LIVERPOOL JMU LIBRARY

3 1111 01391 9277

Guide to Microbiological Control in Pharmaceuticals and Medical Devices

SECOND EDITION

Guide to Microbiological Control in Pharmaceuticals and Medical Devices

SECOND EDITION

Edited by
Stephen P. Denyer
Rosamund M. Baird

CRC Press
Taylor & Francis Group
Boca Raton   London   New York

CRC Press is an imprint of the
Taylor & Francis Group, an informa business

CRC Press
Taylor & Francis Group
6000 Broken Sound Parkway NW, Suite 300
Boca Raton, FL 33487-2742

© 2007 by Stephen P. Denyer and Rosamund M. Baird
CRC Press is an imprint of Taylor & Francis Group, an Informa business

No claim to original U.S. Government works
Printed in the United States of America on acid-free paper
10 9 8 7 6 5 4 3 2 1

International Standard Book Number-10: 0-7484-0615-8 (Hardcover)
International Standard Book Number-13: 978-0-7484-0615-9 (Hardcover)

This book contains information obtained from authentic and highly regarded sources. Reprinted material is quoted with permission, and sources are indicated. A wide variety of references are listed. Reasonable efforts have been made to publish reliable data and information, but the author and the publisher cannot assume responsibility for the validity of all materials or for the consequences of their use.

No part of this book may be reprinted, reproduced, transmitted, or utilized in any form by any electronic, mechanical, or other means, now known or hereafter invented, including photocopying, microfilming, and recording, or in any information storage or retrieval system, without written permission from the publishers.

For permission to photocopy or use material electronically from this work, please access www.copyright.com (http://www.copyright.com/) or contact the Copyright Clearance Center, Inc. (CCC) 222 Rosewood Drive, Danvers, MA 01923, 978-750-8400. CCC is a not-for-profit organization that provides licenses and registration for a variety of users. For organizations that have been granted a photocopy license by the CCC, a separate system of payment has been arranged.

Trademark Notice: Product or corporate names may be trademarks or registered trademarks, and are used only for identification and explanation without intent to infringe.

Visit the Taylor & Francis Web site at
http://www.taylorandfrancis.com

and the CRC Press Web site at
http://www.crcpress.com

Preface to the Second Edition

As opined in the preface to the first edition of this book, microbiological matters continue to exercise considerable influence on product quality. In this second edition, we have tried to follow technological and regulatory change by introducing new chapters. The *Guide to Microbiological Control in Pharmaceuticals and Medical Devices* (the *Guide*) now includes additional material on biotechnological products, medical devices, package integrity testing, and the evaluation of contamination risks in clean rooms. The book also has an amended title, recognizing that similar microbiological considerations apply to the production of pharmaceuticals and of medical devices. It retains its international perspective with contributions from the United Kingdom, Europe, and the United States.

During the period between the first and second editions, we have introduced a companion text, the *Handbook of Microbiological Quality Control in Pharmaceuticals and Medical Devices* (the *Handbook*), edited by R.M. Baird, N.A. Hodges, and S.P. Denyer. This new edition of the *Guide* is intended to be used in partnership with the *Handbook,* offering together a complete theoretical and practical treatment of microbiological control.

ACKNOWLEDGMENTS

Grateful thanks are due to all our contributors, who have been very patient during the gestation period for this second edition. One of us (SPD) would like to make special mention of the hard work of, and support received from, his coeditor (RMB).

Stephen P. Denyer
Rosamund M. Baird

Preface to the First Edition

In the past, consideration of microbiological problems and their means of control occurred only late in pharmaceutical product design; this has led to inevitable compromises in product quality at all stages of development, manufacture, and use. Increasingly sophisticated drug delivery systems are likely to compound further the problems of microbiological integrity, and this coupled with the hardening of regulatory attitudes and increasing consumer awareness means that microbiological problems will assume a new importance. Our current level of understanding does allow us to address many of these problems in advance and thereby to design suitable strategies for their control; to be successful, such strategies must be introduced at the earliest stage of product conception. Recent advances in technology, particularly in clean room design and manufacturing equipment, have provided the means to limit product challenge during manufacture to a minimum. In order to exploit these opportunities to the full, a clear appreciation of all aspects of pharmaceutical microbiology is required. Unfortunately relevant information is often concealed within the literature of several disciplines including those of the engineering, food, cosmetic, and toiletry industries; in this text, we have sought to draw this together in a single volume considering both practical and theoretical aspects.

In recognition of the diverse disciplines involved in pharmaceutical production, a brief introduction to microbiology is presented for the nonmicrobiologist. We have then brought together those principal aspects of microbiology that are relevant to the preformulation, formulation, manufacturing, and licence application stages of pharmaceutical production. Even though attention has been largely focused on the industrial situation much is also of direct relevance to the hospital manufacturing pharmacist. Additionally, we anticipate that this guide will have a wider appeal: scientists from other disciplines frequently have to address similar microbiological problems.

Inevitably in a text such as this, there are areas of common ground between individual chapters. Where these exist, we have attempted to direct the reader's attention through extensive cross-referencing, but without compromising the integrity of individual chapters. Finally, the universal nature of microbiological problems is reflected in the international contributions presented, and although necessarily limited in its size and scope, we hope the text has international appeal.

ACKNOWLEDGEMENTS

The editors would like to record their grateful appreciation to the many contributors who found time in their busy schedules to provide the chapters on which this book is built and to Mrs. J. Woodhouse for her much valued secretarial assistance throughout its preparation. Our thanks and gratitude go also to our families, whose support and encouragement greatly assisted us in the book's final construction; without their patient understanding no publication would have emerged.

S. P. Denyer
R. M. Baird

The Editors

Stephen P. Denyer, Ph.D., received his Bachelor of Pharmacy with First Class Honors from the University of Nottingham in 1975. He obtained his Ph.D. in Pharmaceutical Microbiology in 1979. He is currently Professor of Pharmacy and the Head of the Welsh School of Pharmacy at Cardiff University.

Dr. Denyer's research interests are directed toward clinical application and include microbial pathogenicity and medical device infection, microbial detection, antimicrobial systems, and the design of new biocompatible biomaterials for restorative surgery. He has published more than 150 research articles, including papers, reviews, and invited chapters, and has co-edited five books. He serves on the editorial board of the *Journal of Pharmacy and Pharmacology* and is also a member of the editorial boards of three other scientific journals. Dr. Denyer's contributions to pharmaceutical science have been recognized with a number of awards and distinctions, most recently the British Pharmaceutical Conference Science Chairmanship in 2002.

Dr. Denyer has contributed to the work of a number of national bodies. He has been a microbiology advisor to the U.K. Medical Devices Agency (1993 to 2000), is an expert assessor on active implantable medical devices (1993 to present), is a member of the Health Protection Agency Review Panel for novel infection control agents, and is a member of the National Health Service Engineering and Science Advisory Committee on the Decontamination of Surgical Instruments. He has served on the Veterinary Products Committee (1989 to 1993), is currently a member of the CSM Subcommittee for Pharmacy, Chemistry, and Standards (1999 to present) and is a member of the Panel of Experts on Microbiology, British Pharmacopoeia Commission. He also serves on research council referee panels and as a member of the Engineering and Physical Sciences Research Council Life Sciences College. Dr. Denyer is a current or past consultant to 40 organizations, and was, until recently, the technical director of Biotec Laboratories Limited, a biotechnology company providing diagnostics to the developing world.

Dr. Denyer was elected a Fellow of the Royal Pharmaceutical Society in 2000 and recently received the Society's 2003 Charter Gold Award in recognition of contributions to the profession.

Rosamund M. Baird, Ph.D., is currently Honorary Senior Lecturer in the Department of Pharmacy and Pharmacology at the University of Bath. She graduated from Bath University with a Bachelor of Pharmacy in 1970. She received her Ph.D. in 1975 from London University. Subsequently, she undertook a postdoctoral fellowship at Leiden University and was appointed Honorary Lecturer at St. Bartholomew's Hospital Medical College. From 1975 to 1987, she worked as Principal Pharmacist and Regional Quality Controller at St. Bartholomew's Hospital, London, specializing in pharmaceutical microbiology.

Dr. Baird has specialized in this field for more than 30 years, working initially in the National Health Service and subsequently as a consultant to the pharmaceutical industry. She has published and lectured extensively in this area and has edited a number of related publications and six books on both pharmaceutical and food microbiology. As a consultant, she has provided specialist training on microbiological aspects of production, good manufacturing practices, quality assurance, auditing, and validation to the pharmaceutical, cosmetic, medical device, and herbal industries. She has served on a number of national and international committees, including those of the British Pharmacopoeia Commission. She was awarded a Daphne Jackson Fellowship at the University of Bath in 1996.

Dr. Baird now works for the Daphne Jackson Trust, based at the University of Surrey. She is an honorary member of the Pharmaceutical Microbiology Interest Group. She was elected a Fellow of the Royal Pharmaceutical Society in 2004.

Contributors

Michael J. Akers
Baxter BioPharma Solutions
Bloomington, Indiana, U.S.A.

Haroon A.R.D. Atchia
Quality First International
London, U.K.

Rosamund M. Baird
Department of Pharmacy
 and Pharmacology
University of Bath
Claverton Down, Bath, U.K.

Sally F. Bloomfield
Unilever Research
Bebington, Wirral, U.K.

Allan G. Cosslett
Welsh School of Pharmacy
Cardiff University
Cardiff, U.K.

David J.G. Davies
Centre for Drug Formulation Studies
School of Pharmacy and Pharmacology
University of Bath
Claverton Down, Bath, U.K.

Stephen P. Denyer
Welsh School of Pharmacy
Cardiff University
Cardiff, Wales, U.K.

Eric L. Dewhurst
Teva Pharmaceutical Industries, Ltd.
Runcorn, Cheshire, U.K.

R. Keith Greenwood
Copper Beeches
Wokingham, Berks, U.K.

Klaus Haberer
Compliance Advice and Services in
 Microbiology GmbH
Frankfurt, Germany

Paul Hargreaves
Medicines and Healthcare Products
 Regulatory Agency
London, U.K.

Robin J. Harman
Independent Pharmaceutical and
 Regulatory Consultant
Farnham, Surrey, U.K.

Eamonn V. Hoxey
Johnson and Johnson, Quality and
 Compliance WW
Bracknell, Berks, U.K.

Lee E. Kirsch
College of Pharmacy
University of Iowa,
Iowa City, Iowa, U.S.A.

Bengt Ljungqvist
Safety Ventilation, Building Services
 Engineering
Department of Civil and Architectural
 Engineering
KTH, Royal Institute of Technology
Stockholm, Sweden

Adekunle O. Onadipe
Pfizer, Inc.
Chesterfield, Missouri, U.S.A.

Donald S. Orth
Neutrogena Corporation
Los Angeles, California, U.S.A.

David N. Payne
Reckitt Benckiser
Kingston-upon-Hull, U.K.

Robert A. Pietrowski
David Begg Associates
Kirkbymoorside, York, U.K.

Marie L. Rabouhans
Formerly of The British Pharmacopoeia
  Commission
London, U.K.

Berit Reinmüller
Safety Ventilation, Building Services
  Engineering
Department of Civil and Architectural
  Engineering
KTH, Royal Institute of Technology
Stockholm, Sweden

Nicolette Thomas
Formerly of Medical Devices Agency
London, U.K.

Hans van Doorne
University Centre for Pharmacy
Department of Pharmaceutical
  Technology and Biopharmacy
University of Groningen
Groningen, The Netherlands

Veda K. Walcott
Cook Pharmica, LLC
Bloomington, Indiana, U.S.A.

Contents

1 Introduction to Microbiology

Rosamund M. Baird and Stephen P. Denyer

CONTENTS

1.1 INTRODUCTION

Certain manufactured goods, of which foodstuffs, cosmetics, and pharmaceutical products are the prime examples, can be contaminated with microorganisms during manufacture; this contamination can, at the best, cause spoilage and consequent rejection of the contaminated material and, at the worst, harm or even bring death to the consumer. The culprits are usually bacteria or fungi; they differ fundamentally in subcellular anatomy.

Of contaminants, fungi or molds will be familiar to all as often colored, growth on jams, discarded bread and fruit, and even on the leather of boots stored wet in a garden shed. Bacterial contamination may manifest itself as colonies on foodstuffs but generally is not as visually familiar as molds. A red bacterium, however, later to be named *Chromobacterium prodigiosum* and now *Serratia marcescens,* growing on bread gave rise to many reports of miracles or portents dating back from the third century B.C. The amateur wine maker will also be familiar with the ravages wreaked by bacterial contamination of his product.

1.2 THE GENERAL STRUCTURE OF A BACTERIAL AND FUNGAL CELL

Bacteria are unicellular and can exist as a discrete entity. In the fungi, whereas yeasts are unicellular organisms typically 10 μm in diameter, almost all, if not all, the contaminant molds grow as filaments or hyphae that may be cross-walled (septate) or a continuous tube (coenocytic). Both bacterial and fungal cells possess a cell wall

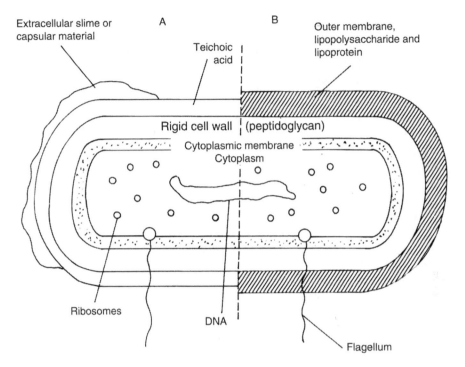

Extracellular slime or capsular material

A

B

Teichoic acid

Outer membrane, lipopolysaccharide and lipoprotein

Rigid cell wall | (peptidoglycan)

Cytoplasmic membrane

Cytoplasm

Ribosomes

DNA

Flagellum

FIGURE 1.1 Diagram of the main features of the bacterial cell. (A) Gram-positive cell; (B) Gram-negative cell.

that is a rigid structure but with differing chemical constitutions. Here any formal anatomical similarity ends.

1.2.1 THE BACTERIAL CELL

Bacteria are small, generally between 0.75 and 4 µm in length. They are characteristically shaped, and those responsible for spoilage come from groups that are either short cylinders with rounded ends (bacilli) or spherical (cocci). Figure 1.1 shows the main features. On, or toward, the outside there is a rigid cell wall that confers the characteristic shape. Chemically this is a complex polymer of sugars, amino sugars, and amino acids. Within that lies a nonrigid structure known as the cytoplasmic membrane. This consists of a raft of phospholipid molecules, which are fatty material containing a phosphate group. In this raft float protein molecules that have structural or enzymic functions (Figure 1.2). The rest of the cell is known as the cytoplasm. It consists of a viscous fluid in which are embedded (a) the nucleus, made up of nucleic acids and responsible for directing enzymic and structural protein synthesis and thus controlling the basic characters of the cell, and (b) ribosomes, which are the sites of the nucleic acid–directed protein synthesis. In addition, enzymes and metabolic reserves (often in polymeric form) are found in the cytoplasm.

Bacterial cells occur in two structural types known as Gram-positive and Gram-negative (Figures 1.1A and 1.1B, respectively), and no one reading a book on

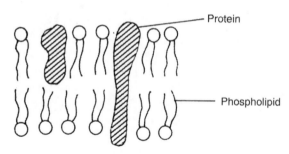

FIGURE 1.2 Phospholipid bilayer with inserted protein molecules: the structure of many biological cytoplasmic membranes.

bacteriology will fail to find these terms. The terms positive and negative refer to a staining reaction and the word Gram refers to the discoverer of the method—Christian Gram. The simplified diagram (Figure 1.1) shows that Gram-negative cells have an additional structure, the outer membrane, which is linked to the outside of the rigid cell wall by divalent cations; this is lacking in Gram-positive cells. This outer membrane confers the differential staining property and in many cases contains toxic material responsible, if ingested or injected, for disease and elevated temperature (pyrogen). It may also contribute to the resistance of some Gram-negative bacteria toward certain antibacterial agents.

1.2.2 THE FUNGAL CELL

Whether unicellular (yeasts) or filamentous with or without cross-walls (molds), both types of spoilage fungi have a rigid cell wall of cellulose and another polymeric material called chitin, chemically related to the shells of crustaceans. Within this rigid wall lies the cell membrane consisting of phospholipids and proteins. In addition, and here is a fundamental difference from bacteria, the fungal membrane contains sterols. Typical sterols found in fungal membranes are ergosterol and zymosterol.

Lying within this membrane is the cytoplasm. This contains the nuclear material surrounded, unlike bacteria, by a pore-containing nuclear membrane. It directs protein synthesis as in the bacterial cell. Also within the cytoplasm are found the ribosomes, as before, sites of the directed protein synthesis. These ribosomes differ in size and structure from those in bacterial cells.

1.2.3 PROKARYOTE AND EUKARYOTE

This very brief outline of the structure of a bacterial and fungal cell has drawn attention to fundamental differences between these cell types. These differences, detected by the techniques of subcellular biology, have enabled biologists to suggest a fundamental division in the living world. Bacteria were named prokaryotic organisms or prokaryotes, a name derived from their unenclosed nucleus, and fungi (and, in fact, all other living plants and animals) were called eukaryotic or eukaryotes; they possessed a nuclear membrane. Some of these differences have been summarized in Table 1.1.

TABLE 1.1
Some Differences between Bacterial (Prokaryotic) Cells and Fungal (Eukaryotic) Cells

Feature	Prokaryote	Eukaryote
Nucleus	No enclosing membrane	Enclosed by a membrane
Flagella	Simple	Complex
Cell wall	Peptidoglycan	Cellulose, chitin, and other polymers
Cytoplasmic membrane	Generally does not contain sterols	Contains sterols
Ribosome	70S[a]	80S[a]
Oxidative phosphorylation	In cytoplasmic membrane	In mitochondria
Mitochondria	Absent	Present

[a] S is a measure of ribosome size (calculated from sedimentation in a centrifuge); 80S ribosomes are larger than 70S.

1.2.4 THE BACTERIAL AND FUNGAL SPORE

Spore is a word common to both bacterial and fungal morphology, but it is in the spore that the largest difference in function between the two groups can be seen. In brief, bacterial spore formation, which is limited to two genera important in contamination, constitutes a survival package that is formed under adverse conditions and from which, when conditions are again suitable, vegetative bacteria arise. Fungal spores, however, are part of the normal life cycle of these organisms.

1.2.4.1 Bacterial Spores

Spore, or endospore, formation in *Bacillus* and *Clostridium* species (the two spore-forming bacterial genera of significance in contamination) is a complicated biochemical and morphological process but the result is a structure (Figure 1.3) that is significantly more resistant to adverse conditions, for example, heat, chemicals, desiccation, and radiation, than the vegetative cell. In fact, it is the heat resistance of spores that dictates the time–temperature relationships of sterilization processes (see Chapter 10) and disinfection regimens. Whatever the factor committing bacteria to sporulation, the stages in the process appear to be the same for all spore-forming bacteria.

An asymmetric cross-wall develops in the bacterial cell and the nuclear material appears in the smaller compartment called the forespore. This becomes the spore protoplast that then forms the cortical membrane, which is followed by the cortex and spore coats. With the formation of the cortex and spore coats is seen the onset of radiation resistance and the onset of resistance to heat and chemicals.

Parallel with the emergence of resistance, complex biochemical changes occur that include synthesis of spore-specific proteins, cysteine-rich structures and dipicolinic acid, and an incorporation of calcium. At the commencement of cortex formation, the cells appear more refractile when viewed under the microscope.

Finally the original cell (mother cell as it is called) lyses and the mature spore is released. This is shown diagrammatically in Figure 1.4. The site and position of spores

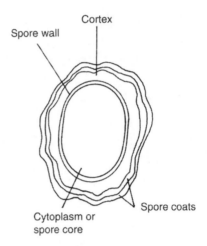

FIGURE 1.3 Cross-section of a bacterial spore.

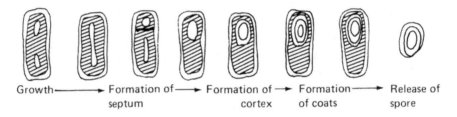

Growth———→ Formation of——→ Formation of —→ Formation———→ Release of
                   septum   cortex   of coats   spore

FIGURE 1.4 Changes occurring during spore formation.

in relation to the mother cell vary from species to species of bacteria. Some are terminal to give a drumstick-like appearance, others are median, and still others are subterminal; the developing spore may be larger than or equal in size to the original mother cell.

The reverse process to spore formation is termed germination. Some divide the process and call it germination and outgrowth. Germination may be initiated by a general improvement in the nutritional environment with a more favorable ambient temperature. It may also be triggered by the presence of a specific chemical such as glucose or L-alanine or by sublethal heating (i.e., at 60°C for 60 min). Germination is accompanied by (a) loss of heat resistance and resistance to chemicals, (b) a loss of refractivity (under the microscope), and (c) a release of dipicolinic acid into the surrounding medium. Finally the vegetative cell so formed grows out from the spore—the outgrowth phenomenon (Figure 1.5).

The biochemistry of bacterial spore formation and regermination and the mechanism of resistance have been the subject of intensive study and publication. For a comprehensive account, see Russell (1982) and Lambert (2004).

1.2.4.2 Fungal Spores

Two types of fungal spores are produced, some originating asexually (arthrospores, chlamydospores, conidiospores, and sporangiospores) and some sexually (ascospores,

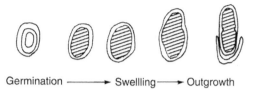

Germination ⟶ Swellling ⟶ Outgrowth

FIGURE 1.5 Stages of spore germination and outgrowth.

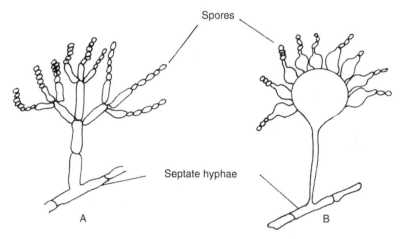

Spores

Septate hyphae

A   B

FIGURE 1.6 Fungal conidiophores bearing asexual reproductive spores of *Penicillium* spp. (A) and *Aspergillus* spp. (B).

basidiospores, and zygospores). These are often colored; furthermore they are resistant to environmental stress and may be carried on air currents in the environment, hence giving rise to further contamination when they alight in favorable situations and germinate. The microscopic structure of fungal fruiting (spore-bearing) bodies is an aid to identification (Figure 1.6).

1.3 BACTERIAL AND FUNGAL GROWTH

It is important to realize from the outset that a large number of bacteria and fungi, including many of those associated with contamination, can grow in what may appear to be nutritionally very simple systems and often at quite low temperatures. Cold is not lethal, although growth is slowed at low temperatures. The notion that bacteria and, to a lesser extent, molds require a rich and often exotic nutritional environment and careful and controlled incubation at 37°C (bacteria) or 25°C (molds) arose because of the dominance of medical bacteriology and mycology where such conditions were almost invariably mandatory.

Process water is an acceptable culture medium for many bacteria and, to a lesser extent, molds and can be a dangerous commodity in the pharmaceutical, food, and cosmetic industries unless carefully handled (see Chapters 2 and 3). The causal organism of Legionnaire's disease grows in water-cooled heat-exchangers (Anonymous 1989).

For each bacterial and fungal species, a set of conditions is necessary for optimum growth and, if correctly balanced, will give maximum yields. It should be realized that although laboratory studies often seek to optimize growth, in the case of contamination, conditions may not be optimal but nevertheless may allow growth to proceed, giving rise to spoilage.

1.3.1 REQUIREMENTS FOR GROWTH

These can be divided conceptually into two categories: first, the range of substrates, the consumables needed; and second, the nature of the environment, that is, temperature, pH, osmotic pressure.

1.3.1.1 Consumables

It would perhaps be salutary and illustrative of the point made in the opening paragraphs of this section to quote a formula that will support the growth of many bacterial and fungal contaminants. It consists of an aqueous solution containing (g.L^{-1}): $(NH_4)_2HPO_4$, 0.6; KH_2PO_4, 0.4; glucose, 10.0. Growth will be slow and can be enhanced by the presence of trace elements (in addition to those present in the laboratory reagents) and carbohydrates, fats, proteinaceous material, amino acids, sugars, and vitamins such as nicotinic acid, riboflavin, and thiamine. Notwithstanding the simple nutritional requirements of some organisms, others exhibit specific needs that form the basis of selective and diagnostic media (see Section 5.1). For a fuller treatment of media design, see Baird (2000).

1.3.1.2 Environmental Factors

Water—The presence of water is essential and dry products or intrinsically anhydrous material are not liable to spoilage. From very early in history, man has exploited the drying of foods to preserve them.

The requirement of a microorganism (bacterial or fungal) for water can be quantified by determining a factor that measures available water in a system and is called its water activity. If the water vapor pressure (P) of the system—a solution or moist semisolid—is measured and compared with that of pure water under the same conditions (P_0) then

$$\text{water activity} = a_w = \frac{P}{P_0} \tag{1.1}$$

When culture media are adjusted by means of solutes to known values of a_w, it is possible to determine, in general terms, the ability of various groups of organisms to grow at these values (Table 1.2). The use of syrups, strong solutions of (usually) sucrose, where the water activity is low, has proved valuable in preservation (Corry 1975).

Gaseous nutrients—Some microorganisms grow in the absence of oxygen and are termed anaerobic; most, however, require oxygen and are called aerobic. Some

TABLE 1.2
Limiting Water Activity (a_w) Values for a Range of Microorganisms

Organism	Limiting a_w for Growth
Bacteria in general	0.95–0.92
Micrococci, lactobacilli	0.90
Staphylococcus aureus	0.86
Halophilic bacteria	0.75
Yeasts	0.94–0.88
Molds	0.93–0.70
Osmophilic yeasts	0.73

bacteria possess the facility to grow in either the absence or presence of oxygen and are termed facultative organisms. In general terms, this means that the exclusion of oxygen (air) may not prevent some spoilage organisms from growing.

pH—There is an optimum pH range over which microorganisms can grow (Figure 1.7A). In particular, bacteria grow best around pH 7 but many molds can tolerate more acid conditions, pH 5 to 6. Some microorganisms are surprisingly tolerant of hostile pH environments and contamination can occur in products where pH is outside the optimum growth value although the growth may be slower.

Temperature—As with pH, there is an optimum temperature range. Low temperatures will slow growth and rising temperatures will increase growth rate. As the temperature rises above optimum, however, growth is inhibited and microorganisms are killed, a phenomenon exploited in heat sterilization (see Chapter 10). The effect of temperature on growth is shown in Figure 1.7B.

It is very important to realize that whereas higher temperatures are lethal as pointed out above, low temperatures are not, and very slow growth may in certain cases occur in refrigerators and even deep-freeze cabinets. Although most human pathogenic bacteria, those of interest in medicine, grow best at the temperature of the human body (37°C), surveys (Skinner 1968) have shown a wide variety of optimum temperature ranges to exist for bacteria. Thus to summarize, bacteria are usually grouped as follows:

- Psychrophilic bacteria or psychrophiles; grow best at 15 to 20°C, can grow at 0 to 5°C, and do not grow above 30°C.
- Mesophilic bacteria or mesophiles; grow best at 25 to 40°C, do not grow above 45°C or below 5°C.
- Thermophilic bacteria or thermophiles; grow best at 55°C, some will grow at 80°C or above; little growth occurs below 25°C.

Many fungi grow best over temperature ranges of 15 to 25°C, although they can grow at refrigeration temperatures and human pathogens will grow well at 37°C.

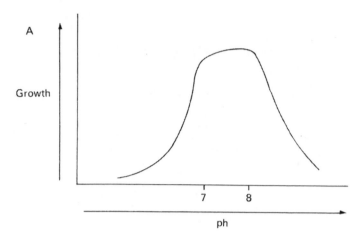

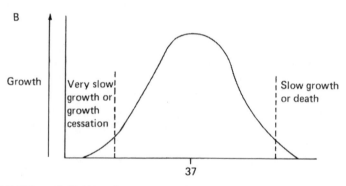

FIGURE 1.7 Effect of pH (A) and temperature (B) on bacterial growth.

1.3.2 MEASUREMENT OF MICROBIAL GROWTH

Bacterial and yeast growth under the differing environmental conditions outlined above may be put on a numerical basis by carrying out counts or by other methods of quantifying growth.

1.3.2.1 Total Counts

The microbial cells are placed on a counting chamber slide that is engraved with squares of known area and a coverslip is supported over the grid to give sections of known volume. By counting cells in these sections, a calculation can be made of the total cell count. This method cannot distinguish between live and dead cells. It is similar to the hemocytometric method of counting red cells in a sample of blood.

1.3.2.2 Viable Counts

Microorganism-containing material, diluted if necessary, is placed in a solid culture medium and the organisms are allowed to grow into visible colonies. The count represents the number of colony-forming units (CFU) in the (diluted) sample. The actual count in the original sample is obtained by multiplying the CFU by the dilution factor.

1.3.2.3 Spectrophotometric Method

A suspension of bacteria or yeasts scatters light. If light is passed through a suspension, it is possible to obtain a measure of the cell numbers present by referring to calibration curves in which optical densities are compared to either total or viable counts.

1.3.2.4 Total Biomass Determinations

In the laboratory situation, microbial growth can be followed by wet or dry weight determinations on samples removed from suspension or by chemical determination of microbial components, for example, protein content. These approaches find the greatest value in measuring growth of filamentous fungi, as these organisms do not form discrete units as do bacterial or yeast cells.

1.3.3 THE PATTERN OF GROWTH

When bacteria enter a new environment, either deliberately in experimental situations or as a contaminant, the relationship of bacterial growth against time shows a very characteristic pattern (Figure 1.8).

At first there is no apparent growth and this is called the lag phase. During this phase, it is thought that bacteria are adapting to their environment. Bacterial numbers remain constant and equal to the inoculum number although a few may die. After this period of lag and adaptation, growth proceeds quickly and, in highly favorable circumstances, bacterial cell numbers may double in 20 to 40 min. Here, the relationship

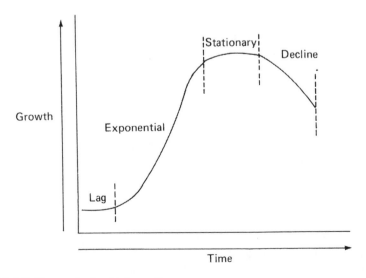

FIGURE 1.8 Phases of microbial growth.

between the log of bacterial numbers and time is linear; this phase of growth is called the logarithmic (log) or exponential phase. If nutrients are limited, or are not renewed, growth eventually ceases due to food exhaustion and possible accumulation of inhibitory waste products. Again, there is a period when bacterial numbers do not change. This period is called the stationary phase. Finally, numbers of viable organisms are found to fall. This is thought to be due to an exacerbation of the forces that induce the stationary phase. This part of the growth curve is referred to as the phase of decline or, sometimes, the death phase.

Fungal growth follows a similar pattern. For filamentous fungi, the increase of total cell dry weight is usually used as a measurement of growth.

It should be remembered that fungal hyphae will make their way through their substrate, environment, or a contaminated product. These hyphae, when they reach the surface of the material in which they are growing, will form fruiting bodies that are often colored and that give rise to the familiar visual sign of mold or fungal growth. It is from these aerial bodies that spores can be disseminated to colonize fresh areas (see Section 1.2.4.2).

Over their growth phase, and following changes in their environment, microorganisms exhibit considerable plasticity that may influence significantly their adaptability and sensitivity to antimicrobial effects.

1.3.4 BACTERIAL REPRODUCTION

Bacteria grow best if all the conditions—water, nutrients, temperature, pH, and osmotic pressure—are optimal. Growth will be inhibited if any of these factors varies from the maximum or if growth inhibitors are present. The process of preservation depends on using suitable growth inhibitors.

Bacteria reproduce by what is called binary fission, that is, one cell becomes two, two become four, four become eight, and so on. This reproductive pattern does not involve the exchange of genetic material and each cell is a copy or clone of the parent. It is possible to calculate the mean time for the bacterial cells in a culture to duplicate by the pattern of binary fission. If the bacterial numbers at the start are N_0 and, after a given time t are determined to be N, then the mean generation time (MGT) is given by the formula

$$\text{MGT} = \frac{t \times 0.301}{\log N - \log N_0} \tag{1.2}$$

It is possible to quantify the various factors affecting growth by calculating the MGT for each parameter being investigated.

1.3.4.1 Genetic Exchange

Originally, binary fission was thought to be the only method by which bacteria were able to reproduce genetic information, but now three methods of reproduction involving genetic exchange are known. In 1928, before the role of deoxyribonucleic acid (DNA) and the genetic code were elucidated, genetic exchange was found to occur in a culture of an organism called *Diplococcus pneumoniae;* this process was called transformation. Subsequently, an active transfer of genetic material (plasmids) via special appendages was discovered and called conjugation. Later, it was discovered that bacterial viruses, known as bacteriophages, could promote genetic exchange; this process was called transduction. These techniques of genetic exchange are responsible for much of the development of antibiotic-resistant microorganisms faced in clinical practice today.

However, from the point of view of growth of contaminants, binary fission can be considered the method of bacterial reproduction.

1.3.5 FUNGAL REPRODUCTION

Most filamentous fungi reproduce by both asexual and sexual processes. The most common method of asexual reproduction is by means of spores (see Section 1.2.4.2), contained in saclike structures called sporangia or borne on hyphal offshoots known as conidiophores (Figure 1.6). Fragmented hyphae may also play a role in propagation. Sexual reproduction requires the union of two compatible nuclei, brought about by the successful elaboration of, and contact between, specialized sex organ structures (gametangia).

Yeast cells commonly reproduce by budding, where a small outgrowth (daughter cell) arises from the parent and, following mitosis, continues to grow until it separates from the parent cell. Occasionally, yeast cells may also reproduce by binary fission as described earlier for bacteria (see Section 1.3.4).

1.3.6 CHEMICAL INHIBITION AND DESTRUCTION
OF MICROORGANISMS

Microbial growth can be inhibited by adverse conditions other than a shift from the optimum growth environment. Thus chemical preservatives and sterilization processes, for example, heat, reactive gas, or irradiation, can also affect growth and survival and are exploited in the pharmaceutical, food, environmental, and medical fields (see Chapters 10, 11, 13, and 14). Table 1.3 indicates the regions in the microbial cell that may be targets for growth inhibitory or lethal action. The majority of mechanism of action studies have been performed on bacterial systems.

The difference in the structures of the Gram-negative and Gram-positive bacteria, insofar that the former have an outer membrane (Figure 1.1), has an interesting bearing on the resistance of these two groups to biocides (antiseptics, disinfectants, and preservatives). Generally, Gram-negative bacteria often show a greater resistance to biocides than do the Gram-positive cells, and this is thought to be due to the protective nature of the Gram-negative outer membrane, which acts as an exclusion barrier to biocidal agents. In this respect, *Pseudomonas aeruginosa* presents a particular challenge to biocidal agents. In much the same manner, the bacterial spore is considered to be more resistant than its corresponding vegetative cell through the presence of its spore coats (see Section 1.2.4.1). We now also recognize that bacterial growth on surfaces, termed a biofilm, results in the elaboration of extracellular material that may act as a barrier to harmful agents.

TABLE 1.3
The Site of Action of Selected Chemical and Physical Antimicrobial Agents

Antimicrobial Agent	Possible Target(s)
Alcohols	Cytoplasmic membrane permeability
Bronopol	Enzymes with thiol groups
Chlorhexidine	Cytoplasmic membrane permeability; general cytoplasm coagulation
Chlorine, chlorine releasers	Bacterial cell wall structure; enzyme thiol groups; amino groups of proteins
Ethylene oxide	Enzyme thiol groups; amino and other groups of proteins
Formaldehyde, formaldehyde releasers	Bacterial cell wall (low concentrations); amino and thiol groups
Glutaraldehyde	Bacterial cell wall; thiol, amino and other reactive groups; general cytoplasm coagulation
Heat	
Dry	Oxidation
Moist	Enzyme denaturation and cytoplasm coagulation
Hydrogen peroxide	Enzyme thiol groups
Irradiation	DNA
Mercury (II) salts, organic mercurials	Bacterial cell wall (low concentrations); enzyme thiol groups; general cytoplasm coagulation
Parabens (p-hydroxybenzoic acid esters)	Cytoplasmic membrane permeability; DNA and RNA synthesis
Phenols, halogenated phenols	Bacterial cell wall (low concentrations); cytoplasmic membrane permeability
Quaternary ammonium compounds	Cytoplasmic membrane permeability
Sorbic acid, other lipophilic weak acids	Cytoplasmic membrane permeability
Sulfur dioxide, sulphites	Enzymes with thiol groups; amino groups of proteins

1.4   METABOLIC ACTIVITY

Bacteria and fungi possess a remarkable ability to metabolize a very wide range of substances. This metabolic versatility is of paramount importance to microbial survival in, and subsequent spoilage of, contaminated products. Spoilage is discussed in detail in Chapter 2. Many of the metabolic pathways are similar to those found in higher animals; some generalized metabolic reactions of bacteria and fungi are given in Figure 1.9. These reactions are catalyzed by enzymes.

1.4.1   CARBOHYDRATES

Starches and celluloses are converted to glucose, which is then metabolized to substances such as organic acids, aldehydes, or alcohols or, in the presence of adequate levels of oxygen, to carbon dioxide and water. The Embden–Meyerhof and Krebs cycles are two well-documented cycles in which living organisms metabolize carbohydrates. Some of the by-products of carbohydrate metabolism during spoilage may acidify or taint the product.

Carbohydrate metabolism

$+C_6H_{10}O_5+_n + nH_2O \rightarrow nC_6H_{12}O_6$ hydrolysis
$C_6H_{12}O_6 \rightarrow 2CH_3 \cdot CH_2OH + 2CO_2$ anaerobic respiration
$C_6H_{12}O_6 + 6O_2 \rightarrow 6CO_2 + 6H_2O$ aerobic respiration

Fat metabolism

$$
\begin{array}{llll}
CH_2OOC(CH_2)_n \cdot CH_3 & & CH_2 \cdot OH & \\
| & & | & \\
CHOOC(CH_2)_n \cdot CH_3 & +3H_2O \rightarrow & CH \cdot OH & + \quad 3CH_3(CH_2)_n \cdot COOH \\
| & & | & \\
CH_2OOC(CH_2)_n \cdot CH_3 & & CH_2 \cdot OH & \quad\quad lipolysis
\end{array}
$$

Protein and amino acid metabolism

$+NH \cdot CHR \cdot CO+_n + nH_2O \rightarrow nNH_2 \cdot CHR \cdot COOH$ proteolysis

$NH_2 \cdot CHR \cdot COOH \rightarrow NH_2 \cdot CH_2R + CO_2$ decarboxylation

FIGURE 1.9 Some examples of generalized metabolic reactions of bacteria and fungi.

1.4.2 FATS

Fats will be converted to glycerol and fatty acids. The latter acidify and taint products containing fatty material when contaminated with lipolytic (fat-splitting) organisms.

1.4.3 PROTEINS AND AMINO ACIDS

Proteins may be cleaved to amino acids and these are then decarboxylated to yield evil-smelling amines that will also tend to alkalinize the mixture. Some of the very bad odors of contaminated products are due to this sequence of reactions.

1.4.4 PRESERVATIVES

The antimicrobial agents included to preserve products are not immune from metabolic attack and if their concentration falls, due to instability in storage, for instance, they may be metabolized, especially if bacteria of the genus *Pseudomonas* form the contaminating organism (see Chapter 2).

1.5 PRINCIPLES OF MICROBIAL IDENTIFICATION

Although bacteria are small, when growing in aggregate (colonies), they are readily visible and as such are seen when they are contaminating solid material or growing on solidified culture media. Similarly, molds just visible with the naked eye are readily seen if growing in colonies, especially if pigmented spores are present.

Direct observations may be made with the microscope and experts will quickly be able to tell whether bacteria are present as rods or spheres, and if molds are being examined, a good idea of the nature of the mold (fungus) will be obtained from the characteristic shape of hyphae and spore-bearing structures. Ultimate identification, often not necessary in contamination studies, falls within the ambit of the specialist mycologist. With bacteria, the performance of special staining techniques, especially the Gram stain, before microscopic examination will allow bacteria to be classified as Gram-positive or Gram-negative.

The metabolic ability of microorganisms can be exploited in identification. Special culture conditions are devised by which bacteria, and to a lesser extent fungi, can be distinguished by the variance in their response to these conditions. In particular, the differing ability of microorganisms to utilize carbohydrates (sugars), that is, their ability to metabolize them via the Embden–Meyerhof or Krebs cycle pathways, can be used in their identification.

In recent years, newer and more rapid methods of identification involving specialist immunological techniques, monoclonal antibodies, bioluminescence, and nucleic acid probes have been exploited (Newby 2000).

1.5.1 SELECTIVE AND DIAGNOSTIC CULTURE MEDIA

As the foregoing account indicates, (a) bacterial growth may be inhibited by adverse conditions including the presence of toxic chemicals, and (b) bacteria, by virtue of their biochemical activities, can effect chemical transformations. Both these factors are exploited in the construction of selective and diagnostic media. Examples of both approaches will be given here.

Whole texts have been devoted to culture media, and *Cowan and Steel's Manual* (see Further Reading) contains some relevant sections. In addition, the catalogs of media manufacturers are very informative. For a detailed account of culture media used in pharmaceutical microbiology, the reader is referred to the companion *Handbook* (Baird 2000).

1.5.1.1 Selective Media

So-called selective media exploit the selective toxicity of added ingredients and also contain agents that aid further identification. A good example of this type of medium is MacConkey's agar. First described in 1905, this medium evolved to detect organisms in drinking water that might be of fecal origin. It was also designed to distinguish bacteria that could, and could not, ferment lactose. The latter property might indicate the presence of salmonellae: dangerous intestinal pathogens that could cause, among other things, outbreaks of food poisoning. The medium consisted of a nutrient base with the addition of bile salts, lactose, and a red dye called neutral red. The bile salts were included to suppress organisms other than those of gut origin, the argument being that bacteria from the gut could survive in the presence of bile.

Those organisms that could ferment lactose acidified the medium and this acidification caused a precipitation of red dye on the colony, which then appeared red. Organisms such as the salmonellae did not ferment lactose and the colonies remained translucent.

1.5.1.2 Diagnostic Media

Bacteria vary in their ability to metabolize a large variety of carbohydrates, typically sugars, but also certain alcohols and glycosides. Diagnostic schemes for bacteria were thus evolved, based on differential metabolic behavior on a range of these substances. It was then necessary to differentiate between those organisms that could utilize the various substrates as shown by their conversion to acid products with the possible production of gas.

The routine method used consisted of a container with a growth medium incorporating the substrate concerned and a dye pH indicator; bromocresol purple has been widely used. In order to detect gas production, a very small inverted test tube (Durham tube) was included in the medium and manipulated to ensure it was full of that medium. Any gas produced would collect in the tube and could be seen easily. Current developments in biochemical tests for identification still utilize these basic principles, while presenting the media in more convenient and miniaturized forms.

A color atlas of pharmaceutically relevant microorganisms is provided in the companion *Handbook* (Hodges 2000) showing their behavior on selective and diagnostic media.

1.6 PATHOGENICITY

Historically, microorganisms have been grouped into pathogenic or nonpathogenic types, according to their ability to cause disease. In recent years, increasing interest has also been shown in the so-called opportunist pathogens, which are capable of causing disease when given the opportunity to do so. These include the free-living Gram-negative bacteria such as pseudomonads and members of the Enterobacteriaceae, for example, *Klebsiella* and *Serratia* spp., all of which have simple nutritional requirements, thus enabling them to survive in some unlikely environments, including disinfectant and antiseptic solutions. Opportunist pathogens pose a particular threat to certain groups of patients at risk, especially neonates, the elderly, and those compromised by trauma, burns, or immunosuppressant therapy.

True pathogens, such as *Clostridium tetani* and *Salmonella* spp., rarely occur in pharmaceutical products but inevitably cause serious problems when present. Opportunist pathogens are more common contaminants in these products, particularly in aqueous preparations where total viable counts may well exceed 10^6 CFU.mL^{-1}. For a more detailed discussion of this topic, the reader is referred to Chapter 2 and Gilbert (2004).

1.7 GENERAL PROPERTIES OF SELECTED MICROORGANISMS

The foregoing sections in this chapter have treated the subject of microbiology in a general way, and little mention has been made of specific microorganisms. The summary below considers some of the microorganisms of particular interest in the field of pharmaceutical microbiology. It includes not only common spoilage or challenge organisms but also some organisms considered to be hazardous if present in pharmaceutical products. This list is not exhaustive. Interested readers should consult the standard microbiological texts identified in the further reading list at the end of the chapter.

1.7.1 GRAM-NEGATIVE ORGANISMS

1.7.1.1 *Pseudomonas aeruginosa*

Pseudomonas aeruginosa has acquired some notoriety as a contaminant of pharmaceutical products, especially eye preparations (because of its ocular pathogenicity), and above all because of its peculiar resistance to biocides, some of which it is able to metabolize. Widely distributed in the environment, it is a free-living opportunist pathogen and can cause particular problems in susceptible groups of patients, especially neonates, the elderly, and the immunocompromised. It is a flagellated rod-shaped organism that does not form spores and grows aerobically. It does not ferment carbohydrates and thus will not produce gaseous products when growing in the presence of carbohydrates. Under appropriate conditions of growth, it may produce a blue or green fluorescent pigment. *Pseudomonas aeruginosa* grows readily on standard laboratory media, but will grow even in distilled water provided trace amounts of organic matter are present. As a mesophile, *P. aeruginosa* will grow over a temperature range from 20 to 42°C, with optimal growth occurring at 37°C. It is commonly found in biofilm growth.

1.7.1.2 *Burkholderia cepacia*

Formerly known as *Pseudomonas cepacia, Burkholderia cepacia* is an aerobic, rod-shaped, motile, free-living organism that can form biofilms in aqueous systems. It causes particular problems for patients with cystic fibrosis. The organism is known to survive in improperly prepared quaternary ammonium disinfectant solutions.

1.7.1.3 *Escherichia coli*

Escherichia coli is a motile, nonsporing rod typically of dimensions 1 μm × 4 μm and is a member of the Enterobacteriaceae. It differs from *P. aeruginosa* in that it is able also to grow anaerobically. When it grows on carbohydrates, it does so by fermentation, producing gaseous products. Media with only a single carbon source are sufficient for promoting *E. coli* growth. It can grow at low temperatures and also at temperatures as high as 40°C. Optimal growth occurs at 37°C. It forms part of the gut flora of man; its presence in products is therefore indicative of fecal contamination and gross defects in hygiene. It has gained particular notoriety through the verotoxigenic strain *E. coli* 0157:H7.

1.7.1.4 *Salmonellae*

Salmonellae are a large group of Gram-negative rods, closely related to *E. coli,* and may be present in the fecal material of animals or of human carriers. *Salmonella typhimurium* and *S. enteritidis* are both causes of poisoning, and if present in oral pharmaceutical products or foods, they can give rise to nausea, vomiting, and diarrhea through endotoxin release.

1.7.1.5 *Klebsiella*

A member of the Enterobacteriaceae, this genus is represented by *Klebsiella pneumoniae,* subsp. *aerogenes,* a short (1 to 2 μm long by 0.5 to 0.8 μm wide), nonmotile rod, frequently encapsulated with carbohydrate material. The *Klebsiella* are aerobic and usually ferment carbohydrates with the production of acid and gas. The optimum temperature for growth is 37°C with limits of 12°C and 43°C. The natural habitat for this genus is the bowel and respiratory tract of humans and animals, and in the environment they are found in soil and water. *Klebsiella* may be a cause of bacteremia or urinary and respiratory tract infections in humans, especially in hospitalized and immunocompromised patients, and *K. pneumoniae* is responsible for a variety of diseases in animals.

1.7.1.6 *Serratia*

Also a member of the Enterobacteriaceae, the genus *Serratia* consists of short (0.5 to 1.0 μm), motile rods. They are typically found in soil and water. Generally, the optimum temperature for growth is between 30 and 37°C, but some strains will show growth at temperatures as low as 1°C. Certain strains produce a red pigment (prodigiosin), the production of which is oxygen-dependent and encouraged at suboptimal growth temperatures. *S. marcescens* may cause bacteremia and localized infections in the urinary and respiratory tracts and in wounds; most infections occur in hospitalized patients.

1.7.2 GRAM-POSITIVE ORGANISMS

1.7.2.1 *Staphylococcus aureus*

This is a spherical organism of approximately 1 μm in diameter. It is nonmotile and does not form spores. It is able to grow aerobically and anaerobically and will grow readily in a chemically defined medium containing glucose, essential salts, selected amino acids, thiamine, and nicotinic acid. It is relatively resistant to antimicrobial preservatives such as phenol, can remain alive at temperatures as cold as 4°C and as warm as 60°C, and will grow in media containing up to 10% sodium chloride.

If present in pharmaceutical products, *S. aureus* and the closely related *S. epidermidis* may indicate contamination from a human source, for example, from the hands, skin, or hair. In particular, *S. epidermidis* is frequently implicated in medical-device-related infection. *Staphylococcus aureus* is a common cause of boils, middle ear infection, pneumonias, and osteomyelitis. It can proliferate in foods, secreting an exotoxin that can give rise to food poisoning.

1.7.2.2 *Bacillus*

These are spore-forming, rod-shaped organisms, usually motile. *Bacillus cereus* can grow anaerobically but *B. subtilis* cannot. These organisms, if present in pharmaceutical products, are indicative of dust contamination. Therefore, any structural alterations to buildings or uncontrolled sweeping may give rise to these contaminants appearing in pharmaceutical formulations. *Bacillus cereus* can cause intestinal poisoning. *Bacillus stearothermophilus* produces particularly heat-resistant spores and these are exploited in the biological monitoring of heat sterilization processes.

1.7.2.3 Clostridia

The clostridia are rod-shaped bacteria that produce spores. They are strict anaerobes; although a few species will tolerate the presence of oxygen, most will ferment carbohydrates. They are found widely distributed in nature, in the soil, and in the alimentary tract of man and animals. *Clostridium perfringens (welchii)*, *C. tetani*, and *C. botulinum* are pathogens causing gas-gangrene wound infection, tetanus, and botulism food poisoning, respectively. Owing to their ubiquitous presence in the environment, clostridia, and in particular clostridial spores, can be found as contaminants in pharmaceutical products.

1.7.3. FUNGI

1.7.3.1 Aspergillus niger

This organism grows only in the filamentous (mycelial) form and is familiar to most people as a white, turning to black, disk of growth on jams and other exposed foodstuffs. Colonies grow over a wide temperature range, up to 50°C, although optimal temperature for growth is 24°C. The characteristic fruiting bodies (Figure 1.6), which when formed are responsible for the color change from white to black, may be seen under the microscope.

Spores of *Aspergillus* are commonly present in air and can infest and germinate in pharmaceutical and cosmetic products, causing discoloration and spoilage. They are generally not as resistant to antimicrobial agents as are bacterial spores. Some *Aspergillus* strains produce characteristic carcinogens, the aflatoxins.

1.7.3.2 Candida albicans

This is a commensal yeast, which may cause oral and vaginal thrush. It has been occasionally implicated in infusion-related infections. It grows readily on conventional mycological media at room temperature (optimal growth at 25°C) or at 37°C. It is dimorphic, growing first as yeast cells, but with aging will form chlamydospores, which are more difficult to destroy. There are no temperature tolerance differences between the two forms. Viewed microscopically, it appears to possess septate hyphae, known as pseudomycelia, among the yeastlike cells. It is unpigmented and colonies have a creamy white appearance.

1.7.3.3 Zygosaccharomyces rouxii

This is another species of yeast that is able to grow in liquids and gels of high osmotic pressure. It is nonpathogenic but can cause spoilage if it infests pharmaceutical and toilet products.

ACKNOWLEDGMENTS

The authors acknowledge the considerable contribution made by W.B. Hugo (deceased) who prepared this chapter in its earlier form for the first edition of the *Guide* (1990).

REFERENCES

Anonymous. (1989). *Report of the Expert Advisory Committee on Biocides.* Her Majesty's Stationery Office, London.

Baird, R.M. (2000). Culture media used in pharmaceutical microbiology. In *Handbook of Microbiological Quality Control: Pharmaceuticals and Medical Devices.* Baird, R.M., Hodges, N.A., and Denyer, S.P., Eds. Taylor & Francis, London, pp. 22–37.

Corry, J.E.L. (1975). The water relationships and heat resistance of micro-organisms. *Prog. Ind. Microbiol.,* 17, 73–108.

Gilbert, P. (2004). Principles of microbial pathogenicity and epidemiology. In *Hugo and Russell's Pharmaceutical Microbiology.* 7th ed. Denyer, S.P., Hodges, N.A., and Gorman, S.P., Eds. Blackwell Publishing, Oxford, pp. 103–114.

Hodges, N.A. (2000). Pharmacopoeial methods for the detection of specified micro-organisms. In *Handbook of Microbiological Quality Control: Pharmaceuticals and Medical Devices.* Baird, R.M., Hodges, N.A., and Denyer, S.P., Eds. Taylor & Francis, London, pp. 86–106.

Lambert, P.A. (2004). Bacterial resistance. Resistance of bacterial spores to chemical agents. In *Russell, Hugo, & Ayliffe's Principles and Practice of Disinfection, Preservation and Sterilization.* Fraise, A.P., Lambert, P.A., and Maillard, J.-Y., Eds. Blackwell Publishing, Oxford, pp. 184–190.

Newby, P. (2000). Rapid methods for enumeration and identification in microbiology. In *Handbook of Microbiological Quality Control: Pharmaceuticals and Medical Devices.* Baird, R.M., Hodges, N.A., and Denyer, S.P., Eds. Taylor & Francis, London, pp. 107–119.

Russell, A.D. (1982). *The Destruction of Bacterial Spores.* Academic Press, London.

Skinner, F.A. (1968). The limits of bacterial existence. *Proc. R. Soc. B.,* 171, 77–89.

FURTHER READING

The following texts provide comprehensive accounts of general microbiology and bacteriology, identification methods, and biochemistry.

Cowan, S.T. (1993). *Cowan and Steel's Manual for the Identification of Medical Bacteria.* 3rd ed. Cambridge University Press, Cambridge, U.K.

Denyer, S.P., Hodges, N.A., and Gorman, S.P., Eds. (2004). *Hugo and Russell's Pharmaceutical Microbiology.* 7th ed. Blackwell Publishing, Oxford.

Dring, G.J., Ellar, D.J., and Gould, G.W. (1985). *Fundamentals and Applied Aspects of Bacterial Spores.* Academic Press, London.

Krieg, N.R. and Holt, J.G., Eds. (1984). *Bergey's Manual of Systematic Bacteriology.* 9th ed. Vol. 1. Williams & Wilkins, Baltimore, MD.

Parker, T. and Collier, L.H., Eds. (1990). *Topley and Wilson's Principles of Bacteriology, Virology and Immunity.* 8th ed. Edward Arnold, London.

Stryer, L. (1988). *Biochemistry.* 3rd ed. W.H. Freeman, San Francisco.

2 Microbial Contamination: Spoilage and Hazard

Sally F. Bloomfield

CONTENTS

2.1 INTRODUCTION

Any pharmaceutical product, whether manufactured in the hospital or industrial environment, has the potential to be contaminated with microorganisms, which may include bacteria, yeasts, or molds. Microbial contamination may originate from the raw materials or may be introduced during manufacture. Products may also become contaminated during storage and use.

Microbial contamination in pharmaceutical products represents a potential hazard for two reasons. First, it may cause product spoilage; the metabolic versatility of microorganisms is such that any formulation ingredient from simple sugars to complex aromatic molecules may undergo chemical modification in the presence of a suitable organism. Spoilage will not only affect therapeutic properties but may also discourage patient compliance. Second, product contamination represents a health hazard to the patient, although the extent of the hazard will vary from product to product and patient to patient, depending on the types and numbers of organisms present, the route of administration, and the resistance of the patient to infection.

The most logical approach to minimizing the microbial hazard would be to specify that all products are manufactured as sterile products in single-dose packs. Although products introduced into sterile areas of the body, or those coming into contact with the eye or products applied to broken skin and mucous membrane, are manufactured in a sterile form, sterility is deemed neither appropriate nor commercially viable for oral or other topical products at the present time. Thus an acceptable level of microbial quality assurance needs to be established for such products.

Similar problems are encountered with medical devices. As these products are intended for use on or in areas of the body where defenses against infection are most likely to be impaired, microbial quality assurance represents a crucial issue.

In attempting to set suitable microbial limits or standards for both pharmaceuticals and medical devices, however, we are faced with the problem that, because of the multiple and varied factors involved, we cannot define precisely what level and types of contamination represent a hazard and what is safe—nor are we ever likely to. Thus by limiting microbial contamination, we are merely reducing the risks of infection and spoilage to a level dictated by our current knowledge and the financial resources available.

This chapter reviews current knowledge of microbial contamination of pharmaceutical products and medical devices and the associated hazards. Earlier reviews on this subject are given by Smart and Spooner (1972), Beveridge (1975, 1987), Baird (1988, 2004), and Spooner (1988).

2.2 CONTAMINATION OF PHARMACEUTICAL PRODUCTS

During the 1960s, mainly due to concern over the growing number of reported outbreaks of infection attributed to contaminated pharmaceuticals and cosmetics, a number of studies were initiated to assess the extent of the problem and to make recommendations. In the United Kingdom, two studies were initiated, one by the Public Health Laboratory Service (PHLS) to investigate hospital medicaments, and the other by the Pharmaceutical Society of Great Britain (PSGB) to study both products and raw materials (Anonymous 1971a, 1971b). The object of these studies was to assess the general microbial quality of manufactured products and to identify the sources of product contamination. Extensive investigations were also undertaken in other European countries (most notably Sweden) and the United States. The overall results of these studies are summarized in Table 2.1. Of the 6700 samples examined, about 27% were found to contain detectable contamination, although contamination rates varied from 2 to 80% according to product type. Contamination levels for individual samples ranged from less than 10 organisms per sample up to 10^5 to 10^6 organisms g^{-1} or mL^{-1}. Typical results from the PHLS survey (Table 2.2) showed that, although the overall contamination rate was similar for aqueous, gel, oily, and dry products, the frequency of heavy contamination was related to water availability in the product.

For the 6000 or more samples where isolates were identified, the majority were found to be Gram-positive bacilli and micrococci; these organisms are generally regarded as nonpathogenic. Yeasts and molds were also commonly found, mainly in creams and ointments. Of the species regarded as potentially pathogenic (Table 2.3) *Pseudomonas aeruginosa* and other pseudomonads were most frequently isolated. Other reported contaminants included species of *Alcaligenes, Flavobacterium, Acinetobacter, Achromobacter, Serratia,* and *Citrobacter.*

TABLE 2.1
Survey of Microbial Quality of Pharmaceuticals, Toiletries, and Cosmetics

Reference		Type of Product	Number Investigated	Number Contaminated
Baker	(1959)	Toiletries	5	2 (40%)
Kallings et al.	(1966)	Pharmaceuticals	134	91 (68%)
Ulrich	(1968)	Pharmaceuticals	696	535 (76%)
Hirsch et al.	(1969)	Pharmaceuticals	57	47 (82%)
Wolven and Levenstein	(1969)	Cosmetics	250	61 (24%)
Dunnigan and Evans	(1970)	Cosmetics	169	33 (20%)
Anonymous	(1971a)	Pharmaceuticals	1220	390 (31%)
Beveridge and Hope	(1971)	Pharmaceuticals	138	58 (42%)
Robinson	(1971)	Pharmaceuticals	279	85 (30%)
Wolven and Levenstein	(1972)	Cosmetics	228	8 (3%)
Ahearn et al.	(1973)	Cosmetics	200	3 (2%)
Baird et al.	(1976)	Pharmaceuticals	499	46 (9%)
Awad	(1977)	Pharmaceuticals	911	109 (13%)
Awad	(1977)	Pharmaceuticals	247	79 (32%)
Awad	(1977)	Pharmaceuticals	110	31 (28%)
Awad	(1977)	Pharmaceuticals	1462	184 (13%)
Baird	(1977)	Cosmetics	147	48 (32%)
Wallhaeusser	(1978)	Pharmaceuticals	17	12 (70%)
Total			6764	1814 (27%)

TABLE 2.2
Contamination Rates for Manufactured Pharmaceutical Products

Product Type	Total Percentage Contaminated	Percentage Contaminated with > 10^5 g^{-1} or mL^{-1}
Aqueous	35%	22%
Gels	34%	15%
Oily	26%	10%
Dry	33%	7%
Spirits	3%	3%
Total	32%	18%

Source: PHLS survey. Anonymous (1971a). *Pharm. J.,* 207, 96–99. With permission.

TABLE 2.3
Contaminants Isolated from 6008 Pharmaceutical, Cosmetic, and Toiletry Products during Surveys Carried Out from 1959 to 1978

Contaminant	Number of Isolates
Pseudomonas aeruginosa	46
Other pseudomonads	36
Escherichia coli	6
Staphylococcus aureus	2
Salmonella spp.	0
Enterobacter spp.	5
Klebsiella spp.	6
Proteus spp.	1

Note: As described in Table 2.1.

2.3 SOURCES OF MICROBIAL CONTAMINATION IN PHARMACEUTICAL PRODUCTS

In order to optimize methods for controlling microbial contamination of pharmaceuticals, it is necessary to understand the sources and routes from which contamination may originate. Microbial contamination from raw materials will invariably be transferred to the product, and further contamination may be introduced from the manufacturing equipment and environment, from the process operators, and from packaging materials.

In general, dry powders of synthetic origin used as raw materials in pharmaceuticals yield low bacterial counts, the organisms present being mainly aerobic spore bearers. In contrast, dry powders of natural origin may be heavily contaminated, often with Gram-positive spore-formers and molds, but on occasions with coliforms and Gram-negative species (Westwood and Pin Lim 1971; Anonymous 1971a,

1971b). By far the most common source of spoilage or pathogenic organisms is water or unpreserved stock solutions; for example, solutions such as peppermint water may become heavily contaminated with Gram-negative organisms if not properly prepared or if incorrectly stored. Typical Gram-negative waterborne genera that may grow even in distilled water include *Acinetobacter, Achromobacter, Enterobacter, Flavobacterium,* and *Pseudomonas;* species of enteric origin such as *Escherichia coli* and *Salmonella* spp., although not free-living, may survive for substantial periods in polluted water. Microbial contamination of raw materials including water is discussed in more detail in Chapter 3.

Contamination arising from a manufacturing and filling plant, which comes into direct contact with the product, must also be carefully controlled. Growth of contaminants, most particularly Gram-negative bacteria, readily occurs in dead spaces (e.g., joints and valves) where water and product residues accumulate; once established this contamination can be very persistent and difficult to eliminate.

Although contamination from the manufacturing environment is considered less important, because it is not in direct contact with the product, there is clear evidence that transfer can occur where control is not properly implemented. In a study of a hospital manufacturing unit (Baird et al. 1976), *P. aeruginosa* was isolated from products manufactured in the pharmacy. Isolation of the same strain of this organism from environmental sites within the pharmacy indicated that this was the source of contamination in many products. When strict environmental control procedures were implemented, the product isolation rate for *P. aeruginosa* fell to 2 to 3%.

Environmental contamination of dry surfaces such as floors and walls comprise mainly Gram-positive rods, cocci, and fungal spores. Gram-negative bacteria are more susceptible to the lethal effects of drying, but small numbers may persist on dry surfaces for substantial periods (Scott and Bloomfield 1990). In wet areas such as sinks and drains, particularly where stagnant water accumulates, typical waterborne species of *Pseudomonas* and *Acinetobacter* can readily survive and grow. Airborne contamination is mainly associated with dust and skin scales and is again mostly mold and bacterial spores and skin cocci. The reader is referred to Chapter 3 for a more detailed discussion of environmental contamination.

Contamination from process operators must be considered a significant hazard. During normal activity, loss of skin scales by shedding is about 10^4 min^{-1}; a large proportion of these skin scales will be contaminated with species of the normal skin flora. These are mainly nonpathogenic micrococci, diphtheroids, and staphylococci but may also include *Staphylococcus aureus* as part of the normal skin flora. Other organisms (e.g., *Salmonella* and *Escherichia coli*), not part of the resident skin flora, may also be carried transiently on the skin surface where poor hygienic practices exist among operators, and may be shed into the product via skin scales or direct contact. Control of manufacturing contamination is further discussed in Chapters 3 through 5.

Product containers and closures must be bacteriologically clean. They should be adequately designed and constructed to protect the product. This is of particular importance with sterile fluids. Felts et al. (1972) reported contamination of intravenous fluids with species of *Erwinia, Enterobacter,* and *Pseudomonas* genera. It was subsequently found that the space between the infusion bottle cap and neck had become contaminated when bottles were spray-cooled with tap water following sterilization. In another reported incident (Phillips et al. 1972), contamination of infusion fluids was traced to spray-cooling water contaminated with *P. thomasii* gaining ingress to infusion bottles. Further experiments demonstrated that small amounts of water became trapped above the rubber bung and in the screw threads of the bottlenecks (Coles and Tredree 1972). Robertson (1970) reported contamination of glucose saline infusion with *Trichoderma* and *Penicillium* spp. caused by use of cracked infusion bottles. Similar problems were also reported by Sack (1970) and Daisy et al. (1979).

In addition to contamination arising from raw materials and during manufacture, products may become contaminated during storage and use. Types and levels of "in-use" contamination are almost impossible to predict and only limited published data are available. Topical products, especially those packed in pots, are at particular risk; Baird et al. (1979a) carried out a study of "in-use" contamination of topical medicaments used for treatment of pressure sores. In samples taken from two hospitals, 20 to 25% were found to be contaminated with *Pseudomonas aeruginosa* (counts ranging from 10^2 to 10^6 CFU.g^{-1}) and 21 to 59% with *S. aureus* (counts ranging from 10^2 to 10^4 CFU.g^{-1}). In a third hospital, these organisms were not isolated from any products. On further investigation, it was found that, in the former hospitals, nurses removed creams from containers with their bare hands, and in the third hospital, medicaments were removed from pots with spatulae and applied with gloved hands. Introduction of this procedure into the other hospitals caused isolation rates to fall to 0.4 and 5% for *P. aeruginosa* and *S. aureus*, respectively. In a similar study of topical medicaments used by patients in a skin diseases hospital, it was found that 73% of 41 samples of emulsifying ointment were contaminated with *P. aeruginosa*, with counts ranging from 10^4 to 10^6 CFU.g^{-1} (Baird et al. 1980). *Proteus* spp. and *S. aureus* were also isolated from some samples. Savin (1967) reported a study of 194 pots of creams and ointments of which 6 were found to contain skin flora and 3 contained coliforms that were not present in samples from the pharmacy.

"In-use" contamination of ophthalmic ointments was studied by comparing open and unused tubes. Only 1 out of 24 unused tubes was contaminated, but 18 of the 50 used tubes contained organisms that included *S. epidermidis, S. aureus,* and fungi (Lehrfeld and Donnelly 1948).

Contamination of oral medicaments in use in the hospital environment was studied by Baird and Petrie (1981). Of 445 products sampled, 25.6% were found to be contaminated, mainly with aerobic Gram-positive rods, although Gram-positive cocci and Gram-negative rods were found in 2.9 and 3.1% of samples, respectively. Most of the Gram-negative isolates were identified as *Pseudomonas* spp. but in this survey *P. aeruginosa* was not found. Viable counts varied between 10^2 to more than 10^6 CFU.mL^{-1}. A survey of microbial contamination of pharmaceutical products in the home was reported by Baird et al. (1979b). Viable organisms were recovered from 14% of 1977 samples examined. One problem encountered in these two latter surveys was the lack of information on microbial quality of samples at the time of issue but it was generally concluded that medicines used in the home were less vulnerable to contamination than those used in hospitals.

Whereas contamination of sterile products during manufacture is now rarely a problem, contamination of intravenous (IV) fluids and apparatus during clinical use is still known to occur. From a literature survey (1971–1983), Denyer (1984) concluded that "in-use" contamination of IV fluids from bottled containers and collapsible plastic containers was of the order of 4.3 and 2.6%, respectively; in general, the organisms were present in low numbers. Where a breakdown in asepsis occurred, the growth-supporting character of the product had a significant influence on the infectious and endotoxin risk (Arduino et al. 1991).

In evaluating the results of surveys of "in-use" microbial contamination, it must be borne in mind that contamination levels in used products reflect not only the bioburden introduced by the patient but also the survival characteristics of the contaminant in the product. More recently (Koerner et al. 1997), an outbreak of septicemia was reported in an acute cardiology ward. The outbreak was attributed to *Enterobacter cloacae* contamination of heparin infusions that were compounded in a nonclinical area adjoining the ward from bags of IV fluids, which remained open for 24 h or more at room temperature, and multiple-use heparin vials.

2.4 FACTORS THAT AFFECT SURVIVAL AND GROWTH OF ORGANISMS IN PRODUCTS

The microbial quality of pharmaceuticals is determined not only by the types and levels of organisms introduced during manufacture, storage, and use, but also by their subsequent behavior within the product. Whereas small numbers of organisms introduced into a product during manufacture or use may be of little consequence to the patient, where these organisms undergo subsequent multiplication within the product the infection hazard will be significantly increased. Guyne (1973) and Felts et al. (1972) demonstrated that when small numbers of species such as *E. coli, P. aeruginosa, Klebsiella pneumoniae, S. aureus,* and *S. epidermidis* were introduced into IV solutions of saline, Ringer lactate, and 5% dextrose, Gram-negative bacilli increased in population over 24 h, whereas Gram-positive cocci remained stationary or died off. In all cases, bacterial counts of 10^6 CFU.mL^{-1} were recorded although the solutions showed no visible turbidity.

Many physicochemical factors can affect the fate of microorganisms entering a product; in practice many products are self-preserving such that residual contamination is reduced to undetectably low levels during storage. Furthermore, such contamination may comprise "stressed" cells, requiring specialized recovery conditions. For a more detailed account of this, the reader is referred to the companion *Handbook* (Baird 2000).

As microorganisms have an absolute requirement for water, water activity and water availability have a profound effect on survival and growth. Although dry products are less susceptible than aqueous products, contamination and spoilage of solid dosage forms have been reported in the literature on a number of occasions (Kallings et al. 1966, Lang et al. 1967, Komarmy et al. 1967, Eikhoff 1967, Beveridge 1975). Factors affecting survival and growth of microbial contamination in solid oral dosage forms have been studied by Blair (1989).

As microorganisms grow optimally at neutral or near neutral pH, formulations that are acid or alkaline in character are obviously less prone to spoilage. Other factors to be considered are nutrient availability, osmotic pressure, surface tension, and oxygen availability (see Chapter 1). The potential for exploitation of the self-preserving properties of pharmaceutical formulations is illustrated by the results of the PHLS survey (Anonymous 1971b); in general, it was found that low microbial counts were associated with products of low water availability, low pH, and high sucrose content.

2.5   MICROBIAL SPOILAGE OF PHARMACEUTICAL PRODUCTS

The ability of microorganisms to produce degradative spoilage in products depends on their ability to synthesize appropriate enzymes. Pharmaceuticals, cosmetics, foods, and other products are so much at risk because microorganisms are extremely versatile and adaptive in their ability to synthesize degradative enzymes. It should be noted, however, that within the whole range of organisms (bacteria, yeasts, and molds), degradative ability varies considerably such that certain species are frequently involved and others only rarely so. Degradative half-lives for individual formulation constituents can vary from a few hours up to several months or even years depending on the nature of the molecule, the product environment, the numbers and types of organisms present, and whether initially produced metabolites support further growth.

Low-molecular-weight substrates such as sugars, amino acids, organic acids, and glycerol are broken down by primary catabolic pathways. The enzymes for these pathways are constitutive in a wide range of organisms. Most compounds of pharmaceutical interest, however, are high-molecular-weight or aromatic molecules and are more resistant to degradation such that the enzymes responsible are produced by a smaller range of organisms.

Breakdown of proteins, polysaccharides, and lipids is brought about by proteinases and peptidases, polysaccharidases, and lipases, respectively (see Chapter 1). Many of these enzymes have low substrate specificity and will attack a wide range of compounds within the group. Polysaccharidases capable of hydrolyzing starch, agar, and cellulose are produced by a range of organisms including *Bacillus, Pseudomonas,* and *Clostridia.* Production of α-amylase is particularly prevalent in *Bacillus* spp. These species, together with *Aspergillus* and *Penicillium* spp., are the most common source of proteinase and peptidase enzymes causing breakdown of compounds such as gelatin. Production of lipase is less widely distributed and occurs most commonly among molds—hence the reason that spoilage of creams and emulsions is generally due to mold growth.

The ability of certain organisms to degrade aromatic compounds and long-chain hydrocarbons is remarkable because these types of molecules are stable to attack by many chemically active agents. This property, however, is restricted to a limited group of bacteria and some fungi, the most notable being the pseudomonads and related Gram-negative species. The capacity to degrade aromatic compounds is due

acetylsalicyclic acid

catechol

OCOCH₃ — esterase → OH COOH — oxygenase → OH OH

oxygenase

CH₃CO~SCoA

acetylcoenzyme A ← COOH COOH

CH₂COOH
|
CH₂COOH

FIGURE 2.1 Probable pathway for breakdown of acetylsalicylic acid by *Acinetobacter lwoffi*.

to the synthesis of oxygenase enzymes that catalyze direct incorporation of oxygen into the substrate molecule. For aromatic molecules, introduction of hydroxyl groups, catalyzed by oxygenase enzymes, affects the stable resonance of the benzene ring, thus making it more susceptible to attack by further oxygenase enzymes that catalyze the ring opening. The ring fission products are then further metabolized, producing intermediates of the tricarboxylic cycle, which are in turn metabolized via that cycle. The probable pathway for degradation of acetylsalicylic acid by *Acinetobacter lwoffi*, as elucidated by Grant et al. (1970), is illustrated in Figure 2.1.

Product spoilage arising from enzymatic degradation can manifest itself in a number of ways. Breakdown of the active ingredient will cause a loss of potency. Alternatively, degradation of formulation components may occur that may be associated with inefficient dosage delivery and reduced bioavailability. Observable formulation breakdown, such as the production of pigments and noxious odors, will make the product unacceptable to the patient. Microbial activity in the product can also result in production of potentially harmful toxins or the degradation of preservatives allowing the growth of other contaminant species.

2.5.1 BREAKDOWN OF ACTIVE INGREDIENTS

Laboratory experiments demonstrate the range of therapeutic agents that may be degraded by microbial enzymes. Degradation of antibiotics, particularly the degradation of penicillin by β-lactamase, has been extensively studied, as reviewed by Franklin and Snow (1989). Microbial enzymes that degrade chloramphenicol have been known since 1949 (Smith and Worrel 1949). Laboratory studies of microbial degradation of compounds such as aspirin, phenacetin, and paracetamol (Grant et al. 1970, Grant 1971, Hart and Orr 1974), alkaloids (Bucherer 1965, Grant and Wilson 1973, Kedzia et al. 1961), thalidomide (Mitvedt and Lindstedt 1970), and steroid esters (Cox and Sewell 1968, Brookes et al. 1982) have been reported in the literature. In some cases, the degradative pathways have been elucidated. Reports

IVERPOOL JOHN MOORES UNIVERSITY
LEARNING SERVICES

TABLE 2.4
Microbial Transformations of Biologically Active Compounds in Formulated Products

Product	Transformation	Organisms	Reference
Syrup of Tolu	Cinnamic acid → toluene-like product	*Penicillium* spp.	Wills (1958)
Atropine eye drops	Loss of atropine	*Corynebacterium* spp. *Pseudomonas* spp.	Kedzia et al. (1961) Beveridge (1975)
Belladonna and ipecacuanha paediatric mixture BPC	Loss of atropine	Mixed flora	Beveridge (1975)
Aspirin mixture	Aspirin hydrolysis	—	Beveridge (unpublished)
Paracetamol mixture	Paracetamol degradation	—	Beveridge (unpublished)
Aspirin and codeine tablets BP	Aspirin hydrolysis	—	Denyer (unpublished)
Prednisolone tablets	Localized steroid transformation	*Aspergillus* spp.	Beveridge (1987, and unpublished)
Hydrocortisone cream	Hydrocortisone androst-4-ene-11-ol-3:17 dione	*Cladosporium herbarum*	Cox and Sewell (1968)
Rose-hip preparation	Destruction of vitamin C	Anaerobic bacteria	Beveridge (unpublished)

Note: BPC = British Pharmaceutical Codex; BP = British Pharmacopoeia. Dashes indicate no organisms identified.

Source: From Denyer, S.P. (1988). In *Biodeterioration 7.* Elsevier Applied Science, London, pp. 146–151. With permission.

of microbial transformations of biologically active molecules in formulated products were reviewed by Denyer (1988), as shown in Table 2.4.

2.5.2 PRODUCTION OF TOXINS

Some species of microorganisms are capable of releasing toxic metabolites that may render the product dangerous to the patient. Probably the most important of these in relation to pharmaceutical products are pyrogens. These are mainly lipopolysaccharide components of Gram-negative bacterial cell walls, which can cause acute febrile reactions if introduced directly into the bloodstream. These toxins are heat stable and may be present even when viable organisms are no longer detectable. They are, however, only poorly adsorbed via the gastrointestinal tract and are therefore of little importance in oral preparations. Bacterial toxins that are associated with acute food-poisoning outbreaks have not been reported in pharmaceuticals, although the presence of toxin-producing *Bacillus cereus* (Arribas et al. 1988) and aflatoxin-producing *Aspergillus flavus* has been detected (Hikoto et al. 1978). It has been shown experimentally that pharmaceutical starches can support the formation of mycotoxins when artificially inoculated with toxigenic strains of *Aspergillus* (Fernandez and Genis 1979). Toxin production in pharmaceuticals is further discussed by Wallhaeusser (1977).

2.5.3 General Formulation Breakdown

Spoilage is usually observed as the result of a general breakdown of the formulation. Complex formulations are particularly prone to this type of spoilage, which may involve phenomena such as cracking or creaming of emulsions, viscosity changes, or separation of suspended material. Practical observations of formulation breakdown, as reviewed by Denyer (1988), are shown in Table 2.5. These changes will not only make the product aesthetically unacceptable but can also affect dosage delivery through separation of suspended particles or an emulsified oily phase, or the alteration of drug release characteristics from surface-damaged tablets. Physicochemical changes may in turn be the effect of many different degradative events within the product.

2.5.3.1 Degradation of Surfactants

Many of the surfactants used in pharmaceutical emulsions are subject to microbial degradation, particularly the nonionic surfactants. Ability to degrade surfactant molecules is again limited to a small range of organisms, particularly the pseudomonads. Susceptibility to biodegradation depends on the chemical structure, biodegradability decreasing as the length of the hydrocarbon chain and the degree of chain branching increases.

Formulations containing anionic surfactants are normally quite well protected because of the alkaline nature of the formulation. Nevertheless, alkyl and alkylbenzene sulphate and sulphonate esters can be hydrolyzed to the corresponding alcohol, which is then converted to a fatty acid. ω-Oxidation of the terminal methyl group is then followed by β-oxidation of the hydrocarbon chain and fission of aromatic rings.

Many nonionic surfactants are found to be biodegradable; although degradation pathways may not have been elucidated, these generally involve hydrolytic and oxidative pathways. Biodegradation of surfactants has been reviewed by Swisher (1987).

2.5.3.2 Degradation of Thickening and Suspending Agents

A wide variety of materials are used in pharmaceuticals for this purpose, many of which are biodegradable by extracellular enzymes. Degradation of these polymeric molecules yields monomers such as sugars, which serve as substrates for growth and further spoilage effects. Starch, acacia, sodium carboxymethylcelluloses, and dextran are all biodegradable. Agar is resistant to attack although agar-degrading organisms do exist. Polyethylene glycols except those of high molecular weight are degraded although polymers used in packaging are generally resistant. Gelatin is hydrolyzed by a wide range of commonly occurring microorganisms.

TABLE 2.5
Products Suffering Observable Degradative Changes Following Microbial Contamination

Type of Product	Examples	Observed Change	Spoilage Organisms	Reference
Solutions and mixtures	Simple aqueous solutions	Turbidity	Algae, molds, bacteria, yeasts	Smart and Spooner (1972)
	Concentrated dill water	Visible growth	Penicillium roqueforti	Parker and Barnes (1967)
	Indigestion mixtures	Color change (white to black)	Sulphate reducers	Spooner (1988)
	Magnesium hydroxide antacids	Ammoniacal odors	Coliforms	Robinson (1971)
Suspending/thickening agents	Carboxymethyl-cellulose gels/lubricants	Loss of viscosity	Trichoderma spp.	Beveridge (1975)
	Starch, tragacanth, acacia mucilages	Loss of viscosity	Unknown	Beveridge (1975)
	Tragacanth mucilage	Fermentation (alcohol and gas)	Yeasts	Denyer (unpublished)
Syrups	Syrup BP	Fermentation	Osmotolerant yeasts	Westwood and Pin-Lim (1971)
Emulsions	Syrup-containing cough remedies	Microbial deposits	Osmotolerant molds	Smart and Spooner (1972)
	Oil-water olive oil emulsions	Separation of phases, discoloration, unpleasant taste and smell	Trichoderma viride, Pseudomonas aeruginosa, Aspergillus flavus, Aspergillus niger	Beveridge (1975)
Creams	Calamine	Visible growth	Mold	Smart and Spooner (1972)
	Cosmetic	Visible surface growth	Penicillium spp.	Parker and Barnes (1967)
Tablets	Aspirin and codeine BP	Surface discoloration and acetic acid smell	Unknown	Denyer (unpublished)
	Nitrazepam	Discoloration, strange smell	Mold	Anonymous (1983)
	Prednisolone	Surface discoloration	Aspergillus spp.	Beveridge (unpublished)
	Sugar-coated tablets	Disintegration and "pitting" of coating	Penicillium spp. Aspergillus spp.	Parker and Barnes (1967)

Source: From Denyer, S.P. (1988). In *Biodeterioration 7*. Elsevier Applied Science, London, pp. 146–151. With permission.

2.5.3.3 Utilization of Humectants and Cosolvents

Glycerol and sorbitol support microbial growth at low concentrations. At low concentrations alcohol can serve as a substrate for growth.

2.5.3.4 Degradation of Sweetening, Flavoring, and Coloring Agents

Sugars and other sweetening agents can act as substrates for microbial attack, particularly by the osmophilic yeasts. Oral suspensions or emulsions containing sugars are liable to ferment with production of gas and acid, which may be sufficient to alter the stability of the formulation. Stock solutions of flavoring and coloring agents, and even simple salt solutions such as potassium citrate and calcium gluconate, will support growth of nutritionally nonexacting bacteria and yeasts.

2.5.3.5 Degradation of Oils and Emulsions

Microorganisms do not grow in a nonaqueous environment, but in an emulsified system, they may grow in the aqueous phase producing lipolytic enzymes that attack the triglyceride oil component at the oil–water interface liberating glycerol and fatty acids. Fatty acids may be further metabolized by β-oxidation of the alkyl chain with the production of ketones. Glycerol is a utilizable substrate for many organisms. Some insight into spoilage of emulsified systems has been gained from studies of the breakdown of engineering cutting oils (Guyne and Bennett 1959).

Ointments and oils are less prone to attack but spoilage may occur where these products contain traces of condensed water or are stored in a humid atmosphere. Spoilage of arachis oil and liquid paraffin has been reported (Rivers and Walters 1966).

2.5.4 Changes in Acceptability of the Product

Changes in product formulation may not only affect therapeutic potency but may also make the product unacceptable to the patient. In addition, other changes involving microbial degradation with the production of metabolic waste materials may make the product aesthetically unacceptable. Examples of such compounds include hydrogen sulphide, amines, ketones, fatty acids, alcohols, or ammonia, which impart a noxious odor or unpleasant taste to the product. Products may become discolored by various microbial pigments; organisms most frequently implicated are the pseudomonads, which produce pigments varying from blue to brown.

The most obvious sign of spoilage is visible growth either as colonies or a pellicle on the surface of a product. Alternatively, polymerization of sugar or surfactant molecules in syrups or shampoos produces a viscous slimy mass within the product, while aggregation of material can produce sediment in a liquid product or a gritty texture in a cream.

2.5.5 Degradation of Preservatives

Although antimicrobial agents are used in pharmaceutical products to prevent spoilage, many commonly used preservatives are themselves degraded by microorganisms, most notably the pseudomonads but also species of *Acinetobacter, Moraxella,* and *Nocardia.* Preservatives known to be susceptible to degradation include chlorhexidine, cetrimide, phenolics, phenylethylalcohol, benzoic acid, benzalkonium chloride, and the *p*-hydroxybenzoates (Beveridge and Hugo 1964a, Beveridge and Hugo 1964b, Hugo and Foster 1964, Beveridge and Tall 1969, Beveridge and Hart 1970, Grant 1970, Hart 1970, Kido et al. 1988).

2.6 INFECTION HAZARDS FROM MICROBIAL CONTAMINATION OF PHARMACEUTICALS

Experience has shown that products may show no sensory evidence of contamination but may still contain a growing microbial population. In the normal healthy adult this represents little problem unless the organism is a primary pathogen; but if the user is immunocompromised in any way, or the product is introduced into a normally sterile area of the body or applied to damaged skin, mucous membrane, or the eye, infection may occur. The clinical significance of microorganisms in pharmaceuticals was reviewed by Parker (1972) and Ringertz and Ringertz (1982). Overall, the infection risk depends on four factors: type of organism, infective dose, host resistance to infection, and route of administration.

2.6.1 TYPE OF ORGANISM

Of the organisms that have been isolated from pharmaceutical products, *Salmonella* is the only one that must be regarded as a primary pathogen, i.e., an organism that produces disease even when unintentionally administered by the normal (in this case oral) route. The majority of primary pathogens do not survive well outside the human or animal body. By contrast, organisms regarded as potential or conditional pathogens have been quite frequently reported as contaminants of pharmaceuticals. These are not generally harmful to the normal healthy adult, but readily become infectious where resistance mechanisms are impaired. Such opportunist pathogens include pseudomonads and enterobacteria and species of *Flavobacterium* and *Staphylococcus.*

Of the Enterobacteriacae, it is the free-living genera (*Enterobacter, Klebsiella,* and *Serratia*) that most usually occur in pharmaceuticals; although they may cause disease, these species are not found as part of the normal body flora. By contrast, species such as *Escherichia coli, Proteus* spp., and perhaps also some species of *Klebsiella* are commonly found as part of the normal body flora; although harmless in the bowel, they readily cause disease if transferred to other areas. These organisms have only a limited ability to survive outside the body and are less frequently found as contaminants of pharmaceuticals.

Pseudomonas aeruginosa is a free-living organism. Although it is present as part of the normal bowel flora in only about 4 to 6% of the normal population, it will readily cause infection of wounds or burns. *Burkholderia cepacia* (previously known as *Pseudomonas cepacia*) is another free-living species that has caused infection on a number of occasions (Levey and Guinness 1981, Berkelman et al. 1984). *Flavobacterium meningosepticum* is a pigmented waterborne species that can cause severe generalized sepsis in infants (Parker 1972). Up to 30% of the population may be either persistent or intermittent carriers of *S. aureus* (Armstrong-Esther and Smith 1976); although this organism does not thrive outside the body, it is quite resistant to drying and can persist in products for substantial periods of time.

Other species most frequently implicated in opportunistic nosocomial infections include *Acinetobacter anitratus, S. epidermidis,* and *Streptococcus pyogenes.* However, none of these organisms has yet been definitely implicated in outbreaks of infection due to contaminated pharmaceuticals.

2.6.2 INFECTIVE DOSE

Observations recorded in the literature indicate that the "infective dose" of an organism may vary considerably and may depend on several factors. Administered orally, quite large doses may be required; experiments with healthy adults suggest that the infective dose of *Salmonella* or *E. coli* may be as high as 10^6 to 10^7 organisms but may be as low as 10^2 to 10^3 (McCullough and Eisele 1951, Ferguson and June 1952). Outbreaks involving chocolate and Cheddar cheese suggest that the infective dose for some species of *Salmonella* may be as little as 50 to 100 organisms, and less than 10 organisms, respectively (Gill et al. 1983, Greenwood and Hooper 1983, D'Aoust 1985). In very small children, the infective dose for *Salmonella* may be as low as 200. Buck and Cooke (1969) found that ingestion of at least 10^6 *P. aeruginosa* may be necessary for intestinal colonization of normal persons. As far as topical preparations are concerned, experiments with healthy volunteers showed that an inoculum up to 10^6 *Staphylococcus aureus* may be required to produce pus, but as little as 10^2 may be sufficient where the skin is traumatized or occluded (Marples 1976). According to Crompton et al. (1962), intraocular injection of as little as 60 cells of *P. aeruginosa* was sufficient to cause infection in rabbits. Even relatively trivial abrasions of the eye can substantially increase the risk of infection.

2.6.3 HOST RESISTANCE TO INFECTION

Although, in most situations, the normal healthy adult has adequate resistance to infection, pharmaceutical products are frequently administered to people whose body defenses to infection are impaired. This situation may arise in patients with preexisting disease such as leukemia or diabetes. It may also be associated with immunosuppressive drug therapy, which may include steroid treatment or chemotherapy. Impaired local resistance is associated with surgical and accident wounds, burns, and pressure sores. It is also frequently associated with insertion of catheters or other surgical devices. Geriatric patients and babies, particularly during the first few days of life, show increased susceptibility to infection.

2.6.4 ROUTE OF ADMINISTRATION

Infection risks associated with contaminated pharmaceuticals will depend largely on their intended use. In general, the risk of infection will be much reduced for a drug given orally or applied to intact skin compared with a formulation used for treatment of abraded skin or mucous membrane, or a damaged eye.

For products that are introduced into normally sterile areas of the body, the potential risks are considerable. A review of microbial contamination of IV fluids during use suggested that the small numbers of mainly nonpathogenic organisms were generally well tolerated and that infection was a rare occurrence (Denyer 1984).

As reviewed earlier in this chapter, although some controlled studies with healthy adult volunteers have been used to assess infection risks, most of our knowledge regarding the clinical significance of contamination in medicinal products in fact comes from published reports of infection outbreaks resulting from use of such products. These are reviewed in the following sections.

2.6.4.1   Oral Medicines

The literature contains several reports of outbreaks of *Salmonella* infection resulting from contaminated oral products. In 1966, Kallings et al. (1966) reported 202 patients in Sweden who became infected. The infection was subsequently traced to thyroid tablets that were found to contain *Salmonella bareilly, S. muenchen,* and other fecal organisms in excess of 10^6 CFU.g^{-1} (Kallings 1973). In 1967, a number of cases of gastrointestinal tract infection in the United States were traced to carmine capsules contaminated with *S. cubana* (Lang et al. 1967, Komarmy et al. 1967, Eikhoff 1967). Infection from *S. agona* associated with contaminated pancreatin was reported in two children (Glencross 1972). *Salmonella schwarzengrund* and *S. eimsbuettel* were also found in batches of pancreatin causing infection in two children with cystic fibrosis (Rowe and Hall 1975).

Shooter et al. (1969) reported bowel colonization (although not infection) of a number of hospital patients following inadvertent administration of peppermint water containing 10^6 CFU.mL^{-1} *P. aeruginosa.*

An outbreak of septicemia caused by use of a thymol mouthwash contaminated with *P. aeruginosa* was recorded by Stephenson et al. (1984). Millership et al. (1986) also reported colonization of patients from communal ward use of mouthwashes and feeds that were heavily contaminated with coliforms.

2.6.4.2   Topical Administration

Although intact skin presents an efficient barrier against infection, for damaged skin the infection risks are substantially increased. Multidose containers, particularly those used by more than one person, are prone to contamination, especially in the hospital situation, and there are reports of infection outbreaks from the application of such products.

One of the earliest reports was a fatal outbreak of *Clostridium tetani* infection in babies caused by contaminated talcum powder (Tremewan 1946, Hills 1946). In 1966, Noble and Savin reported a number of cases of skin lesions that were traced to betamethasone cream contaminated with *P. aeruginosa* (Morse et al. 1967). McCormack and Kunin (1966) attributed an outbreak of umbilical sepsis to *Serratia marcescens* in the saline solution used to moisten the cord stump. In 1967, hand cream contaminated with *K. pneumoniae* was implicated in an outbreak of septicemia in an intensive therapy unit (Morse et al. 1967). Bassett et al. (1970) isolated

P. multivorans from infected operation wounds in nine hospital patients; the source of infection was subsequently traced to a diluted chlorhexidine-cetrimide disinfectant found to be contaminated with 10^6 pseudomonads per milliliter. Colonization and infection with *P. aeruginosa* arising from contaminated detergents, disinfectants, and topical medicaments including zinc-based barrier creams and emulsifying ointment were also reported by Victorin (1967), Cooke et al. (1970), Baird and Shooter (1976), Baird et al. (1979a), and Baird et al. (1980). Salveson and Bergan (1981) reported contamination from a range of organisms in a chlorhexidine hand cream used to prevent infection in catheterized patients; use of the contaminated cream was associated with urinary tract infection in a number of these patients. Clinical infection following application of an iodophor antiseptic solution contaminated with *P. aeruginosa* and *Burkholderia cepacia* was also reported by Parrot et al. (1982) and Berkelman et al. (1984).

2.6.4.3 Respiratory Preparations

Nosocomial respiratory tract infections involving Gram-negative bacteria have been reported quite frequently and in some instances this has been traced to nebulizers and humidifiers used in inhalation therapy (Favero et al. 1971, Pierce and Sanford 1973, Petersen et al. 1978). In two instances, aerosol solutions contaminated with *Klebsiella* spp. and *Serratia marcescens* (Sanders et al. 1970) have been implicated as the cause of pulmonary infection. An outbreak of respiratory infection was also traced to endotracheal tubes lubricated with contaminated lignocaine jelly (Kallings et al. 1966).

2.6.4.4 Disinfectants

Although disinfectants are used to kill microorganisms, some Gram-negative species, most notably the pseudomonads, can acquire resistance and grow in disinfectant solutions, utilizing the antimicrobial agent as a carbon source. Investigations associated with the use of contaminated disinfectants have been reviewed by Bloomfield (1988). Of the 42 incidents investigated, 24 were reported to be associated with, or implicated in, colonization of patients or outbreaks of infection. The majority of reports related to phenolics, quaternary ammonium, or chlorhexidine formulations. The contaminants were mainly pseudomonads, particularly *P. aeruginosa* and *Burkholderia* (formerly *Pseudomonas*) *cepacia*. Other Gram-negative species included *Alcaligenes faecalis, Enterobacter cloacae, E. agglomerans, Escherichia coli, Serratia marcescens,* and *Flavobacterium meningosepticum.* Gram-positive isolates were not recorded, but were reported in a later study of hospital-prepared disinfectants (Nkibiasala et al. 1989).

2.6.4.5 Ophthalmic Preparations

Although sterility has been a pharmacopoeial requirement for ophthalmic drops and lotions since 1966, there are a number of reports in the literature, particularly prior to 1966, of infections resulting from the use of contaminated eye preparations. The organism of major concern was again *P. aeruginosa,* which produces corneal ulceration and can cause blindness.

In 1952, Theodore and Feinstein published a paper indicating strong correlation between *Pseudomonas* infections in clinical practice and contaminated eye preparations. Allen (1959) estimated that in the United States at least 200 people each year were suffering loss of eyesight from use of contaminated eye drops. Crompton et al. (1964) reported evidence of eye infections from use of eye drops contaminated with a species of *Aerobacter.* Ringertz and Ernerfeldt (1965) and Kallings et al. (1966) reported *P. aeruginosa* infections arising from contaminated batches of neomycin and cortisone eye ointments, respectively. Ayliffe et al. (1966) reported infections causing blindness from the use of saline contaminated with *P. aeruginosa* during intraocular operations. More recently, a number of cases of keratitis were traced to eye drops contaminated with *S. marcescens*, although the organisms appeared to have originated from the dropper caps (Templeton et al. 1982).

A number of serious infections from the use of contaminated eye cosmetics have also been reported (Wilson et al. 1971, Ahearn et al. 1973, Wilson et al. 1975, Wilson and Ahearn 1977, Reid and Wood 1979). Infecting organisms included not only bacterial species but also various yeasts and fungi such as *Candida albicans, Fusarium,* and *Penicillium* spp.

2.6.4.6   Irrigating Fluids and Dialysis Fluids

The literature contains several reports of urinary tract infections caused by contaminated irrigating fluids (Last et al. 1966, Mitchell and Hayward 1966). An outbreak of *Pseudomonas* infection from a contaminated urinary catheter kit was traced to the germicidal cleansing solution, which was contaminated with 10^3 bacteria per milliliter (Hardy et al. 1970).

Hemodialysis represents a particular problem. Favero et al. (1974) reported two cases of septicemia from dialysis fluid contaminated with *P. aeruginosa.* Curtis et al. (1967) described an outbreak of bacteremia due to *Bacillus cereus* contamination of the dialysis fluid.

2.6.4.7   Injections and Infusions

Undoubtedly the most serious infection outbreaks are those associated with contaminated injection fluids where bacteremic shock and, in some cases, death of the patient may result. In 1970, Robertson reported two cases of infection arising from glucose-saline solution contaminated with *Trichoderma* and *Penicillium* spp. A number of cases of sepsis following the administration of fluids were reported by Felts et al. (1972). Species of *Erwinia, Enterobacter cloacae,* and *Pseudomonas stutzeri* were isolated from the fluids. Phillips et al. (1972) reported an outbreak of infection involving 40 hospital patients following administration of infusions contaminated with *P. thomasii.* The most notorious outbreak in the United Kingdom resulted in the death of six hospital patients following the administration of contaminated dextrose solution (Meers et al. 1973). The contents of 155 bottles of the suspect batch were examined, of which approximately one third were cloudy. *Klebsiella aerogenes, E. cloacae, Erwinia herbicola, P. thomasii,* coryneforms, and other species of Enterobacteriacae were isolated from the fluids. Lapage et al. (1973) identified

80 bacterial strains, chiefly enterobacteria, pseudomonads, and coryneforms associated with two outbreaks of infection from contaminated IV fluids.

At the present time, the preparation of total parenteral nutrition (TPN) solutions probably represents the area of greatest concern (Freund and Rimon 1990). In 1994, contaminated TPN solutions prepared in a hospital pharmacy caused the death of two children (Anonymous 1994). The contaminating organism was *E. cloacae*. DNA typing showed that the organism isolated from the children was the same as an organism isolated from the disposable tubing used during the compounding process. During the same year, an outbreak of *Serratia odorifera* septicemia associated with contaminated parenteral nutrition fluids was reported in South Africa (Frean et al. 1994). The organism was recovered from surfaces inside the flexible film isolator system in which the solution had been prepared, despite routine decontamination procedures having been performed shortly before use.

Even during use, lipid-based medications carry significant microbiological risk. For instance, during the early 1990s, a series of bacteremic episodes was traced to the administration of propofol (2,6 di-isopropylphenol) anesthetic formulated in a soybean oil emulsion (Bennett et al. 1995, Kuehnert et al. 1997). This sterile product was unpreserved and proved a vehicle for bacterial growth in circumstances of poor asepsis and delayed administration (Arduino et al. 1991, Farrington et al. 1994, Crowther et al. 1996).

2.7 CONTAMINATION OF MEDICAL DEVICES

As stated previously, because medical devices by definition are intended for use on or in areas of the body where defenses against infection are most likely to be impaired, microbial quality assurance for these items represents a crucial issue. The high-risk nature of medical devices suggests that use of sterile single-use items is preferable in all cases, but in many situations, most particularly for highly sophisticated endoscope equipment, this is not an option. For many of these items, the issues of "in-use" contamination represent the greatest problem; the 1981 survey of the prevalence of nosocomial infection (Meers et al. 1981) indicated that 9.2% of all hospital patients acquire some sort of infection and that 30.3% of these infections are urinary tract infections, mainly associated with indwelling catheters.

The factors that determine the likelihood of either colonization or clinical infection following contact with contaminated devices are the same as those for pharmaceutical products—namely the type and number of organisms, the health of the patient, and the route of administration or introduction. These factors have already been considered in Sections 2.6.1 through 2.6.4. As with pharmaceutical products, however, most knowledge regarding the clinical significance of contaminated medical devices comes from published reports of infection outbreaks from the use of such products.

Flexible endoscopy is now a common clinical procedure. However, because endoscopy procedures carry a high risk of cross-infection and because these fragile heat-sensitive items can only be decontaminated by disinfection rather than sterilization procedures, outbreaks of infection associated with contaminated endoscopes are relatively more frequent than those associated with medical devices where heat or irradiation sterilization is routinely applied. Transmission of infection by gastrointestinal endoscopy and bronchoscopy was reviewed by Spach et al. (1993). This review covered 281 infections transmitted by gastrointestinal endoscopy and 96 infections transmitted by bronchoscopy. The infections ranged from asymptomatic colonization to serious disease and fatality. Agents most usually identified with transmission of infection included *Salmonella* spp. and *P. aeruginosa* in gastrointestinal endoscopy, and *Mycobacterium tuberculosis*, atypical mycobacteria, and *Pseudomonas* spp. in bronchoscopy. In general, *M. tuberculosis and Salmonella* spp. tended to contaminate endoscopes from an infected patient, whereas *Pseudomonas* spp. and atypical mycobacteria originated from aqueous reservoirs such as those used in cleaning and disinfection. Single incidents of infection associated with *Serratia marcescens, Staphylococcus epidermidis, Citrobacter freundii, Enterobacter aerogenes, Helicobacter pylori, Trichosporon beigelli,* and *Strongyloides stercoralis* were also reported. The reviewers reported little evidence of viral agents transmitted by endoscopy apart from a single case of hepatitus B virus (HBV) (Birnie et al. 1983). Two studies, however, showed the potential for endoscope transmission of HBV. In one study, iodophor alcohol was ineffective in removing the virus from cytology brushes and biopsy forceps (Bond and Moncada 1978), and in the other, contaminated biopsy forceps exposed to chlorhexidine and cetrimide were found to be positive for HBV (McClelland et al. 1978). A further case of hepatitis associated with endomyocardial biopsy was subsequently reported (Drescher et al. 1994).

Contact lens wear is often associated with a range of corneal infections (Schein et al. 1989, Dart et al. 1991, Stapleton et al. 1993). Gram-negative bacteria, particularly *P. aeruginosa,* have been frequently implicated as causative agents in up to 70% of contact lens–associated microbial keratitis as compared with non-contact lens–related infections where Gram-positive organisms are most usually involved (Galentine et al. 1984). In addition, Gram-negative species in large numbers including *P. aeruginosa, Serratia marcescens,* and *Haemophilus influenzae* have been recovered from the contact lenses of wearers with acute corneal inflammatory response (Holden et al. 1996). By contrast, high numbers of Gram-negative bacteria have rarely been isolated from contact lenses of asymptomatic wearers. Contamination normally appears to be infrequent and typically involves less than 30 CFU per lens (Mowrey-McKee et al. 1992, Hart et al. 1993). Coagulase-negative staphylococci are most usually isolated from lenses following wear although isolation of *Staphylococcus aureus*, streptococci, and Gram-negative bacteria are occasionally reported (Fleiszig and Efron 1992, Gopinathan et al. 1997). Contact lens contamination is usually attributed to lens handling but may also be derived from contaminated storage cases (Mowrey-McKee et al. 1992). Where lenses have been sampled following storage, higher rates of contamination with Gram-negative bacteria are reported (Lipener et al. 1995). *Acanthamoeba* contamination of contact lenses was reviewed by Larkin et al. (1990).

For medical devices such as catheters, which are supplied to the patient in a sterile condition, the main hazard is contamination in use, which is largely then a clinical problem. Clinical studies suggest that infections associated with catheterization can include a broad range of Gram-negative and Gram-positive species, but there is general agreement that there is a close correlation between infection and the duration of catheterization. Studies and reviews of intravenous catheter-associated infections given by Knudsen et al. (1993), Raad et al. (1993), and Egebo et al. (1994) indicated that reported colonization rates for catheters varied from 3.8 to 47%, with estimated incidence of septicemia between 2.5 and 25%. The major route of contamination is not fully established. Although some studies suggest that infections mainly originate from the insertion site, other studies implicate catheter-administration set junctions as the most important source. Other studies of infections associated with urinary drainage catheters and epidural catheters have been done by Wille et al. (1993), Stickler and Zimakoff (1994), and Holt et al. (1995).

2.8 OVERVIEW

Substantial improvements in the microbial quality of manufactured products have been seen in the last 30 or so years. For pharmaceutical products, this can be illustrated by comparing the results of the 1971 PHLS survey (Anonymous 1971a) with the results of a more recent survey of hospital-manufactured products carried out over the period 1975–1983 by Baird (1985). Whereas the PHLS survey indicated that 2.7% of 1220 product samples were contaminated with *P. aeruginosa* and 15% with other Gram-negative rods, the more recent survey (13,000 samples) showed only two isolates of *P. aeruginosa* with only 0.24% of samples containing Gram-negative rods. Despite these improvements, however, there is no room for complacency because outbreaks of infection continue to be reported from time to time (Baird et al. 1980, Salveson and Bergen 1981, Stephenson et al. 1984, Millership et al. 1986, Anonymous 1994).

REFERENCES

Ahearn, D.G., Wilson, C.A., Julian, A.J., Reinhardt, D.J., and Ajello, G. (1973). Microbial growth in eye cosmetics: contamination during use. In *Developments in Industrial Microbiology*. Vol. 15. Plenum Press, New York, pp. 211–219.

Allen, H.F. (1959). Aseptic technique in ophthalmology. *Trans. Am. Ophthalmol. Soc.*, 57, 377–472.

Anonymous. (1971a). Microbial contamination of medicines administered to hospital patients. *Pharm. J.*, 207, 96–99.

Anonymous. (1971b). Microbial contamination in pharmaceuticals for oral and topical use. *Pharm. J.*, 207, 400–402.

Anonymous. (1983). Editorial. *Pharm. J.*, 230, 300.

Anonymous. (1994). Two children die after receiving infected TPN solutions. *Pharm. J.*, 252, 596.

Arduino, M.J., Bland, L.A., McAlliser, S.K., Aguero, S.M., Villarino, M.E., McNeil, M.M., Jarvis, W.R., and Favero, M.S. (1991). Microbial growth and endotoxin production in the intravenous anesthetic propofol. *Infect. Control Hosp. Epidemiol.*, 12, 535–539.

Armstrong-Esther, C.A. and Smith, J.E. (1976). Carriage patterns of *Staphylococcus aureus* in a healthy non-hospital population of adults and children. *Ann. Hum. Biol.*, 3, 221–227.

Arribas, M.L.G., Plaza, C.J., de la Rosa, M.C., and Mosso, M.A. (1988). Characterisation of *Bacillus cereus* strains isolated from drugs and evaluation of their toxins. *J. Appl. Bacteriol.*, 64, 257–264.

Awad, Z.A. (1977). In Use Contamination of Pharmaceutical Products and Their Possible Role in Hospital Cross Infection. Ph.D. thesis, University of London.

Ayliffe, G.A.J., Barry, D.R., Lowbury, E.J.I., Roper-Hall, M.J., and Walker, W. M. (1966). Postoperative infection with *Pseudomonas aeruginosa* in an eye hospital. *Lancet*, i, 1113–1117.

Baird, R.M. (1977). Microbial contamination of cosmetic products. *J. Soc. Cosmet. Chem.*, 28, 17–20.

Baird, R.M. (1985). Microbial contamination of pharmaceutical products made in a hospital pharmacy: a nine year survey. *Pharm. J.*, 231, 54–55.

Baird, R.M. (1988). Microbial contamination of manufactured products: official and unofficial limits. In *Microbial Quality Assurance in Pharmaceuticals, Cosmetics and Toiletries*. Bloomfield, S.F., Baird, R.M., Leak, R.E., and Leach, R., Eds. Ellis Horwood, Chichester, U.K., pp. 61–76.

Baird, R.M. (2000). Culture media used in pharmaceutical microbiology. In *Handbook of Microbiological Quality Control: Pharmaceuticals and Medical Devices*. Baird, R.M., Hodges, N.A., and Denyer, S.P., Eds. Taylor & Francis, London, pp. 22–37.

Baird, R.M. (2004). Microbial spoilage, infection risk and contamination control. In *Hugo and Russell's Pharmaceutical Microbiology*. 7th ed. Denyer, S.P., Hodges, N.A., and Gorman, S.P., Eds. Blackwell Publishing, Oxford, pp. 263–284.

Baird, R.M., and Petrie, P.S. (1981). A study of microbiological contamination of oral medicaments. *Pharm. J.*, 226, 10–11.

Baird, R.M., and Shooter, R.A. (1976). *Pseudomonas aeruginosa* infections associated with use of contaminated medicaments. *Br. Med. J.*, 2, 349–350.

Baird, R.M., Brown, W.R.L., and Shooter, R.A. (1976). *Pseudomonas aeruginosa* in hospital pharmacies. *Br. Med. J.*, i, 511–512.

Baird, R.M., Sturgiss, M., Awad, Z.A., and Shooter, R.A. (1979a). Microbial contamination of topical medicaments used in the treatment and prevention of pressure sores. *J. Hyg. Camb.*, 83, 445–450.

Baird, R.M., Crowden, C.A., O'Farrell, S.M., and Shooter, R.A. (1979b). Microbial contamination of pharmaceutical products in the home. *J. Hyg. Camb.*, 83, 277–283.

Baird, R.M., Awad, Z.A., and Shooter, R.A. (1980). Contaminated medicaments in use in a hospital for diseases of the skin. *J. Hyg. Camb.*, 84, 103–108.

Baker, J.H. (1959). That unwanted cosmetic ingredient: bacteria. *J. Soc. Cosmet. Chem.*, 10, 133–143.

Bassett, D.C.J., Stokes, K.J., and Thomas, W.R.G. (1970). Wound infection due to *Pseudomonas multivorans*: a water-borne contaminant of disinfectant solutions. *Lancet*, i, 1188–1191.

Bennett, S.N., McNeil, M.M., Bland, L.A., Arduino, M.J., Villarino, M.E., Perrotta, D.M., Burwen, D.R., Welbel, S.F., Pegues, D.A., Stroud, L., Zeitz, P.S., and Jarvis, W.R. (1995). Postoperative infections traced to contamination of an intravenous anesthetic, propofol. *New Engl. J. Med.*, 333, 147–154.

Berkelman, R.L., Anderson, R.L., Davis, B.J., Highsmith, A.K., Petersen, N.J., Bono, W.W., Cook, E.H., Mackel, M.S., Favero, M.S., and Martone, W.J. (1984). Intrinsic bacterial contamination of a commercial iodophor solution. *Appl. Environ. Microbiol.*, 47, 752–756.

Beveridge, E.G. (1975). The microbial spoilage of pharmaceutical products. In *Microbial Aspects of the Deterioration of Materials*. Lovelock, D.W. and Gilbert, R.J., Eds. Academic Press, London, pp. 213–235.

Beveridge, E.G. (1987). Microbial spoilage and preservation of pharmaceutical products. In *Pharmaceutical Microbiology*. 4th ed. Hugo, W.B. and Russell, A.D., Eds. Blackwell Scientific, Oxford, pp. 360–380.

Beveridge, E.G. and Hart, A. (1970). The utilization for growth and the degradation of *p*-hydroxybenzoate esters by bacteria. *Int. Biodet. Bull.*, 6, 9–25.

Beveridge, E.G. and Hope, I.A. (1971). Microbial content of pharmaceutical solutions. *Pharm. J.*, 207, 102–103.

Beveridge, E.G. and Hugo, W.B. (1964a). The resistance of gallic acid and its alkylesters to attack by bacteria able to degrade aromatic ring structures. *J. App. Bacteriol.*, 27, 304–311.

Beveridge, E.G. and Hugo, W.B. (1964b). The metabolism of gallic acid by *Pseudomonas convexa* X.1. *J. Appl. Bacteriol.*, 27, 448–459.

Beveridge, E.G. and Tall, D. (1969). The metabolic availability of phenol analogues to bacterium NCIB 8250. *J. App. Bacteriol.*, 32, 304–311.

Birnie, G.G, Quigley, E.M., Clements, G.A., Follet, E.A., and Watkinson, G. (1983). Endoscopic transmission of Hepatitis B virus. *Gut*, 24, 171–174.

Blair, T.A. (1989). Some Factors Influencing the Survival of Microbial Contamination in Solid Oral Dosage Forms. Ph.D. thesis, University of London.

Bloomfield, S.F. (1988). Biodeterioration and disinfectants. In *Biodeterioration 7*, Houghton, P.R., Smith, R.N., and Egins, H.O.W., Eds. Elsevier Applied Science, London, pp. 135–145.

Bond, W.W. and Moncada, R.E. (1978). Viral hepatitis B infection risk in flexible fibreoptic endoscopy. *Gastrointest. Endosc.*, 24, 225–230.

Brookes, F.L., Hugo, W.B., and Denyer, S.P. (1982). Transformation of betamethasone 17-valerate by skin microflora. *J. Pharm. Pharmacol.*, 34, 61P.

Bucherer, H. (1965). Microbial degradation of toxicants. IV. Microbial degradation of phenyl acetate, strychnine, brucine, vomicine and tubocurarine. *Zentbl. Bakt. Parasitkde. Abt. II*, 119, 232–238.

Buck, A.C. and Cooke, E.M. (1969). The fate of ingested *Pseudomonas aeruginosa* in normal persons. *J. Med. Microbiol.*, 2, 521–525.

Coles, J. and Tredree, R.L. (1972). Contamination of autoclaved fluids with cooling water. *Pharm. J.*, 207, 193–195.

Cooke, E.M., Shooter, R.A., O'Farrell, S.M., and Martin, D.R. (1970). Faecal carriage of *Pseudomonas aeruginosa* by newborn babies. *Lancet*, ii, 1045–1046.

Cox, P.H. and Sewell, B.A. (1968). The metabolism of steroids by *Cladosporium herbarum*. *J. Soc. Cosmet. Chem.*, 19, 461–467.

Crompton, D.O., Anderson, K.F., and Kennare, M.A. (1962). Experimental infection of the rabbit anterior chamber. *Trans. Ophthalmol. Soc. Aust.*, 22, 81.

Crompton, D.O., Murchland, J.B., and Anderson, K.F. (1964). Sterility of eye drops: a rare ocular pathogen. *Lancet*, i, 1391.

Crowther, J., Hrazdil, J., Jolly, D.T., Galbraith, J.C., Greacen, M., and Grace, M. (1996). Growth of microorganisms in propofol, thiopental, and a 1:1 mixture of propofol and thiopental. *Anesth. Analg.*, 82, 475–478.

Curtis, J.R., Wing, A.J., and Coleman, J.C. (1967). *Bacillus cereus* bacteraemia: a complication of intermittent dialysis. *Lancet*, i, 136–138.

Daisy, J.A., Abrutyn, E.A., and MacGregor, R.R. (1979). Inadvertent administration of intravenous fluids contaminated with fungus. *Ann. Intern. Med.*, 91, 563–565.

D'Aoust, J.Y. (1985). Infective dose of *Salmonella typhimurium* in Cheddar cheese: brief report. *Am. J. Epidemiol.*, 122, 717–720.

Dart, J.K.G., Stapleton, F., and Minassian, D. (1991). Contact lenses and other risk factors in microbial keratitis. *Lancet*, 338, 650–653.

Denyer, S.P. (1984). Microbial contamination of intravenous fluids during use. *Br. J. Pharm. Practice*, 6, 122–126.

Denyer, S.P. (1988). Clinical consequences of microbial action on medicines. In *Biodeterioration 7*, Houghton, D.R., Smith, R.N., and Egins, H.O., Eds. Elsevier Applied Science, London, pp. 146–151.

Drescher, J., Wagner, D., Haverich, A., Flick, J., Stachan-Kunstyr, R., and Vahegan, W. (1994). Nosocomial hepatitis B virus in cardiac transplant recipients transmitted during transvenous myocardial biopsy. *J. Hosp. Infect.*, 26, 81–92.

Dunnigan, A.P. and Evans, J.R. (1970). Report of a special survey: microbiological contamination of topical drugs and cosmetics. *Toilet Goods Assoc. J. Cosmet.*, 2, 39–41.

Egebo, K., Toft, P., Christensen, E.F., Steenson, P., and Jakobsen, C.-J. (1994). Contamination of central venous catheters: use of infusion lines does not increase catheter contamination. *J. Hosp. Infect.*, 26, 105–109.

Eikhoff, T.C. (1967). Nosocomial salmonellosis due to carmine. *Ann. Intern. Med.*, 66, 813–814.

Farrington, M., McGinnes, J., Matthews, I., and Park, G.R. (1994). Do infusions of midzaolam and propofol pose an infection risk to critically ill patients? *J. Anaesthesia*, 72, 415–417.

Favero, M.S., Carson, L.A., Bond, W.W., and Peterson, N.J. (1971). *Pseudomonas aeruginosa*: growth in distilled water from hospitals. *Science*, 173, 836–838.

Favero, M.S., Peterson, N.J., Boyer, K.M., Carson, L.A., and Bond, W.W. (1974). Microbial contamination of renal dialysis systems and associated health risks. *Am. Soc. Artif. Int. Organs*, 20, 175–183.

Felts, S.K., Schaffner, W., Melly, M.A., and Koenig, M.G. (1972). Sepsis caused by contaminated intravenous fluids: epidemiologic, clinical and laboratory investigation of an outbreak in one hospital. *Ann. Intern. Med.*, 77, 881–890.

Ferguson, W.W. and June, R.C. (1952). Experiments on feeding adult volunteers with *Escherichia coli* 111, B4, a coliform organism associated with infant diarrhoea. *Am. J. Hyg.*, 55, 155–169.

Fernandez, G.S. and Genis, M.J. (1979). The formation of aflatoxins in different types of starches for pharmaceutical use. *Pharm. Acta. Helv.*, 54, 78–81.

Fleiszig, S.M.J. and Efron, N. (1992). Microbial flora in the eyes of current and former contact lens wearers. *J. Clinical Microbiol.*, 30, 1156–1161.

Franklin, T.J. and Snow, G.A. (1989) in collaboration with Barrett-Bee, K.J. and Nolan, R.D., *Biochemistry of Antimicrobial Action*. Chapman and Hall, London.

Frean, J.A., Arntzen, L., Rosekilly, I., and Isaacson, M. (1994). Investigation of contaminated parenteral nutrition fluids associated with an outbreak of *Serratia odorifera* septicaemia. *J. Hosp. Infect.*, 27, 263–273.

Freund, H.R. and Rimon, B. (1990). Sepsis during parenteral nutrition. *J. Parenter. Enterol Nutr.*, 14, 39–41.

Galentine, P.G., Cohen, E.J., Laibson, P.R., Adams, C.P., Michaud, R., and Arentsen, J.J. (1984). Corneal ulcers associated with contact lens wear. *Arch. Ophthalmol.*, 102, 891–894.

Gill, O.N., Barlett, C.L.R., Sockett, P.N., and Vaile, M.S.B. (1983). Outbreak of *Salmonella napoli* infection caused by contaminated chocolate bars. *Lancet*, i, 574–575.

Glencross, E.J.G. (1972). Pancreatin as a source of hospital-acquired Salmonellosis. *Br. Med. J.*, 2, 376–378.

Gopinathan, U., Stapleton, F., Sharma, S., Willcox, M.D.P., Rao, G.N., and Holden, B.A. (1997). Microbial contamination of hydrogel contact lenses. *J. Appl. Microbiol.*, 82, 653–658.

Grant, D.J.W. (1970). The oxidative degradation of benzoate and catechol by *Klebsiella aerogenes (Aerobacter aerogenes)*. *Antonie van Leeuwenhoek*, 36, 161–177.

Grant, D.J.W (1971). Degradation of acetylsalicyclic acid by a strain of *Acinetobacter lwoffii*. *J. Appl. Bacteriol.*, 34, 689–698.

Grant, D.J.W. and Wilson, J.V. (1973). Degradation and hydrolysis of amides by *Corynebacterium pseudodiphtheriticum* NCIB 10803. *Microbios*, 8, 15–22.

Grant, D.J.W., De Szocs, J.C., and Wilson, J.V. (1970). Utilisation of acetylsalicyclic acid as sole carbon source and the induction of its enzymatic hydrolysis by an isolated strain of *Acinetobacter lwoffii*. *J. Pharm. Pharmacol.*, 22, 461–463.

Greenwood, M.J. and Hooper, W.L. (1983). Chocolate bars contaminated with *Salmonella napoli*: an infective study. *Br. Med. J.,* 286, 1394.

Guyne, C.J. (1973). Growth of various bacteria in a variety of intravenous fluids. *Am. J. Hosp. Pharm.,* 30, 321–329.

Guyne, C.J. and Bennett, E.O. (1959). Bacterial deterioration of emulsion oil. *Appl. Microbiol.,* 7, 117–125.

Hardy, P.C., Ederer, G.M., and Matsen, J.M. (1970). Contamination of commercially packed urinary catheter kits with *Pseudomonad* EO-1. *New Engl. J. Med.*, 282, 33–35.

Hart, A. (1970). The Stability of Some Esters of *p*-Hydroxybenzoic Acid to Microbial Attack. Ph.D. thesis, Sunderland Polytechnic.

Hart, A. and Orr, D.L.J. (1974). Degradation of paracetamol by a Penicillium species. *J. Pharm. Pharmacol.*, 26, suppl., 70P.

Hart, D.E., Reindel, W., Proskin, H.M., and Mowrey-McKee, M.F. (1993). Microbial contamination of hydrophilic contact lenses: quantitation and identification of microorganisms associated with contact lenses while on the eye. *Optometry and Visual Science,* 70, 185–191.

Hikoto, H., Morozumi, S., Wauke, T., Sakai, S., and Kurata, H. (1978). Fungal contamination and mycotoxin detection of powdered herbal drugs. *Appl. Environ. Microbiol.*, 36, 252–256.

Hills, S. (1946). The isolation of *Cl. tetani* from infected talc. *N.Z. Med. J.,* 45, 419–423.

Hirsch, J.I., Canada, A.T., and Randall, E.L. (1969). Microbial contamination of oral liquid medications. *Am. J. Hosp. Pharm.*, 23, 625–629.

Holden, B.A., Grant, T., La Hood, D., Willcox, M.D.P., Sweeney, D.F., Baleriola-Lucas, C., and Newton-Howes, J. (1996). Gram negative bacteria can induce acute red eye (CARE). *CLAO J.,* 22, 47–52.

Holt, H.M., Anderson, S.S., Anderson, O., Gahrn-Hansen, B., and Siboni, K. (1995). Infections following epidural catheterisation. *J. Hosp. Infect.,* 30, 253–260.

Hugo, W.B. and Foster, J.H. (1964). Growth of *Pseudomonas aeruginosa* in solutions of esters of *p*-hydroxybenzoic acid. *J. Pharm. Pharmacol.,* 16, 209–210.

Kallings, L.O. (1973). Contamination of therapeutic agents. In *Contamination in the Manufacture of Pharmaceutical Products*. Secretariat of the European Free Trade Association, Geneva, pp. 17–23.

Kallings, L.O., Ringertz, O., Silverstolpe, L., and Ernerfeldt, F. (1966). Microbiological contamination of medical preparations. *Acta Pharma. Suec.,* 3, 219–228.

Kedzia, W., Lewon, J., and Wisniewski, T. (1961). The breakdown of atropine by bacteria. *J. Pharm. Pharmacol.*, 13, 614–619.

Kido, Y., Kodama, H., Uraki, F., Uyeda, M., Tsuruoka, M., and Shabata, M. (1988). Microbial degradation of disinfectants. II. Complete degradation of chlorhexidine, *Eisei Kagaku,* 34, 97–101.

Knudsen, A.M., Rosdahl, V.T., Espersen, F., Frimodt-Moller, N., Skinhoj, P., and Bentzon, M.W. (1993). Catheter-related *Staphylococcus aureus* infections. *J. Hosp. Infect.,* 23, 123–131.

Koerner, R.J., Morgan, S., Ford, M., Orr, K.E., McComb, J.M., and Gould, F.K. (1997). Outbreak of Gram-negative septicaemia caused by contaminated continuous infusions prepared in a non-clinical area. *J. Hosp. Infect.,* 36, 285–289.

Komarmy, L.E., Oxley, M., And Brecher, G. (1967). Acquired salmonellosis traced to carmine dye capsules. *New Engl. J. Med.,* 276, 850–852.

Kuehnert, M.J., Webb, R.M., Jochimsen, E.M., Hancock, G.A., Arduino, M.J., Hand, S., Currier, M., and Jarvis, W.R. (1997). *Staphylococcus aureus* blood stream infections among patients undergoing electroconvulsive therapy traced to breaks in infection control and possible extrinsic contamination by propofol. *Anesth. Anagl.,* 85, 420–425.

Lang, D.J., Kunz, L.J., Martin, A.R., Schroeder, S.A., and Thompson, L.A. (1967). Carmine as a source of nosocomial salmonellosis. *New Engl. J. Med.,* 276, 829–832.

Lapage, S.P., Johnson, R., and Holmes, B. (1973). Bacteria from intravenous fluids. *Lancet,* ii, 284–285.

Larkin, D.F.P., Kilvington, S., and Easty, D.L. (1990). Contamination of contact lens storage cases by *Acanthamoeba* and bacteria. *Br. J. Opthalmol.,* 74, 133–135.

Last, P.M., Harbison, P.A., and Marsh, J.A. (1966). Bacteraemia after urological instrumentation. *Lancet,* i, 284–285.

Lehrfeld, L. and Donnelly, M.D.G. (1948). Contaminated ophthalmic ointments. *Am. J. Ophthalmol.,* 31, 470–471.

Levey, J.M. and Guinness, M.D.G. (1981). Hospital microbial environment: need for continued surveillance. *Med. J. Aust.,* 590–592.

Lipener, C., Nagoya, F.R., Zamboni, F.J., Lewinski, R., Kwitko, S., and Uras, R. (1995). Bacterial contamination rates in soft contact lens wearers. *CLAO J.,* 21, 122–124.

Marples, R.R. (1976). Local infections: experimental aspects. *J. Soc. Cosmet. Chem.,* 27, 449–457.

McClelland, D.B., Burrell, C.J., Tonkin, R.W., and Heading, J.C. (1978). Hepatitis B: absence of transmission via gastrointestinal endoscopy. *Br. Med. J.,* 1, 23–24.

McCormack, R.C. and Kunin, C.M. (1966). Control of a single source nursery epidemic due to *Serratia marcescens*. *Paediatrics,* 37, 750–752.

McCullough, N.B. and Eisele, C.W. (1951). Experimental human salmonellosis. *J. Infect. Dis.,* 88, 278–289.

Meers, P.D., Calder, M.W., Mazher, M.M., and Lawrie, G.M. (1973). Intravenous infusion of contaminated dextrose solution: the Devonport incident. *Lancet,* ii, 1189–1198.

Meers, P.D., Ayliffe, G.A.J., and Emmerson, A.M. (1981). Report on the national survey of infection in hospitals. *J. Hosp. Infect.,* 2, 1–51.

Millership, S.E., Patel, N., and Chattopadhyay, B. (1986). The colonization of patients in an intensive treatment unit with Gram-negative flora: the significance of the oral route. *J. Hosp. Infect.,* 7, 226–235.

Mitchell, R.G. and Hayward, A.C. (1966). Postoperative urinary-tract infections caused by contaminated irrigating fluid. *Lancet,* i, 793–795.

Mitvedt, T. And Lindstedt, G. (1970). Metabolism of thalidomide in *Pseudomonas aeruginosa* NCTC A7244. *Acta Path. Microbiol. Scand.,* 78, Sec. B, 488–494.

Morse, L.J., Williams, H.I., Grenn, F.P., Eldridge, E.F., and Rotta, J.R. (1967). Septicaemia due to *Klebsiella pneumoniae* originating from a handcream dispenser. *New Engl. J. Med.,* 277, 472–473.

Mowrey-McKee, M.F., Sampson, H.J., and Proskin, H.M. (1992). Microbial contamination of hydrophilic contact lenses. Part II: quantitation of microbes after patient handling and after aseptic removal from the eye. *CLAO J.,* 18, 240–244.

Nkibiasala, S.M., Devleeschouwer, M.J., Van Gansbeke, B., Rost, F., and Dony, J. (1989). Disinfectants prepared in a hospital pharmacy: assessment of their microbiological purity and antimicrobial effectiveness. *J. Clin. Pharm. Ther.*, 14, 457–464.

Noble, W.C. and Savin, J.A. (1966). Steroid cream contaminated with *Pseudomonas aeruginosa. Lancet*, i, 347–349.

Parker, M.S. and Barnes, M. (1967). Microbiological quality control of cosmetic and pharmaceutical preparations. *Soap Perfum. Cosmet.*, (December), 1–4.

Parker, M.T. (1972). The clinical significance of the presence of micro-organisms in pharmaceutical and cosmetic preparations. *J. Soc. Cosmet. Chem.*, 23, 415–426.

Parrot, P.L., Terry, P.M., Whitworth, E.N., Frawley, L.W., Coble, R.S., Wachsmith, I.K., and McGowan, J.E. (1982). *Pseudomonas aeruginosa* associated with contaminated polaxamer-iodine solutions. *Lancet*, ii, 683–684.

Petersen, N.J., Carson, L.A., Favero, M.S., Marshall, J.H., and Bond, W.W. (1978). Microbial contamination of mist therapy units of six paediatric wards. *Health Lab. Sci.*, 12, 41–46.

Phillips, I., Eykyn, S., and Laker, M. (1972). Outbreak of hospital infection caused by contaminated autoclaved fluids. *Lancet*, i, 1258–1260.

Pierce, A.K. and Sanford, J.P. (1973). Bacterial contamination of aerosols. *Arch. Intern. Med.*, 131, 156–159.

Raad, I., Umphrey, J., Khan, A., and Bodey, G.P. (1993). The duration of placement as a predictor of peripheral and pulmonary arterial catheter infections. *J. Hosp. Infect.*, 23, 17–26.

Reid, F.R. and Wood, T.O. (1979). *Pseudomonas* corneal ulcer: the causative role of contaminated eye cosmetics. *Arch. Ophthalmol.*, 97, 1640–1641.

Ringertz, O. and Ernerfeldt, F. (1965). Microbiological Contamination of Medical Products. Report to the Royal Medical Board of Stockholm.

Ringertz, O. and Ringertz, S. (1982). The clinical significance of microbial contamination in pharmaceutical and allied products. *Adv. Pharm. Sci.*, 5, 201–226.

Rivers, S.M. and Walters, V. (1966). The effect of benzoic acid, phenol and hydroxybenzoates on the oxygen uptake and growth of some lipolytic fungi. *J. Pharm. Pharmacol.*, 18, 45S.

Robertson, P.R. (1970) Fungi in fluids: a hazard of intravenous therapy. *J. Med. Microbiol.*, 3, 99–102.

Robinson, E.P. (1971) *P. aeruginosa* contamination of liquid antacids: a survey. *J. Pharm. Sci.*, 60, 604–606.

Rowe, B. And Hall, M.L. (1975). *Salmonella* contamination of therapeutic panel preparations. *Br. Med. J.*, 2, 51.

Sack, R.A. (1970). Epidemic of Gram-negative organism septicaemia subsequent to elective operation. *Am. J. Obstet. Gynecol.*, 107, 394–399.

Salveson, A. and Bergen T. (1981). Contamination of chlorhexidine cream used to prevent ascending urinary tract infections. *J. Hyg. Camb.*, 86, 295–301.

Sanders, C.V., Luby, J.B., Johnson, W.G., Barnett, J.A., and Sanford, J.P. (1970). *Serratia marcescens* infections from inhalation therapy medications: nosocomial outbreak. *Ann. Intern. Med.*, 73, 15–21.

Savin, J.A. (1967). The microbiology of topical preparations in pharmaceutical practice. 1. Clinical aspects. *Pharm. J.*, 199, 285–288.

Schein, O.D., Glynn, R.J., Poggio, E.C., Seddon, J.H., and Kenyon, K.R. (1989). The relative risk of ulcerative keratitis among users of daily wear and extended wear soft contact lenses. A case control study. *New Engl. J. Med.*, 321, 773–778.

Scott, E. and Bloomfield, S.F. (1990). The survival and transfer of microbial contamination via cloths, hands and utensils. *J. Appl. Bacteriol.*, 68, 271–278.

Shooter, R.A., Cooke, E.M., Gaya, H., Kumar, P., Patel, N., Parker, M.T., Thom, B.T., and France, D.R. (1969). Food and medicaments as possible sources of hospital strains of *Pseudomonas aeruginosa. Lancet*, i, 1227–1231.

Smart, R. and Spooner, D.F. (1972). Microbiological spoilage in pharmaceuticals and cosmetics. *J. Soc. Cosmet. Chem.*, 23, 341–346.

Smith, G.M. and Worrel, C.S. (1949). Studies on the action of chloramphenicol on enzymatic systems. *Arch. Biochem.*, 23, 341–346.

Spach, D.H., Silverstein, F.E., and Stamm, W.E. (1993). Transmission of infection by gastrointestinal endoscopy and bronchoscopy. *Anns. Intern. Med.*, 118, 117–128.

Spooner, D.F. (1988). Hazards associated with the microbiological contamination of non-sterile pharmaceuticals, cosmetics and toiletries. In *Microbial Quality Assurance in Pharmaceuticals, Cosmetics and Toiletries*. Bloomfield, S.F., Baird, R.M., Leak, R.E. and Leech, R., Eds. Ellis Horwood, Chichester, U.K., pp. 15–34.

Stapleton, F., Dart, J.K.G., and Minassian, D. (1993). Risk factors with contact lens associated keratitis. *CLAO J.*, 19, 204–210.

Stephenson, J.R., Head, S. R., Richards, M.A., and Tabaqchali, S. (1984). Outbreak of septicaemia due to contaminated mouthwash. *Br. Med. J.*, 289, 1584.

Stickler, D.J. and Zimakoff, J. (1994). Complications of urinary tract infections associated with devices used in long term bladder management. *J. Hosp. Infect.*, 28, 177–194.

Swisher, R.D. (1987). *Surfactant Biodegradation*. 2nd ed. Surfactant Science Series no. 18. Marcel Dekker, New York.

Templeton, W.C., Eiferman, R.A., Snyder, J.W., Melo, J.C., and Raff, M.J. (1982). *Serratia* keratitis transmitted by contaminated eye droppers. *Am. J. Ophthalmol.*, 93, 723–726.

Theodore, F.H. and Feinstein, R.R. (1952). Practical suggestions for the preparation and maintenance of sterile ophthalmic solutions. *Am. J. Ophthalmol.* 35(1959), 656–658.

Tremewan, H.C. (1946). Tetanus neonatorum in New Zealand. *N.Z. Med. J.*, 45, 312–313.

Ulrich, K. (1968). Microbial content in non-sterile pharmaceuticals. *Dansk. Tidss Farm.*, 42, 1–4, 50–55, 71–83, 257–263.

Victorin, L. (1967). An epidemic of otitis in newborns due to infection with *Pseudomonas aeruginosa. Acta Paediatr. Scand.*, 56, 344–348.

Wallhaeusser, K.H. (1977). Microbiological aspects on the subject of oral dosage forms. *Pharm. Ind.*, 39, 491–497.

Wallhaeusser, K.H. (1978). Microbial quality control of skin care preparations. *Cosmet. Toiletries*, 93, 42–48.

Westwood, N. and Pin Lim, B. (1971). Microbial contamination of some pharmaceutical raw materials. *Pharm. J.*, 207, 99–102.

Wille, J.C., Blusse, van Oud Alblas, A., and Thewesson, E.A.P.M. (1993). Nosocomial catheter-associated bacteriuria: a clinical trial comparing two closed urinary drainage systems. *J. Hosp. Infect.*, 25, 191–198.

Wills, B.A. (1958). Fungal growth in syrup of Tolu. *J. Pharm. Pharmacol.*, 10, 302–305.

Wilson, L.A. and Ahearn, D.G. (1977). Pseudomonas-induced corneal ulcers associated with contaminated eye mascaras. *Am. J. Ophthalmol.*, 84, 112–119.

Wilson, L.A., Kuehne, J.W., Hall, S.W., and Ahearn, D.G. (1971). Microbiol contamination in ocular cosmetics. *Am. J. Ophthalmol.*, 71, 112–119.

Wilson, L.A., Julian, A.J., and Ahearn, D.G. (1975). The survival and growth of microorganisms in mascara during use. *Am. J. Ophthalmol.*, 79, 596–601.

Wolven, A. and Levenstein, I. (1969). Cosmetics: contaminated or not. *Toilet Goods Assoc. Cosmet. J.*, 1, 34.

Wolven, A. and Levenstein, I. (1972). Microbiological examination of cosmetics. *Am. Cosmet. Perfum.*, 87, 63–65.

3 Microbial Ecology of the Production Process

David N. Payne

CONTENTS

3.1 INTRODUCTION

It is necessary to understand the ecology of the total production process in order to be in control of the microbiological quality of the finished pharmaceutical product. It is not sufficient to know only the quality of the finished product, because, particularly in the case of a preserved or sterile product, this gives no information on the bioburden inflicted on that product at its earlier stages of manufacture, and therefore, the safety margin afforded by any preservative or sterilizing process cannot be assessed.

Furthermore, it is essential not only to be aware of the microbial quality of the raw materials, but also of the packaging materials and the production environment itself. The relative importance of the raw materials and the environment will depend largely on the type and quantity of raw material used and on the type of finished product. The microbiology of the environment will thus be of little significance for a tablet or dry powder product whose major constituent is heavily contaminated; at the opposite extreme, an aseptically prepared product, made from sterile ingredients and packaged into sterile containers, will be highly susceptible to the environment quality. There are thus three major areas contributing to the ecology of the production process, namely the raw materials, the primary packaging, and the environment.

3.2 RAW MATERIALS

Untreated raw materials derived from natural sources may be expected to be heavily contaminated, whereas those of an essentially synthetic nature are usually free from all but incidental microbial contamination. In 1976, Grigo reviewed the reports of contamination in 282 raw materials and was able to group them into five categories based on their total count (Table 3.1).

   Category 1 raw materials were principally chemically synthesized or highly purified extracts of natural materials. Category 2 contained synthetic and naturally derived materials and category 3 contained predominantly plant extracts. Category 4 and category 5 were exclusively animal and plant products that had little or no processing. Of the 282 raw materials, 188 had details of the organisms isolated (Table 3.2). *Bacillus* species, Enterobacteriaceae, *Staphylococcus aureus,* and molds occurred in all categories, and the Enterobacteriaceae predominated in categories 3 through 5, i.e., those with raw materials of natural origin.

   Bonomi and Negretti, in 1977, surveyed 100 types of raw material in Italy with similar results. Substances of either animal or plant origin were most heavily contaminated, often with counts exceeding 10^4 CFU.g^{-1}, but inorganic chemicals, such as sodium bicarbonate, had counts of less than 10 CFU.g^{-1}. Undesirable or pathogenic bacteria were isolated from 34 of the 100 raw materials tested, all 34 being of natural origin. Of the 1038 samples tested from these 34 types of raw material, 1.2% contained sulphate-reducing *Clostridia,* 2.9% *Pseudomonas aeruginosa,* 2.6% *S. aureus,* 3.1% *Streptococcus faecalis,* 8.5% *Escherichia coli,* 0.48% salmonellae, and 4.0% other Enterobacteriaceae comprising *Proteus* (23 samples), *Citrobacter* (11 samples), *Klebsiella* (3 samples), *Enterobacter* (1 sample), *Erwina* (1 sample), *Providencia* (1 sample), and *Yersinia* (1 sample).

   Favet (1992), examining 20 drugs of plant origin, found total aerobic counts to range from 10^1 to 10^7 CFU.g^{-1} with 52% in excess of 10^5 CFU.g^{-1}. Aerobic spores and facultatively anaerobic bacteria often represented more than 50% of the aerobic count. Fungal counts were generally a factor of 10 lower than the bacterial count. Enterobacteriacae (but never salmonellae) were often present and in some cases at levels in excess of 10^4 CFU.g^{-1}.

   Baggerman and Kannegieter (1984) examined the inorganic salts sodium chloride, potassium chloride, calcium chloride, and sodium bicarbonate, together with glucose, fructose, sorbitol, mannitol, and glycine, all used in large-volume parenteral production, and found all to have minimal contamination, with total counts ranging from 0.1 to 10 CFU.g^{-1}.

   Crude herbal drugs such as powdered coptis may contain fungi at levels up to 10^4 CFU.g^{-1}, and bacteria at levels up to 10^5 CFU.g^{-1} (Hitokoto et al. 1978, Yokoyama et al. 1981). *Aspergillus* and *Penicillium* are the main components of the mycoflora

TABLE 3.1

Quality Categories of Raw Materials Based on Total Counts

Category	Maximum Average Total Count (CFU.g^{-1} or CFU.mL^{-1}) for Each Category
1	10
2	10^3
3	10^4
4	10^5
5	10^6

Source: From Grigo, J. (1976). *Zentbl. Bakt. Hyg. Abt. I. Orig. B.*, 162, 233–287. With permission.

TABLE 3.2

Organisms Isolated from Raw Materials

Microorganism	Percentages of Raw Materials Containing the Specified Organisms in Each Indicated Category				
	1	2	3	4	5
Sterile (no organisms detected)	34	0	0	0	0
Bacillus spp.	42	67	55	55	33
Molds	20	50	30	36	50
Staphylococcus aureus	4	18	15	18	33
Sarcina spp.	0	5	5	9	16
Streptococcus spp.	0	1	0	27	0
Escherichia coli	9	30	55	27	66
Salmonellae	0	4	10	18	83
Enterobacteriaceae (excluding *Escherichia coli* and salmonellae)	1	16	65	73	33
Pseudomonads	0	3	20	0	50
Flavobacterium spp.	0	1	0	0	0
Achromobacter spp.	1	1	5	0	0
Alcaligenes spp.	0	4	10	0	16
Clostridia	0	1	0	9	0

Note: See Table 3.1 for total count in category.

Source: Adapted from Grigo, J. (1976). *Zentbl. Bakt. Hyg. Abt. I. Orig. B.*, 162, 233–287. With permission.

of herbal drugs, with *Aspergillus niger, A. glaucus,* and *A. flavus* being the most prevalent species among the *Aspergillus. Mucor, Rhizopus, Cladosporium,* and *Aureobasidium* may also be isolated. The bacteria isolated are predominantly sporing *Bacillus* with micrococci and staphylococci also being present in some instances.

Ku et al. (1994) reported crude Chinese drugs such as Ginseng Radix, Glycyrrhizae Radix, and Zingiber Rhizome to be seriously contaminated. In a survey of

19 crude drugs analyzing a total of 119 samples, Ku et al. showed 47% to have total counts in excess of 10^4 CFU.g^{-1} with 23% having yeast and mold counts in excess of 10^4 CFU.g^{-1}. In addition, 66% of samples contained coliforms, 59% *Bacillus cereus,* 1% *Salmonella* spp., and 3% *Staphylococcus aureus.*

Data from our own surveys over the period 1984–1996 (Table 3.3) is in agreement with that of earlier published information. Untreated plant material such as senna pods and ispaghula husk may contain up to 10^6 CFU.g^{-1} of bacteria and 10^4 CFU.g^{-1} of molds with Enterobacteriaceae and pseudomonads present; whereas the chemically synthesized or highly refined materials are essentially free of contamination. As in earlier surveys, spore-forming *Bacillus* spp. are the predominant organisms in almost all raw materials.

3.3   WATER

Water is often the major component of pharmaceutical preparations and can be a very significant source of contamination. Product contamination may arise directly from the process water, indirectly from cleaning operations, or by cross-contamination from the wet areas of floors, sinks, and drains to the processing equipment.

3.3.1   METHODS OF WATER PREPARATION

3.3.1.1   Distillation

Distilled water, immediately after condensation, is sterile and with proper aseptic collection and storage precautions may be kept so. In the parenteral industry, distilled water is often maintained at 80°C and circulated around a ring system. This practice maintains the quality of the water by destroying any vegetative organisms that gain access to the system at outlet points, thereby preventing the generation of pyrogens. It is, therefore, important in the construction of circulating ring systems that no dead legs are created and that outlets are as close as possible to the ring to minimize local areas of lower temperature, where adventitious contamination can survive and grow, giving rise to pyrogens in the final product. Although proving effective for the parenteral industry, the maintenance of 80°C throughout the water system is prohibitively expensive for large-scale use in nonsterile pharmaceutical production.

3.3.1.2   Reverse Osmosis

Reverse osmosis (RO) involves forcing water, at a pressure in excess of its natural osmotic pressure, across the surface of a semipermeable membrane. Pure water is driven from the more concentrated upstream feed water side to the downstream permeate side. By configuring the latest generation of spirally wound RO membranes so that the product water passes through two consecutive stages (twin pass), virtually all the microorganisms, endotoxins, and other organic macromolecules can be removed from the feed water, together with up to 98% of the inorganic ions (Cross 1994). Microorganisms can nevertheless be recovered downstream of RO membranes, usually as a result of their subsequent ingress into the storage vessel or distribution system. RO membranes may be sanitized and depyrogenated with various chemicals,

TABLE 3.3
Microbiological Contamination of 34 Raw Material Types Representing 1527 Deliveries over the Period 1984–1996

Substance	Number of Deliveries with Total Counts (CFU.g^{-1} or CFU.mL^{-1}) in the Indicated Ranges						
	<10	10^1–10^2	10^2–10^3	10^3–10^4	10^4–10^5	10^5–10^6	>10^6
Gum acacia powder	—	—	1	8	—	—	—
Alginic acid	—	99	5	3	—	—	—
Aluminium hydroxide	28	3	—	—	—	—	—
Banana flavor	5	—	—	—	—	—	—
Carotene	—	3	—	—	—	—	—
Calcium carbonate	10	6	—	—	—	—	—
Calcium phosphate	11	—	—	—	—	—	—
Caramel flavor	6	1	—	—	—	—	—
Carmoisine soluble	2	2	2	—	—	—	—
Cocoa powder	—	7	17	14	—	—	—
Dextrose anhydrous	32	4	—	—	—	—	—
Ginger powder[a]	—	—	—	—	—	—	9
Grapefruit flavor	5	5	1	3	—	—	—
Hydrolyzed gelatin	9	—	—	—	—	—	—
Hydrogenated glucose syrup	15	32	1	—	—	—	—
Ispaghula husk[b]	—	—	—	6	50	33	—
Lactose	24	1	—	—	—	—	—
Light kaolin	—	2	—	—	—	—	—
Locust bean gum	—	1	6	1	—	—	—
Lime flavor	1	2	1	—	—	—	—
Magnesium stearate	—	2	—	—	—	—	—
Magnesium trisilicate	16	27	5	—	—	—	—
Maize starch	6	36	26	—	—	—	—
Malt extract	1	6	13	1	—	—	—
Mannitol	62	26	4	—	—	—	—
Orange flavor	5	—	—	—	—	—	—
Paracetamol/polyvinyl pyrrolidone	36	2	—	—	—	—	—
Raspberry flavor	3	—	—	—	—	—	—
Senna pods	—	—	—	—	4	11	5
Sodium alginate	—	59	83	39	4	—	—
Sodium bicarbonate	3	2	—	—	—	—	—
Sugar	95	262	164	6	—	—	—
Sunset yellow	6	—	—	—	—	—	—
Sterilized talc	45	2	—	—	—	—	—

Note: Dashes indicate zero deliveries.

[a] *Salmonella* spp. in one delivery.

[b] *Escherichia coli*, one delivery; pseudomonads, two deliveries.

depending on the polymer formulation of the membrane. RO is accepted by both the *European Pharmacopoeia* (2004) and the *United States Pharmacopeia 29* (2005) as a method of preparing purified water but not by the *European Pharmaopoeia* for water for injection (WFI). It is quite common in purified water systems to operate a RO unit upstream of a final deionization unit.

3.3.1.3 Deionization

Purified water prepared by deionization is the most common form of process water used for nonsterile pharmaceuticals, yet this has perhaps the highest potential for contamination. The source water for the production of purified water is normally town water of potable quality. Water suppliers chlorinate the water as it leaves the treatment works to maintain its microbiological quality throughout the distribution network. However, manufacturing plants at the extremities of the distribution network may receive water with little or no residual chlorine; under these circumstances, the potable water may be contaminated with pseudomonads and other Gram-negative waterborne organisms (Rhodes, personal communication). In this case, on-site chlorination or ozonization of the water supplied to the storage tanks may be necessary. Water authority regulations do not allow water to be connected to other equipment directly from a rising main and insist on a break tank. This should be as small as possible as it will tend to remain static at night, weekends, and during plant shutdowns, leading to proliferation of microorganisms in the tank and thus present a high loading to the deionizer of the purified water plant.

Deionized water systems may involve some combination of the following, all of which can harbor microorganisms: carbon filter, water softeners, cation- and anion-exchangers as either twin or mixed beds, and a storage and distribution system (Chapman et al. 1983a).

Carbon filters—Carbon filters are effective in removing chlorine and lower-molecular-weight hydrocarbons, but are less effective in removing high-molecular-weight organic materials, such as humic acid, which are common to surface water supplies. These filters are often used to minimize irreversible fouling of deionizing resins. Microbial growth is supported by organic molecules absorbed onto, and retained within, the activated carbon particles. The highest numbers of microorganisms occur toward the bottom of the bed because the residual chlorine is removed in the top portion. Organisms leave the filter on particle fines or when they are sloughed off by the local flow conditions and shear forces. Coliform organisms together with *Arthrobacter, Alcaligenes, Acinetobacter, Micrococcus, Corynebacterium,* and *Pseudomonas* species may be recovered from carbon filters (Camper et al. 1986).

Water softeners—Softeners, which are required when the source water has a high mineral content (especially Ca^{2+} and Mg^{2+}) are generally more susceptible to microbial contamination than cation- or anion-exchange resins, because the latter are generated with acid and alkali, which have a bactericidal effect (Chapman et al. 1983b). Sodium chloride regeneration solutions for softeners do not provide the necessary periodic bactericidal effect. Consequently, situations can develop in which the microbial population steadily increases over the course of the softener's operating life. Some bacteria will be removed or reduced during a backwash operation, but many softeners are designed with minimal headspace and there is insufficient backwash flow (Chapman et al. 1983b).

The brine makeup tanks associated with softeners can allow proliferation of halophiles and other salt-tolerant microorganisms unless adequate precautions are taken. These include agitation or recirculation of the tank contents to prevent stratification, which leads to high levels of microorganisms in the lower salt concentrations at the top of the tank. Maintaining saturated solutions, minimizing the brine holding time, and providing a clean closed system are worthwhile precautions to reduce the risk of microbial contamination and proliferation.

Ion-exchange resins—The cation or anion deionizers themselves, if correctly sized, may well present less of a problem than the other major elements of the system. Even though the deionizers will be inoculated repeatedly with bacteria from upstream in the system, the frequent regeneration by strong acid and alkali usually sanitizes them. Problems may arise with deionizers when either physical pockets of stagnation occur or after prolonged periods of inactivity, for example, when a standby unit comes into operation during regeneration of the primary unit. If regeneration is infrequent, the standby unit may be seriously contaminated by the time it is called into duty. The adhesive interactions between the microorganisms and the ion-exchange resins have been classified by Wood (1980) as nonspecific, pH-dependent, and salt-dependent.

High-molecular-weight organic molecules and their chlorine-oxidized degradation products will be concentrated onto the resin bed from the supply water. This occurs partly due to molecular absorption onto the bead surface and partly due to the migration of the organics into the matrix of the resin beads. Insoluble organic material will also be retained in the bed due to simple physical filtration. The resin beads do not, of themselves, support the growth of microorganisms; however, microbial proliferation on and between the resin beads is encouraged by the retained organic material (Dudderidge 1988). As the biofilm increases throughout the resin bed, so the chemical and microbiological quality of the water gradually declines. In addition, sudden surges through the bed, caused by shutdowns or start-ups, can lead to sloughing off of portions of the biofilm from the bead surfaces into the stream.

The extent to which microbial growth will occur within an ion-exchange bed will be a function of the type and level of organic material in the input water, the temperature of the bed, and the operating characteristics of the plant. Typically, viable counts made at 22°C can be expected to be 10^3 CFU.mL^{-1} from a deionizer prior to regeneration, if that regeneration is undertaken approximately every 24 h. If the system is so oversized as to need regenerating only weekly, then counts of 10^5 to 10^6 CFU.mL^{-1} are not uncommon. The organisms most often associated with deionizers are *Acinetobacter* spp., the *Alcaligenes* group, and *Pseudomonas* spp., although Gram-positive rods and cocci may also be found. The organisms originate mainly from the supply water, but may also be present in the resin beds as they are supplied. Karavanskaya et al. (1980) found ion-exchange columns to be heavily contaminated with *Moraxella* and *Acinetobacter* species, and Stamm et al. (1969) isolated and identified 44 different bacterial and fungal genera, including pathogens, from the effluent of cation exchangers. Schubert and Esanu (1972) reported

P. fluorescens, putida, lemoignei, acidovorans, cepacia (now *Burkholderia cepacia*), *maltophilia*, and *carophylii*, together with *Alcaligenes* group and Gram-positive rods and cocci, as being present in ion exchangers.

Continuous electrodeionization (CDI) combines alternating anion and cation ion exchange membranes with mixed ion exchange resin to remove ions from water under the influence of a direct electric field applied via two large electrodes. The applied DC current causes ions to migrate from the feed water in the resin filled area through the cationic or anionic membrane and away to waste thereby flushing the membranes. When the electrical potential that builds up between the cation and anion beads reaches 0.8 V, it discharges causing water to dissociate into hydrogen (H^+) and hydroxyl (OH^-) ions, which then act to regenerate the resin continuously. Water dissociation occurs to a much larger extent when the CDI unit is sited downstream of an RO unit because the water is already substantially depleted of ions by the RO before it is fed to the CDI (Ganzi and Parise 1990). The outer surface of most bacteria being electronegative (Mittleman and Geesey 1987), they are attracted to the anion exchange surfaces where the greatest degree of ionization occurs. These surfaces appear to be bactericidal and result in an environment that reduces the rate of bacterial proliferation. Ganzi and Parise (1990) reported operating a RO-CDI system for at least 10 weeks without sanitization continuing to produce water bacterial counts of <0.1 CFU.mL^{-1} and endotoxin levels of <0.25 IU.mL^{-1}. When required, effective cleaning or sanitization of the CDI can be achieved with 1% sodium percarbonate applied for 60 min (Ganzi and Parise 1990).

Once microorganisms are in the distribution system, a proportion will attach themselves to the surface of pipework, filters, storage tanks, and general fittings. Flow rates of 1 to 2 m.s^{-1} are normally recommended for water systems to minimize adhesion. In general, the slower the flow rate and the rougher the internal surfaces, the faster adhesion will occur. The rate of adhesion and subsequent growth will be system specific, but such biofilms can be difficult to remove and the system may require thorough cleaning followed by disinfection.

If stagnant water is allowed to occur in dead legs within a distribution system, counts of 10^6 CFU.mL^{-1} are quite easily attained, and this culture can effectively seed the main body of recirculating water. Ingress of environmental contaminants will occur if tank vents are not properly protected by bacterial filters and takeoff points are poorly handled. The practice of leaving flexible hoses connected to takeoff points is a particular example where stagnant water in the hose may contaminate the main ring via the valve or meter. Hoses need to be removed and drained immediately after use.

Distribution system—Deionizers may be situated between the raw or pretreated input water and the main storage tank, that is, outside the distribution ring, or within the distribution ring. In either case, accepting that contamination from the deionizers is almost inevitable, there must be some system of controlling the microbial load in the circulating ring. Two main approaches are used:

1. Ultraviolet (UV) light—UV lamps operating at a wavelength of 254 nm are most often used in the circulating ring main, but there may be occasions where they are better sited at the point of use. UV lamps are effective at

controlling the microbial load provided they are correctly sized to cope with the flow rate, optical clarity, and expected bioburden of the water. Correct maintenance of UV "sterilizers," particularly in the regular cleaning of the quartz sleeves covering the UV lamp, is essential if effective light penetration is to be maintained. Lamps need to be changed at the end of their rated hours as the UV output of a lamp decays with time to an ineffective level.

2. Filtration—Filtering water through 0.45- or 0.2-μm pore size membrane filters is a method widely used in the pharmaceutical industry for reducing the microbial burden. Prefiltration may be necessary to protect the expensive bacterial filters from fouling with larger particles or colloids in the source water and fines shed from the ion exchangers. Some waterborne bacteria will pass through a 0.45-μm pore size filter, and for microbiologically pure water, a maximum pore size of 0.2 μm should be used. The correct installation of filters within the housings is crucial if the microbiological quality of the water is to be maintained. Poorly seated filters allow bacteria to pass around the seal and contaminate the rest of the distribution system. Wherever possible, it is advisable to test the integrity of a filter in its housing immediately after installation by, for example, a bubble point or forward flow pressure hold test (see Chapter 11).

3.3.2 MICROBIOLOGICAL QUALITY

The microbiological purity required from a water system depends largely on its usage. The alert levels normally used are 100 CFU.mL for purified water and 10 CFU per 100 ml for water used for injection. An endotoxin limit of <0.25 IU.mL^{-1} also applies to water for injection. For further information on water sampling and water testing methods, including testing for endotoxins, the reader is referred to the companion *Handbook* (Baines 2000, Baird 2000, Millar 2000).

3.4 PACKAGING

Primary packaging should not be ignored in the total quality equation. It has a dual role in containing the product and in preventing contamination with microorganisms and the ingress of moisture that may lead to subsequent spoilage (see Chapter 16). The packaging can also act as a source of microbial contamination. In practice, when used for nonsterile dry products or preserved liquids, the packaging seldom contributes significantly to the total bioburden of the product. However, in the case of sterile products, particularly those aseptically prepared, any contamination in the container becomes significant and sterilization of the container is essential.

At the time of manufacture, glass containers are sterile and molded plastic containers are likely to contain only very low levels of contamination. Incidental contamination may arise during packing, storage, and shipment, particularly when cardboard or paperboard is used between the layers of stored bottles or unlined cardboard boxes are employed. Contamination is predominantly with spores of *Penicillium* spp., *Aspergillus* spp., and *Bacillus* spp. Laminates, metal foils, and blister-pack materials all have smooth impervious surfaces with a high-temperature stage employed in their manufacture and, therefore, have low surface microbial counts. Because they are stored and transported as wound reels, incidental contamination is restricted to the outer surface of the reel.

Negretti (1981) examined 4200 samples of pharmaceutical packaging materials and accessories (glass and plastic bottles, flexible metal tube, droppers, blisters, and cap liners). The samples most often contaminated were cap liners (96.4%), plastic bottles (94.8%), droppers (93.6%), and flexible metal tubes (89.7%); less frequently contaminated were the glass bottles (54.6%) and blisters (33.5%). All those contaminated had low microbial counts (only 2.54% in excess of 50 CFU per container) consisting mainly of sporing bacilli (42.4%) and molds (25%), but pathogenic or hygienically undesirable organisms were occasionally found (Table 3.4).

3.5 ENVIRONMENT

3.5.1 WAREHOUSE

Having obtained good quality raw materials and packaging components, it is important to maintain that quality throughout storage. Although most dry raw materials are nowadays supplied in polythene-lined sacks or kegs, possibly with pallet covers, there is still a potential for contamination if the standard of warehousing is inadequate. Warehouses with racking to hold individual pallets are ideal; air can then circulate around the pallets, allowing the outer packaging to dry out if it has become damp in transit. In unracked warehousing, the close proximity of uncovered pallets to each other may result in damp areas persisting, thereby allowing mold spores present on the outer paper sacks or the cardboard packaging components to germinate and grow. Pallets stored in contact with outer walls may become damp if the wall is at all porous, a situation particularly prevalent in older buildings, with the result that mold growth may occur. The contamination of the outer packaging can be transferred in the dispensaries to the raw material itself as the bags are handled. Small pockets of macroscopic mold on the outer bags can easily be missed by the operators in a busy dispensary. Care needs to be exercised if pallet covers are added at the receiving warehouse because goods must be dry before covers are applied. Conditions ideal for mold growth may be provided if polythene covers are used to overwrap damp materials.

It is essential that pest control measures, against rodents and insects for instance, are maintained, as droppings will be heavily contaminated with enteric pathogens. Strip curtaining or automatic doors should be used at all entrances to warehousing to reduce the risk of birds entering, perching, and nesting within the warehouse. Bird droppings, which are not only unsightly, are heavily contaminated.

Liquid raw materials preserved by virtue of their high osmotic pressure, for example, sugar solutions and malt extract, can be susceptible to mold growth if condensation forms on the inner surfaces of the storage vessels and falls back into the liquid, thereby causing surface areas of reduced osmotic pressure. Trace heating of tanks to prevent condensation or continuous stirring to mix in any condensation can prevent the problem in large tanks, but small tubs (25 kg) can sometimes be a

TABLE 3.4
Microflora Present on Containers

Type of Container	Number Examined	Number Contaminated	Bacilli	Mold	Staphylococcus aureus	Pseudomonas aeruginosa	Streptococcus faecalis	Enterobacteriaceae spp.
Glass bottle	1,000	546	308	103	25	4	0	K. pneumoniae 8, Escherichia coli 6
Plastic bottle	1,000	948	466	353	40	0	2	Escherichia coli 19, K. pneumoniae 7, Serratia 6, Enterobacter 5, Yersinia 5, Citrobacter 3
Metal tubes	1,000	897	521	248	64	0	5	Escherichia coli 11, Proteus 7, K. pneumoniae 4, Citrobacter 3, Enterobacter 2, Hafnia 2
Droppers	500	468	187	189	11	0	0	Escherichia coli 10, Proteus 8, K. pneumoniae 5, Enterobacter 5
Cap liners	500	482	257	151	14	0	0	K. pneumoniae 9, Escherichia coli 8, Proteus 7, Serratia 6
Blister	200	67	42	9	4	0	0	0

Source: Adapted from Negretti, F. (1981). *Boll. Chim.-Farm.*, 120, 193–201.

problem. In our experience, malt extract delivered in the summer and stored in a cool warehouse developed condensation on the lids and surface of the malt, allowing the mold spores present to germinate and grow.

3.5.2 MANUFACTURE OF NONSTERILE DRY PRODUCTS

In a manufacturing facility dealing with only dry powder mixing, granulation and drying, and final sacheting or tabletting, contamination of the product from the environment is minimal. The contamination that does occur will be predominantly *Bacillus* and mold spores from the environment dust, together with micrococci and staphylococci from skin scales shed by the operators. The major source of contamination is cross-contamination from the outer packaging to the raw materials during dispensing. However, in practice, with good handling procedures and suitable ventilation for dust control, cross-contamination is minimal and normally undetectable. Blending and mixing operations are now largely enclosed and ingress of airborne contamination is minimal.

Fluid bed driers and mixer granulator driers pass large quantities of air through the product during the drying and granulation stage and can potentially be a source of contamination depending on the siting of the input air duct. However, in practice, coarse filtration of the input air appears to be all that is required, and submicron filtration is rarely necessary to prevent large increases in microbial numbers during the drying stage. Bed temperatures of 80 to 100°C achieved in fluid bed driers and granulator driers cannot be relied upon to destroy vegetative organisms such as the Enterobacteriaceae that may, for instance, be present in a raw material of natural origin because the heat is essentially dry heat.

Aqueous granulation and drying can become a problem if drying is not carried out immediately or if temperature tray drying is carried out over an extended time. Proliferation of the flora originating from the raw materials may occur during the tray drying stage, which then dies out as the water activity is reduced. High spore counts may be all that remain in the finished product to indicate such a problem.

Tabletting machines exert some antimicrobial effects. The shearing forces and localized heat involved in pressing tablets is sufficient to destroy many mold spores and vegetative organisms, although *Bacillus* spores appear to survive. Reductions of 67 to 93% in total viable counts of a dry blended product have been reported to be produced by tabletting (Chesworth et al. 1977). Increasing compression pressures and tabletting speeds increase the antimicrobial effect (Fassihi and Parker 1987). Generally tablets produced by direct compression give lower microbial levels than those of the wet granulation method (Ibraham and Olurinola 1991).

Sachet or tablet filling operations have no effect on the microbiological quality of the product. The accumulation of product dust in vacuum transfer lines, hoppers, tabletting, and filling machines is largely a problem of good manufacturing practices (see Chapter 6), and does not pose a microbiological hazard due to the low water activity of the powder preventing growth from occurring.

The method used for cleaning equipment often has a major influence on the potential for microbial proliferation. Dry vacuuming is the best method as this does not produce conditions conducive to growth, but it may not always be appropriate for product changeovers. On occasions, washing with water may be the only effective form of cleaning, but carries with it the risk of presenting conditions ideal for microbial proliferation. Unless equipment is thoroughly cleaned and dried, growth is almost inevitable. Poorly cleaned equipment can result in small pockets of product, dried on the outside, but remaining damp in the center, where growth can occur. Subsequent use of the equipment results in sporadic incidences of contamination, which may include Gram-negative organisms originating from the cleaning water supply or molds from the previous product. Any areas of the plant that are not dry at the end of the cleaning operation must be regarded as potential areas for microbial growth. In addition, cleaning solutions themselves require careful attention. Dilute solutions of detergents may well support growth (see Chapter 2). Dilute citric acid solutions, used to return stainless steel equipment to a polished appearance, can become heavily contaminated with mold if not freshly prepared (unpublished observations).

3.5.3 Production of Nonsterile Liquids, Creams, and Ointments

Cross-contamination from the outer packaging of raw materials is possible in the dispensary, although in practice, with good air handling and dust extraction, it is minimal. *Bacillus* spp., mold spores from environmental dust, and micrococci and staphylococci from skin scales are the commonest contaminants at this stage.

Because water is involved in the production of liquids, creams, and ointment products and in the cleaning of the plants, there is a potential for microbiological contamination of the production environment. The greatest danger lies in cross-contamination from the manufacturing environment to the product. Water on the floor, in the drains and gullys of the manufacturing environment, and the wash areas, enables *P. aeruginosa,* other *Pseudomonas* spp., *Enterobacter* spp., and other Gram-negative bacteria to grow profusely. Counts of 10^6 to 10^7 CFU.mL^{-1} are easily attained in the wet areas if regular disinfection and drying is not carried out. However, the total elimination of contamination can never be effected, as there is always a reservoir of contamination in the drains. In addition, the feet of the operators and the wheels of pallet trucks are probably a major vector for transferring contamination from one area to another and it is, therefore, important to keep the floors and surfaces as dry as possible.

Similar situations were described in a study of environmental contamination in hospital manufacturing pharmacies, where *P. aeruginosa* was isolated from a variety of wet sites, including sinks, drains, taps, draining boards, tank water supplies, label dampers, and cleaning equipment (Baird et al. 1976). The same strains of *P. aeruginosa* were invariably isolated from contaminated products made in these pharmacies and subsequently from patients who used the products (Baird and Shooter 1976).

Good manufacturing practices are particularly important in preventing cross-contamination from the plant environment to the product (see Chapter 6). Particularly at risk are those operations that may have to be carried out in close proximity to the floor, such as hose connections.

Cleaning equipment can also be a significant source of contamination. Mops, buckets, cloths, scrubbing machines, and other items, particularly when stored wet, may provide suitable conditions for the proliferation of Gram-negative bacteria, especially *Pseudomonas* spp. The use of contaminated cleaning equipment has contributed to the spread of infections in hospitals (Maurer 1985) and there is no doubt that contamination in pharmaceutical factories has occurred in the same way.

Cleaning the processing equipment, storage tanks, pipelines, and filling machines creates its own potential for contamination because it inevitably has a water rinse as its final stage. If the plant remains moist, then growth of Gram-negative organisms is to be expected. Ideal equipment is crevice free, smooth surfaced, easily drained or dried, and is readily accessible. Vessels, mixers, and homogenizers with their associated shafts may easily create obstacles to cleaning lances leaving shadows of unclean equipment areas. Inspection ports and inlets for probes, often recessed into the sides of the vessel, can be particularly difficult to clean and dry. Pumps, especially those with pressure release valves, need careful attention. Pipelines, with minimal dead legs that are self-draining, are essential if cleaning is to be achieved without extensive dismantling. Complex filling machines, although having smooth inner surfaces and being easily cleaned in place, can be difficult to dry without dismantling. Those areas of the plant that are difficult to clean have the potential for creating areas of dilute product, where organisms may become adapted to preservative levels present in that product.

Compressed air coming into contact with product or primary packing, for example, bottle blowers, unless sterilized by filtration, can be expected to be a source of contamination. Gram-negative organisms may colonize the condensate in the air reservoir tanks.

In a well-maintained manufacturing facility operating with good manufacturing procedures, environmental air is not a significant source of contamination during processing or filling of preserved products. Covered mixing vessels prevent significant ingress of air during the mixing stage and air turbulence created both within and around container openings appears, in practice, to be insufficient to cause significant numbers of microorganism-laden particles to enter the container.

Chemical disinfectants used throughout the plant must be prepared to the correct use concentration. Overdilute solutions may become contaminated and thereby represent a source of contamination for the environment to be cleaned (see Chapter 2). Most chemical disinfectants, in particular the halogens, some phenolics, and quaternary ammonium compounds, are inactivated in the presence of organic matter and it is essential that all cleaning materials such as buckets are kept clean. Biofilms may arise on the surfaces of dirty containers in which organisms, normally sensitive to the disinfectant solution, may survive and grow. The practice of storing bulk concentrated disinfectant, and preparing use concentrations daily in clean containers, significantly reduces the chances of contamination. The rotation of disinfecting agents is a wise precaution against the development of resistant strains.

For a more detailed discussion of cleaning practices, their validation and role in contamination control programs, the reader is referred to Hodges and Baird (2000) in the companion *Handbook*.

3.5.4 STERILE PRODUCTS

Many precautions need to be taken in the design and operation of a sterile product-manufacturing facility to minimize both particulate and microbial contamination (see Chapters 6 and 7). The production suite needs to be of a high standard of construction, having impervious smooth walls, floor and ceiling covering, and all dust traps eliminated. All surfaces and equipment should be easy to clean and constructed in materials capable of withstanding repeated chemical disinfection. High efficiency particulate absorbing (HEPA)-filtered air supplies either to provide conventional airflow in the room space or localized laminar flow at the critical sites of operation are normal in the pharmaceutical industry (see Chapter 7). Regular chemical disinfection of the surfaces, walls, and floors, together with maintenance of the air filter integrity and air change rates, can produce areas well within the standards in the unmanned state. Fumigation of areas with formaldehyde is commonly carried out, but provided chemical surface disinfection is thorough, this is probably necessary only after the integrity of the area has been broken during maintenance. Microbial burdens of 0 to 3 CFU per 24-cm^2 contact plate immediately after cleaning are easily attained and counts of <10 CFU per 24 cm^2 can be expected even during operation. The level of microorganisms present in the air is dependent on the type of operation carried out, the number of operators, and the size of the room. The greater the complexity of operation and number of operators, the higher the level of contamination to be expected in a given space.

The quality of gowning and level of awareness and training of operators is very important in determining the levels of environmental contamination in clean aseptic areas. Full coveralls with boots, hood, face mask, and sterile gloves afford the maximum protection to the environment. Close-weave fabrics that prevent skin scales from escaping yet allow a degree of air movement give operator comfort and, therefore, high procedure compliance rates. Air-impervious garments are extremely uncomfortable and become counterproductive, as the only areas from which air can escape are the cuffs and collar. Thus, each time a movement is made there is a tendency to puff skin scales and organisms out into the environment. In addition, hot, uncomfortable, and irritable operators are less likely to comply strictly with procedures.

3.6 OVERVIEW

An understanding of the microbial ecology of production processes and environments enable the industrial microbiologist to appreciate the critical control points, set sensible quality limits and quickly identify reasons for changes in quality. Extended treatments of environment standards and good manufacturing practices that contribute to good microbiological control are given in Chapters 6 and 7.

REFERENCES

Baggerman, C. and Kannegieter, L.M (1984). Microbiological contamination of raw materials for large volume parenterals. *Appl. Environ. Microbiol.,* 48, 662–664.

Baines, A. (2000). Endotoxin testing. In *Handbook of Microbiological Quality Control: Pharmaceuticals and Medical Devices*. Baird, R.M., Hodges, N.A., and Denyer, S.P., Eds. Taylor & Francis, London, pp. 144–167.

Baird, R.M. (2000). Sampling: principles and practice. In *Handbook of Microbiological Quality Control: Pharmaceuticals and Medical Devices*. Baird, R.M., Hodges, N.A., and Denyer, S.P., Eds. Taylor & Francis, London, pp. 38–53.

Baird, R.M. and Shooter, R. A. (1976). *Pseudomonas aeruginosa* infections associated with use of contaminated medicaments. *Br. Med. J.*, 2, 349–350.

Baird, R.M., Brown, W.R.L., and Shooter, R.A. (1976). *Pseudomonas aeruginosa* in hospital pharmacies. *Br. Med. J.*, 1, 511–512.

Bonomi, E. and Negretti, F. (1977). Studies on the microbial content of raw materials used in pharmaceutical preparations. *Ann. 1st Super. Sanita*, 13, 805–832.

Camper, A. K., Le Chevallier, M. W., Broadaway, S.C., and McFeters, G.A. (1986). Bacteria associated with granular activated carbon particles in drinking water. *Appl. Environ. Microbiol.*, 52, 434–438.

Chapman, K.G., Heinze, G.E., Flemming, C.V., Kochling, J., Croll, D.B., Kladko, M., Lehman, W.J., Smith, D.C., Adair, F.W., Amos, R.M., and Soli, T.C. (1983a). Protection of water treatment systems, Part II a: potential solutions. *Pharm. Technol.*, 7(9), 86–92.

Chapman, K.G., Alegnani, W.C., Heinze, G.E., Flemming, C.V., Kochling, J., Croll, D.B., Kladko, M., Lehman, W.J., Smith, D.C., Adair, F.W., Amos, R.L., Enzinger, R.M., Grant, D.E., and Soli, T.C. (1983b). Protection of water treatment systems, Part 1: the problem. *Pharm. Technol.*, 7(5), 48–57.

Chesworth, K.A.C., Sinclair, A., Stretton, R.J. and Hayes, W.P. (1977). Effect of tablet compression on the microbial content of granule ingredients. *Microbios Lett.*, 4, 41–45.

Cross, J. (1994). Removing microbial contaminants from process water. *Pharm. Manufact. Rev.*, (September), 2.

Dudderidge, J.E. (1988). Biofilm growth in water for cosmetics. *Manuf. Chem.*, 59(5), 42–44.

European Pharmacopoeia. (2004). *European Pharmacopoeia*. 5th ed. Council of Europe, Strasbourg, France.

Fassihi, A.R. and Parker, M.S. (1987). Inimical effects of compaction speed on microorganisms in powder systems with dissimilar compaction mechanisms. *J. Pharm. Sci.*, 76, 466–470.

Favet, J. (1992). The microbial contamination of twenty herbal drugs. *Pharm. Acta Helv.*, 67, 250–258.

Ganzi, G.C. and Parise, P.L. (1990). The production of pharmaceutical grades of water using continuous deionization post reverse-osmosis. *Parent. Sci. Technol.*, 44, 231–241.

Grigo, J. (1976). Micro-organisms in drugs and cosmetics: occurrence, harms and consequences in hygienic manufacturing. *Zentbl. Bakt. Hyg. Abt. I. Orig. B.*, 162, 233–287.

Hitokoto, H., Morozumi, S., Wauke, T., Sakai, S., and Kurata, H. (1978). Fungal contamination and mycotoxin detection of powdered herbal drugs. *Appl. Environ. Microbial.*, 36, 252–256.

Hodges, N.A. and Baird, R.M. (2000). Disinfection and cleansing. In *Handbook of Microbiological Quality Control: Pharmaceutical and Medical Devices*. Baird, R.M., Hodges, N.A., and Denyer, S.P., Eds. Taylor & Francis, London, pp. 205–220.

Ibrahim, Y.K.E. and Olurinola P.F. (1991). Comparative microbial contamination levels in wet granulation and direct compression methods of tablet production. *Pharm. Acta Helv.*, 66, 298–301.

Karavanskaya, N.A., Yakimenko, A.I., Zemlanskii, V.V., and Listetskaya, N.F. (1980). Degree of microbial contamination of certain types of ion exchange resins and activated carbon. *Mikrobiol. Zh. (Kiev),* 42, 428–431.

Ku, Y.R., Chou, L.M., Jang, C.F., Liu, Y.C., Lin, J.H., and Wen, G.C. (1994). Study of microbial contamination in concentrated Chinese medicine. *Food Drug Anal.,* 1, 49–62.

Maurer, I. M. (1985). Cleaning the hospital: care with water. In *Hospital Hygiene.* 3rd ed. Edward Arnold, London, pp. 42–57.

Millar, R. (2000). Enumeration of micro-organisms. In *Handbook of Microbiological Quality Control: Pharmaceuticals and Medical Devices.* Baird, R.M., Hodges, N.A., and Denyer, S.P., Eds. Taylor & Francis, London, pp. 54–68.

Mittleman, M.W. and Geesey, G.G. (1987). *Biological Fouling of Industrial Water Systems: A Problem Solving Approach.* Water Micro Associates, San Diego, pp. 138–193.

Negretti, F. (1981). Findings on the microbiological characteristics of pharmaceutical containers. *Boll. Chim.-Farm.,* 120, 193–201.

Schubert, R.H.W. and Esanu, J. (1972). On bacterial aftergrowth in drinking and industrial water. I. The influence of ion-exchange plants. *Zentbl. Bakt. Hyg. Abt. I Orig. B.,* 155, 488–501.

Stamm, J.M., Engelhard, W.E., and Parsons, J.E. (1969). Microbiological study of water-softeners resins. *Appl. Microbiol.,* 18, 376–386.

United States Pharmacopeia. (2005). *United States Pharmacopeia 29.* 2006 USP Convention. Rockville, Maryland.

Wood, J.M. (1980). The interaction of micro-organisms with ion-exchange resins. In *Microbial Adhesion to Surfaces.* Berkley, R.C.W., Lynch, J.M., Melling, J., Rutter, P.R., and Vincent, B., Eds. Ellis Horwood, Chichester, U.K., pp. 163–185.

Yokoyama, H., Yamasaki, K., Sakagami, Y., Nunoura, Y., Umezawa, C., and Yoneda, K. (1981). Investigating on quality of pharmaceutical products containing crude drugs (I). *J. Antibact. Antifung. Agents,* 9, 421–428.

4 The Design of Controlled Environments

Allan G. Cosslett

CONTENTS

4.1 INTRODUCTION

4.1.1 DEVELOPMENT OF ENVIRONMENTAL CONTROL

The early development of cleanrooms during World War II resulted from the identification of a need to improve the quality and reliability of the instruments of war. The relationship between dirty work environments and the failure of equipment during battle was quickly noted, and manufacturers started to consider methods they could use to clean up their working environments to improve the quality control of their products. Included among their developments was the high efficiency particulate air (HEPA) filter, which allowed air to be "filter cleaned" so that contamination from both microorganisms and particles could be removed, thus providing the mechanism by which very clean working environments could be established. In 1962, the Sandia Laboratories' development of the "unidirectional" airflow system heralded a major landmark in cleanroom technology, providing a mechanism and series of methodologies that future "clean" manufacturing industries would rapidly follow (Whyte 1999).

As a result of the developments during the early 1960s, a need for standardization across those industries utilizing cleanroom technology was realized. In the United States, the development and release of Federal Standard 209 (FS209) A, B, C, D, and E (1962, 1976, 1987, 1988, 1992) heralded the provision of documentation that could be used to classify both conventional and unidirectional cleanrooms, as well as detailing cleanroom design principles, appropriate mechanisms of working within the clean environments, requirements for the cleaning of the rooms, information on restriction of certain types of equipment, and details of personnel clothing. Although this and many other country-specific standards [e.g., British Standard 5295 (BS5295) 1976, 1989] have now been superseded by the International Organization for Standardization's (ISO) document ISO 14644–1 (1999), the basic concepts and thoughts outlined still very much hold true today.

4.1.2 CLEANROOM CLASSIFICATION

Different manufacturing industries and processes require different levels of cleanliness within their manufacturing environment. The original Federal Standard 209A (1962) introduced the concept of room classification, by which particulate levels of contamination were restricted at different size levels and numbers, depending upon the room's purpose. Today the current harmonized ISO definition of "classification" is as follows:

> Level (or process of specifying or determining level) of airborne particulate cleanliness applicable to a cleanroom or clean zone, expressed in terms of an ISO Class N, which represents maximum allowable concentrations (in particles per cubic metre of air) for considered sizes of particles.

Importantly, the use of the term *classification* is now related to one or more of three occupancy states: "as-built," "at-rest," and "operational." As-built occurs only once and thereafter the other two states are tested and defined. When designating the classification of a cleanroom or clean zone, the ISO classification number, occupancy state, and particle size(s) and concentration(s) must all be used to provide the user with full knowledge of the appropriateness of the classification.

In order to determine the classification number of a cleanroom or clean zone and, hence, the particle size and concentration the room or zone should achieve, the ISO standard uses the following equation:

$$C_n = 10^N \times (0.1/D)^{2.08}$$

where C_n is the maximum permitted concentration (in particles per cubic meter of air) of airborne particles that are equal to or larger than the considered particle size. C_n is rounded to the nearest whole number, using no more than three significant figures. N is the ISO classification number, which shall not exceed a value of 9. Intermediate ISO classification numbers may be specified, with 0.1 the smallest permitted increment of N. D is the considered particle size, in micrometers; 0.1 is a constant, measured in micrometers.

Using the ISO classification equation, selected airborne particulate cleanliness classes for cleanrooms and clean zones can be determined (see Table 4.1).

Although cleanroom classifications are required to be used in the pharmaceutical industry, they only specify the requirements of the maximum level of particulate contamination, with no consideration of the presence and level of microorganisms. Specifications regarding microbial contamination detection and limits can be found in the European Community's "Guide to Good Manufacturing Practice for Medicinal Products" (EC GGMP) (Eudralex 2004) or the U.S. Food and Drug Administration's (FDA's) "Guidance for Industry Sterile Drug Products Produced by Aseptic Processing—Current Good Manufacturing Practice" (FDA 2004). These guides define the grades of environmental quality in terms of both the particulate concentration and the microbiological contamination levels acceptable for pharmaceutical manufacture of medicinal products.

The design of cleanroom pharmaceutical manufacturing facilities is very much dependent upon the manufacturing processes that are to be performed during the production of a medicinal product. The major critical determining factor is the requirement of aseptic manipulation steps as opposed to the terminal sterilization of the finished product. EC GGMP defines four grades of cleanliness. Grade A is the

---

TABLE 4.1
ISO Classification of Cleanroom Particulate Limits

ISO Classification Number (N)	Maximum Concentration Limits (Particles.m⁻³ of Air) for Particles Equal to and Larger Than the Considered Sizes Shown Below					
	0.1 μm	0.2 μm	0.3 μm	0.5 μm	1 μm	5 μm
ISO class 1	10	2	—	—	—	—
ISO class 2	100	24	10	4	—	—
ISO class 3	1,000	237	102	35	8	—
ISO class 4	10,000	2,370	1,020	352	83	—
ISO class 5	100,000	23,700	10,200	3,520	832	29
ISO class 6	1,000,000	237,000	102,000	35,200	8,320	293
ISO class 7	—	—	—	352,000	83,200	2,930
ISO class 8	—	—	—	3,520,000	832,000	29,300
ISO class 9	—	—	—	35,200,000	8,320,000	293,000

Note: Dashes indicate not applicable.

TABLE 4.2
Cleanroom Classification According to the EC GGMP

	At Rest		In Operation	
	Maximum Permitted Number of Particles.m⁻³ Equal to or Above			
Grade	0.5 μm	5 μm	0.5 μm	5 μm
A	3,500	1	3,500	1
B	3,500	1	350,000	2,000
C	350,000	2,000	3,500,000	20,000
D	3,500,000	20,000	not defined	not defined

TABLE 4.3
Recommended Limits for Microbial Contamination According to EC GGMP Cleanroom Classification

	Recommended Limits for Microbial Contamination			
Grade	Air Sample CFU/m³	Settle Plates CFU/4 h	Contact Plates CFU/plate	Glove Print CFU/glove
A	<1	<1	<1	<1
B	10	5	5	5
C	100	50	25	—
D	200	100	50	—

Note: CFU are colony-forming units. Dashes indicate no limit specified.

Source: From Eudralex. (2004). *Volume 4—Medicinal Products for Human and Veterinary Use: Good Manufacturing Practice*. Office for Official Publications of the European Communities, Luxembourg. With permission.

cleanest with regard to both particulate and microbiological contamination. For appropriate conditions for aseptic manufacture, the level of "clean zone" should be achieved both "at rest" and "in operation" (see Table 4.2 and Table 4.3).

The ISO standards and the FDA and EC GGMP guidelines have allowed the continued development of pharmaceutical medicinal products even though some of today's newer pharmaceutical technologies (e.g., nanotechnology and gene therapy) require extremely stringent manufacturing conditions that 30 to 40 years ago would not have been attainable.

4.2 SOURCES OF ENVIRONMENTAL CONTAMINATION

Environmental contaminants are typically referred to either as viable, resulting from particles that harbor culturable microorganisms, or nonviable, resulting from inanimate particles. The ratio of viable to nonviable particles has been variously estimated at between 1:500 and 1:20,000, in a total concentration of approximately 10^8 to 10^{13} m^{-3}, with the particle size distribution of a typical atmospheric air sample shown in Table 4.4. It can be seen that particles up to 1 μm constitute almost 99% of the numerical particulate load, but account for only 2% of the total contaminant load weight. This factor has important implications for the filtration systems employed in reducing airborne particulate contamination (De Vecchi 1986). It should also be noted that the levels of contamination are very dependent upon the environment location (e.g., highly populated or industrial environments being high in background levels of contamination), as well as being influenced by prevailing climatic conditions.

Within the cleanroom environments the pharmaceutical industry has available for manufacture, the major source of particulate contamination arises from the operator because the equipment and work surfaces generate very few particles, and the extremely clean air supplies now used provide very few particles to the total room particulate contamination load. The human body sheds approximately 10^9 skin cells per day (Heuring 1970, Whyte 2001), as well as aerosolizing particles from the nose and mouth during exhalation. Humans also generate particles from undergarments and from specialist cleanroom clothing. The concentration of particles released by the first two sources is very much dependent upon the level of operator activity, and the type and quality of garment material control the level of release of particles from the latter two sources. Typically, as operators increase their levels of activity within the cleanroom, the release of particulate contamination into the surrounding work environment is increased, with a close association between viable and nonviable particles (1:4 average) (MacIntosh et al. 1978). The use of synthetic undergarments will help reduce the level of released particles by retaining the shed skin cells and by not releasing any contaminating particles themselves (Whyte 2001). The level of particulate retention effectiveness of cleanroom garments is related to their air permeability, their ability to retain particles, and the pore size of the fabric; all of these can be assessed using "air-shower/body box" devices, allowing pharmaceutical manufacturers to determine the extent of operator particle release protection they need to provide for each manufacturing activity (Institute of Environmental Sciences and Technology's RP-CC003; IEST 2003).

TABLE 4.4
Size Distribution of a Typical Atmospheric Dust Sample

Particle Size Range (μm)	Proportionate Particle Count	Percent by Particle Count	Percent by Weight
10–30	1,000	0.005	28
5–10	35,000	0.175	52
3–5	50,000	0.25	11
1–3	214,000	1.07	6
0.5–1	1,352,000	6.78	2
0–0.5	18,280,000	91.72	1

4.3 CLEANROOM TECHNOLOGY

4.3.1 AIR QUALITY

The development of the HEPA filtration systems allowed high quality air to be supplied to pharmaceutical facilities. Since their mechanism of operation (filtration) results in the removal of significant numbers of large particles, that is, an absence of 5-μm particles and extremely low concentration of 0.5-μm particles, then for practical purposes, the resulting air stream will be free from viable contamination because microorganisms are almost invariably associated with larger inanimate particles that protect them from desiccation in the air stream.

4.3.1.1 Mechanisms of Filtration

In HEPA filtration systems, the overall filtration mechanism is the combined effect of several different filtration processes, whose relative contributions vary markedly with particle size and the velocity of air passing through the filter medium. For large particles, a "sieving effect" by coarse prefilters would normally remove all or most of these particles; an additional three mechanisms are responsible for retaining smaller particles on the glass fibers that make up the HEPA filter medium, as shown diagrammatically in Figure 4.1.

Besides the three filter mechanisms described below, the minimum efficiency of a filter will depend upon a number of variables, including:

- Particle density
- Particle velocity and mean free path through the filter media
- The depth of the filter medium
- The environmental conditions of the carrier gas (i.e., velocity, pressure, and temperature)
- The size and distribution of the fibers within the filter media

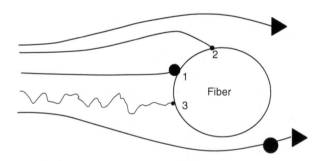

1   Inertial Impaction
2   Electrostatic Retention
3   Diffusive Retention

FIGURE 4.1 Schematic representation of mechanisms of filtration.

4.3.1.1.1   Inertial Impaction

As air flows around the randomly orientated fibers, the inertia of suspended particles precludes rapid changes of direction and they become embedded on the fibers. This mechanism is of major significance for particles of >1 μm and becomes more effective as air velocity increases through the filter medium.

4.3.1.1.2   Electrostatic Retention or Interception

Historically, particles of 0.5 to 1.0 μm, which have very low mass, were thought to become trapped by electrostatic attraction to those fibers with which they come into close proximity. It is now thought that, most commonly, these particles strike the filter fibers or previously trapped particles as they pass (i.e., tangentially) and are thus retained against the surface, a mechanism referred to as "interception."

4.3.1.1.3   Diffusive Retention

Particles of <0.5 μm are principally retained by impaction resulting from randomized particle movement (i.e., Brownian motion), and this mechanism operates most efficiently at lower flow velocities and very small particle sizes.

4.3.1.2   HEPA Filters

New air filtration technology available today has allowed nonpharmaceutical industries to benefit from the use of ultra-low particulate air (ULPA) filters, which are 99.99996% efficient against 0.12-μm particles. For even the most stringent pharmaceutical purposes, however, filters that are 99.97% efficient against 0.3-μm particles are entirely adequate for final filtration of air entering controlled areas (Whyte 2001).

HEPA filters have traditionally been made from a nonflammable, water-repellent microglass material with individual fiber diameters of around 0.1 μm. A pleated construction system (see Figure 4.2) provides a large surface area over which particulate filtration can take place. In order to support each pleat, a corrugated separator (usually crinkled aluminium foil) is used to maintain even packing across the surface that results in optimum and consistent airflow off the surface. The filter material and separators are normally bonded to a rigid frame using low-solvent adhesive to minimize shrinkage. More recent HEPA filters have used a "minipleat" design where construction dispenses entirely with separators and uses bonded glass threads, ribbons, or molded media to maintain close, regular packing of the pleated filter material. The minipleat construction method provides a considerably greater volume flow capacity, more uniform face velocities, and a significant reduction in weight in comparison to the traditional construction method.

The modern minipleat HEPA filter has now evolved into an extremely dependable component, providing consistently high performance and good durability, and being virtually free from "pinhole" leaks. The typical final HEPA filter for pharmaceutical aseptic zones would have a flow-rate face velocity of 0.45 m.s^{-1}; however, because any filter material subjected to a particulate load will eventually become clogged, prefilters are used to remove the gross contamination found in untreated air in order to extend the useful life of the expensive final filters and minimize cleanroom downtime. For all pharmaceutical manufacturing units, the cleanroom design process

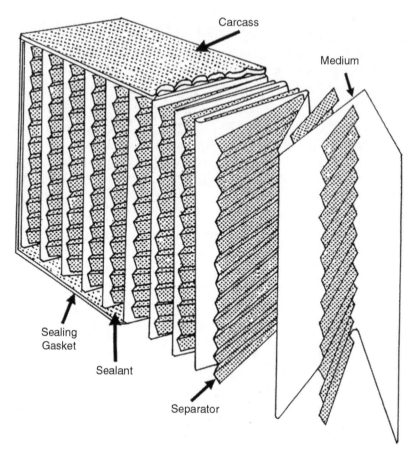

FIGURE 4.2 Construction of conventional HEPA filter panel.

must include the appropriate level of prefilters for the particular manufacturing unit because the rate of filter clogging will be dependent upon the extent of the particulate challenge, which, as previously noted, will vary with environmental location and prevailing weather conditions.

4.3.1.3   HEPA Filter Installation

The only air system component downstream from a final HEPA filter should be a diffuser grill, if required, or a simple protective grill or splash guard. The frame housing in which the filter panels are mounted must be securely fixed and sealed to their apertures in the ceiling or walls of the cleanroom or clean zone equipment. The joint between the filter housing frame and the ducting system that feeds air to the filter is particularly critical, as the contaminated, high-pressure air zone built behind the filter media surface has the capability of forcing extremely large numbers of particles into the room or zone through the smallest of cracks or leaks.

Wherever considerations of space and access permit, the filter housing should be designed for rearward replacement of the filter panels. This arrangement results

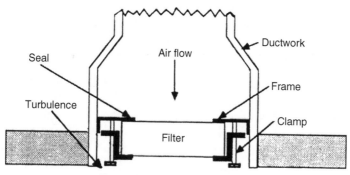

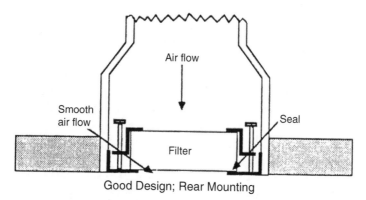

FIGURE 4.3 HEPA filter installation.

in a neater, easily cleanable installation, as shown in Figure 4.3; it also allows the cleanroom to be protected from the dirty interior of the ductwork during filter replacement. Furthermore, the detection and diagnosis of edge leaks become more straightforward due to the absence of air turbulence, and *in situ* sealing of minor leaks is more likely to be successful.

4.3.2 AIRFLOW MANAGEMENT

The basic aim of any cleanroom design process is to provide the manufactured product with protection from airborne contamination. In the most simplistic terms, this is achieved by continuously sweeping away any operator-derived particulates from the immediate work zone, thereby reducing the surrounding particulate challenge to the protected work zone to the lowest achievable level. Overall the ability of a well-designed aseptic suite to maintain low equilibrium levels of particulates (operator-generated vs. room cleaning) under fully operational conditions is a direct measure of the suite's contribution to sterility assurance for the manufactured product. In practice, particulate contamination removal is usually achieved by one of two mechanisms, conventional or unidirectional "laminar" airflow, although more commonly by a combination of both processes. These mechanisms, together with room pressurization, which is achieved by supplying more air into the room than is released, allow the high cleanliness classifications required for pharmaceutical aseptic manufacturing to be achieved.

4.3.2.1   Conventional (Turbulent) Flow

The mechanisms of particle removal in a conventional cleanroom are those of mixing and dilution of the contaminated air. The greater the amount and the better the quality of air supplied to the room, combined with a high efficiency of mixing of the fresh air within the room, the better the cleanliness achieved. The level of cleanliness of a conventional cleanroom can be approximately determined by the following equation:

$$\frac{\text{Airborne concentration}}{(\text{count.m}^{-3})} = \frac{\text{Number of particles generated.min}^{-1}}{\text{Air volume supplied (m}^3.\text{min}^{-1})}$$

Even though a conventional flow room can reduce the particulate load within the cleanroom, in order to achieve suitable levels of cleanliness in the zone of operation unidirectional airflow [or laminar airflow (LAF)] cabinets or workstations have traditionally been utilized, although increasingly manufacturing facilities are using isolator technology. These items of LAF equipment are usually self-contained and constructed to provide a high quality of clean air from a HEPA filter, which may be installed either vertically or horizontally, as shown schematically in Figure 4.4. The operator performs the critical manufacturing steps directly in front of, or beneath, the HEPA filter, within an envelope of controlled airflow. As well as rigidly constructed workstations, aseptic areas may be provided with LAF modules for local protection of equipment such as filling lines. Down-flow modules may employ flexible plastic sheeting or even peripheral high-velocity air curtains to contain the LAF over the working area.

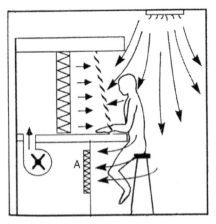 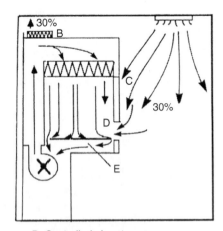

A: Prefilter
B: Exhaust HEPA filter
C: Glazed panel

D: Controlled air entry
E: Dished work-top,
   with peripheral slots

FIGURE 4.4 Schematic diagram of horizontal and vertical LAF workstations and interaction with room air inlets.

When considering cleanroom airflow options, the appropriate choice between vertical or horizontal airflow protection should always be made to ensure that an undistorted, uncontaminated airstream flows over critical points, such as the open necks of vials, and operatives' hands should ideally be downstream, or at least to one side of the vulnerable components. Regardless of airflow direction, mechanical or human movement within the working area should be slow, deliberate, and as infrequent as possible to minimize the generation of vortices and eddy currents.

In the conventional airflow room, supplied air should be directed to induce turbulent conditions throughout the room, thereby attempting to sweep away the particulate contamination. In order to achieve this, the clean air supply is normally via a small number of wall or ceiling inlet diffusers with appropriately positioned air outlets or leakage points optimizing the air's flushing potential and attempting to prevent "dead space" effects.

Significantly, a potential problem when designing aseptic suites, which consist of a conventional room with sited LAF units, is the possibility of interference between the room's air supplies and the protective airflow within LAF workstations. Typically, air input velocities to the room are significantly above the LAF unit's output velocity ($0.45 \ m.s^{-1}$) and, as shown in Figure 4.4, contamination can potentially be transported off the operator directly into the critical working zone of a poorly sited workstation.

4.3.2.2   Unidirectional (Laminar) Flow

Unidirectional flow or laminar flow is the description given to mechanisms of particulate contamination removal where appropriate levels of air moving at a velocity in the range 0.3 to $0.45 \ m.s^{-1}$ in a horizontal or vertical direction remove large particles by preventing them from settling onto any surface, thereby systematically flushing them away from the working area.

Any physical obstruction (e.g., equipment, operator's hand) to the laminar airflow will create downstream turbulence, which may compromise the manufactured product's protection. However, at laminar flow velocity there is little tendency for turbulent flow to develop upstream of an obstruction, and uniform flow is usually reestablished downstream at a distance equivalent to 3 to 6 times the cross-stream dimension of the obstruction (Figure 4.5). Historically, the selection of a horizontal or vertical protective airstream has been partly dependent upon the product's physical and chemical characteristics; for example, where operator protection is as equally important as protection of product from contamination, as in the case of parenteral cytotoxic drug powder reconstitution, then a down-flow setup would have been utilized. Today the use of an isolator unit has to some degree reduced the choice of laminar flow device to one of strict operator protection (isolator) vs. product protection (traditional LAF unit).

4.3.2.3   Unidirectional Flow Rooms (UFRs)

The design objective for a UFR is to move filtered air evenly and uniformly through the entire volume of the room, so that the air displaces any particulate matter from

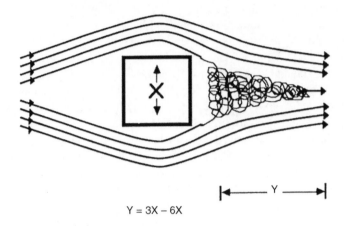

$$Y = 3X - 6X$$

FIGURE 4.5 Effect of obstruction on airflow patterns.

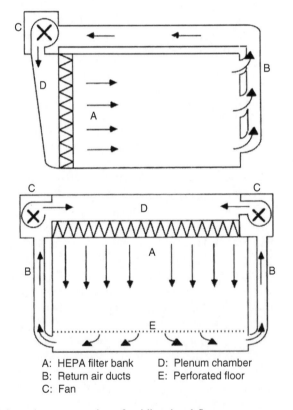

A: HEPA filter bank   D: Plenum chamber
B: Return air ducts   E: Perforated floor
C: Fan

FIGURE 4.6 Schematic representation of unidirectional flow rooms.

the room in a continuous manner. In practice, a UFR will have one complete wall or entire ceiling made from a bank of HEPA filters, producing horizontal or vertical flow rooms, as shown in Figure 4.6.

Any critical work surface must be designed to be positioned directly in front of or beneath the filter bank, where the particulate challenge to the critical zone is much reduced because the airflow is essentially uniform throughout the entire room and, therefore, stagnant areas and high-velocity turbulence are eliminated.

As with conventional cleanrooms, the appropriate selection of horizontal or vertical airflow very much depends upon the work tasks to be performed; however, the relative work positions of the operators assume increased importance. For large aseptic facilities, with critical locations distributed throughout the floor plan, down flow may be the only realistic option because this arrangement should provide good environmental control at any point within the room, regardless of adjacent personnel. The degree of protection offered to products on an open bench within a vertical-flow room may not be entirely comparable with that afforded by a LAF cabinet sited in a conventional cleanroom, due to the absence of a physical barrier to exhaled particles and debris shed from the operator's upper body. Various studies (Whyte et al. 1979, Howorth 1987) have demonstrated that partial vertical-flow rooms are unsatisfactory, as entrainment of contamination into the critical area is inevitable as personnel move into or out of the protected zone.

For smaller facilities, horizontal flow provides outstanding product protection, with no cross-contamination between adjacent operatives working in front of the filter wall. However, debris shed from each individual forms a well-defined "wake" of contaminated air, rendering the downstream environment unsuitable for unpro-tected aseptic work. In contrast to down-flow rooms, partial horizontal-flow rooms, as shown in Figure 4.7, can provide excellent contamination control in a very cost-effective manner.

Another important consideration when deciding on whether to utilize UFR technology is the running cost of the unit. Because the air supply to the UFR must

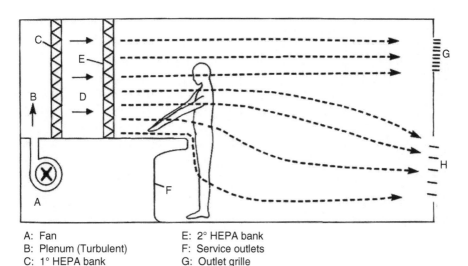

A: Fan
B: Plenum (Turbulent)
C: 1° HEPA bank
D: Plenum (Uniform flow):

E: 2° HEPA bank
F: Service outlets
G: Outlet grille
H: Pressure control valves

FIGURE 4.7 Partial horizontal-flow room.

maintain the laminar flow air velocity target, then the volumetric flow rate of air entering the room will consequently depend on the dimensions of the wall or ceiling HEPA filter bank. The financial implications of single-pass temperature or humidity control and filtration system for a UFR has led to most facilities being designed with recirculation of exhaust air together with about 10% fresh air to maintain adequate ventilation requirements for the operators.

4.3.2.4 Validation and Monitoring of Cleanrooms

Once a cleanroom has been built, the commissioning involves a series of comprehensive checks to confirm that the quality and quantity of air supplied to each area is appropriate and that the cleanroom achieves the standards laid down in ISO 14644–1. These initial checks on compliance with the standards also confirm that the design specification and characteristics for the room are appropriate; these become the benchmark for subsequent tests demonstrating that the suite continues to perform satisfactorily. For the pharmaceutical industry, tests on the facilities would also include detailed analysis of the level of microbiological contamination that may be present prior to and, if appropriate, during production runs. For a more detailed discussion on monitoring microbiological contamination, the reader is referred to Chapter 8 and also Bill (2000) in the companion *Handbook*.

For continued compliance with ISO 14644 classifications (see Table 4.1), particulate counts are performed in accordance with annex B of ISO 14644-1 (1999) at a maximum time interval of 6 months for ISO classes ≤5 and of 12 months for ISO classes >5. In addition, for all classes, airflow velocity tests (ISO 14644-3, clause B.4) and air pressure differences (ISO 14644-3, clause B.5) must be undertaken at a maximum time interval of 12 months. Further optional tests, upon agreement with the customer and supplier, are usually undertaken within a maximum period of 24 months; these include installation filter leakage, airflow visualization, recovery of working environment, and containment leakage.

By undertaking these tests, the cleanroom is monitored for continued compliance with the international standards throughout its working life. Furthermore, the tests are proving in simple terms that the basic principles of cleanliness are being achieved, that is, sufficient quantity of high-quality air is being supplied in such a way that air movement should prevent high concentrations of contamination occurring within a room and that the overall movement of air within the suite is from clean to less-clean areas. Besides particulate tests, good manufacturing practice guidelines indicate that appropriate microbiological testing is undertaken; in terms of the European Community's guidelines (Eudralex 2004), this involves several methods of testing, including air sampling, the use of settle and contact plates, and operator glove prints ("finger dabs") (see Table 4.3).

4.3.3 Design Principles

In practice, the cleanroom design will develop a suite of connecting rooms of various cleanliness classifications to accommodate the different parts of the manufacturing processes. For example, the EC GGMP guidelines (Eudralex 2004) indicate that the requirements for aseptic and terminally sterilized parenteral products are different in terms of the classification of the environmental cleanliness in the area surrounding the critical zone of work (grade C around a grade A zone for terminally sterilized products compared with grade B around a grade A zone for aseptically produced parenteral products).

In order to achieve the appropriate level of cleanliness in a cost-efficient method, the design objectives must attempt to ensure:

- The exclusion of the surrounding environment
- The removal or dilution of contamination generated during the manufacturing processes (including that from operators)
- A mechanism by which operators are protected from the product and the product protected against environmental hazards
- The operators have optimum working conditions
- That effective monitoring of room conditions can be achieved at pre-defined time intervals

In order to ensure these objectives, the control and management of the flow of product material throughout the manufacturing process is determined by means of layout and configuration diagrams that define the exact number of steps required for the particular products to be made. Alongside this flow design, reviews of the control and management of operator movement within the rooms are performed and then optimized; thus the arrangement and connections to the individual rooms and zones within the suite, and the control of the entry and exit of operators and materials, provides security to the overall processes and prevents cross-contamination from occurring.

4.3.3.1 Design Features

The overriding requirement of all types of cleanrooms is the ability to clean the equipment and critical work surfaces easily. In order to achieve this, every effort is made to have smooth, flat, impervious surfaces, with all equipment and fixtures flush fitting wherever possible to prevent potential contamination traps from developing. All junctions between floors, walls, and ceilings should be seamless and coved, and, whenever practicable, major items of equipment should either be movable for cleaning, or be built into the structure of the room to present smooth, cleanable surfaces and eliminate inaccessible gaps and recesses.

Careful consideration must be given to the design of windows and communication mechanisms; appropriately sealed and implanted window units and "speech panels" provide suitable approaches for integrating these objects. Light fittings, as with all other cleanroom objects, should be flush fitting, in permanently sealed units with access for lamp replacement gained from a service void above the cleanroom; maintenance can then be achieved without the need to enter the cleanroom suite itself. Internal doors should be positioned and hinged such that their opening produces minimum airflow disturbance at critical locations within the cleanroom, and that air flows smoothly out into the lower classification areas. Doors should normally open into the more stringently controlled area to allow personnel to back through the doorway without contaminating gloved hands and to ensure that room overpressure will assist prompt door closure. It is vital that, in order to reestablish environmental control in the critical areas as quickly as possible, doors are set to self-close efficiently. In addition, all doors communicating with a changing area, technically the dirtiest area of the cleanroom suite, should be interlocked to prevent more than one door being open simultaneously, thereby preserving overall pressure differentials as personnel enter or leave the critical areas.

4.3.3.2   Construction Materials and Surface Finishes

Cleanroom construction has typically developed in one of two ways. The first means is adaptation of traditional building techniques, finishing interior surfaces with plasterboard, which is then sprayed with epoxy-based paints to different thicknesses, which provide a good impact-resistant surface finish. The alternate, and now more common, form of construction involves the use of modular components; here either a studless wall system (wall panels made from laminated cleanroom–compatible material are mounted in anodized aluminium tracks) or a framed wall system (where similar wall panels are attached to joined aluminum extrusions that form the stud and cross members of the room) is employed.

All interior construction surfaces should provide smooth, nonshedding, washable surfaces, free from cracks, crevices, or unsealed joints. Over the years, various polymeric paints have been developed that possess enhanced elasticity, which theoretically should accommodate any relative movement in their substrate. Additionally, some of these purpose-developed paints also incorporate fungicides and "cobwebbing agents" that allow the applied paint layer to bridge across junctions between dissimilar materials.

The importance attached to ease of cleaning throughout a facility's service life has led to the adoption of durable constructional materials and surface finishes. Floors are usually constructed of a concrete base on which a smooth, durable, and nonpenetrable material (e.g., epoxy resin) is added to provide a surface that is slip-resistant to the operator and also withstands disinfectant cleaning solutions.

For all materials used within the cleanroom, a design consideration based on cost vs. effectiveness is required. This needs to be achieved while also following the guidelines provided by the authorities, for example, the EC GGMP and the ISO standards, which provide informative documentation on suitable materials and how they should be used (ISO 14664-4: Annex E).

4.4   ISOLATOR TECHNOLOGY

Isolation technology provides the manufacturer with the ability to have close control of the environmental zone immediately surrounding a manufacturing process, thus avoiding operator-derived product contamination and protecting the operator from hazardous materials. Isolation technology has been used for many years in certain areas of the pharmaceutical industry; however, only in the last decade have the potential financial and operational advantages of the technique been widely recognized, particularly within the hospital sector. The ability to have isolator units ranging from small glove boxes, through single free-standing workstations with multiple glove access, to fully integrated industrial units where a series of stations are interlinked to form the production line for a product, means flexibility in process design can now be more readily achieved than with traditional LAF workstations or zones.

The availability of reasonably priced flexible-film or rigid-walled units with options on internal sterilization with gas sterilants or validated disinfection techniques means manufacturers can now select the most appropriate isolator specification for their needs. That choice depends upon the nature, extent, and predictability of manufacturing work to be undertaken, the quality of the environment in which the equipment will be installed, and the degree of confidence required in the microbial integrity of the final products. As an example, for radiopharmaceutical production work, only the use of rigidly constructed isolators with lead glass panels would be appropriate because operator protection would be of paramount consideration.

The manufacturing tasks that operators are to perform must also be considered because this may determine the type of operator access to the isolator. The two principal methods of gaining manipulative access to the interior of an isolator are through two or more sleeves or by means of a half-suit. For sensitive glove-access work, the use of multicomponent sleeves that allow the use of conventional latex gloves, together with sleeves fabricated from material such as nylon-lined polyvinyl chloride (PVC), offer the operator comfort while presenting an impervious, wipeable surface within the work zone. However, continual movement by the operator within the gloves increases the potential risk for leakage at many points within the glove, sleeve, and fixtures and so routine testing of the system must take place.

Half-suits are fabricated from the same material as sleeves and use the same cuff rings for glove attachment. A transparent helmet is sealed to the neck of the suit, which is normally ventilated by its own pressurized air supply, maintaining operator comfort over prolonged work sessions. The advantage of a half-suit is that the operator has around 80% greater reach radius, and can also maneuver larger or heavier items. Half-suit isolators are generally flexible film types and are usually employed as dedicated production equipment or for commercial aseptic compounding of products such as parenteral nutrition formulations.

As with all cleanroom manufacturing processes, workflow-related planning diagrams must determine the appropriate mechanisms by which the operator can manipulate the product under manufacture safely and without risk to that product or operator. Recent increases in incidences of repetitive stress injuries have been linked to the increased use of gloved isolators where the operator's working position is greatly influenced by the size of the unit and the tasks to be performed inside the isolator. Strict time limits on staff working in the isolators and rotation of operational staff through other manufacturing processes normally can reduce these problems (Abbot and Johnson 2002).

Within the cleanroom facility, validation of working practices, including equipment usage, must be undertaken. Isolator environmental integrity must be routinely checked, and besides checking the glove or suit access points, additional testing of the walls of the unit must also be undertaken. This task may be performed either by using traceable gases and chemical sniffer units or by releasing a mixture of ammonia and air into the isolator's internal environment and then testing the surface of the unit by passing a cloth impregnated with a pH indicator over the surface and observing color changes; this latter method has found favor with users of flexible-film isolator units (Burnett 2001).

USEFUL SOURCES OF CLEANROOM INFORMATION ON THE INTERNET

EU Guide to Good Manufacturing Practice, available at http://pharmacos.eudra.org/F2/eudralex/vol-4/home.htm.

FDA Guidance for Industry on Sterile Drug Products Produced by Aseptic Processing—Current Good Manufacturing Practice, available at http://www.fda.gov/cder/guidance/5882fnl.htm.

Institute of Environmental Sciences and Technology—Guides and Recommended Practices, available at http://www.iest.org/.

International Confederation of Contamination Control Societies, available at http://www.icccs.net.

International Organization for Standardization—Standards ISO 14644—"Cleanrooms and Associated Controlled Environments" and ISO 14698—"Cleanrooms and Associated Controlled Environments—Biocontamination Control," available at http://www.iso.org.

REFERENCES

Abbott, L. and Johnson, T.N. (2002). Minimising pain resulting from the repetitive nature of aseptic dispensing. *Hosp. Pharm.,* 9, 77–79.

Bill, A. (2000). Microbiology laboratory methods in support of the sterility assurance system. In *Handbook of Microbiological Quality Control: Pharmaceuticals and Medical Devices.* Baird, R.M., Hodges, N.A., and Denyer, S.P., Eds. Taylor & Francis, London, pp. 120–143.

British Standard 5295 (1976, 1989). *Environmental Cleanliness in Enclosed Spaces.* British Standards Institution, London.

Burnett, S.A. (2001). Pharmaceutical isolator leak testing—proposal for specification of leak test rates which allows a comparison of quoted values. *Eur. J. Parent. Sci.,* 6, 13–20.

De Vecchi, F. (1986). Environmental control in parenteral drug manufacturing. In *Pharmaceutical Dosage Forms: Parenteral Medications.* Avis, K.E., Lachman, L., and Lieberman, H.A., Eds. Vol. II. Marcel Dekker, New York, pp. 309–359.

Eudralex. (2004). *Volume 4—Medicinal Products for Human and Veterinary Use: Good Manufacturing Practice.* Office for Official Publications of the European Communities, Luxembourg.

FDA (2004). *Food and Drug Administration's Guidance for Industry on Sterile Drug Products Produced by Aseptic Processing—Current Good Manufacturing Practice.* Food and Drug Administration, Rockville, MD.

Federal Standard 209A, B, C, D, and E (1962, 1976, 1987, 1988, 1992). *Cleanroom and Workstation Requirements: Controlled Environment.* Government Printing Office, Washington, D.C.

Heuring, H. (1970). People: the key to contamination control. *Contamin. Control* 9, 18–20.

Howorth, F.H. (1987). Prevention of airborne infection in operating rooms. *J. Med. Eng. Tech.*, 11, 263–266.

IEST. (2003). *Institute of Environmental Sciences and Technology's Recommended Practices and Guides RP-CC003*. Institute of Environmental Sciences and Technology, Rolling Meadows, IL.

ISO. (1999–2004). Cleanrooms and associated controlled environments, In *International Standards Organization 14644 (Parts 1 to 8)*. International Organisation for Standardisation, Geneva.

MacIntosh, C.A., Lidwell, O.M., Towers, A.G., and Marples, R.R. (1978). The dimensions of skin fragments dispersed into the air during activity. *J. Hyg.*, 81, 471–479.

Whyte, W. (1999). An introduction to the design of clean and containment areas. In *Cleanroom Design*. W. Whyte, Ed. 2nd ed. John Wiley & Sons, New York, pp. 1–20.

Whyte, W. (2001). Cleanroom clothing. In *Cleanroom Technology: Fundamentals of Design, Testing and Operation*. W. Whyte, Ed. John Wiley & Sons, New York, pp. 237–262.

Whyte, W., Bailey, P.V., and Hodgson, R. (1979). Monitoring the causes of cleanroom contamination. *Man. Chem. Aer. News*, 50, 65–81.

5A Microbiological Considerations for Biotechnological Products

Adekunle O. Onadipe

CONTENTS

5A.1 INTRODUCTION

Biotechnological products differ from conventional pharmaceuticals in that biological processes and materials are employed in their production. The biological basis of these processes is inherently variable; thus, in comparison to conventional pharmaceutical production, there is greater potential for variation in the quantity and quality of the resultant product.

Biotechnology exploits a variety of host organisms to manufacture therapeutic products and these can be broadly classified into prokaryotic cell-derived and eukaryotic (often mammalian) cell-derived. The range of therapeutic products from these processes includes both recombinant and nonrecombinant proteins (e.g., antibodies, hormones, enzymes, and receptors), viral and bacterial vaccines, and viral vectors for gene therapy (Birch 1997). Potential therapeutic products from eukaryotic cell biotechnology include hematopoietic progenitor cells for bone marrow transplantation (Meissner et al. 1998), *in vitro* cultured skin cells for grafting (Kolokoltsova et al. 1998), islet cell cultures for bioartificial pancreases (Handa-Corrigan et al. 1998), hepatocytes for artificial livers (Bratch et al. 1998), and transgenic plants and animals (Cartwright 1997).

The general microbiological considerations required for the production of therapeutic products using biotechnological processes are broadly similar, irrespective of the host organism. However, mammalian cell biotechnology processes present some additional peculiar issues, which are not found in microbial processes. The most important of these is that the production cells themselves may contain adventitious agents, such as mycoplasmas and viruses, or they may contain potentially oncogenic material or may produce undesirable proteins. The use of a number of animal-derived raw materials in the production process also presents a potential source of product contamination with adventitious agents. All these represent quality and safety issues and, therefore, potential hazards for the patient. Thus, this chapter discusses therapeutic protein production using mammalian cell biotechnology as the most exacting process to illustrate the range of microbiological considerations for biotechnological products.

Historically, biological products, including certain vaccines, hormones, and clotting factors from human blood have been implicated in the transmission of infectious diseases (Minor 1994). In order to prevent the possibility of contamination, biotechnology processes used for the production of therapeutic products are highly controlled and regulated. This is particularly evident for products derived from mammalian, as opposed to microbial, cell culture resulting from the additional perceived risks associated with the culture process for these cell lines. To date, and as a measure of success, there has been no reported transmission of viruses from more than 30 recombinant proteins and monoclonal antibodies approved in the United States and Europe since 1982.

5A.2 PRODUCTION OF THERAPEUTIC PROTEINS FROM CELL CULTURE

The production of a typical therapeutic protein from mammalian cell culture is illustrated in Figure 5A.1. The cell lines are created using either hybridoma or

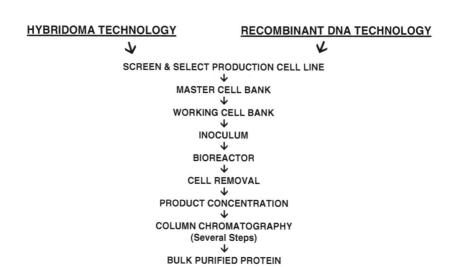

FIGURE 5A.1 Development and production of a therapeutic protein from cell culture.

recombinant deoxyribonucleic acid (DNA) technology. Following a screening of the cell lines created, one or more rounds of cloning may be performed to obtain clonal cell lines. In hybridoma and recombinant DNA technologies, the clones are then screened with respect to growth and productivity and a production cell line is selected. The next stage is the preparation of a master cell bank (MCB) of typically 50 to 400 ampoules and a working cell bank [(WCB), sometimes called a manufacturer's working cell bank (MWCB)] of typically 100 to 400 ampoules prepared in a similar procedure from an ampoule of the MCB (Froud 1999). For the manufacture of the product, an ampoule of the WCB is expanded to provide enough cells to seed the production bioreactor. Following a successful fermentation, the cells and cell debris are separated from the culture supernatant by filtration, centrifugation, or both. When the product is in the supernatant fluid, the cells are discarded and the supernatant material may be concentrated by ultrafiltration. The product is then purified by multistep column chromatography. Depending on the product, chromatographic methods available include affinity, ion exchange, hydrophobic interaction, and size exclusion chromatography (Racher et al. 1999). The bulk purified product is the outcome of a successful purification process; this may then be formulated to produce the final drug product for clinical use.

5A.3 CONSEQUENCES OF CONTAMINATION

From the previous discussion, it is evident that the manufacture of a therapeutic protein by mammalian cell culture is a complex multistep process and hence there is a potential risk of contamination at every stage. Consequences of contamination include the introduction of pathogens into the product and the introduction of toxins (e.g., endotoxins) or enzymes (e.g., proteases which may degrade the product). Ultimately, all these may lead to the loss of the product batch, as product safety or efficacy may be compromised.

Furthermore, in contrast to traditional pharmaceuticals, therapeutic biotechnology products do not generally involve "harsh" downstream processes, which make in-process monitoring and prevention of microbial contamination essential. It is, therefore, important that in addition to the testing of the final bulk purified product, the production process is monitored at every stage to ensure that adventitious agents are not introduced from the raw materials (including the production cells), the environment, or the process operators.

5A.4 SOURCES OF CONTAMINATION

Everything that interfaces with the production process for a biopharmaceutical product can potentially contaminate that product. This includes the raw materials, the process equipment, water, air, the production cells, and the process operators.

5A.4.1 PROCESS MATERIALS

It is important to ensure that all raw materials used in the manufacture of biopharmaceutical products are evaluated by established laboratory procedures and that they meet stringent quality specifications for current good manufacturing practice (CFR21 2000).

The use of disposable plasticware (e.g., cell culture flasks, centrifuge tubes, pipettes, filter units, and storage bags) is commonplace in cell culture and is effective in reducing the potential for contamination. Once packaged, these plasticware items are usually sterilized by the manufacturers using a gamma or ionizing radiation source (see Chapter 10). Even though the sterility of the product can be guaranteed if the sterilization procedure is performed properly, care should be taken when opening or resealing the packaging to maintain the integrity of these items and to avoid contaminating the contents.

Although the majority of materials can be obtained as disposable plastic items, some operations require reuseable glassware. These may be sterilized by autoclaving or dry heat (see Chapter 10). It is important that the sterilizing process is validated and the equipment is properly maintained and operated. For example, the sterilizing chamber must not be overloaded and the appropriate validated sterilization cycle and sterilizing time must be used to ensure sterility is achieved (Allwood 1999). Prior to sterilization, the glassware must be packaged in such a way that all the exposed orifices are covered. After sterilization, the glassware must be stored appropriately to avoid contamination.

Most cell culture media, sera, and supplements are heat labile and, therefore, cannot be sterilized by autoclaving; these are usually filter-sterilized to remove contaminating microbiological agents. High-temperature, short-time treatment strategies have been developed to inactivate adventitious agents (e.g., bacteria and viruses) while sparing small solutes (e.g., vitamins and amino acids). This is possible due to differences in heat inactivation rates between the larger biological agents and the small solutes (Foreman et al. 1995). When obtained from commercial sources, these products are routinely sterility-tested by the manufacturer. When media are prepared in house using nonsterile materials and components, they should be sterile-filtered and then tested for sterility before use. Although filtration through a 0.22-μm pore size filter removes most microbiological contaminants, the removal of mycoplasmas, nanobacteria, and viruses cannot be guaranteed, especially from sera (McGarrity et al. 1985, Roche and Levy 1992). Manufacturers use 0.1-μm membrane filtration (or smaller pore size), gamma, or ultraviolet irradiation to remove or inactivate adventitious agents from sera (Wyatt et al. 1994, Hodgson 1995).

5A.4.2 PROCESS EQUIPMENT

The design of process equipment such as bioreactors, centrifuges, filters and housings, purification equipment, medium and buffer tanks must allow for their cleaning, sterilization, aseptic additions, and removal of in-process samples (Vranch 1992). Where possible, the equipment design should include automation to minimize human intervention. Efficient cleaning regimes should be developed and validated to ensure that residual materials from a previous process are removed from the equipment (Sherwood et al. 1996). In addition, each piece of equipment and associated pipework should either be sterilized or treated to minimize potential bioburden. Due to the size of some of these items, a steam-in-place method (for pressure vessels) or "sanitization" (for nonpressure vessels) may be appropriate. Where steam sterilization is used, the procedure must be validated for the equipment to ensure that sterility is achieved (Hodges and Baird 2000).

Nonpressure vessels, such as ultrafiltration rigs and chromatographic columns, which are not steam sterilizable, are chemically disinfected (or sanitized) with solutions of sodium hydroxide, sodium hypochlorite, hydrochloric acid, peracetic acid, or guanidine hydrochloride. However, the sanitization process is not a sterilization process and not all microorganisms will be killed. Validation of the sanitization process for each item of equipment is important to ensure adequate bioburden reduction.

5A.4.3 WATER

The microbiological quality of process water has been discussed elsewhere in this book (see Chapter 3). The microbiological requirements for the monitoring and control of purified water systems are presented in the *United States Pharmacopeia* (USP) chapter "Water for Pharmaceutical Purposes" (USP 2005). Microorganisms that can be isolated from industrial water systems include, among others, *Pseudomonas vesicularis, P. mesophilica, Moraxella phenylpyruvica, Flavobacterium meningosepticum, Burkholderia pickettii, Sphingomonas paucimobilis, Bacillus licheniformis,* and *Xanthomons maltophilia* (Wills et al. 1998).

The regulatory authorities recommend that water meeting pharmacopoeial standards is used and its quality is monitored routinely (e.g., USP 2004). No single microbiological test method is capable of detecting all potential microbial contaminants. Classical methods such as membrane filtration and most probable number rely on the growth of the microorganisms and results may not be available for up to 7 days, by which time the water may have been used. Rapid microbiological test methods have been developed based on bioluminescence and fluorescent labeling (Wills et al. 1998, Jones et al. 1999). These methods provide results faster than the traditional method and, as they approach real-time monitoring, they offer the potential for remedial actions to be taken when specifications are not met. These methods are gaining acceptance in the pharmaceutical industry and it is anticipated that in due course they will be accepted by the regulatory authorities as valid alternative methods for detecting microorganisms and sterility testing of materials and processes (see also Chapter 9). For a more detailed discussion on the use of rapid methods for enumeration and identification, the reader is referred to the companion *Handbook* (Newby 2000).

5A.4.4 AIR

The amount of airborne contamination found in pharmaceuticals has been shown to be proportional to the concentration of microorganisms in the air (Bradley et al. 1991) and the probability that microbes will gain access into the container (Whyte 1986, 1996). During the production of biopharmaceuticals, some stages are at a higher risk from airborne contamination than others. Airborne contaminants are mainly spore-formers including the *Bacillus* spp. and fungi. However, organisms shed by individuals and from dried liquid spills can also be found in the air. Therefore, guidelines exist for the air classification of different areas of biological manufacturing facilities based on the activities being performed and the risk to the product [CFR21 2000, European Commission (EC) 1991]. In general, the highest grade of air supply is required in critical operations such as inoculum preparation and aseptic filling. Thus, these operations are performed in laminar flow cabinets (see Chapter 4). In production areas where the product is contained (e.g., in bioreactors) the environmental air may be of a lower classification. Unclassified air may be used in areas that are not critical, such as nonsterile solution preparation rooms and storage areas for raw materials kept in closed containers (Kaplan 1995).

Regulatory authorities request the monitoring of all manufacturing areas to control viable and nonviable particles. Where aseptic operations are performed, monitoring should be frequent, using a variety of methods such as settle plates, volumetric air sampling, and contact plates (for work surfaces and personnel monitoring), especially before and after critical operations (Agalloco 1999). The results should be reviewed and assessed prior to product release. It is important to analyze trends, set appropriate alert and action limits for the monitoring results, and, where these limits are exceeded, establish a plan of corrective measures (see also Chapter 8). It is worth noting that the majority of organisms isolated from contaminated cultures can also be isolated from the air in the production environment.

Consideration must also be given to the gas inlets and outlets from the bioreactors and other process equipment. Appropriate air vents and gas filters should be used and their integrity tested before and after use (Denyer and Hodges 2004). The air-handling units for the facility should be validated and the integrity of the filters should be tested at regular intervals.

5A.4.5 PERSONNEL

As microbe-carrying particles originate mainly from human skin scales (Delattin 1998), it is important to reduce the shedding of microorganisms into the manufacturing environment by appropriate gowning. A typical gowning regime includes the removal of outdoor clothing, wearing non-particle-shedding clothing, gloves, and dedicated footwear, with hair, beards, and moustaches suitably covered. As additional safeguards, the wearing of cosmetics, watches, and jewelry should be prohibited within the facility. Personnel suffering from any infectious illnesses (e.g., respiratory tract infections) which could result in the shedding of abnormally high numbers of contaminating agents should be discouraged from entering the manufacturing facility.

5A.4.6 Cells

The cells used in the production of a biopharmaceutical product must be free from contamination with adventitious agents that may cause harm to humans, the product, or the environment. The manufacturing process starts from an ampoule of a cell bank, which should be prepared and stored according to current good manufacturing practice (cGMP) as described in guidelines issued by the regulatory authorities (Center for Biologics Evaluation and Research (CBER) 1993, EC 1994, International Conference on Harmonisation 1997, World Health Organization 1997). It is important that cells are obtained from a reliable source (e.g., cell culture collections). The source of the cells and the materials used for their cultivation must be well documented. These measures will reduce the risk of cellular contamination (e.g., with HeLa cells) previously observed (Stulburg et al. 1976, Nelson-Rees et al. 1981). Where cells are obtained from sources outside the facility, they should be quarantined and tested for bacteria, fungi, mycoplasma, and specified viruses.

5A.4.7 Summary

In summary, all items that come into contact with the product or the manufacturing process represent potential sources of contamination. Sources of bacterial and fungal contamination include air, raw materials, water, process equipment, and the operators. Mycoplasma and viruses can be introduced into the process and the product from raw materials, other cell lines, or the process operators.

5A.5 POTENTIAL BIOLOGICAL CONTAMINANTS

5A.5.1 Bacteria and Fungi

In the absence of antibiotics in a cell culture, it is usually easy to detect contamination by bacteria and fungi. Cultures may become unusually turbid, pH changes may be observed, poor cell growth or even cell death may occur, and the contaminating organisms may be visible by direct microscopy. In bioreactors, contamination may also be accompanied by an increase in the observed oxygen uptake rate of the culture. However, bacteria and fungi can be more difficult to detect when antibiotics have been routinely used during cell culture or if the microorganisms are slow growing or intracellular. In these cases, the contamination may persist without obvious evidence, resulting in subtle but significant changes to the cell cultures. Some bacteria and fungi commonly isolated from contaminated cell cultures are listed in Table 5A.1.

TABLE 5A.1
Bacteria and Fungi Commonly Isolated from Contaminated Cell Cultures

Organism	Source
Corynebacterium spp.	Usually of human origin
Lactobacillus spp.	
Micrococcus spp.	
Staphylococcus spp.	
Streptococcus spp.	
Comamonas spp.	Usually liquid-borne
Flavobacterium spp.	
Pseudomonas spp.	
Sphingomonas spp.	
Bacillus spp.	Usually airborne, associated with dust and dirt particles
Candida spp.	
Aspergillus spp.	
Penicillium spp.	

The procedures for testing cell cultures for bacteria and fungi include the inoculation of soya bean–casein digest (tryptone soya broth) for the detection of aerobes, facultative anaerobes and fungi, and fluid thioglycolate for the detection of aerobic, microaerophilic, and anaerobic microorganisms (see Chapter 8). Other media, such as blood agar, may be included for the detection of more fastidious microorganisms (Hay and Cour 1995). The inclusion of blood agar and other media have been found to be useful for the detection of low-level, slow-growing bacteria that were undetected using the media recommended in the pharmacopoeias.

5A.5.2 ENDOTOXINS

Bacterial contamination can introduce endotoxins and other pyrogens. An endotoxin is a complex lipopolysaccharide that is a major component of the outer membrane of many Gram-negative bacteria, shed by both living and dead organisms (Raetz 1990, Morrison et al. 1994). Endotoxins can be found in water and some raw materials, such as sera. Endotoxins are exceptionally potent and cause pyrogenic responses ranging from fever and chills to irreversible and fatal septic shock if administered to patients in injected products. In cell cultures, endotoxin concentrations of 10 to 20 ng/ml have been shown to affect the growth and productivity of cells and higher concentrations may result in experimental variability (Sibley et al. 1988, Epstein et al. 1990, Wille et al. 1992).

Endotoxins are detected and quantified using the very sensitive Limulus amoebocyte lysate (LAL) assay. The LAL gel clot assay detects down to 0.03 endotoxin units (EUs) mL^{-1}, while the LAL kinetic turbidimetric or chromogenic assays are more sensitive, with sensitivities down to 0.005 EU.mL^{-1}.

Endotoxins, and pyrogens in general, may be very difficult to remove. They are not destroyed by autoclaving; however, exposure of glassware to 250°C for more than 30 min or 180°C for 3 h will destroy endotoxins (Weary and Pearson 1988). It is also known that electron beam or gamma irradiation will not destroy all endotoxins (Roslansky et al. 1991). Endotoxins may be introduced in unclosed systems (e.g., during purification). Environmental monitoring and sterilization (or sanitization) of the equipment will monitor and control the bioburden to which the product may be exposed and, in turn, will reduce the potential for contamination of the product with endotoxins. Some purification process steps (e.g., ion exchange chromatography) may remove endotoxins.

5A.5.3 MYCOPLASMAS

Mycoplasmas are small Gram-negative prokaryotes (0.3 to 0.5 µm in diameter), which form minute colonies (10 to 100 µm in diameter) on appropriate agar with a characteristic "fried egg" morphology. Mycoplasmas lack cell walls, are highly pleomorphic, and may be difficult to remove by filtration. They grow in aerobic or anaerobic conditions with optimum growth at 37°C and pH 7.0, that is, the same conditions required for mammalian cell culture.

Mycoplasmas deserve special consideration due to their ability to infect cultures to high densities without showing visible signs of their presence. They cannot be detected by the naked eye, usually show no macroscopic alteration of the cells or culture media, grow slowly, and do not destroy the host cells. They may induce chromosome aberrations (Aula and Nichols 1967), morphological alterations, cytopathology (Butler and Leach 1964), changes in cell growth rate (McGarrity et al. 1980), and cell metabolism (Stanbridge et al. 1971). Signs of chronic infection include reduced cell growth rate, reduced cell concentration, and increased clumping in suspension culture. They are very difficult to eradicate from cell cultures and for these reasons, mycoplasmas are, possibly, the most serious and devastating contaminants of cell culture. Mowles and Doyle (1995) reported that 12% of cultures sent to a testing laboratory were infected with mycoplasmas. Regulatory authorities recommend that cell cultures used for the production of biologicals are free from infection with mycoplasma (e.g., CBER 1993).

Five species of mycoplasma commonly implicated in the infection of cell cultures are *Acholeplasma laidlawii* (bovine origin), *Mycoplasma arginini* (bovine origin), *M. orale* (human origin), *M. fermentans* (human origin), and *M. hyorhinis* (porcine origin). Sources of mycoplasma contamination include other contaminated cell lines, animal-derived raw materials (including bovine serum and porcine trypsin), and humans.

A variety of techniques are available for the detection of mycoplasma contamination. These include DNA staining, culture, DNA probes, polymerase chain reaction (PCR), and cocultivation. It is advisable to use two different methods in the detection of mycoplasmas to allow the differentiation between false-positive and false-negative results. The methods commonly used are direct culture (with enrichment broth and mycoplasma agar medium), PCR, and the Hoechst 33258 DNA staining technique (Doyle and Bolton 1994, Mowles and Doyle 1995). Figure 5A.2 shows an uninfected and a mycoplasma-infected culture stained with Hoechst 33258.

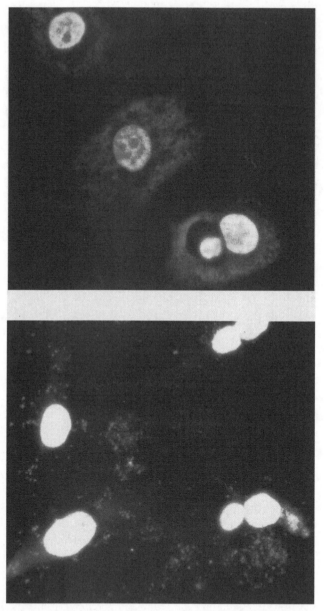

FIGURE 5A.2 Hoechst 33258–stained DNA of Vero cells (top) uninfected culture and (bottom) mycoplasma-infected culture. (Reproduced with permission of BioReliance Ltd.)

5A.5.4 Cross-Contamination

Cell lines can potentially be contaminated by other cell lines (Fogh et al. 1971). A survey by Nelson-Rees et al. in 1981 showed that more than 90 human cell lines examined were contaminated by HeLa cells (a human tumor cell line), 16 other human cell lines were contaminated with non-HeLa cells, and 13 cases of interspecies contamination were also found. Cross-contamination may be very difficult to detect, especially in suspension culture, because the cells look similar by microscopy. Cross-contamination results mainly from human error, such as incorrect labeling or using medium contaminated with one cell line for other cell lines. The ramifications of cross-contamination in the production of biologicals would be catastrophic, especially for a multiproduct manufacturing facility.

Tests for detecting intra- and interspecies contamination include chromosome analysis, isoenzyme analysis, detection of species-specific antigens, and DNA fingerprinting (Doyle and Griffiths 1998). However, for a facility with many recombinant products from the same host and similar vectors, a product-specific identity test will be more appropriate.

5A.5.5 Transmissible Spongiform Encephalopathies

Transmissible spongiform encephalopathies (TSEs) are fatal neurodegenerative diseases that have prolonged incubation periods. They are thought to result from the deposition of the prion protein (PrP) in the brain of infected individuals. Agents have been identified that infect cattle [bovine spongiform encephalopathy (BSE)], sheep (scrapie), and humans [Creutzfeldt-Jakob disease (CJD), variant CJD and Kuru]. Other species are also affected. These transmissible agents have extreme resistance to physicochemical inactivation including routine sterilization procedures. Suggested procedures for inactivation of TSEs include prolonged exposure to 1 M sodium hydroxide, porous load autoclaving at 134 to 138°C for 18 min, exposure to a solution of sodium hypochlorite containing 2% available chlorine and exposure to guanidine thiocyanate (Manuelidis 1997, Garland 1999). BSE has been reported to be transmissible to a range of animals including pigs and humans (Olson 1998). The infective agent, which can be found in blood, lymphoid, and other tissues, has been shown to be transmissible via blood (Manuelidis et al. 1985, Dealler 1996). The risk of transmission of TSEs through animal-derived raw materials such as fetal calf serum (FCS), bovine serum albumin (BSA), porcine trypsin, bovine transferrin, and human transferrin must, therefore, be considered. Such products are widely used in cell culture media.

The detection of TSE contamination involves direct intracranial injection of PrP transgenic mice with the test material, followed by prolonged incubation periods (usually many months) to allow the disease to develop. An alternative method of detection of prions is by Western blot (Beekes et al. 1995). At present, such tests are not performed routinely. Sourcing animal-derived materials from BSE-free regions is currently used to avoid contamination with TSEs.

5A.5.6 Viruses

Viral contamination of cell lines used for the production of biopharmaceuticals presents potential hazards to the process operators as well as the patient. Viral contamination may originate from the cell line, the raw materials, or the process operators. The cell line may contain endogenous viruses; for example, infectious mouse retroviruses are present in all cell lines derived from mouse myelomas (Froud et al. 1997). Cells also may be contaminated with viruses from other infected cultures handled in the same laboratory. The raw materials that may introduce viruses include FCS (bovine viruses), porcine trypsin (porcine viruses), and cell feeder layers. Garnik (1996) reported a series of contaminations of fermenter cultures of Chinese hamster ovary cells with minute virus of mice (MVM). MVM was not detected in the MCB and as the fermentation processes used large quantities of raw materials, the actual source of the contamination was not traceable. The investigation concluded that the available evidence was consistent with the virus being introduced into the manufacturing process through a contaminated raw material. This has led to the request for more rigorous screening for MVM by the regulatory authorities.

In order to test a cell line adequately, its detailed history must be examined to determine the viruses with which it could potentially be infected. For rodent cell lines (mouse, rat, and hamster) commonly used in the production of biopharmaceutical products, the viruses of concern include, among others, viruses from the host species and viruses from the media components used (Doyle and Bolton 1994, Lees and Darling 1995). Some of these viruses are listed in Table 5A.2.

The detection of viruses in cell lines and their products is performed using a variety of methods. Specific *in vitro* viral assays use detector cell lines to detect viruses of interest by cytopathic effect, morphological changes, hemadsorption and hemagglutination. The detector cells usually include human, primate, and cells from the same species. The MAP, RAP, and HAP (mouse, rat, and hamster antibody production assays, respectively) tests are species-specific tests for rodent cell lines that involve injection of the test material into mice, rats, or hamsters. The serum from the inoculated animal is then tested for the presence of a range of viruses. Up to 16 different murine viruses can be detected using the MAP test. Other viruses that may not be detectable using *in vitro* assays can be detected using *in vivo* assays performed by inoculating the test materials into embryonated chicken eggs, suckling

TABLE 5A.2
Viruses of Concern for Products from Rodent Cell Lines

Virus	Source
Lymphocytic choriomeningitis virus (LCMV)	Host species
Hantaan virus	
Sendai virus	
Reovirus-3	
Rat rotaviruses	
Bovine viral diarrhea virus (BVDV)	Bovine serum
Bovine rhinotracheitis virus	
Bovine polyoma virus	
Bovine rotaviruses	
Parainfluenza	
Bovine retroviruses	
Porcine viruses (e.g., porcine parvoviruses)	Porcine trypsin

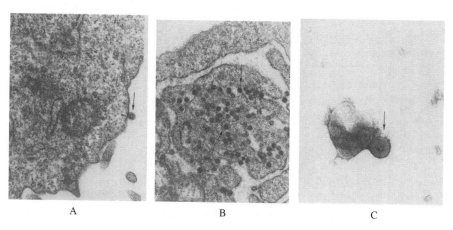

A   B   C

FIGURE 5A.3 Electron micrographs of retroviruses. (A) C-type retrovirus (arrowed) (total magnification × 101,600); (B) intracisternal A-particles (arrowed) (total magnification × 76,800); (C) negatively stained C-type retrovirus particle (arrowed) (total magnification × 280,000). (Reproduced with permission of BioReliance Ltd.)

mice, guinea pigs, or rabbits, which are subsequently examined for signs of detrimental effects. The reverse transcriptase assay detects retroviruses based on the presence of the retroviral enzyme reverse transcriptase. However, this assay does not distinguish between infectious and noninfectious retroviruses. Transmission electron microscopy (TEM) is a quantitative assay that can detect and identify a broad range of viruses. However, TEM cannot differentiate between infectious and defective viruses and is relatively insensitive. TEM is also capable of detecting mycoplasma, bacteria, and fungi. Figure 5A.3 shows electron micrographs of rodent retroviruses. The cocultivation assays are used to test for viral infectivity. In these assays, the test cell line is incubated with one or more cell lines susceptible to a variety of viruses. The cocultivation is maintained by passage for a number of weeks, during which the presence of viruses is detected, for example, by cytopathic effects, hemadsorption, or reverse transcriptase assays.

Other assays include PCR assays for viruses such as bovine viral diarrhea virus, rat XC cell plaque assays for the detection of ecotropic mouse leukemia viruses (which multiply in mouse or rat cells), S^+L^- (sarcoma positive, leukemogenic negative) focus forming assay for the detection of xenotropic mouse leukemia viruses (which can multiply in cells of other species), and the *Mus dunni*–immunofluorescence assay (a sensitive assay for the detection of murine retroviruses). Cell lines of human or primate origin should be exhaustively tested for known human and primate retroviruses and adventitious viruses. The recommended tests have been summarized by Froud (1999).

5A.5.7 AVOIDANCE OF BIOLOGICAL CONTAMINATION

No single approach will necessarily ensure the safety of a biopharmaceutical product. Therefore, the following combined approaches are required to minimize the risk of contamination.

It is most strongly recommended that cell lines are obtained from well-characterized cell banks prepared at reputable cell culture collections. Where this is not possible or verifiable, the cells should be quarantined until properly tested for contaminants before being used routinely. In addition, cell cultures in use should be regularly tested for mycoplasma contamination. All cell banks prepared should be tested following the detailed protocols provided by the regulatory authorities (e.g., CBER 1993). It may be necessary to include other media to detect fastidious and slow-growing contaminants. The identification of a contaminating organism can help trace its source.

The use of good aseptic techniques in the handling of cell cultures cannot be overemphasized (Freshney 1987). The use of antibiotics should be avoided as much as possible to prevent the maintenance of a low-level contaminant and to reduce the risk of selecting resistant strains. It is important to ensure sterility of media (e.g., by filtration) and all equipment to be used (the method employed will depend on scale). It is good practice to test media routinely before use by incubating a sample for up to 14 days in various broths and agars at 30 and 37°C (McGarrity 1979). There is no substitute for regular visual examination of the cell cultures.

As animal- and human-derived raw materials contribute to the risk of viral and mycoplasma contamination, many manufacturers of biopharmaceutical products are moving toward using media free of animal- and human-derived components. If non-animal-derived replacements are not available, the appropriate certification should be provided to confirm the origin and the health of the animals before slaughter. In general, animal-derived materials should be obtained from certified herds from areas known to be free of BSE, such as New Zealand. All animal-derived raw materials (especially serum and trypsin) should be tested for mycoplasma and the relevant viruses of concern. Sera and trypsin should be treated to remove or inactivate potential contaminants. Sera may be heat inactivated at 56°C for 30 min or filtered through 0.1-μm pore size filters. Trypsin should be filtered through a 0.1-μm filter. Some manufacturers and users of bovine sera and porcine trypsin use gamma irradiation for the inactivation of viruses. Although reports suggest that prions can be removed using high-molecular-weight exclusion limit cross-flow ultrafiltration (Michaels et al. 1995, Olson 1996) and monoclonal antibodies to PrP27-30 (Gabizon et al. 1989), the procurement of animal-derived materials from areas free of BSE is presently used to avoid contamination with TSEs. Currently, validation studies for the removal or inactivation of TSE agents are not generally required because of the difficulty in performing *in vivo* studies with these agents. However, Bailey (1998) has reported the development of strategies for validating the removal of TSEs in biopharmaceutical processes.

In addition to using cell lines obtained from a reliable source, cross-contamination can be avoided by adequate segregation of cell lines, by using good aseptic techniques and by segregating media, reagents, and supplements for each cell line. It is advisable to check cultures regularly and to treat as suspicious any sudden change in culture characteristics (e.g., in morphology or growth rate). Procedures such as physical separation, storage, and standard methods for labeling that are understood and accepted by everyone in the facility should be adopted for solutions, cultures, cryopreserved cells, and other components to avoid cross-contamination.

TABLE 5A.3
A Testing Strategy for a Rodent Cell Bank and In-Process Samples

	Master Cell Bank	Working Cell Bank	End of Production Cells	Bulk Harvest	Purified Bulk
Sterility	+	+	+		+
Mycoplasma	+	+	+	+	
In vitro virus assays	+	+	+	+	
In vivo virus assays	+		+		
MAP/RAP/HAP	+		+		
Retrovirus assays	+		+	+	+
Electron microscopy	+		+		
Cocultivation	+		+		

Note: The "+" indicates which tests should be performed. All test results should be reported. If an adventitious agent is detected in any of the tests with the shaded boxes, the cell line (or the product batch) may not be suitable for the production of a biopharmaceutical product.

Regular disinfection of the laboratory and the work area is strongly recommended (Smith 2000). All contaminated cell cultures must be discarded immediately to prevent the spread of infection. Irreplaceable stocks may be "rescued" using antibiotics for eradicating the contamination as described by Mowles (1995). These eradication methods vary in their degrees of success. The cell line may be considered contaminant free after extended maintenance in antibiotic-free culture and rigorous testing. An alternative approach is to use molecular biology techniques to re-express the product in a noninfected host cell line.

Cell banks should be tested for endogenous viruses, viruses that may be derived from the raw materials used, and other adventitious agents using a scheme similar to that in Table 5A.3. Broad viral screens, such as the general *in vitro* assays and retroviral assays, can be used for in-process detection of viruses.

Due to the limits of detection of these assays, the nondetection of viruses in a cell line does not prove that the cell line is virus free. Therefore, additional safeguards are required to minimize the potential risk of infection of the patient who receives a product derived from the cell line contaminated with an undetected virus. The regulatory authorities expect to see data showing virus removal during the purification process (virus clearance studies) and the introduction of specific virus removal or virus inactivation steps in the process (e.g., low pH inactivation, use of solvents, detergents, or virus filters).

5A.5.7.1 Virus Clearance Validation

The demonstration of virus clearance combines the following five steps:

1. Scaling down the purification process by considering various methods of virus inactivation.

2. Selecting appropriate model viruses (Table 5A.4).
3. Spiking selected steps of the process with high titers of the viruses.
4. Calculating the virus reduction factors for each step.
5. Adding the separate virus reduction factors to obtain an overall clearance factor for the whole of the purification process

The limitations of virus validation studies include the use of model viruses that may not behave exactly as contaminating viruses and the inability to scale down completely while maintaining the exact equipment and conditions of the full-scale manufacturing process. Table 5A.5 shows data obtained from the virus validation studies of a murine monoclonal antibody purification process. The overall clearance factor gives an indication of the capacity of the purification process to remove and inactivate viruses. However, to ascertain whether this clearance factor is sufficient, it is essential to relate this figure to the estimated virus load derived from the starting material (e.g., by electron microscopy) and to take account of the proposed use of the product.

TABLE 5A.4
Some Viruses Selected for Virus Clearance Validation Studies

Virus	Size (nm)	Envelope	Nucleic Acid
Murine leukemia virus	80–120	Yes	RNA
Poliovirus	20–30	No	RNA
Herpes simplex virus type 1	120–150	Yes	DNA
Adenovirus	70–90	No	DNA
Bovine viral diarrhea virus	50–70	Yes	RNA
Bovine parvovirus	18–24	No	DNA

TABLE 5A.5
Retrovirus Clearance for a Monoclonal Antibody Purification Process

Purification step	Log Reduction
Protein A affinity chromatography	4.7 to > 6.2
Low pH inactivation	4.4 to > 5.9
Ion exchange chromatography	5.4 to 7.2
Size exclusion (gel permeation) chromatography	1.5
Virus filtration	> 6.6
Cumulative reduction	> 21.1

Note: The cumulative reduction does not include the value obtained from size exclusion chromatography, because the value was very low in comparison to the other reductions.

5A.6 OVERVIEW

Bacteria, fungi, mycoplasmas, other cell lines, and viruses from various sources can contaminate cell lines, the production process, and the biopharmaceutical products obtained from them. There is also a potential risk of contamination with prion proteins. The sources of contamination include raw materials of animal origin, air, water, equipment, the cells themselves, and the process operators.

A variety of techniques exist for the detection of the various biological contaminants in the culture raw materials, the cell lines, and their products. In general, bacteria and fungi can be detected by microscopy and culturing in appropriate media, and mycoplasma can be detected using PCR, the Hoechst 33258 DNA stain, and culturing in selective media. Chromosome analysis, isoenzyme analysis, detection of species-specific antigens, genetic fingerprinting, and detection of specific proteins can be used to detect cross-contamination with other cell lines. Testing for prions is rarely done, but they can be detected by Western blot analysis and prolonged incubation in mice. Electron microscopy, various cell culture tests, and *in vivo* assays are used to detect the presence of a variety of viruses. The test regime will vary depending on the history of the cell line and the proposed use of the product.

In order to avoid contamination of cell culture processes in the production of therapeutic biotechnological products, it is very important to use cell lines from well-characterized cell banks. If it is unavoidable to receive cell lines from other laboratories, it is essential to quarantine these cell lines until they are tested for the absence of mycoplasma, bacterial, and fungal contamination. Segregating cell lines in time and space is also essential to avoid cross-contamination between cell lines. The use of good aseptic techniques cannot be overemphasized. The use of antibiotics is not a substitute for good aseptic technique and should be avoided where possible: it is better to know of the presence of a contaminant than to suppress it with antibiotics. All raw materials and equipment should be adequately tested and sterilized before use. As much as possible, the design of the production equipment should ensure aseptic operation and where possible include automation to minimize human intervention. The environmental and process bioburden should be monitored to ensure they are both within acceptable limits. Finally, where known risks are apparent, specific process steps should be included to reduce these potential risks to a level consistent with the treatment required for the patient.

This chapter has concentrated on mammalian cell biotechnology because it captures most of the microbial quality issues encountered in biopharmaceutical production. Where recombinant prokaryotes are employed, good fermentation practices must be applied to ensure production integrity (Primrose 1998). Here, the special problem of identifying in-process microbial contamination against a high-level recombinant background demands novel solutions (Delaney et al. 1996).

ACKNOWLEDGMENT

The author would like to thank Mr. E.W. Milne of BioReliance Ltd. for permission to reproduce the photographs used in this chapter.

REFERENCES

Agalloco, J. (1999). Barriers, isolators and microbial control. *PDA J. Pharm. Sci. Technol.,* 53, 48–53.

Allwood, M.C. (1999). Medical applications of thermal processes. In *Principles and Practice of Disinfection, Preservation and Sterilization,* 3rd ed. Russell, A.D., Hugo, W.B., and Ayliffe, G.A.J., Eds. Blackwell Science, Oxford, pp. 657–664.

Aula, P. and Nichols, W.W. (1967). The cytogenetic effects of mycoplasma in human leucocyte cultures. *J. Cell Physiol.,* 70, 281–290.

Bailey, A. (1998). Strategies for the validation of biopharmaceutical processes for the removal of transmissible spongiform encephalopathies (TSEs). *Eur. Biopharm. Rev.,* September, pp. 62–66.

Beekes, M., Baldauf, E., Casen, S., Dirringer, H., Keyes, P., Scott, A.C., Wells, G.A., Brown, P., Gibbs, Jr., C.J., and Gajdusek, D.C. (1995). Western blot mapping of disease-specific amyloid in various animal species and humans with transmissible spongiform encephalopathies using a high-yielding purification method. *J. Gen. Virol.,* 76, 2567–2576.

Birch, J.R. (1997). Review of biotechnology-derived products in use and in development. *Eur. J. Parent. Sci. Biotechnol. Special Issue 1997,* 3–10.

Bradley, A., Probert, S.P., Sinclair, C.S., and Tallentire, A. (1991). Airborne microbial challenges of blow/fill/seal equipment: a case study. *J. Parent. Sci. Technol.,* 45, 187–192.

Bratch, K., Strain, A.J., and Al-Rubeai, M. (1998). High density perfusion culture of primary rat hepatocytes for potential use as a bioartificial liver device. In *New Developments and New Applications in Animal Cell Technology.* Merten, O.W., Perrin, P., and Griffiths, J.B., Eds. Kluwer Academic Publishers, Dordrecht, the Netherlands, pp. 661–667.

Butler, M. and Leach, R.H. (1964). A mycoplasma which induces acidity and cytopathic effect in tissue culture. *J. Gen. Microbiol.,* 34, 285–294.

Cartwright, T. (1997). Biotechnology-based manufacturing methods for bulk-purified pharmaceutical proteins. *Eur. J. Parent. Sci. Biotechnol. Special Issue 1997,* 11–18.

CBER. (1993). Points to consider in the characterisation of cell lines used to produce biologicals. Center for Biologics Evaluation and Research, U.S. Food and Drug Administration, Washington, D.C.

CFR21. (2000). Code of Federal Regulations, Title 21, Part 210—211 and Part 600—680, U.S. Food and Drug Administration, Washington, D.C.

Dealler, S. (1996). A matter for debate: The risk of bovine spongiform encephalopathy to humans posed by blood transfusion in the UK. *Trans. Med.,* 6, 217–222.

Delaney, J., Wright, G.E., Michaels, T., Tsai, L., and Sitney, K. (1996). Bioburden testing of recombinant *Escherichia coli* K-12 fermentation samples using bacteriophages T4 and P1. *Eur. J. Parent. Sci.,* 1, 83–87.

Delattin, R. (1998). EU status of GMP for sterile products. *PDA J. Pharm. Sci. Technol.,* 52, 82–88.

Denyer, S. and Hodges, N.A. (2004). Filtration sterilization. In *Russell, Hugo, & Ayliffe's Principles and Practice of Disinfection, Preservation and Sterilization.* Fraise, A.P., Lambert, P.A., and Maillard, J.-Y., Eds. 4th ed. Blackwell Publishing, Oxford, pp. 436–472.

Doyle, A. and Bolton, B.J. (1994). The quality control of cell lines and the prevention, detection and cure of contamination. In *Basic Cell Culture—A Practical Approach.* Davies, J.M., Ed. Oxford University Press, Oxford, pp. 243–271.

Doyle, A. and Griffiths, B.J. (1998). *Cell and Tissue Culture: Laboratory Procedures in Biotechnology.* John Wiley & Sons, Chichester, U.K., pp. 3–52.

EC. (1991). EC Directive 91/356/EEC. European Commission, Brussels.

EC. (1994). Production and quality control of monoclonal antibodies. European Commission, Brussels.

Epstein, J., Lee, M.M., Kelly, C.E., and Donahue, P.K. (1990). Effect of *E. coli* endotoxin on mammalian cell growth and recombinant protein production. *In Vitro Cell. Dev. Biol.,* 26, 1121–1122.

Fogh, J., Holmgren, N.B., and Ludovici, P.P. (1971). A review of cell culture contaminations. *In Vitro,* 7, 26–41.

Foreman, L.W., Choi, T., Ihrig, T.J., Kiss, R.D., and Goodnight, M.V. (1995). Prototype system for heat inactivation of virus in culture medium. *Cytotechnology,* 17, Suppl. 1 TH18 (abstract).

Freshney, R.I. (1987). *Culture of Animal Cells—A Manual of Basic Techniques.* 2nd ed. Alan R. Liss Inc., New York.

Froud, S.J. (1999). Cell bank preparation and characterization. In *Methods in Biotechnology.* Vol. 8: *Animal Cell Biotechnology.* Jenkins, N., Ed. Humana Press Inc., Totowa, NJ, pp. 99–115.

Froud, S.J., Birch, J.R., McLean, C., Shepherd, A.J., and Smith, K.T. (1997). Viral contaminants found in mouse cell lines used in the production of biological products. In *Animal Cell Technology.* Carondo, M.J.T., Griffiths, B., and Moreira, J.L.P., Eds. Kluwer Academic Press, Dordrecht, the Netherlands, pp. 681–686.

Gabizon, R., McKinley, M.P., Groth, D., Westaway, D., DeArmond, S.J., Carlson, G.A., and Prusiner, S.B. (1989). Immunoaffinity purification and neutralization of scrapie prions. *Prog. Clin. Biol. Res.,* 317, 583–600.

Garland, A.J.M. (1999). A review of BSE and its inactivation. *Eur. J. Parent. Sci.,* 4, 86–93.

Garnik, R.L. (1996). Experience with viral contamination in cell culture. In *Developments in Biological Standardization.* Proceedings of the IABS International Scientific Conference, Bethesda, MD, June 14–16, 1995, on viral safety and evaluation of viral clearance from biopharmaceutical products. Kargar, Basel.

Handa-Corrigan, A., Green, I.C., Mabley, J., Hayavi, S., Kass, G.N., Hinton, R.H., Morgan, I.M., and Wright, J. (1998). Novel mini bioreactors for islet cell culture. In *New Developments and New Applications in Animal Cell Technology.* Merten, O.W., Perrin, P., and Griffiths, J.B., Eds. Kluwer Academic Publishers, Dordrecht, the Netherlands, pp. 669–672.

Hay, R.J. and Cour, I. (1995). Testing for microbial contamination—bacteria and fungi. In *Cell and Tissue Culture: Laboratory Procedures.* Doyle, A., Griffiths, J.B., and Newell, D.G., Eds. John Wiley & Sons, Chichester, U.K., module 7A:2.

Hodges, N.A. and Baird, R.M. (2000). Disinfection and cleansing. In *Handbook of Microbiological Quality Control: Pharmaceuticals and Medical Devices.* Baird, R.M., Hodges, N.A., and Denyer, S.P., Eds. Taylor & Francis, London, pp. 205–220.

Hodgson, J. (1995). To treat or not to treat: that is the question for serum. *Bio/Technology,* 13, 333–343.

International Conference on Harmonisation (1997). Quality of biotechnological products: derivation and characterization of cell substrates for the production of biotechnological/biological products.

Jones D.L., Brailsford M.A., Drocourt, J.-L. (1999). Solid-phase, laser-scanning cytometry: a new two-hour method for the enumeration of micro-organisms in pharmaceutical water. *Pharmacopoeial Forum,* 25(1), 7626–7645.

Kaplan, R. (1995). Current GMP considerations for biotechnology facilities, Part 2: process validation, contamination control and compliance auditing. *BioPharmaceuticals,* 8, 26–30.

Kolokoltsova, T.D., Yurchenko, N.D., Kolosov, N.G., Shumakova, O.V., and Nechaeva, E.A. (1998). Cultivation of skin cells suitable for recovery of burn wounds. In *New Developments and New Applications in Animal Cell Technology.* Merten, O.W., Perrin, P., and Griffiths, J.B., Eds. Kluwer Academic Publishers, Dordrecht, the Netherlands, pp. 673–675.

Lees, G. and Darling, A. (1995). Biosafety considerations. In *Monoclonal Antibodies: Principles and Applications.* Birch, J.R. and Lennox, E.S., Eds. Wiley-Liss, New York, pp. 267–298.

Manuelidis, E.E., Kim, J.H., Mericngas, J.R., and Manuelidis, L. (1985). Transmission to animals of CJD from human blood. *Lancet,* ii, 896–897.

Manuelidis, L. (1997). Decontamination of Creutzfeldt-Jakob disease and other transmissible agents. *J. Neurovirol.,* 3, 62–65.

McGarrity, G.J. (1979). Detection of contamination. In *Methods in Enzymology: Cell Culture.* Jacoby, W.B. and Pasten, I.H., Eds. Vol. 58. Academic Press, New York, pp. 18–29.

McGarrity, G.J., Phillips, D., and Vaidya, A. (1980). Mycoplasmal infection of lymphocyte cultures: infection with *M. salivarium. In Vitro,* 16, 346–356.

McGarrity, G.J., Sarama, J., and Vanaman, V. (1985). Cell culture techniques. *ASM News* 51, 170–183.

Meissner, P., Werner, P., Schroder, B., Herfurth, C., Wandrey, C., and Biselli, M. (1998). Expansion of human hematopoietic progenitor cells in a fixed bed bioreactor. In *New Developments and New Applications in Animal Cell Technology.* Merten, O.W., Perrin, P., and Griffiths, J.B., Eds. Kluwer Academic Publishers, Dordrecht, the Netherlands, pp. 635–636.

Michaels, S.L., Antoniou, C., Goel, V., Keating, P., Kuriyel, R., Michaels, A.S., Pearl, S.R., de Los Reyes, G., Rudolph, E., and Siwak, M. (1995). Tangential flow filtration. In *Separations Technology, Pharmaceutical and Biotechnology Application.* Olson, W.P., Ed. Interpharm Press, Buffalo Grove, IL, pp. 57–194.

Minor, P.D. (1994). Ensuring safety and consistency in cell culture processes: viral screening and inactivation. *Trends Biotechnol.,* 12, 257–261.

Morrison, D.C., Dinarello, C.A., Munford, R.S., Natanson, C., Danner, R., Pollack, M., Spitzer, J.J., Ulevitch, R.J., Vogel, S.N., and McSweegan, E. (1994). Current status of bacterial endotoxins. *ASM News,* 60, 479–484.

Mowles, J. (1995). Testing for microbial contamination—elimination of contamination. In *Cell and Tissue Culture: Laboratory Procedures.* Doyle, A., Griffiths, J.B., and Newell, D.G., Eds. John Wiley & Sons, Chichester, U.K., module 7C:1.

Mowles, J. and Doyle, A. (1995). Testing for microbial contamination—detection of Mycoplasma. In *Cell and Tissue Culture: Laboratory Procedures.* Doyle, A., Griffiths, J.B., and Newell, D.G., Eds. John Wiley & Sons, Chichester, U.K., module 7A:1.

Nelson-Rees, W.A., Daniels, D.W., and Flandermeyer, R. (1981). Cross-contamination of cells in culture. *Science,* 212, 446–452.

Newby, P. (2000). Rapid methods for enumeration and identification in microbiology. In *Handbook of Microbiological Quality Control: Pharmaceuticals and Medical Devices.* Baird, R.M., Hodges, N.A., and Denyer, S.P., Eds. Taylor & Francis, London, pp. 107–119.

Olson, W.P. (1996). Sterilization of small-volume parenterals and therapeutics by filtration. In *Aseptic Pharmaceutical Manufacturing Technology for the 1990s.* Olson, W.P. and Groves, M.J., Eds. Interpharm Press, Buffalo Grove, IL, pp. 101–149.

Olson, W.P. (1998). Prions: a review, theories and proposals. *PDA J. Pharm. Sci. Technol.,* 52, 134–146.

Primrose, S.B. (1998). Production of therapeutically useful substances by recombinant DNA technology. In *Pharmaceutical Microbiology.* Hugo, W.B. and Russell, A.D., Eds. 6th ed. Blackwell Scientific, Oxford, pp. 453–468.

Racher, A.J., Tong, J.M., and Bonnerjea, J. (1999). Manufacture of therapeutic antibodies. In *Biotechnology.* Rehm, H.J. and Reed, G., Eds. 2nd ed. Vol. 5A. Wiley-VCH, Weinheim, Germany, pp. 245–274.

Raetz, C.R.H. (1990). Biochemistry of endotoxins. *Ann. Rev. Biochem.,* 59, 129–170.

Roche, K.L. and Levy, R.V. (1992). Methods used to validate microporous membranes for the removal of mycoplasma. *BioPharmaceuticals,* 5, 22–23.

Roslansky, P.F., Dawson, M.E., and Novitsky, T.J. (1991). Plastics, endotoxins and the Limulus amebocyte lysate test. *J. Parent. Sci. Technol.,* 45, 83–87.

Sherwood, D., Fisher, D., Clifford, J., and Slade, S. (1996). Experiences with clean-in-place validation in a multi-product biopharmaceutical manufacturing facility. *Eur. J. Parent. Sci.,* 1, 35–41.

Sibley, C.H., Terry, A., and Raetz, C.R.H. (1988). Induction of κ light chain synthesis in 70Z/3 B lymphoma cells by chemically defined lipid A precursors. *J. Biol. Chem.,* 263, 5098–5103.

Smith, A.W. (2000). Safe microbiological practices. In *Handbook of Microbiological Quality Control: Pharmacueticals and Medical Devices.* Baird, R.M., Hodges, N.A., and Denyer, S.P., Eds. Taylor & Francis, London, pp. 1–21.

Stanbridge, E.J., Hayflick, L., and Perkins, F.T. (1971). Modification of amino acid concentrations induced by mycoplasmas in cell culture medium. *Nat. (London) New Biol.,* 232, 242–244.

Stulburg, C.S., Petersen, Jr., W.D., and Simpson, W.F. (1976). Identification of cells in culture. *Am. J. Hematol.,* 1, 237–242.

USP. (2005). Water for Pharmaceutical Purposes. *United States Pharmacopeia 29.* 2006 U.S. Pharmacopeial Convention Inc., Rockville, MD.

Vranch, S.P. (1992). Plant design and process development for contract biopharmaceutical manufacture. *Ann. N.Y. Acad. Sci.,* 646, 367–375.

Weary, M. and Pearson, F. (1988). A manufacturer's guide to depyrogenation. *BioPharmaceuticals,* 1, 22–29.

World Health Organization. (1997). Requirements for use of animal cells as *in vitro* substrates for the production of biologicals. World Health Organization, Geneva.

Whyte, W. (1986). Sterility assurance and models for assessing bacterial contamination. *J. Parent. Sci. Technol.,* 40, 188–197.

Whyte, W. (1996). In support of settle plates. *J. Pharm. Sci. Technol.,* 50, 201–204.

Wille, J.J., Park, J., and Elgavish, A. (1992). Effects of growth factors, hormones, bacterial lipopolysaccharides and lipotechoic acids on the clonal growth of normal urethral epithelial cells in serum-free culture. *J. Cell. Physiol.,* 150, 52–58.

Wills, K., Woods, H., Gerdes, L., Hearn, A., Kyle, N., Meighan, P., Foote, N., Layte, K., and Easter, M. (1998). Satisfying microbiological concerns for pharmaceutical purified waters using a validated rapid test method. *Pharmacopeial Forum,* 24(1), 5645–5664.

Wyatt, D.E., Keathley, C.W. and Broce, R. (1994). Inactivation of serum contaminants. In *Animal Cell Technology Products of Today, Prospects for Tomorrow.* Spier, R.E., Griffiths, J.B., and Berthold, W., Eds. Butterworth-Heinemann, Oxford, pp. 140–141.

5B Microbiological Considerations in the Production of Medical Devices

Haroon A.R.D. Atchia

CONTENTS

5B.1 INTRODUCTION

Given the complex nature of medical device production processes and the considerable range of medical device types, it is not surprising that significant effort is required to assess and control microbial contamination. In this context, it is important to recognize that the microbial ecology of medical devices concerns not only the raw materials used but also the influence of the wider production program. This includes an appreciation of processes at the principal manufacturer's site, the characteristics of subcontracted operations, and the quality of the total production environment.

5B.2 THE NATURE OF MEDICAL DEVICES

Medical devices are often of complex construction reflecting their medical or surgical role. Although not exhaustive, Table 5B.1 describes many of the raw materials used in the construction of medical devices.

TABLE 5B.1
Selected Materials Used in the Construction of Medical Devices

Material	Typical Device Use
Plastics, polymers, and elastomers	Infusion devices (administration sets, blood bags), urinary catheters, fluid collection devices, hemodialysis tubing sets, bulking agents
Ceramics	Orthopedic implants, dental restorative products, cements
Metals	External orthopedic fixation devices, internal orthopedic fixative devices and other orthopedic implants, intravascular guide wires, peripheral and coronary stents, other coronary implants
Biological and chemical coatings	Device surface lubrication, reduction of thrombus deposition, improved biological tolerance by the body
Animal tissues	Reconstruction; urethral devices, heart valves
Human tissues	Reconstruction

The use of these very different materials determines the production process, with a concomitant influence on the production environment and microbial ecology.

The microbial flora associated with medical device production does not appear to show uniformity in the quantity or type of microorganism recovered. Although this may reflect population nonhomogeneity, it is more probably related to the nature of sampling for microbiological attributes (Clontz 1998). Sampling presents two difficulties:

1. The relatively small sample sizes used.
2. The lack of uniformity in the distribution of contaminants.

The distribution and type of microbial contaminants are partly related to the raw material being converted into a medical device (see Chapter 8), but subsequently they are also influenced considerably by the production process. The data presented in Table 5B.2, collected from 48 manufacturers of medical devices as a snapshot of their production operations, illustrate the inherent variation in contamination for the material types used. The range of device complexity, production operations, and manufacturing environment are reflected in the very wide *infra*-material type distribution of microbial loading. As all the sampling, cultivation, and enumeration were performed according to the same methodology, variation in organism count reflects true product differences.

Bioburden data reveal a predictable tendency for increasing contamination from low levels on metallic items to high burden on fibrous devices such as cotton wool swabs. Most of the microorganisms constituting contaminants of metal products are introduced onto the device by the production process. This contrasts with natural organic materials in which the microbial contamination is indigenous.

TABLE 5B.2
Bioburden Data Taken from a Number of Anonymous United Kingdom Customers and Grouped by Product Type

Bioburden* by Product Type					
Tubing		Steel Items	Catheter/Guidewire		Swabs
22	304	7	5	4	16
20	8	9	24	5	37
21	5	4	22	12	68
2	6	5	0	0	97
5	4	10	0	0	13
11	1	4	37	2	250+
11	1	1	11	1	13
2	0	8	34	2	250+
62	0	0	3	4	250+
37	2	2	1	3	250+
19	2	3	3	0	250+
8	5	5	3	2	250+
2	2	3	1	1	—
62	11	1	3	2	—
37	172	0	3	5	—
19	20	1	4	2	—
8	0	2	1	2	—
2	108	1	2	0	—
29	1	0	2	1	—
20	2	0	0	0	—
15	56	0	1	1	—
5	112	0	36	5	—
7	0	0	0	3	—
12	0	0	3	2	—
7	0	0	3	1	—
14	0	0	5	1	—
14	1	0	3	7	—
60	6	—	3	5	—
16	11	—	10	1	—
9	4	—	10	1	—
40	4	—	11	3	—
27	4	—	11	2	—
2	4	—	14	4	—
7	1	—	14	1	—
22	0	—	2	27	—
47	2	—	1	3	—
10	3	—	0	1	—
5	3	—	1	5	—
14	0	—	0	2	—

(continued)

TABLE 5B.2 (continued)
Bioburden Data Taken from a Number of Anonymous United Kingdom Customers and Grouped by Product Type

Bioburden* by Product Type					
Tubing		Steel Items	Catheter/Guidewire		Swabs
5	10	—	0	2	—
15	28	—	0	1	—
56	27	—	2	1	—
3	13	—	17	0	—
38	5	—	3	8	—
42	6	—	2	1	—
2	3	—	11	1	—
33	2	—	5	3	—
75	31	—	2	0	—
Average					
	21	2	5		N/A
Minimum					
	0	0	0		13
Maximum					
	304	10	37		250+

* *Note*: All data are expressed as colony forming units (CFUs) per product except swabs at $CFU.g^{-1}$. Dashes indicate no data. N/A indicates not applicable.

Source: From Sterigenics UK, personal communication. With permission.

Production processes are also known to be a major contributory factor for contamination of medical devices made from plastic polymers (Figure 5B.1).

An understanding of the sources and anticipated distribution of microbial contaminants is crucial in developing an effective microbiological control programme. Measures for control have been reflected in sterile medical device regulation since the 1960s, culminating in a variety of recognized standards for environmental monitoring. Most of the standards have avoided specifying microbial contamination in terms of the identify of the microorganisms. Instead, they remain confined to the determination of inanimate particulate contamination (Anonymous. 1976, 1989) or simply as to *how* microorganisms can be detected through the use of screening tests.

Overtly pathogenic microorganisms are rarely found on non-organic materials; microorganisms found in the medical device production environment are mainly Gram-positive bacilli and cocci. The presence of Gram-negative bacilli is almost invariably associated with water in the production process. It is important to understand the production processes adequately in order to implement an effective microbiological control program.

FACTORS

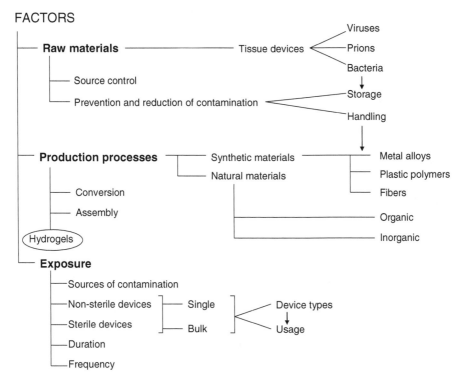

FIGURE 5B.1 Factors influencing the microbial ecology and nature of medical devices.

5B.3 FACTORS INFLUENCING THE MICROBIAL ECOLOGY

Factors influencing the microbial ecology of medical devices may be grouped into four main areas:

1. Raw materials
2. Production processes
3. Extent of exposure
4. Scale of production

Although the influence of both the raw materials and the production processes may be obvious as sources of contamination, the extent of exposure is often overlooked. For devices made on a small batch basis, the intensity of handling by production workers and repeated handling are the most prolific sources of contamination. For this reason, the environmental microbiological control program requires a clear appreciation of how the production process affects the distribution of contamination at each step in the process. End-stage monitoring of microbial contamination of the finished device (traditionally termed "bioburden") is insufficient for characterizing contamination and does not permit the detection of "hot spots" in the production process. Comprehensive profiling is implied through process control requirements of recognized quality management systems applicable to the production of medical devices. In these situations, manufacturers are obliged to know the capability of their processes; specific operations are qualified to yield a consistently manufactured product conforming to the specified requirement for that product (or to established specifications given by appropriate standards).

The features of such process qualification must naturally take into account the following:

- Materials
- Processing fluids or agents with which the materials will come into contact
- Exposure of the materials to air in the production environment
- Exposure of materials to surfaces
- Handling

The logical corollary of process qualification is the implementation of appropriate storage areas such as receiving or inspection, intermediate storage during the production of the device, and finished goods storage and distribution to the primary user. Controls to minimize the introduction of contamination due to incoming raw material as well as equipment should then be developed. Typical controls include placing consignments of raw materials in double plastic bags to facilitate transfer from storage into the production environment in the case of devices manufactured partly or fully in a controlled environment.

5B.4 SUITABILITY OF MICROBIAL METHODS FOR MONITORING CONTAMINATION

The absence of statutory or regulatory specifications for the level of microbial contamination has led to the emergence of industry norms. For example, products such as blood contact medical devices tend to be produced in clean rooms (areas of higher specified cleanliness; see Chapter 4). Hence, the microbiological control program should be commensurate with the anticipated level of contamination to be controlled (Pendlebury and Pickard 1997).

Since the importance of correct methods is crucial, appropriately detailed documented procedures are needed. These should cover the methodologies and techniques to be used for each test, together with the frequency of testing. Corrections for the efficiency of the techniques to recover microorganisms from the test devices should be included. These may be supplemented by control measures such as indicator strains to be used for comparisons or other reference measures.

Detailed guidance on bioburden estimation in medical devices is given in BS EN 1174 (1994) (BS EN are European standards accepted by British Standards Institution). Five distinct stages are involved: sample selection, removal of microorganisms from sample, transfer of microorganisms to recovery conditions, enumeration of microorganisms with specific characteristics, and interpretation of data, involving application of appropriate correction factors determined during validation studies. Considerable significance is attached to bioburden testing; it is regarded as a critical step in the manufacturing process of sterile devices. An under-estimation of the bioburden population could result in a miscalculation of the sterilizing requirements for a given product; in contrast, an overestimation could result in excessive exposure to the sterilizing agent, which in turn could affect the function of the device. For a more detailed discussion on sampling and testing of medical devices and on the recovery conditions used, the reader is referred to Baird (2000) and Millar (2000) in the companion *Handbook*.

5B.5 USE OF INFORMATION FROM MICROBIOLOGICAL MONITORING

The most frequent use of information from microbiological monitoring of medical device production is for determination or confirmation of the achievement of the *specified quality* of the product. Information gained from microbiological monitoring may also be required as part of the regulatory scrutiny needed for obtaining marketing approval or clearance from government organizations responsible for safety.

Further uses of information may be in the compilation of quality profiles of the production process. In this regard, characterization of the microbiological contamination allows determination and identification of potential deleterious sources that could compromise quality.

In certain situations, such as devices to be labeled "sterile," manufacturers are obliged by convention to prepare a documented programme for accomplishing the performance qualification requirements specified in standards or monographs. It is noteworthy that the use of many of the applicable standards is voluntary. Nevertheless, convention and regulatory pressure have led to the emergence of certain norms, which has conferred a technological status to such references (Table 5B.3).

The status of the various standards currently available creates confusion for manufacturers. For example, convention requires the use of standard methods, yet manufacturers intending to place product on the market within the European Union conceivably achieve the necessary approvals through the use of non-harmonized European Standards. Although this may at first appear attractive, the emphasis on

---

TABLE 5B.3
Typical Uses of Microbiological Monitoring Information

Determination of contamination distribution (profile)
Control of routine production quality
Investigation of excursions from set limits
Remedy of contamination problems
Demonstration of achievement of specified quality
Accomplishment of applicable standards for performance qualification (validation)
Regulatory approval
Investigation of postproduction events

---

good quality management becomes paramount: equivalent controls would be needed within the manufacturers' quality management system.

Confusion also arises because of the different geographical expectations for microbiological control. A particularly revealing example concerns the U.S. expectation for bacteriostasis and fungistasis tests. Manufacturers of medical devices supplying the U.S. market are required to perform such tests, even though the quality of the methods used is not necessarily microbiologically defensible. In Europe, however, their use would be optional, with the manufacturer obliged only to determine whether to test and to determine the suitability of the test method accordingly. There is a tendency in certain medical markets to rely increasingly on microbiological test methods originally devised for control of pharmaceutical products.

In terms of the routine control of medical device production, by far the most important concern for the manufacturer is whether the batch in question can be released to the market. There remains no universal consensus on the best method to use and so it is important to determine both the local and national requirements. Despite any such variation, the process for release of products remains uniform:

- Confirmation that the sterility assurance level specified for the medical device has been achieved according to qualified (validated) methods
- Confirmation that all assembly steps have written specifications
- Confirmation that microbiological contamination levels are within specifications
- Demonstration or confirmation that sterilization process cycle conditions are within specification

In certain situations, manufacturers may need to perform additional testing prior to release. These include detection of endotoxins on devices that come into contact with water during production and tests for sterility. Typically, however, such tests do not add weight to the demonstration of microbiological quality, as they tend not to be required for regulatory clearance without good justification.

5B.6 DYNAMIC MICROBIOLOGICAL QUALITY ASSURANCE FOR MEDICAL DEVICE PRODUCTION

Manufacturers of medical devices could easily be forgiven for wondering how to respond to the bewildering array of voluntary and statutory requirements that may determine how to effect microbiological control of production. Clearly, these standards, particularly in the Western world, can be expected to increase in the future. This, along with the continued appeal of formal, recognized quality management systems demonstrating good manufacturing practice, suggest that the development of dynamic assurance for microbiological contamination control should constitute an essential part of the quality system (Table 5B.4).

A clear understanding of the contamination sources, influence of each production step, and the regulatory requirements are obviously crucial but so too is the *purpose* of any testing that is integrated within the quality system.

TABLE 5B.4
Typical Components of a Dynamic Microbiological Control Program

Policies for control of contamination
Evidence for selection of methods and approaches, with justification
Recognized use of standard methods, with justification
Comprehensive work instructions, e.g., laboratory methods
Empirically qualified cultivation techniques
Empirically qualified enumeration and identification techniques
Control limits

Fully documented policies, procedures, and practices are implied by the recognized quality management system standards, notably EN ISO 13485:2000 and EN ISO 13488:2000 (Anonymous. 2000a, 2000b). This is further supported by the continued drive towards international harmonization of requirements, in which the prevailing regulations and standards all require some level of documentary evidence for achievement of specified requirements, including those concerning microbiological contamination. The message for the industry is clear: the structure and purpose of any microbiological control program not only must be good microbiology, reflecting in-depth theoretical academic knowledge and empirical evidence, but must permit the achievement of consistent production quality to be communicable to regulatory authorities.

ACKNOWLEDGMENT

Grateful appreciation is extended to Sterigenics UK Ltd., Alperton, United Kingdom, for kindly providing the bioburden data used in this chapter.

REFERENCES

Anonymous. (1976). *British Standard 5295—Environmental Cleanliness in Enclosed Spaces.* British Standards Institution, London.
Anonymous. (1989). *British Standard 5295—Environmental Cleanliness in Enclosed Spaces.* British Standards Institution, London.
Anonymous. (2000a). *EN ISO 13485: 2000—Quality Systems: Medical Devices.* International Organization for Standardizations, Geneva.
Anonymous. (2000b). *EN ISO 13488: 2000—Quality System: Medical Devices.* International Organization for Standardizations, Geneva.
Baird, R.M. (2000). Sampling: principles and practice. In: *Handbook of Microbiological Quality Control: Pharmaceuticals and Medical Devices.* Baird, R.M., Hodges, N.A., and Denyer, S.P., Eds. Taylor & Francis, London, pp. 38–53.
British Standards Institution. (1994). *BS EN 1174 (1994). Part 11. Sterilization of Medical Devices—Estimation of the Population of Micro-Organisms on Product.* British Standards Institution, London.
Clontz, L. (1998). Microbial limit and bioburden tests. In: *Microbial Contamination and Control.* Interpharm Press, Buffalo Grove, IL, p. 11.

Millar, R. (2000). Enumeration of micro-organisms. In: *Handbook of Microbiological Quality Control: Pharmaceuticals and Medical Devices*. Baird, R.M., Hodges, N.A., and Denyer, S.P., Eds. Taylor and & Francis, London, pp. 38–53.

Pendlebury, D.E. and Pickard, D. (1997). Examining ways to capture airborne micro-organisms. In: *Cleanrooms*. PennWell Publishing Company, Tulsa, OK.

6 Good Manufacturing Practice in the Control of Contamination

Paul Hargreaves[*]

CONTENTS

---

[*] The views and opinions expressed are personal ones and are not necessarily those of the Medicines Inspectorate and should not be taken as GMP requirements.

6.1 INTRODUCTION

Good manufacturing practice (GMP) enshrines the principle that products are consistently manufactured to a quality appropriate to their intended use. GMP is thus an all-embracing philosophy that covers premises, services, equipment, processes, documentation, purchasing, distribution, and staff. The control of microbiological contamination is an important aspect of GMP and should form an integral part of it.

Traditionally, microbiological control has been associated particularly with the manufacture of sterile products. However, experience has shown that microbiological contamination can also cause problems with a wide range of nonsterile products, such as tablets, creams, ointments, and oral liquids. The traditional response to this problem has been the addition of preservatives, but there is now increasing consumer and medical pressure to remove or reduce the levels of preservatives in products. In turn, the pharmaceutical manufacturer has had to respond by increasing the level of microbiological monitoring and control. One of the main consequences of this has been the introduction of microbiological control of existing premises and practices in a piecemeal fashion that is often expensive and ineffective. An alternative and preferred course of action involves a thorough review of all aspects of the operation with a team consisting of representatives from the microbiology, engineering, manufacturing, and training departments. This should ensure that there is a consistent approach throughout the factory and that no important aspects are missed. The food industry is aware of the need for an integrated approach and has developed a system called hazard analysis of critical control points (HACCP). This system ensures that the proper amounts of resources, monitoring, and control are in place at the points where microbiological control is required. This is not only cost effective but also gives a high degree of assurance that the finished product will be suitable for its intended use. A possible HACCP system for cleanroom contamination control has been proposed for pharmaceutical manufacturing (Whyte 2002).

GMP has been interpreted in many different ways. The most succinct definition is "getting it right first time, every time." One of the best ways of achieving this is to prevent mistakes from occurring at the source. Most companies are good at identifying possible hazards for which they then introduce checks or tests at subsequent manufacturing stages to determine whether the hazard has occurred. This is not good manufacturing practice. The correct approach is to identify the hazard and prevent it from occurring in the first place. In the long term, this is also probably the most cost-effective approach, but it may take time and effort on the part of the people involved in solving the problem. It is recognized that going from one crisis to another is a common problem in manufacturing industries and that lack of time and resources often prevents the ideal problem-solving approach from being taken. Indeed, there can be considerable commercial pressure to maintain the status quo rather than develop better, more secure systems or working practices. In such a situation, a cost-benefit analysis may be necessary to demonstrate that the benefits from an initial investment of staff and resources can often be significant in terms of longer shelf life (especially in warmer climates), lower rejection rate, and fewer customer complaints. It is important, there-fore, that scientific and manufacturing personnel are able to present their requirements in such a way that they are readily understandable to the commercial arm of the company. GMP must evolve from, and involve, senior management. Without their commitment, resources may be unduly limited.

6.2 PRINCIPLES OF FACTORY OR HOSPITAL HYGIENE

6.2.1 ZONING CONCEPT

A factory will naturally be divided into specific areas for different activities. Each activity will have different requirements in terms of air handling and filtration, temperature and humidity control, cleaning and protective clothing. By taking this natural segregation of activities one stage further, the zoning concept is achieved. When designing a new factory or renovating an existing one, it is relatively easy to zone activities according to cleanliness requirements and to ensure that these zones are different colors. Access would be restricted to those people who were correctly clothed, trained, and authorized to be in that zone.

   Although this practice is common in cleanroom suites, its extension to cover the whole factory is now becoming more widespread as the need to manufacture non-sterile products to a high microbiological quality is recognized. Traditionally, zones have been classified as black, gray, and white. However, with the numerous activities that occur in nonsterile manufacture, a more varied classification is needed. The zoning concept enables the correct number of resources to be allocated to an area. For example, an oral liquid manufacturing area may require a more frequent and thorough sanitization program than a nonaqueous ointment manufacturing area. In the same way, clothing, air handling, and air filtration should reflect the necessary degree of protection that the products require.

6.2.2 CLEANROOM CLASSIFICATION

Whyte et al. (1982) have demonstrated that virtually all contamination occurring during aseptic manufacturing results from poor operator technique or airborne viable particles; however, the crucial factor for achieving low contamination levels with a trained workforce will be viable particles in the air. It might therefore seem logical to define environmental standards on the basis of colony-forming units (CFUs) per unit air volume; however, the relatively low levels of airborne microbial contamina-tion would render measurements imprecise and inconsistent. Instead, the quality of cleanroom air is defined by total particulate counts that can be continually or periodically monitored. Thus sterility assurance relies upon the exclusion from the work zone of all particles that could be viable, rather than attempting to define an

TABLE 6.1
Cleanroom Classifications Based on Airborne Particulates

Classification System							
Federal Standard 209E Class		1	10	100	1,000	10,000	100,000
EU GMP Class		—	—	A/B	—	C	D
ISO 14644–1 Class		3	4	5	6	7	8
Performance specifications for the control of airborne particulates defined as limits on the number of particles of given size (µm) which may be present in a cubic foot of air	0.1 µm	35	350	—	—	—	—
	0.2 µm	7.5	75	750	—	—	—
	0.3 µm	3	30	300	—	—	—
	0.5 µm	1	10	100	1,000	10,000	100,000
	5.0 µm	—	—	—	7	70	700

Note: Dashes indicate no standard requirement.

acceptable limit for viable particle concentration. A cleanroom therefore can be classified by the quality of air supplied to the facility. Some unification of standard has been achieved through the global International Organisation for Standardisation (ISO) classification system; this is compared to national standards in Table 6.1. For a more detailed discussion on the design of cleanrooms and controlled environments, the reader is referred to Chapter 4.

6.2.3 CLOTHING

Ideally, all personnel entering or working in pharmaceutical manufacturing areas should be clothed in dedicated factory clothing. This should consist of white shirt, white trousers, and appropriate footwear. The clothes should be freshly laundered. It is not unusual to find that people are issued with insufficient changes of clothing, perhaps having to wear the same shirt and trousers for up to 1 week in the interest of economy. For products that are prone to adventitious microbiological contamination, for example, oral liquids and creams, and are not manufactured in closed systems, this can prove to be an expensive economy. It is absurd to invest money and time on performing bioburden tests on raw materials when the manufacturing staff is not suitably clothed.

Much has been written on the design of, and the textiles used for, cleanroom clothing (Redlin and Neale 1987, Anonymous 1987a, Miller 1987, Clemens 1988, Goodwin 1988, Dixon 1989, Clayton 2002) to which the reader is referred. The purpose of cleanroom clothing is to reduce the shedding by personnel of microorganisms and nonviable particulate matter into the environment. If cleanroom clothing is to be effective, therefore, it should be demonstrated to be so, prior to its being worn in the cleanroom. In this respect, the laundering and sterilization (or sanitization) of clothing can damage it to the extent that it is no longer suitable for its intended use. There are a number of methods available for testing cleanroom clothing, and one of these methods or a suitable alternative should be used (Anonymous 1968, Australian Standard AS 2013 1977a, Australian Standard AS 2014 1977b,

British Standard 3211 1986). The approximate life of the garments can then be determined so that they can be replaced before becoming defective. This will require stability testing whereby a garment will undergo successive laundering and sterilization cycles until it fails. The rationale behind this is that it is extremely difficult for cleanroom monitoring programs to detect defective garments in a timely manner. A defective garment may be worn only once or twice a month and it may be many months before the monitoring program will pinpoint the problem. It helps, of course, if all garments are identified and logged so that high microbial counts in the cleanroom can be correlated to the specific garments worn. Apart from the standard cleanroom clothing worn by the regular personnel, there is usually a requirement for clothing for people who are infrequent visitors, for instance, maintenance or engineering personnel and self-inspection audit teams. Disposable cleanroom clothing is a popular choice in these circumstances, but this clothing also needs to be vigorously tested for adequacy before a particular make is used; in particular, some types do not have sealed seams over the stitching.

6.2.4 CLEANING AND DISINFECTION

The *Rules and Guidance for Pharmaceutical Manufacturers and Distributors* (Anonymous 2002) contains some general advice with regard to cleaning and disinfection (see sections 3.36, 3.9–3.11, 5.19, A1.37–A1.39). Disinfection and sanitization of specific manufacturing areas will be examined in greater detail in subsequent sections of this chapter. Some problems associated with cleaning of equipment are illustrated in Chapter 3. Seiberling (1987) gives an extensive treatment of cleaning-in-place and sterilizing-in-place.

A cleaning and disinfection policy should be prepared that covers the whole factory or hospital manufacturing suite. This should be drawn up in conjunction with a microbiologist and manufacturing and janitorial staff. The cleaning and disinfection policy should cover written cleaning schedules and procedures including the name of cleaning or disinfection agent, concentration of agent, quality of water to be used in preparation of use-dilutions, and shelf life of the diluted agent. The procedures should also include details of how the disinfectant is measured and diluted. For instance, it is not uncommon to find small volumes being measured in large measuring cylinders, a bad practice that arises particularly when disinfectants are changed, especially to ones that are used at concentrations of 0.5% or below. Both above and below certain concentrations, biocidal activity may decrease dramatically.

Validation of disinfectant activity is an area that has been somewhat ignored in the past. Retrospective validation can no longer be accepted, especially where products that are susceptible to microbial contamination are involved. A validation protocol should be prepared and used whenever a new disinfectant is proposed; this should include an initial assessment of the disinfectant's properties (Table 6.2).

If the results of this initial survey and prospective validation are satisfactory, then the next stage is to consider the current operating procedures. Procedures may need to be rewritten to incorporate such changes as the name of the new disinfectant, its concentration and frequency of use, rotation with other agents, and so forth. Dilutions may well be different, requiring new measures or containers for diluting

TABLE 6.2
Disinfectant Validation Protocol

Review manufacturer's data sheet. Determine whether work was performed in-house or independently.
 Obtain copies of references.
Literature search. Determine whether further information is available.
Determine minimum contact time.
Determine compatibility of disinfectants or cleaning agents in use.
Determine compatibility of disinfectant with surface finishes in manufacturing areas, for example,
 stainless steel, glass, plastics, vinyl.
Determine in-use concentration range.
Determine compatibility with any product residues to ensure neutralization does not occur.
Determine spectrum of activity.

and storing the new disinfectant. Another important factor to take into account is user acceptance. If the disinfectant causes irritation of the eyes and throat in sensitive individuals, strict compliance with the disinfection regime may prove difficult. Therefore, user acceptance trials may be of benefit at this stage.

The next stage is the most difficult one: a decision has to be made to accept or reject the new disinfectant. It is important that the rationale for accepting or rejecting the agent is well documented. If problems are encountered in the future, the documentation could prove to be very useful, especially if there has been a change in staff. When the new disinfectant is brought into use, the level of environmental monitoring must be increased. The extent of this will depend upon the results obtained during the prospective validation study. Once the relevant amount of historical data has been obtained and found to be satisfactory, then environmental monitoring can revert to its previous frequency.

Disinfectants can become contaminated with microorganisms and special precautions need to be taken to prevent this problem arising (Anonymous 1958, Burdon and Whitby 1967, Ayliffe et al. 1969, Bassett et al. 1970, Berkelman et al. 1984; see also Chapter 2). Disinfectants and cleaning agents should be monitored for microbial contamination. Dilutions should be kept in previously cleaned containers and should not be stored unless sterilized. Part-empty containers should not be refilled (Whyte and Donaldson 1989). For a more detailed discussion on the role of disinfection and cleaning in a contamination control program, the reader is referred to the companion *Handbook* (Hodges and Baird, 2000).

The type and condition of mops and buckets used are a frequent source of embarrassment to companies during audits. It is not unusual to find dirty, dilapidated mops stored in a damp condition in a cupboard with inadequate ventilation. This presents a microbiological hazard and generates an excessive challenge for disinfectants. This situation can arise when cleaning and disinfection are performed outside of business hours and also when outside contractors are used. The examination of cleaning equipment should be part of the self-inspection program and also one of the area supervisor's routine checks.

6.2.5 TRAINING

Training of personnel is an area to which little practical attention in both time and financial terms is paid, although this would form an integral part of any microbiological audit (see Chapter 21). Training programs tend to be unimaginative and ineffective as they do not reflect real-life situations. Commercially available programs may have a professional finish to them but it can be difficult to relate their content to actual work situations. Training programs produced in house may not provide a highly polished presentation but, if well constructed, they will at least be relevant to the participants. Initial GMP induction programs usually have an audience willing and eager to learn; ongoing GMP programs are much more difficult to produce.

Fortunately, microbiological control is one area in which the theory and practice can be both educational and entertaining. The effects of the operator and the environment on product quality can easily be demonstrated by use of settle plates, swabs, bioburden plates, and particle counters. It is important to remember that people need to see bacterial colonies growing on nutrient agar plates in order to appreciate the significance of the training. The theory of microbiological control can be presented in an interesting manner because it can be related to both home and work situations. For example, disinfection of work surfaces in the cleanroom to prevent microbial contamination of the product can be linked with disinfection of work surfaces in the kitchen to prevent food poisoning. It is essential that personnel understand why they are performing a particular task and what the significance to the end user of the product may be if that task is not performed correctly. All staff should be encouraged to attend relevant meetings, perhaps in conjunction with managerial staff in order to promote discussion of current working practices.

Training programs should be documented and each person should have a training manual that details what training has been undertaken, and by whom, and the skills or knowledge attained. Ongoing training and reinforcement are important aspects with regard to microbial control. Periodic assessment of the effectiveness of training programs should be made. All too frequently, the outcome of an investigation into defective medicinal products that have reached the marketplace reveals that human error was involved.

6.2.6 MOTIVATION

Motivation of staff is an integral part of GMP and is the mainstay of microbial control. No matter how modern the facility, there will always be human interaction. A well-motivated and knowledgeable workforce will be alert to possible contamination risks to the product and will inform the relevant people. If the workforce is not motivated, there may be some telltale signs such as water filters not being changed according to schedule, equipment not being thoroughly cleaned, damage to fabric not being reported, and many other small, but important, items being overlooked.

Disillusionment soon sets in if the principles of GMP learned in training courses are overridden for commercial reasons. It is important to explain to staff if a change is to be made that might appear to be in conflict with their idea of GMP. For instance, the environmental monitoring program might show that one area is historically well within limits and hence the monitoring is to be decreased, thus releasing staff to work elsewhere. The resource realized by this action might be put into an area that is not so well within limits. The staff in the good area might believe that the decreased monitoring is part of a cost-cutting exercise rather than a fine-tuning of the monitoring program if they are not told the rationale behind it. Communication between the people involved in monitoring microbial contamination and the manufacturing staff is essential if demotivation is not to occur. A typical example of demotivating action is to report only out-of-limit results to the manufacturing department. Even worse is to name individuals associated with these results. This not only creates bad feeling but also could ultimately lead to obstructive practices developing between manufacturing and monitoring departments. A feedback system should be developed so that the manufacturing department receives the good news as well as the bad news. If a problem should arise with certain groups or individuals, retraining or reinforcement of GMP principles should be the first step rather than disciplinary action.

6.2.7   SELF-INSPECTION

A self-inspection or audit team can be a most powerful tool in microbiological control (see Chapter 21). The team has two main functions: to provide a fresh pair of eyes and to ensure a regular unannounced visit to all manufacturing areas by management. It is a good idea to include on the team someone who has little knowledge of manufacturing or of the area. It is usually this person who asks the most questions and can prove to be an effective challenge to current practices and systems. Regular visits by managerial staff demonstrate that they are interested in the day-to-day life of a manufacturing unit and this helps reinforce training and increases motivation. Much care needs to be taken with the reporting and feedback of audit results so that they are presented in a positive and constructive manner. For a more detailed discussion of microbiological auditing in practice, the reader is referred to the companion *Handbook* (Lush 2000).

6.2.8   PROCEDURES

Written procedures have two main functions: they allow all activities to be performed in a reproducible way and to the same standard every time. They can also encourage the person who writes them to examine current working practices and, perhaps, to improve them.

Written procedures should be kept as simple as possible, should be in easily followed steps, and should be written in the imperative. The use of diagrammatic flowchart procedures should be considered for complex operations as these can be very effective. All too frequently, working practice does not reflect the written procedure because the person performing the task did not write it. Furthermore, the procedures may be held inaccessible in the supervisor's office and not available for use in practice. Indeed, the operator may not have seen the procedures because they may have been prepared for presentation to regulatory authorities or certifying bodies only.

There is a simple rule of thumb to determine whether written procedures are satisfactory: if an experienced person who does not work in that department can follow the procedure without having to ask questions, the procedure works. If questions have to be asked, this indicates that important steps are missing from the procedure.

6.2.9 MONITORING

Monitoring is the assessment or measurement of a number of parameters to estimate the condition of an area, product, or process. The purpose is to confirm that the chosen parameters are within defined limits or whether the limits have been exceeded and corrective action is required. Table 6.3 summarizes the features required in a microbiological monitoring program.

It is important that the results of the monitoring program are documented in such a way that trends are discernible at the earliest possible moment. Any unusual results should be investigated immediately and not assumed to arise from contaminated plates or operator error.

6.2.10 DOCUMENTATION

All too frequently, work is performed but not adequately documented. This is particularly true for microbiological investigations of contamination problems. Much work may be done to identify and solve the problem but it may not be written up at the time. Subsequently, it can be very difficult to reconstruct the work performed and hence convince an outside person that the problem has been adequately resolved. Furthermore, should a similar problem arise, it may again need to be tackled from first principles in the absence of adequate documentation from the first occasion.

It is now common practice to enter environmental monitoring results into a computer to aid trend analysis. However, in certain instances it has been found that actual results are converted to averages or means in this process. Although this practice may be satisfactory for the purpose of trend analysis, the raw data must be kept, and any reviewer of the data must be informed about software manipulation of the raw information.

TABLE 6.3
Essential Features of a Microbiological Monitoring Program

Reflects the risk to the product
Provides meaningful results (e.g., "no colony-forming units on a settle plate" does not provide meaningful information)
Initially be all-embracing before concentrating on those areas that show highest risk
Has both warning and action limits
Is flexible, allowing on-the-spot decisions to be taken by monitoring staff
Has a known percentage recovery rate so that results can be converted to actual levels (particularly important for swabs)

6.3 INFLUENCE OF PRODUCTION ENVIRONMENT DESIGN

The environment of the manufacturing and storage areas may have an impact on the microbial quality of the finished product. The extent of this impact will be mainly dependent upon equipment design, operating practices, and the process. If closed systems are used, then the environmental standards may not be required to be as stringent as those for open systems. However, for this relaxation in standards to be acceptable, the operating procedures and practices must also be appropriate. Two different companies manufacturing the same products, using identical equipment, might have to operate to different levels of environmental standards to achieve the same product quality because of differences in operating procedures and practices. This might at first seem to be unwarranted, but it is important to remember that it is not possible to validate bad manufacturing practices.

A number of key areas need to be addressed when examining the effect of the environment in manufacturing and storage areas on product quality. These are surface finishes; heating, ventilation, and air conditioning (HVAC) systems; drains; equipment; operating procedures; and clothing. All should be carefully reviewed when designing a new, or developing an existing, production environment (Paley 2002).

Each of the above points will have a different significance in different manufacturing areas. As a general guide, Table 6.4 summarizes the environmental standards for typical production processes; further details are given in the following sections.

6.3.1 DRY PRODUCTS

There is no doubt that tablets, capsules, and powders are, on occasion, susceptible to microbial contamination and this can result in spoilage of the product and, in some cases, harm to the patient (Flatau et al. 1996). Although it is accepted that the main source of contamination is usually the raw material, once an undesirable microorganism enters a dry products area there are usually a number of places where it can become established to form a reservoir. Subsequent batches of the same or different products may then become contaminated. The correct design, construction, and cleaning of manufacturing areas will diminish the risk of microbial contamination. The reader's attention is drawn to the many papers, articles, and monographs that have been written on this subject (e.g., Anonymous 1987b, 1989, 1997, 2002; British Standard 5295 1989); some of the more important aspects will be examined here.

Drains, if present, should be regularly monitored and sanitized. Floors should slope toward the drain and there should be no dead areas where water can accumulate. Wash bays should be designed with proper drainage and an adequate air supply and extraction system. Inadequate air extraction often arises because the amount of steam used in cleaning manufacturing equipment is frequently underestimated at the outset. Surface finishes must be robust because of the aggressive nature of steam and some cleaning agents. All too often, insufficient time, effort, and expertise are put into designing wash bays because they are perceived not to be directly involved in manufacturing the product.

TABLE 6.4
Environmental Standards for Typical Production Activities

Environmental Standard	Production Activity
Containment workstation	Sterile radiopharmaceuticals
Class A/B	Aseptic preparation and filling of vials, ampoules, and aerosol containers
	Solution preparation before aseptic fill
	Handling or filling of aseptically prepared products
	Filling product into open vials, ampoules, or aerosol containers before terminal sterilization
	Manufacture of implanted medical devices
Class C	Solution preparation before sterile filtration
	Open solution preparation and filling prior to terminal sterilization
	Preparation of containers and parts for transitory indwelling devices
Class D	Handling of cleaned components for use in aseptic products
	Closed solution and component preparation prior to filling of terminally sterilized products
	Dispensing, blending, and preparation of tablets, oral liquids, and powders
	Open preparation of nonsterile liquids, creams, and ointments

Water may be used as an ingredient, such as in granulation and coating solutions, and as a cleaning agent. The microbial quality of the water may be poor because of the storage conditions or because of unsuitable pipework and valves. Flexible hoses should be removed, drained after use, and stored vertically (see also Chapter 3). Some water systems are designed to have point-of-use filters. Unless these filters are changed and sterilized on a frequent basis, they can be a source of microbial contamination. It is usually better practice to design a system that does not require point-of-use filtration to obtain water of good microbial quality.

Manufacturing staff can also be a source of undesirable microorganisms. In particular, fungal spores carried on clothing during the summer months may cause a problem with certain dry products, such as paracetamol tablets. Adequate changing facilities should be provided to enable staff to change quickly at the end of working sessions and without contaminating their factory clothing. Another source of fungal spores can be the HVAC system. The air supply to a dry products area should be adequately filtered to prevent the ingress of spores. The system should be so designed that it is easy to examine and replace filters. When filters become blocked there is an increased risk of air bypassing the filters and hence unfiltered air entering the manufacturing areas. Where air extraction systems are used, for example, to remove local heat, the source of makeup air must be considered.

6.3.2 Liquids, Creams, and Ointments

This section refers to nonsterile products only. Many of the comments made in Section 6.3.1 regarding dry product manufacturing areas are also applicable to areas devoted to the manufacture of liquids, creams, and ointments. However, these products may be more prone to microbial contamination as discussed in detail in

Chapter 3, and require more stringent manufacturing environments. If closed manufacturing systems are not used, then a grade D room (Anonymous 2002) should be considered for manufacturing these types of products. This approach has two advantages: staff members are constantly aware that the products are at risk from microbial contamination, and the products are protected to some degree. These preventative measures may seem extreme but the time and cost involved in eliminating a contamination problem in a manufacturing environment and equipment can be considerable, as can the cost of a product recall.

6.3.3 TERMINALLY STERILIZED PRODUCTS

The manufacturing environment for the bulk solution for this type of product should meet at least a grade C standard (Anonymous 2002). However, if the solution is sterile filtered via a closed system and then collected in sterile receiving vessels, or filled into the final containers and terminally sterilized within a specified short time period, then a grade D environment may be acceptable. The solution should, however, be filled in a room that meets grade C.

The main source of microbial contamination in these manufacturing areas will be water. It is essential that the area be so designed and built to prevent the accumulation of water on floors. Any flexible hoses used should have quick-release fittings to encourage their removal and suitable hose storage facilities should be available. Water outlets should not have dead legs exceeding six pipe diameters in length. Where large volumes of solutions are prepared, the environment may become extremely humid, especially if the solutions are heated. This can cause structural damage and allow molds to become established. Adequate dehumidification systems should be installed in the air-handling units to cope with this problem.

6.3.4 ASEPTIC MANUFACTURE

The manufacture of aseptically prepared products is one of the most challenging areas within the pharmaceutical industry. The production of a sterile product by this means requires a total quality system. The failure of just one part of the system may result in a batch of product not having a sufficient level of sterility assurance and thus causing its rejection. The sterility test cannot be used as an alternative to the concept of sterility assurance (see Chapter 12). An important aspect of sterility assurance is parametric rejection, that is, if a critical parameter is not met during the manufacturing, sterile filtration, or filling process then the batch may have to be rejected or reprocessed.

Sterility assurance is highly dependent upon the manufacturing environment. Solutions should be prepared in rooms designed and built to grade C (Anonymous 2002) and it should be demonstrated that they meet this classification on a regular basis (see also Chapter 7). The main sources of contamination are as described in Section 6.3.3.

Aseptic filling rooms should be designed and built to meet a grade B standard (Anonymous 2002). They should not just meet the particulate count requirements but should be well within those limits. In addition, where products are exposed, the zone around the product should meet grade A and the room grade B standards (Anonymous 2002). It is important to note that the filling zone must meet a grade A in the *manned* state. In order to achieve the necessary environmental standards, much thought and care must be put into the design, construction, and operation of the area. The advent of computer-aided design can take much of the guesswork out of the design of air-handling systems. The number and locations of air inlets and extracts in relation to the proposed layout can be predicted to achieve the most effective air flow (see also Chapter 7). The reader is referred to Denyer (1998) and Whyte (2001) for more detailed discussions on cleanroom design, including the location of services and equipment.

Components and equipment should be sterilized during passage into the aseptic area via a double-door autoclave or oven. It is no longer acceptable to build an aseptic area without a suitable double-door sterilizer. Certain components such as plastic eye-drop bottles will not withstand heat sterilization; in these cases, it is acceptable to transfer presterilized components into the aseptic area. These components should be triple wrapped prior to sterilization so that the first layer can be removed in the air lock for entry into the grade C area and the second layer removed in the air lock of the aseptic area. It is standard practice to spray or wipe small items with 70% alcohol as they pass into cleanrooms. The alcohol used for this must be sterile and the minimum contact time determined. As alcohol is not sporicidal, there are inherent risks in this practice. Large items of equipment cannot always be readily sterilized during passage into the area. This problem can be overcome by the use of fumigating air locks. The equipment is fumigated within the air lock with formaldehyde or similar agents. Although this system is expensive to fit, it does provide an effective means of transferring equipment into and out of aseptic areas. It can be justified on the basis of reduced downtime of the area.

Maintaining differential air pressures between areas of different risk plays an important role in the design and operation of cleanrooms (see Chapter 7). Pressures are usually monitored with manometers, which should be identifiable and easily read. The limits should be indicated on them. It is recommended that all manometers be in a central bank and be located in ascending or descending order so that it is immediately obvious that the suite is satisfactory. The practice of referencing between rooms and not to ambient can cause confusion and requires a considerable amount of time to calculate actual and total air pressure differentials. Manometers should be zeroed regularly. Errors in zeroing are frequently found during inspections or audits.

6.3.5 Form-Fill-Seal

The design of form-fill-seal equipment can be so varied that it is difficult to determine the influence of production environment design on the product without knowing the exact specification of the equipment. The production environment will have an effect on product quality as form-fill-seal machines are not closed systems and operators will be in close contact with the filling zone especially at start-up and also during filling to some extent. If the machine is used for aseptic filling, then it should have HEPA-filtered air blown over the filling zone. However, the quality of the environment will influence the amount of contamination on the operator and hence the filling zone. For terminally sterilized products, form-fill-seal machines should be located in at least a grade D cleanroom. If form-fill-seal machines are to be used for aseptic manufacture, they should be located in grade C cleanrooms with the operators in grade A/B cleanroom clothing. In this way, the risk of the operator contaminating the product during start-up and during filling is minimized. Location of machines in lower classes of environment may still yield product with a contamination rate that complies with the Parenteral Drug Association guideline of 0.1% for aseptically filled products (Anonymous 1980a). Nevertheless, because modern high-speed conventional filling lines can achieve significantly lower contamination rates, it seems to be a backward step to accept a lower standard when it is not absolutely necessary. It should be remembered that a terminally sterilized product is expected to achieve a contamination rate of less than 1 in 10^6 units.

6.3.6 ISOLATORS

Isolators can be classified into two main types: those where goods are sterilized *in situ* or are transferred into the unit within special transfer containers without sterility being broken, and those where sterile goods are transferred into the unit double wrapped. In the first case, the production environment can contaminate the goods and therefore the interior of the isolator. If the isolator is used for aseptic processing, it should be located in at least a grade D cleanroom. Microbiological monitoring should be carried out routinely and there should be frequent leak testing of the isolator and glove-sleeve system.

6.4 MANUFACTURING PROCESSES AND THEIR INFLUENCE ON CONTROL OF CONTAMINATION

Manufacturing processes can either increase or decrease the level of microbial contamination of the product. It is important to identify which stages of manufacture may cause a significant change and to ensure adequate control at these stages.

6.4.1 CHEMICAL RAW MATERIALS

Chemical raw materials have to be sampled and tested prior to their use in manufacture. Sampling should be performed in a dedicated room that is easily cleanable. It is a common misunderstanding that the rationale behind this is solely to prevent cross-contamination of stored materials. Even if there is no risk or hazard of cross-contamination, sampling should not be undertaken in an open warehouse. Extraneous contamination of the material may occur; flies, other insects and even birds may be found in warehouses.

The next stage of the manufacturing process is the dispensing of chemical raw materials. As this process is carried out in the manufacturing area, the environment is controlled and so the area of risk lies in the scoops and containers used in the dispensary. All equipment which comes into direct contact with the chemical raw materials should be adequately washed, dried, and stored. Inadequate drying of containers and unsuitable storage conditions, for example, in the wash bay, are items frequently missed in self-inspection audits.

6.4.2 DRY PRODUCTS

The manufacturing process for dry products will usually start with a dry blending process of the active ingredient and the excipients. This process will not usually result in a significant change to the bioburden. The total microbial bioburden of the product will usually consist of the sum of the individual bioburdens of the ingredients used.

The wet granulation process used in the manufacture of some tablets can result in a significant change to the bioburden (see also Chapter 3). For aqueous granulation, the quality of the water used is important. Also, the storage conditions of the granulation solution and the granules, prior to and after drying, should be controlled to minimize adventitious contamination. The drying process, whether by tray drying or fluid bed drying, can alter the bioburden depending upon the conditions used (Flatau et al. 1996). Drying at low temperatures, for example, 45°C, may cause an increase in bioburden in certain susceptible products. Drying at high temperatures may result in the bioburden decreasing. Unfortunately, there is no general rule regarding the effect on bioburden during the granulation process and so individual studies on each product and process need to be performed.

The effect of tablet compression on bioburden is also variable (Flatau et al. 1996). Only individual studies can ascertain what the effect of compression will be. If tablets are sugar-coated or film-coated, the coating solution should be monitored microbiologically and expiry dates determined.

Whatever the effect of the manufacturing process on the bioburden, the process should not be relied upon to exercise control over it.

6.4.3 LIQUIDS, CREAMS, AND OINTMENTS

These products may be manufactured in open or closed systems. The use of closed systems for manufacture and transfer, together with the use of microbiologically clean raw materials, will help prevent microbiological problems. However, it should be remembered that such systems can be difficult to monitor, for instance, with swabs and contact plates, and so extensive validation of the sanitization process is required (Sieberling 1986, Berman et al. 1986, Myers et al. 1987, McClure 1988). Valves and pipework joints should be of the sanitary type as it is important that there should be no sites of attachment available for microorganisms. With open systems, there are a number of common problems. In particular, the product may remain exposed during stirring, inadequately sloped pipework may impede drainage, and leaking valves and joints may deposit product on the floor. All of these situations can cause unnecessary microbial contamination of the product.

The manufacturing process for many creams, ointments, and some liquids includes a heating step that may decrease the bioburden. However, care must be taken during cooling to minimize the ingress of air and prevent excessive condensation on vessel walls and lids, for example, by ensuring that batch sizes are suitable for the vessel used. If the manufacturing vessels are jacketed, they can be sanitized by using the jacket to heat water in the vessels to 100°C. The filling process for these products needs to be monitored and controlled. Hoppers for creams and ointments should have close-fitting lids. Break tanks for liquids may require bacterial air vent filters to be fitted, particularly for extended filling runs. This is necessary because the quality of air in the packing hall may be of a substantially lower standard than that in the manufacturing areas. Filling equipment in contact with the product, such as hoses, nozzles, pumps, and so forth, should ideally be sterilized before use; failing that, a validated sanitization program should be used.

Final containers should be cleaned prior to being filled unless the containers are manufactured, delivered, and stored under an approved quality management system. Upon delivery, containers may be contaminated with insects, dust, glass, swarf, and numerous other items; supplier audits are therefore desirable. Ideally, glass containers should be washed before use. Plastic containers should be subject to a vacuum and blowing cycle. Aluminium and laminate tubes usually cannot be cleaned in this way and so supplier audits should be performed in these cases. A number of manufacturers are now manufacturing containers under controlled conditions in cleanrooms and are seeking registration to ISO 9000 (British Standard 5750 1987). All suppliers should be encouraged to achieve registration to this standard.

6.4.4 STERILE PRODUCTS

One of the main aims of sterile product manufacture is to maintain as low a bioburden as possible during preliminary stages to present a minimal challenge to the sterilizing process. This can be achieved by the use of manufacturing vessels and equipment specifically designed for that purpose. Improvisation and modification of basically unsuitable equipment is unacceptable. This is not to say that manufacturers of sterile products need to have state-of-the-art equipment, but the processes and equipment used should be capable of consistently producing a product of the required quality.

Process validation is an area of utmost importance. However, there is still evidence that "validation" is being misapplied. It is impossible to validate equipment or processes that are inherently unsuitable for their purpose. As it is not possible to prove that a product is sterile, then much reliance is placed on the concept of sterility assurance. This implies that if one or more parts of the process break down or deviate significantly from the norm, then there is an increased possibility that the product may not be sterile. It is sobering to realize that if only a small percentage of a batch of product is not sterile, there is little likelihood of it ever being detected and removed from the marketplace (Ringertz and Ringertz 1982).

6.4.4.1 Water for Injection

One of the processes that is regularly highlighted as causing problems is the manufacture of water for injection for use as an ingredient. In the *European Pharmacopoeia* (2004), this water must be prepared by distillation, although some non-European countries do allow the use of reverse osmosis for its preparation. Whatever the means of preparation, most problems are caused by inadequate storage and distribution. The more satisfactory systems maintain the water at about 80°C, recirculate, and use only sanitary-type valves and fittings (Artiss 1982, Anonymous 1983, Jackman 1988). Other types of systems, for example where water is freshly prepared daily and used within a short time, are acceptable but require more stringent control. Microbiological aspects of water are discussed in Chapters 2 and 3.

6.4.4.2 Vessels

Manufacturing, holding, and filling vessels can cause problems. Lids need to be examined routinely for damage. A warped lid may form an inadequate seal and allow ingress of air during the transfer to other vessels or to the filling line. Damaged or missing O rings from vessels is another common problem. These rings have a finite life and should be replaced on a regular basis and not upon gross failure as they may let air pass for a long period of time before detection. Records should be kept of O ring replacement; alternatively, they should be included in a planned preventative maintenance program.

Bacterial air-vent filters should be used wherever possible on vessels. Integrity testing of these filters can be difficult and time-consuming but they do provide an effective means of protecting the product in vessels.

6.4.4.3 Filtration

Most sterile solutions undergo a filtration step. Terminally sterilized products may be filtered through a 5-µm pore size filter to remove particulates and then through a 0.45- or 0.2-µm pore size filter to yield a solution with a low bioburden. The second filtration stage may be performed to enable a longer holding time before the terminal sterilization process provided that the pipework and receiving vessels have been sterilized or sanitized. Aseptic filtration is a very demanding process, especially the making of the aseptic connection to the receiving vessel. Operator training and equipment maintenance play vital roles in successful aseptic filtration. It is not unusual for two sterilizing filters to be used in line such that if one filter fails the integrity test then adequate sterility assurance is still guaranteed.

The number and type of filters available can be confusing, especially as different terminology and test methods for pore size distribution are used. Users should be aware of the difference between absolute and nominal ratings of filters (Denyer and Hodges 2004). Filters should be integrity tested before and after filtration. There are three main test methods available: bubble point, pressure hold, and flow volume (forward flow or diffusion rate; see Chapter 12). Each method has advantages and disadvantages and these should be considered in detail and the limitations of the chosen method understood. Limits should be based upon the solutions being filtered and not based upon water. The bubble point is sometimes incorrectly performed as the compressed air line is regulated to a maximum pressure below the bubble point.

If cartridge filters are used, the style of cartridge and holder is important. A straight, single O ring push fit with a spring to hold the center is not satisfactory, as under moderate pulsing, the filter will bounce or flex in the housing causing random bypass from the nonsterile to the sterile side. A bayonet lock with two O rings, or similar, is a more secure system.

6.4.4.4 Bioburden

Bioburden studies are very useful in determining the effect that the manufacturing equipment and processes have on the microbial content of the product. These studies should initially be performed at each stage of the manufacturing process so that potential problem areas can be identified, rectified, or monitored. These studies should then be repeated at regular intervals to ensure that no significant changes have occurred. When sufficient data have been collected, bioburden limits may then be set. For a more detailed discussion on bioburden testing studies, the reader is referred to the companion *Handbook* (Bill 2000).

6.4.4.5 Filling

The wide variety of filling equipment available ranges from fully automated, microprocessor-controlled lines to simple hand-filling lines. Contamination during the filling process ranges from as high as 1 in 1,000 for hand-filling to better than 1 in 10,000 for automated-filling lines. Consequently, hand-filling should only be performed when absolutely necessary, that is, when the benefit outweighs the risk. The level of sterility assurance of aseptic filling is commonly determined by running broth fills (media fills, system suitability tests). This particular test can provide much useful information if run correctly. The test should challenge both the aseptic filtration and the filling process. The filling of previously autoclaved broth provides little useful information on the sterility assurance of the total manufacturing process. Again, the broth fill cannot be used to validate inadequate or unsuitable manufacturing equipment or processes. Detailed guidance on broth fills and powder fills is provided in Parenteral Drug Association monographs (Anonymous 1980a, 1984b) and the U.K. Parenteral Society monograph (1993).

6.4.4.6 Sterilization Methods

There are four main methods of sterilizing components and finished product: autoclaving, hot air, ethylene oxide, and irradiation (see Chapter 11). There are many texts available on the validation and operation of these processes (Anonymous 1980b, 1981, 1984a, 1988a, 1988b; see also Chapters 11 and 12).

Double-ended sterilizers should be used for components entering aseptic areas and for terminally sterilized products leaving the filling area. If double-ended sterilizers are not used, there is an increased risk of product mix-up between sterilized and nonsterilized product. The use of indicator tape on its own does not provide sufficient evidence that a product has undergone a sterilization cycle. In one particular instance, rolls of indicator tape were subject to a sterilizing cycle to obtain the color change and pieces of this tape were then attached to the product after it had completed its sterilization cycle. The rationale for this was that if the tape was attached to bottles before sterilization, then it was very difficult to remove the tape afterward. This one example illustrates just how easy it is for a system to be made inherently unsafe.

The cooling media in autoclaves can be sources of microbial contamination. Cooling water should be sterile and air should be filtered through a sterilizing-grade filter during the cooling cycle. One hundred percent leak testing of containers should detect any leakers but this is a check system, not a prevention system. There have been instances of sterility failure owing to the necks of vials becoming contaminated during the cooling process. The dye solution used for leak testing should be monitored microbiologically and appropriate limits set. It is important that the sensitivity of detection be determined. It should be remembered that some products will decolorize certain dyes.

Sterilizers play such an important role in the manufacture of sterile products that it is surprising to find that their maintenance and upkeep is sometimes left to unqualified and untrained personnel. Unless the person understands the theory and practice of sterilization and has adequate knowledge of the sterilizer, potentially hazardous alterations to the equipment may unwittingly be made. Any work performed on a sterilizer should only be done by authorized personnel. It is not unusual to find that sterilizers have been modified over the years; these modifications should be well documented and engineering drawings prepared to illustrate the total system and highlight modifications.

6.5 OVERVIEW

In both hospital and industrial pharmaceutical environments, microbiological control is an area that requires a total understanding of the manufacturing systems, constant vigilance, and an inquiring mind. One company's problem may be another company's problem in the future and so a regular review of specialist journals is recommended. In this way, it is possible to recognize potential problems and prevent them from occurring. Lack of time is no excuse for ignorance.

The cost of monitoring and checking is an ongoing expense. The cost of prevention is usually one-off. Application of modern rapid microbiological detection methods should be considered (see Chapter 9) in an attempt to eliminate the retrospective nature of traditional microbiological culture methods. Historical information could lead to difficult decisions having to be made.

REFERENCES

Anonymous. (1958). Bacteria in antiseptic solutions. *Br. Med. J.*, 2, 436.

Anonymous. (1968). *Sizing and Counting Particulate Contamination in and on Clean Room Garments*. No. ASTM F51-68. American Society for Testing and Materials, West Conshohocken, PA.

Anonymous. (1980a). *Validation of Aseptic Filling for Solution Drug Products*. Technical Monograph no. 2. Parenteral Drug Association Inc., Philadelphia.

Anonymous. (1980b). *Validation of Steam Sterilization Cycles*. Technical Monograph no. 1. Parenteral Drug Association Inc., Philadelphia.

Anonymous. (1981). *Validation of Dry Heat Processes Used for Sterilization and Depyrogenation*. Technical Report no. 3. Parenteral Drug Association Inc., Philadelphia.

Anonymous. (1983). *Design Concepts for the Validation of a Water for Injection System*. Technical Report no. 4. Parenteral Drug Association Inc., Philadelphia.

Anonymous. (1984a). *Process Control Guidelines for Gamma Radiation Sterilization of Medical Devices*. Association for the Advancement of Medical Instrumentation, Arlington, VA.

Anonymous. (1984b). *Validation of Aseptic Drug Powder Filling Processes.* Technical Report no. 6. Parenteral Drug Association Inc., Philadelphia.

Anonymous. (1987a). *Garments Required in Clean Rooms and Controlled Environment Areas.* Institute of Environmental Sciences no. RP-CC-003-87T. Institute of Environmental Sciences, Rolling Meadows, IL.

Anonymous. (1987b). *Pharmaceutical Premises and Environment.* Pharmaceutical Quality Group. Institute of Quality Assurance, London.

Anonymous. (1988a). *Process Design, Validation, Routine Sterilization and Contract Sterilization.* Association for the Advancement of Medical Instrumentation, Arlington, VA.

Anonymous. (1988b). *Validation and Routine Monitoring of Sterilization by Ionizing Radiation.* U.K. Panel on Gamma and Electron Irradiation, London.

Anonymous. (1989). *The Rules Governing Medicinal Products in the European Community. Vol. IV. Guide to Good Manufacturing Practice For Medicinal Products.* Her Majesty's Stationery Office, London.

Anonymous. (1997). *Rules and Guidance for Pharmaceutical Manufacturers and Distributors.* The Stationery Office, London.

Anonymous. (2002). *Rules and Guidance for Pharmaceutical Manufacturers and Distributors.* The Stationery Office, London.

Artiss, D.H. (1982). Materials, surfaces and components for WFI and other sanitary piping systems. *Pharm. Technol.,* (August), 37–48.

Australian Standard AS 2013. (1977a). *Clean Room Garments.* Standards Association of Australia, North Sydney, NSW.

Australian Standard AS 2014. (1977b). *Code of Practice of Clean Room Garments.* Standards Association of Australia, North Sydney, NSW.

Ayliffe, G.A.J., Barrowcliff, D.F., and Lowbury, E.J.L. (1969). Contamination of disinfectants. *Br. Med. J.,* 1, 505–511.

Bassett, D.C.J., Stokes, K.J., and Thomas, W.R.G. (1970). Wound infection due to *Pseudomonas multivorans.* A water-borne contaminant of disinfectant solutions. *Lancet,* i, 1188–1191.

Berkelman, R.L., Anderson, R.L., Davis, B.J., Highsmith, A.K., Petersen, N.J., Bono, W.W., Cooke, E.H., Mackel, M.S., Faucio, M.S., and Martone, W.J. (1984). Intrinsic bacterial contamination of a commercial iodophor solution. *Appl. Environ. Microbiol.,* 47, 752–756.

Berman, D., Myers, T., and Chrai, S. (1986). Factors involved in cycle development of a steam-in-place system. *J. Parenter. Sci. Technol.,* 40, 119–121.

Bill, A. (2000). Microbiology laboratory methods in support of the sterility assurance systems. In *Handbook of Microbiological Quality Control: Pharmaceuticals and Medical Devices.* Baird, R.M., Hodges, N.A., and Denyer, S.P., Eds. Taylor & Francis, London, pp. 120–143.

British Standard 3211. (1986). *Method for the Measurement of the Equivalent Pore Size of Fabrics (Bubble Pressure Test).* British Standards Institution, London.

British Standard 5750. (1987). *Quality Systems (ISO 9000–1987).* British Standards Institution, London.

British Standard 5295. (1989). *Environmental Cleanliness in Enclosed Spaces.* British Standards Institution, London.

Burdon, D.W. and Whitby, J.L. (1967). Contamination of hospital disinfectants with *Pseudomonas* species. *Br. Med. J.,* 2, 153–155.

Clayton, N. (2002). Contamination control clothing—selecting a system to meet your requirements. *Eur. J. Parent. Sci.,* 7, 49–54.

Clemens, S.W. (1988). Packaging people to protect products from contamination in the
hospital pharmacy. *Aust. J. Pharm.,* 18, 74.

Denyer, S.P. (1998). Factory and hospital hygiene and good manufacturing practice. In
Pharmaceutical Microbiology. Hugo, W.B. and Russell, A.D., Eds. 6th ed. Blackwell
Science, Oxford, pp. 426–438.

Denyer, S.P. and Hodges, N.A. (2004). Filtration sterilization. In *Russell, Hugo, & Ayliffe's
Principles and Practice of Disinfection, Preservation and Sterilization.* Fraise, A.P.,
Lambert, P.A., and Maillard, J.-Y., Eds. 4th ed. Blackwell Publishing, Oxford, pp.
436–472.

Dixon, A.M. (1989). Garments: a clean approach. *Med. Dev. Diagnost. Ind.,* February, 44–61.

European Pharmacopoeia. (2004). *European Pharmacopoeia.* 5th ed. Council of Europe,
Strasbourg, France.

Flatau, T.C., Bloomfield, S.F., and Buckton, G. (1996). Preservation of solid oral dosage
forms. In *Microbial Quality Assurance in Cosmetics, Toiletries and Non-sterile Pharmaceuticals.* Baird, R.M. with Bloomfield, S.F., Eds. 2nd ed. Taylor & Francis,
London, pp. 113–132.

Goodwin, B.W. (1988). Cleanroom garments and fabrics. In *Handbook of Contamination
Control in Micro-Electronics.* Noyes Publications, Park Ridge, NJ, pp. 110–135.

Hodges, N.A. and Baird, R.M. (2000). Disinfection and Cleansing. In *Handbook of Microbiological Quality Control: Pharmaceuticals and Medical Devices.* Baird, R.M.,
Hodges, N.A., and Denyer, S.P., Eds. Taylor and Francis, London, pp. 205–220.

Jackman, D.L. (1988). Troubleshooting your pharmaceutical liquid filling equipment at point
of contact. *Pharm. Eng.,* 8, 14–17.

Lush, M. (2000). Microbiological hazard analysis and audit: the practice. In *Handbook of
Microbiological Quality Control: Pharmaceuticals and Medical Devices.* Baird,
R.M., Hodges, N.A., and Denyer, S.P., Eds. Taylor & Francis, London, pp. 221–238.

McClure, H. (1988). Sterilization in place: how to sterilize liquid filling equipment at point
of contact. *Pharm. Eng.,* 8, 14–17.

Miller, W.F. (1987). Clean room garments: a day in the life. *J. Soc. Environ. Eng.,* 26, 12–13.

Myers, T., Kasica, T., and Chrai, S. (1987). Approaches to cycle developments for clean-in-
place processes. *J. Parent. Sci. Technol.,* 41, 9–15.

Paley, R. (2002). Cleanroom design: biopharm and pharmaceutical. *Eur. Pharm. Rev.,* 2,
64–73.

Parenteral Society. (1993). *The Use of Process Simulation Tests in the Evaluation of Processes
for the Manufacture of Sterile Products.* Technical Monograph no. 4. The Parenteral
Society, Swindon, U.K.

Redlin, W.J. and Neale, R.M. (1987). Garments for clean room operators: the demands of the
1990s. *J. Soc. Environ. Eng.,* 26, 17–19.

Ringertz, O. and Ringertz, S. (1982) The clinical significance of microbial contamination in
pharmaceutical and allied products. *Adv. Pharm. Tech.,* 5, 201–226.

Seiberling, D.A. (1986). Clean-in-place and sterilize-in-place applications in the parenteral
solution process. *Pharm. Eng.,* 6, 30–35.

Seiberling, D. (1987). Clean-in-place/sterilize-in-place (CIP/SIP). In *Aseptic Pharmaceutical
Manufacturing: Technology for the 1990's.* Olson, W.P. and Groves, M.J., Eds. Interpharm Press, Prairie View, IL, pp. 247–314.

Whyte, W. (2001). *Cleanroom Technology.* John Wiley & Sons, London.

Whyte, W. (2002). A cleanroom contamination control system. *Eur. J. Parent. Sci.,* 7, 55–61.

Whyte, W., Bailey, P.V., Tinkler, J., McCubbin, I., Young, L., and Jess, J. (1982). An evaluation of the routes of bacterial contamination occurring during aseptic pharmaceutical manufacturing. *J. Parent. Sci. Technol.,* 36, 102–106.

Whyte, W. and Donaldson, N. (1989). Cleaning a clean room. *Med. Dev. Diagnost. Ind.,* February, 31–35.

7 Dispersion of Airborne Contaminants and Contamination Risks in Cleanrooms

Bengt Ljungqvist and Berit Reinmüller

CONTENTS

7.1   INTRODUCTION

Environmental microbiological control is aided by good production environment design (see Chapter 6). An important feature of this design, especially in aseptic manufacture, is the integrated development of cleanroom facilities, air-handling systems, and process methodologies. Ideally, air movements must be controlled to eliminate or minimize particulate, and hence microbial, presence in critical product exposure areas. In particular, airborne contaminants must be dispersed in such a way as to limit risks of deposition leading to product microbial contamination. This chapter provides a mathematical basis by which contamination risks may be established.

7.2   GENERAL MATHEMATICAL PRINCIPLES

Air may move in two different ways. One is characterized by a smooth flow, free of any disturbances such as small and temporary vortices or eddies. This is known as *laminar flow*. The other type of flow is characterized by small and temporary fluctuations caused by instabilities. The flow velocity is no longer constant but fluctuates around an average value. This is known as *turbulent flow* and the disturbances are often interpreted as small, temporary eddies.

An organized air flow pattern implies that directional flow can be characterized by means of streamlines, that is, the paths taken by assumed weightless particles in the room as they follow the air stream if turbulent fluctuations are ignored. The transport of contaminants due to streamline flow is often described as "convective transport."

The simplest system for an analysis of the transport of contaminants by ventilation is, therefore, convective transport along streamlines. The disturbances caused by turbulence (turbulent diffusion) are superimposed on this. Obviously, if there is no turbulence, turbulent diffusion is replaced by molecular diffusion or Brownian motion. It can generally be assumed, in regions with well-defined air flow fields, that the settling velocity of contaminants is negligible, which implies that gravitation plays a minor role.

With the assumption of a constant value for the diffusion coefficient, the diffusion equation in a velocity field in rectangular coordinates becomes

$$\frac{\partial c}{\partial t} + v_x \cdot \frac{\partial c}{\partial x} + v_y \cdot \frac{\partial c}{\partial y} + v_z \cdot \frac{\partial c}{\partial z} = D \cdot \left(\frac{\partial^2 c}{\partial x^2} + \frac{\partial^2 c}{\partial y^2} + \frac{\partial^2 c}{\partial z^2} \right) \qquad (7.1)$$

where c = concentration (for particles, number.m^{-3}; for a gas, m^3.m^{-3} or ppm); v_x, v_y, v_z = air flow velocities in x, y, and z directions (m.s^{-1}); D = diffusion coefficient (m^2.s^{-1}).

In still air, the diffusion coefficient for gases (molecular diffusion coefficient) and for particles (particle diffusion coefficient) can be calculated theoretically, see for instance Bird et al. (1960) and Hinds (1982). The molecular diffusion coefficient depends, among other things, on temperature, pressure, and molecular weight and, under ambient conditions, has a value of about 2×10^{-5} m^2.s^{-1}. The particle diffusion coefficient increases with temperature and is inversely proportional to particle size. For example, a 1.0-µm particle has a diffusion coefficient that is about six orders of magnitude smaller than that for molecular diffusion in air.

In laminar air flow, the diffusion coefficients have similar values to those in still air. This means that in laminar air flow, the dispersion of gases is faster than that of particles. However, in turbulent air flow, gases and particles have similar values for their diffusion coefficient and it increases with higher air velocities. Values for the turbulent diffusion coefficient are mostly determined experimentally. The particular case of isotropic-turbulent parallel air flow has been investigated experimentally, using gas and particles (5.5-µm mean diameter), by Ljungqvist (1979) for a room with a cross section of 1.7 m × 1.7 m at two air velocities of 0.20 m.s^{-1} and 0.45 m.s^{-1}. The diffusion coefficients obtained for the two velocities were approximately 1.4×10^{-4} m^2.s^{-1} and 2.4×10^{-4} m^2.s^{-1}, respectively. Thus, in a turbulent unidirectional air flow, the diffusion coefficient is about 10 times higher than that for gases in a laminar unidirectional air flow.

Equation 7.1 gives the simplest possible mathematical model to describe the concentration of contaminants in a velocity field, when those contaminants are emitted from a source at an arbitrary position.

However, the risk of contamination does not only depend on the concentration of airborne contaminants, which is of critical importance, but also on the motion of those contaminants. If particulates are being considered, then it is the rate of incidence (impact) of the particles on the pharmaceutical product that characterizes the risk to that product. When considered mathematically, this incidence is of vector character and the term *flux vector* or *impact vector* is used (see, for instance, Friedlander 1977, Ljungqvist 1979).

The risk will to a great extent, but not entirely, be dependent on the vector (K)

$$K = -D \operatorname{grad} c + v \cdot c \qquad (7.2)$$

where c = particle concentration (number.m^{-3}), v = velocity vector (m.s^{-1}).

In principle, the numerical value of K indicates, per unit time, the number of particles passing a supposed unit area, placed perpendicular to the direction of particle flow.

In a one-dimensional case, for example, in the x direction, the expression of particle impact is

$$K_x = -D \frac{\partial c}{\partial x} + v_x \cdot c \qquad (7.3)$$

For a more thorough description of the interaction between air movements and dispersion of contaminants and contamination risks, see Ljungqvist and Reinmüller (1997).

Air supplied to a cleanroom is filtered and therefore does not usually contribute to airborne contamination. The cleanroom is additionally pressurized to prevent contamination from adjacent areas and, in order to maintain air quality, a sufficient air flow of at least 20 air changes per hour is normally required (controlled areas). In the controlled area, the concept of mixing air is used, whereby complete mixing of air is considered to be achieved. Therefore, the sources of airborne particles within the room are people and machinery. People are the main source of airborne bacteria. The fact that the airborne microbial contaminants are transported on particles means that the same conditions of airborne dispersion are valid for particles with and without microorganisms.

In order to avoid particulate and microbiological contamination of critical process regions (critical areas), unidirectional air flow with high efficiency particulate air (HEPA)-filtered air (0.45 ± 0.1 m.s^{-1}) is used as protection from ambient air. In the following sections, contamination risk in still air, mixing air, and unidirectional air flow will be discussed.

7.3 STILL AIR AND COMPLETELY TURBULENT MIXING AIR

Consider a chamber of height H containing, at time zero, a specified initial concentration c_0 of uniformly distributed monodisperse particles. If there is no motion of the air and diffusion is ignored, all the particles will be settling with the same constant settling velocity v_s and the particle impact becomes $v_s \cdot c_0$. The concentration becomes zero everywhere in the chamber after a time equal to H/v_s has elapsed.

At the other extreme, with completely turbulent mixing air, it is assumed that the concentration is uniform throughout the chamber at all times. Diffusion and deposition on the walls are assumed to be negligible. The particle settling velocity is superimposed on the vertical components of convective velocity. Because the up and down components of convective velocity are equal, every particle will have an average net velocity equal to v_s. The concentration of particles decays exponentially with time and as such never reaches zero. The concentration reaches $1/e$ (~37%) of the original concentration in the same time (H/v_s) that is required for complete removal in the previous case where there is no air motion (Fuchs 1964, Hinds 1982).

The expression, in this case, for the particle impact (K_s) at the time t from a monodisperse aerosol is

$$K_s = v_s \cdot c = v_s \cdot c_0 \cdot e^{-\frac{v_s \cdot t}{H}} \tag{7.4}$$

where c_0 = initial concentration (number.m^{-3}), v_s = settling velocity (m.s^{-1}), H = height of chamber (m).

The number of particles of a monodisperse aerosol settling in a time t on the unit area of the bottom of the chamber becomes

$$N_s = \int_0^t K_s dt = c_0 H \left(1 - e^{-\frac{v_s \cdot t}{H}} \right) \tag{7.5}$$

As mentioned earlier, the expression of particle impact for the case with still air at the bottom of the chamber is

$$K_s = v_s \cdot c_0, \text{ when } t < H/v_s$$

and

$$K_s = 0, \text{ when } t \geq H/v_s \tag{7.6}$$

Figure 7.1 shows schematically the particle impact for the two idealized situations.

When diffusion is neglected and the velocity vector depends only on the settling velocity, a one-dimensional special case of Equation 7.2 applies. When an equivalent mean diameter of bacteria-carrying particles can be established, the settling velocity becomes a constant value. If the concentration of bacteria-carrying particles in the air, the area of exposed surface, and the exposure time are all known, the number of bacteria-carrying particles deposited can be calculated. When the concentration

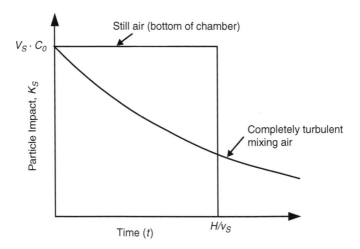

FIGURE 7.1 Particle impact for monodisperse particles in still air and completely turbulent mixing air.

of bacteria-carrying particles is uniformly distributed and constant during exposure time, the expression for the number of particles deposited (N_d) is

$$N_d = v_s \cdot c_b \cdot A_e \cdot t_e \tag{7.7}$$

where c_b = concentration of bacteria-carrying particles in the air (number.m^{-3}), A_e = area of exposed surface (m^2), t_e = exposure time (s).

Airborne particles carrying microorganisms are shed from people, and the particle size is mainly in the range of 4 to 20 μm. An equivalent mean diameter of between 12 and 14 μm (spherical particle with the unit density of 1 g.cm^{-3}) has been established (Noble et al. 1963, Clark et al. 1985, Whyte 1986). Assuming the average size of bacteria-carrying particles is 12 μm, and hence the settling rate is 4.62×10^{-3} m.s^{-1}, the contamination risk will be equivalent to the contamination rate given by Whyte (1986).

With a settle plate of 90 mm diameter and 4 h exposure time, the concentration in Equation 7.7 expressed in colony-forming units (CFU) per cubic meter becomes

$$c_b = 2.36 N_d \tag{7.8}$$

In the EU GMP Annex, *Manufacture of Sterile Medicinal Products* (1997, revised 2003) and the Medicines Control Agency (MCA) *Rules and Guidance for Pharmaceutical Manufacturers and Distributors* (2002), guidance values for micro-biological monitoring of cleanrooms in operation are given, as in Table 7.1. The values for settle plates (N_d) have been used to calculate c_b from Equation 7.8 for comparison with the air sample guidance values. In view of the intrinsic experimental variation in all sampling, this theoretical approach can give a good approximation to the actual situation.

TABLE 7.1

Guidance Values for Microbiological Monitoring of Cleanrooms

	Maximum Number of Viable Organisms		
Cleanroom Classification	Air Sample (CFU.m^{-3})	Settle Plate (90 mm) CFU in 4 h	Airborne Concentration (c_b) Calculated from Equation 7.8 (CFU.m^{-3})
A	<1	<1	—
B	10	5	12
C	100	50	118
D	200	100	236

Note: Dash indicates not applicable.

Source: From EU GMP. (1997). *The Rules Governing Medicinal Products in the European Community. Vol. IV—Guide to Good Manufacturing Practice. Annex 1—Manufacture of Sterile Medicinal Products.* Rev. Ed. European Commission, Brussels. With permission.

Equation 7.7 and Equation 7.8 can be used to estimate airborne contamination when there is no air motion (still air), for completely mixing air (fully turbulent air), and in the vertical direction of horizontal unidirectional air flow where there is little influence of turbulent fluctuations (laminar air flow). Because Equation 7.4 through Equation 7.8 require uniformly distributed particles in the air, a fact not always fulfilled in real case situations, alternative theoretical or experimental methods must be taken into consideration when estimating the contamination risk. An experimental method for microbiological assessment of potential risks in cleanrooms is described in Section 7.6.

7.4   UNIDIRECTIONAL AIR FLOW

Dispersion from a fixed source in a uniform parallel flow has been described theoretically and experimentally by Bird et al. (1960), Fuchs (1964), Hinze (1975), Ljungqvist (1979), and Ljungqvist and Reinmüller (1997). For a continuous point source situated at the origin in a parallel flow with a constant velocity v_0 in the x direction, the solution of the diffusion equation in a velocity field (Equation 7.1) when $\partial c / \partial t = 0$ and after simplification becomes

$$c = \frac{q}{4\pi Dx} \cdot e^{-\frac{v_0\left(y^2 + z^2\right)}{4Dx}} \tag{7.9}$$

where q = outward particle flow from the point source (number.s^{-1}), v_0 = constant velocity in the x direction (m.s^{-1}).

The dispersion pattern in the x, y plane ($z = 0$) is schematically shown in Figure 7.2.

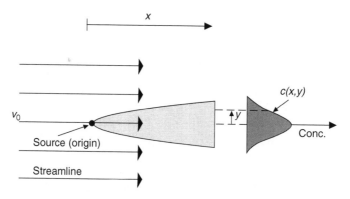

FIGURE 7.2 A schematic of the particle dispersion pattern caused by a continuous point source situated in a unidirectional flow with constant velocity in the x direction, where *conc.* is particle concentration and $c(x,y)$ is the concentration at the coordinates x, y.

The solution for a continuous line source situated along the z axis with a strength, q_l per unit of length can, in a simplified form, be expressed as

$$c = \frac{q_l}{2\left(\pi D\, v_0 x\right)^{1/2}} \cdot e^{-\frac{v_0 y^2}{4Dx}} \tag{7.10}$$

where q_l = outward particle flow per unit of length from the line source (number.s^{-1}.m^{-1}).

The derivations of particle impact for the case with the point source in a unidirectional air flow are described by Ljungqvist and Reinmüller (1997). It is shown that the expressions for particle impact are proportional to air velocity and concentration at a certain distance from the source.

In a horizontal unidirectional air flow, the settling velocities of bacteria-carrying particles cannot always be neglected, and the particle impact due to gravitation will be proportional to the settling velocity and concentration ($v_s \cdot c$). According to Whyte (1986), the average size of bacteria-carrying particles is 12 μm, which gives a settling velocity of 4.62×10^{-3} m.s^{-1}. This value is close to 1% of the air velocity in the unidirectional flow, that is, the particle impact due to gravitation will be about 1% of the value of the particle impact in the main flow direction.

In conclusion, in a horizontal turbulent unidirectional air flow with particles of an average size of 12 μm, the particle impact caused by diffusion will be in approximately the same range as the particle impact due to gravitation, but much less than the particle impact in the main air flow direction.

7.5 FACTUAL SITUATIONS

In a vortex, Ljungqvist (1979) has shown that the mean value of the particle concentration over the entire region inside the streamline, where the point of emission is situated, is considerably higher than that of the outside. This allows us to use the concept of contamination accumulation in the context of vortices. It has also been shown that accumulation can occur in the wake created by people or objects placed in a parallel flow provided that the contaminants are emitted in the wake region. Special consideration must be taken with instabilities and vortices generated by the working person.

Personal vortices are of two kinds: relatively stable and stationary wakes created by the body or unstable and nonstationary vortices that arise as a consequence of body movements. In this latter respect, it is obvious that the movements of the hands and arms play a significant part in creating unstable situations. With visual illustrative methods, it is easy to demonstrate that each of these two kinds of vortices is capable of destroying the intended beneficial effect of the ventilation system (Ljungqvist 1979, 1987; Flynn and Ljungqvist 1995; Ljungqvist and Reinmüller 1997).

In a unidirectional airflow, wakes and vortex streets are easily created behind obstacles and in these regions an increased contamination risk will occur (Ljungqvist and Reinmüller 1997). The length of the reversed region of a wake can be estimated to be two to three times the characteristic length of the obstacle, and can be double this when the obstacle is situated just beside a side wall. Parallel flow often leads to stagnation regions in front of machinery and working surfaces situated perpendicular to the main flow direction. The air movements in this case are mostly irregular, and in real case situations, air movements are not easy to predict. These regions can have an impact on the contamination risk, and in regions with increased turbulence, the diffusion component of Equation 7.2 becomes important.

Side walls are used as restricted access barriers around the critical zone and as physical barriers for the protection of operators as well as product. These side walls can be either rigid or flexible. From the standpoint of controlled air movements, rigid walls are always preferred. In order to allow necessary access, there can be specially designed openings. The length of the side walls should be chosen with regard to necessary product safety, protection from moving parts, necessary access, and correct air movements.

The velocity of the air flow out from the critical zone should be high enough to give protection from ingress of room air during operating production conditions. The velocity depends on the area of the outlet opening created by the side walls. Higher protection will be obtained with higher outflow velocity, that is, smaller outlet openings. However, if the openings are too small, stagnation regions will occur in the critical zone. In these stagnation regions, the air movements will be highly unpredictable and uncontrolled with accompanying contamination risks (Figure 7.3). If the side walls are not directly connected to the HEPA filters, there will be a certain air flow escaping from the clean zone into the cleanroom. The amount of escaped air depends on the total resistance of air flow over the critical zone within the side walls. This arrangement will usually produce less well-defined air movements in the critical zone, and contaminants might not be directly transported out from this zone.

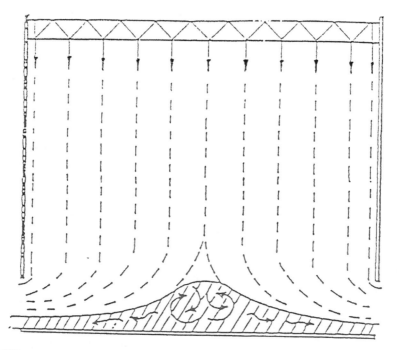

FIGURE 7.3 An example of the observed stagnation region in a unit with vertical unidirectional flow of HEPA-filtered air.

7.6 LIMITATION OF RISKS

Environmental monitoring in critical zones is used as an indirect method for assessing the sterility of the product during aseptic production. With present microbiological methods, it is difficult to measure and evaluate how the performance of single operations or interventions affects the microbiological risks. The limitation of risks method (LR method), described in detail by Ljungqvist and Reinmüller (1995, 1997) presents a reliable method for evaluation of microbial safety and is a useful tool in the hazard analysis critical control points (HACCP) approach. The concept incorporating visualization of air movements, particle challenge tests, and calculation of risk factors presents an effective method for limitation of risk and is a valuable complement to standard microbiological methods for monitoring and evaluation. It can be used for tracing the dispersion routes of airborne contamination, for evaluation of single process steps or interventions, for detection and assessment of potential risks, and for evaluating the efficacy of sampling sites in critical regions.

The LR method is performed in three steps. The first step is to visualize the main air movements and identify critical vortex or turbulent regions by using the smoke technique. The second step, the challenge test, is to place the probe for particle counting in the critical area, and during measurements to generate particles (by using air current test tubes) in the vicinity of the critical work areas to a challenge level of more than 10^7 particles per m^3 equal to, and larger than, 0.5 μm (300,000 particles.ft^{-3}). These measurements should be carried out during simulated production activity. This simulation should, to some extent, exaggerate the human interference and interventions during the measuring period. The third step is to calculate the risk factor that is defined as the ratio between measured particle concentration in the critical region and the challenge level in the surrounding air. Owing to limitations in measuring accuracy at high particle concentrations, a value of 10^7 particles (0.5 μm) per cubic meter (300,000 particles.ft^{-3}) is used as the challenge level at all calculations of risk factors. According to the authors' experience, when the risk factor is less than 10^{-4} (0.01%) during the challenge test, there should be no microbiological contamination from the air to the process during normal operational conditions.

Microbiological problems in the production cleanroom often depend on the interaction between air movements and dispersion of contaminants. In the following examples, risk evaluation at in-feeding tables, stopper bowls, and out-feeding stations will be discussed.

According to the experience of the authors, rotating container feeding tables, due to their large horizontal surface perpendicular to the main air flow, cause wake regions that under certain conditions allow passage of ambient air into critical zones. Furthermore, the rotating movements of the table surface can increase the risk. When linear feeding tables with smaller horizontal surfaces are used, the contamination risk can be reduced. If the process requires that the product vials are only partly closed when passing from the filling line to the out-feeding station or accumulation table, it is critical that the microbiological safety is not endangered. Here again, knowledge regarding wake regions and their accumulation of contaminants plays an important role in the evaluation of microbiological risk.

Table 7.2 shows that the risk factors for the rotating table are not acceptable but that the linear feeding table has satisfactory values. This is in agreement with results from media fills and other microbiological tests. The high risk factor value below the rotating table depends on the accumulating effect of a large wake vortex.

Different risk factor values for stopper bowls (container closure hoppers) are also presented in Table 7.2. The differences mainly depend on the shape of the bowl (slanting sides or straight), the size of the table surface on which the bowl is located, and whether the side of the bowl is perforated to allow air flow through.

Each of the different shapes has its advantages and disadvantages. The shape of the bowl, the location in combination with the filling line, the movements of the ambient air, and the procedure for transferring stoppers into the bowl, are all factors that must be evaluated separately.

The risks associated with the out-feeding station are low and show no potential entrainment of surrounding ambient air (IQ values), when the permitted interference region is limited to a distance of 0.15 m from the device (PQ values). This limitation arises because of a wake region close to the feeding device.

In conclusion, the LR method gives valuable information regarding the probability of contamination, but does not take into account the exposure time. For example, open vials on a feeding table, or stoppers in a stopper bowl with long

TABLE 7.2
Calculated Risk Factor Values during Installation Qualification (IQ) and Performance Qualification (PQ)

Equipment	Location of Particle Sensor	Risk Factor		Media Fill Result Contamination Rate (%)
		IQ	PQ	
Circular rotating in-feeding table	Above the table	6×10^{-1}	—	>>0.1
	Below the table	1.6	—	—
Linear in-feeding table	Above the table	$<10^{-4}$	$<10^{-4}$	<0.1
Stopper bowl	Along the edge; slanting shape, large table	$\sim 10^{-3}$	—	—
	Along the edge; slanting shape, smaller table	$<10^{-4}$	$<10^{-4}$	<0.1
	Along the edge; straight shape, perforated bowl	$<10^{-4}$	$<10^{-4}$	<0.1
Out-feeding table	Along the edge, 5 cm below	$<10^{-5}$	$\sim 10^{-3}$	—
	5 cm below and 15 cm away from the edge	$<10^{-5}$	$<10^{-4}$	<0.1

Note: A challenge exceeding 10^7 particles.m^{-3} of ≥ 0.5 μm diameter was generated in ambient air (see text).

exposure time can, even if the risk factor is low, have the same or higher total risk than a process stage with a short exposure time and higher risk factor.

The LR method gives fast information about potential risks and can be used as an engineering tool during design and development of aseptic processes as well as during preliminary surveys and performance qualifications. The LR method does not replace the final microbiological evaluation of an aseptic process and must be complemented with media stimulation runs.

REFERENCES

Bird, R.B., Steward, W.E., and Lightfoot, E.N. (1960). *Transport Phenomena*. John Wiley and Sons, Inc., New York.

Clark, R.P., Reed, P.J., Seal, D.V., and Stephenson, M.L. (1985). Ventilation conditions and air-borne bacteria and particles in operating theatres; proposed safe economies, *J. Hyg. Camb.*, 95, 325–335.

EU GMP. (1997). *The Rules Governing Medicinal Products in the European Community. Vol. IV—Guide to Good Manufacturing Practice. Annex 1—Manufacture of Sterile Medicinal Products*. Rev. ed. 2003. European Commission, Brussels.

Flynn, M. and Ljungqvist, B. (1995). A review of wake effects on workers exposure. *Ann. Occup. Hyg.*, 39, 211–221.

Friedlander, S.K. (1977). *Smoke, Dust and Haze, Fundamentals of Aerosol Behaviour.* John Wiley & Sons, Inc., New York.

Fuchs, N.A. (1964). *The Mechanics of Aerosols*. Pergamon Press, Oxford.

Hinds, W.C. (1982). *Aerosol Technology*. John Wiley & Sons, Inc., New York.

Hinze, J.O. (1975). *Turbulence*. McGraw-Hill, Inc., New York.

Ljungqvist, B. (1979). Some observations on the interaction between air movements and the dispersion of pollution. Document D8: 1979, Swedish Council for Building Research, Stockholm.

Ljungqvist, B. (1987). Air movements—the dispersion of pollutants: studies with visual illustrative methods. *ASHRAE Trans.*, 93(1), 1304–1317.

Ljungqvist, B. and Reinmüller, B. (1995). Hazard analyses of airborne contamination in clean rooms—application of a method for limitation of risks. *PDA J. Pharm. Sci. Tech.*, 49, 239–243.

Ljunqvist, B. and Reinmüller, B. (1997). *Clean Room Design. Minimizing Contamination through Proper Design.* Interpharm/CRC, Boca Raton.

MCA (2002). Rules and Guidance for Pharmaceutical Manufacturers and Distributors. The Stationery Office, London.

Noble, W.C., Lidwell, O.M., and Kingston, D. (1963). The size distribution of airborne particles carrying microorganisms, *J. Hyg. Camb.*, 61, 385–391.

Whyte, W. (1986). Sterility assurance and models for assessing airborne bacterial contamination. *J. Parenter. Sci. Technol.*, 40, 188–197.

8 Monitoring Microbiological Quality: Conventional Testing Methods

Rosamund M. Baird

CONTENTS

8.1 INTRODUCTION

Microbiological control begins with the design of plant and premises and ends with testing of the finished product for microbial contamination. As in any production chain, the quality of that finished product hinges upon the weakest link in the chain. Quality must therefore be built in at every production stage; by definition it cannot be inspected into the final product (see Chapter 6).

Microbiological control is applied at several key points during the manufacture of both sterile and nonsterile products. Raw materials, including water, must meet the required specifications before being approved for use. In-process monitoring will indicate whether product quality has been compromised during manufacture, abnormal results often providing the first indication of an earlier problem in processing. Finished product tests will demonstrate whether release specifications have been met. These results, combined with other accumulated data, provide the vital documentary evidence that the product is fit for its intended use.

Regardless of whether the product under examination is a raw material or a sterile or a nonsterile product, all samples withdrawn for testing should be selected on a statistical basis, with the sampling method taking account of the individual product type. Similarly the test method, whether quantitative or qualitative, should be adapted to the product characteristics. Various contributory factors are known to influence the outcome of the test, including the choice and sensitivity of the test method itself, the effectiveness of antimicrobial neutralization techniques and the choice of resuscitation and enrichment media. Test methods used in pharmaceutical microbiology have traditionally been based upon the techniques developed for use in food, water, and dairy microbiology; newer techniques involving rapid methods have also been introduced in recent years with some success, as discussed in Chapter 9.

As with other control laboratories, the quality of service provided by the pharmaceutical microbiology laboratory should be monitored by a process of continuous self-inspection and regular external audit. Factors such as the choice, standardization, and validation of test methods will clearly need to be reviewed.

8.2 SAMPLING

In any sampling scheme, the quality of a given batch is assessed on the basis of test results of samples drawn from that batch. Any sampling process therefore involves the selection of appropriate indicators of quality, known as attributes, which are assumed to be homogeneously distributed throughout the batch. In microbiological sampling, however, contaminants may not always be randomly distributed within a product. In the case of heat-sterilized products, samples selected for sterility testing are normally taken from the coolest part of the autoclave, where the risk of sterilization failure is deemed to be higher. Likewise nonaqueous products may not always be homogeneously contaminated because pockets of contaminants may occur. For example, surface contamination of an antibiotic eye ointment was reported to have occurred through the condensation of water onto the ointment surface, thereby enabling local proliferation of *Pseudomonas aeruginosa* in an otherwise hostile environment (Kallings et al. 1966). Similarly, aerobic contaminants may grow preferentially on the surface of topical products. In such cases, sampling schemes should take account of this difference in distribution. Aqueous products are assumed to be homogeneously contaminated; nevertheless, the contents of individual containers should be mixed thoroughly before samples are withdrawn. Sampling schemes should also be adapted to reflect the nature of the product. For a more detailed discussion, the reader is referred to Baird (2000a).

8.2.1 Sampling Techniques

The likelihood of accidental contamination occurring while taking samples should clearly be minimized. Personnel involved in sampling will not necessarily be microbiologists but they should have been properly trained in the use of aseptic techniques and should be conversant with relevant microbiological principles.

Samples taken should obviously be representative of the batch as a whole. Sampling procedures should document the method of sampling, the equipment and type of sample container to be used, the quantity of sample to be taken, and how it should be treated thereafter. Equipment should be properly cleaned after use and reserved only for sampling purposes.

Microbiological sampling of products should precede other quality control tests. Previously unopened packs should be sampled and each pack should be suitably marked so that it can be identified in the case of a failure. The interval between sampling and examination should be kept to a minimum, particularly when sampling water supplies. Contaminants may multiply during this time, leading to erroneously high counts. If necessary, samples should be refrigerated.

In the case of aseptically prepared products, both random and nonrandom sampling is required; one-quarter of the required samples should be taken from the beginning of the filling run and similarly from the end of the run, the remaining half being randomly selected from the finished batch. As mentioned before, for heat-sterilized products, samples should be taken from specified sites in the load, previously identified as the slowest to reach sterilizing temperatures. Samples for pyrogen testing should be collected from among the last filled containers. Oily products, creams, and ointments frequently present sampling problems. This may be overcome to some extent by homogenizing the product with a suitable emulsifying agent, such as polysorbate (Tween 80). The *British Pharmacopoeia* (BP 2005) and the *United States Pharmacopeia* (USP) *27* (2005) then recommend warming the resulting homogenate to not more than 40°C and 45°C, respectively, for an unspecified period of time. By so doing, however, misleading information on product quality may be generated; preservative action is temperature dependent and so the microbial count may well be reduced by such heating.

8.2.2 Sampling Schemes

In formulating any sampling scheme, a decision must first be made as to the level of assurance required. This in turn will determine the size of the sample and the number required. Sampling rates can then be found in British Standard (BS) 6001 (1972). Clearly, for greater assurance, the number of samples must increase accordingly. In those instances where test results indicate a deterioration in product quality, sampling plans can be switched from a normal inspection to a tightened inspection, as stated in BS 6000 (1972) and BS 6002 (1979). Sampling will then remain at this level until test results show that normal inspection levels are again warranted.

8.2.2.1 Single Sampling

Conventional sampling plans are usually based upon the examination of single samples taken from a predetermined number of containers in a batch. Traditionally this number has been $\sqrt{n} + 2$, where n is the number of containers in the batch. Such samples are mixed together and a further sample from this is then analyzed, resulting in the acceptance or rejection of the batch concerned. Results on individual products may be examined for trends over a period of time.

An alternative but somewhat time-consuming approach involves the plotting of results using cumulative sum (CUSUM) charts, whereby trends in microbiological quality can easily be observed. Such a system can provide an early warning that product quality is not being maintained. It has also been successfully used in monitoring the quality of water supplies as shown in Figure 8.1 and in environmental monitoring programs (Russell et al. 1984, Russell 1996).

8.2.2.2 Attribute Sampling

An alternative approach to sampling has utilized two- and three-class attribute sampling schemes. As before, several samples are taken but these are then analyzed separately. Although more costly in terms of staff resources, such schemes are considered to offer a greater degree of assurance and to provide rather more useful information on the batch concerned. By definition, such schemes incorporate so-called tolerances, allowing a small proportion of samples to show slight deficiencies; clearly, however, they are unsuitable for the testing of sterile products.

Two-class attribute plan—In this plan, both the number of samples to be taken and the maximum permitted number of positive results are stated. Such a scheme is used in the examination of water supplies where *Escherichia coli* is not allowed in any sample and only two out of five samples may contain Enterobacteriaceae (Anonymous 1982). Thus two quality levels are defined by the required absence of *E. coli* and by the limited number of samples that may contain Enterobacteriaceae.

Three-class attribute plan—Here three levels of quality are acknowledged: the fully acceptable, often known as the acceptable quality level (AQL); the marginally acceptable; and the unacceptable. Such schemes are closely associated with the establishment and use of microbiological reference values and have been used successfully in food and cosmetic microbiology for many years (Mossel 1982, Anonymous 1996a). In pharmaceutical microbiology, their use has in the past been

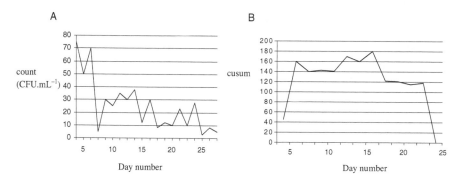

FIGURE 8.1 Variations in bacterial counts of deionized water over a period of time (A) plotted traditionally and (B) depicted as a cusum chart, where upward trends indicate a loss of control and vice versa. In the case of the latter, trends are more easily discernible, thus enabling corrective action to be taken at an early stage, if required. From Russell, M. (1996) In *Microbiological Quality Assurance in Cosmetics, Toiletries and Pharmaceuticals*. Baird, R.M. and Bloomfield, S.F., Eds. Taylor & Francis, New York, pp. 31–47. With permission.)

largely restricted to the in-house testing of raw materials and nonsterile products, although reference is now made to their use in the BP (2005). When used, they are considered to provide invaluable, detailed information on the batch concerned and can also forewarn of any potential problems at an earlier stage. Additional information can then be obtained by comparing data amassed from different batches.

8.3 CULTURE MEDIA

Over the years, a wide range of culture media has been developed for the isolation and identification of microorganisms from different ecological niches. The range of available media has expanded considerably through the introduction of new selective agents and the adaptation of existing formulae, particularly in the field of food microbiology. Both nonselective media, rich in nutrients, and highly selective, specific media are used for the detection of the increasingly large number of recognized microbial types.

In contrast, the range of media used in pharmaceutical microbiology is comparatively narrow and has altered very little in recent times. Media used in the pharmacopoeial tests are required to recover a relatively small number of microbial types or to demonstrate that contaminants cannot be recovered from the product in question (Hodges 2000). In-house tests on pharmaceutical products, however, frequently employ a less restricted range of culture media.

Culture media used in the identification and isolation of microbial contaminants may be in a liquid or solid form. They are generally formulated to contain a source of carbon, nitrogen, buffers, sulphur, phosphorous, inorganic salts, and a significant proportion of water. In addition, nutritional supplements such as serum, blood, carbohydrates, vitamins, and certain salts (calcium, magnesium, manganese, sodium, and potassium, for example) may be added. Selective agents (antibiotics and dyes) may also be included to inhibit unwanted organisms, thereby increasing the selective properties of a given medium. Media are usually formulated with a pH close to neutral, thereby providing optimum conditions for growth.

In the past, a wide range of culture media would have been prepared and stocked for identification purposes. The often laborious task of identifying microbial isolates has nowadays become much simpler through the introduction of commercially available kits; these contain dehydrated reagents to which suspensions of individual bacterial cultures are added. Following incubation, the biochemical reactions are recorded according to observed color changes and these are then converted to a score that corresponds to a certain bacterial type and species.

8.3.1 PREPARATION AND USE OF CULTURE MEDIA

Culture media may be purchased in a ready-to-use form with a limited shelf life or in a more stable dehydrated form to which water is added. Less commonly, culture media may be prepared from the component raw materials. Good laboratory practice requires that culture media are prepared to pharmaceutical and not "cook-book" standards. Thus culture media and their components should be purchased from approved suppliers and dated on receipt; media should be prepared according to master formulae and written procedures; sterilization methods should be carefully followed so that the resulting product is sufficiently processed, but not overheated, thereby affecting its nutritive properties; the finished product should be given a batch number and an expiry date. Culture media and their component raw materials should be stored in appropriate conditions, as recommended by the manufacturer.

Quality control tests on the final product should include not only tests on sterility and pH but also performance tests on the medium's selectivity, productivity, and ability to produce distinctive colonies. Significant differences in productivity may occur between brands of media and even between different batches of the same brand.

With the exception of fertility tests carried out on sterility test media, the importance of performance tests has been largely overlooked in pharmaceutical microbiology. In food microbiology, however, their importance has been recognized through the publication of an international pharmacopoeia, detailing test methods, test strains, and monographs on the most commonly used media (Baird et al. 1985, Baird et al. 1987, Corry et al. 1995, Corry et al. 2003).

The performance of a given culture medium is known to be affected by three factors (Mossel et al. 1980). The so-called "intrinsic" factors are of primary importance because they include not only the available nutrients, the water activity (a_w), the redox potential (e_h), and pH of the medium at the time of inoculation and during incubation, but also any antimicrobial agents that may be present. "External" factors such as the temperature and gaseous environment can also markedly affect the performance of media during the incubation period. Other "implicit" factors may be significant during this period, depending upon the vitality of the inoculum itself and the likelihood of synergism or antagonism between competing cells. For a more detailed discussion on culture media, the reader is referred to Baird (2000b).

8.3.2 Culture Methods

Recovery of microbial contaminants from pharmaceutical products involves both an isolation and identification stage. Additionally, an initial resuscitation stage may be required to recover sublethally injured cells. Such cells may result from exposure to preservatives, antimicrobial compounds or other toxic components, extremes of osmotic pressure or pH, or from the injurious effects of processing generally, including freezing, heating, and dehydration. Microorganisms surviving such conditions carry sublethal lesions that inhibit their development on selective media, although quite suitable for nonstressed cells of the same species. The use of selective media for counting purposes may therefore result in a gross underestimation of the number of cells present.

Food microbiologists have long recognized the damaging consequences of exposure to sublethal effects (Hurst 1977, Ray 1979). There is no universal resuscitation medium, although complex, rather than simple media are more likely to provide suitable substrates for alternative metabolic pathways to aid the recovery of damaged organisms (Mossel and Corry 1977). Recovery methods frequently therefore incorporate a resuscitation step when cell repair can take place. This may take the form of recovery in a liquid medium at room temperature for a period of 2 h. In other cases, a longer period may be required, for example, up to 6 h. In such cases, liquid medium repair is inappropriate because it would allow multiplication of both uninjured and repaired cells, leading to erroneously high counts. In these instances, a solid medium repair technique is used, where dilutions of the product are spread onto a rich nonselective agar and held at room temperature for 4 to 6 h. Subsequently, a suitable selective agar at 44°C is overlayered, allowed to set, and then incubated at an appropriate temperature.

Apart from the recommended techniques used to recover specified organisms from nonsterile products (see BP 2005), resuscitation techniques have in the past been largely overlooked in pharmaceutical microbiology. This clearly illustrated the need for closer cooperation between microbiologists of different disciplines. In recent times, however, the significance of these stressed cells has become apparent to pharmaceutical microbiologists and increasing attention has been paid to the attempted recovery of these so-called viable nonculturable cells (Barer et al. 1993, Barer 1997, Newby 2001).

An additional stage in the examination of pharmaceutical products is a neutralization step to inactivate any antimicrobial agents that may be present (see Chapter 17). This should take place at an early stage of culture and before any attempt at resuscitation is made. In some cases, for example, alcohols and phenols, the antimicrobial activity may simply be diluted out. The success of this method depends upon the product having a high dilution coefficient (concentration exponent, see Chapter 14). In the case of phenol, with a dilution coefficient of six, a threefold dilution will result in a decrease of activity of 3^6-fold. Where dilution is inappropriate, a filtration and rinsing technique may be used or, alternatively, various inactivating agents may be added to the culture media, as shown in Table 8.1. Where a specific

TABLE 8.1
Methods of Inactivating Antimicrobial Compounds

Antimicrobial Compound	Method/Addition of
Alcohols	Dilution (plus Tween 80)
Ampicillin, benzylpenicillin	β-Lactamase from *Bacillus cereus*
Antibiotics, other	Membrane filtration
Bronopol	Cysteine hydrochloride
Chlorhexidine	Lecithin plus Tween 80 or Lubrol W
Cresols	Dilution plus Tween 80
Halogens	Sodium thiosulphate
Hexachlorophane	Tween 80
Mercurials	Sodium thioglycollate or cysteine
Parabens	Dilution plus Tween 80
Phenolics	Dilution plus Tween 80
Quaternary ammonium compounds	Lecithin plus Tween 80 or Lubrol W
Sorbic acid	Dilution plus Tween 80
Sulphonamides	*p*-Aminobenzoic acid

inactivator is not available, a multipurpose neutralizing agent, such as 3% Tween 80 or 4% Lubrol W, may be used. In sampling pharmaceutical products of unknown origin, a nonspecific neutralization method should be used. An all-purpose neutralizing medium is now commercially available. Neutralization methods are further discussed in Chapter 14.

8.4 MICROBIOLOGICAL TESTS

8.4.1 Control of Raw Materials

Raw materials, including water, used in the manufacture of both sterile and nonsterile pharmaceuticals may contain a variety of microbial contaminants. During processing, these may be reduced in number or conversely they may be given a suitable opportunity to multiply. Hence, the purpose of microbiological quality control tests is, first, to establish whether contaminants are present and in what sort of numbers and, second, to determine which organisms are present and whether they represent a potential hazard either from a health or from a spoilage point of view.

The microbiological quality of the final product is determined by the quality of the starting materials; materials with a known low bioburden should be purchased whenever possible. Poor-quality raw materials will not only compromise the quality and life of the final product but also contribute to the overall level of background contamination in the manufacturing plant. Particularly in the case of free-living opportunist organisms, reservoirs of contamination may be established in the manufacturing environment, thus providing a potential source of contamination to successive products. Gram-negative bacilli found in water supplies have often caused notable problems in the past (see also Chapters 2 and 3).

Microbiological standards for raw materials will vary according to the type of product and its susceptibility to contamination. Table 8.2 summarizes the microbiological test requirements for selected raw materials of the BP (2005) and the USP 27 (2004). Although a wide range of raw materials is used in pharmaceutical production, it is not necessary to monitor the microbiological quality of all starting materials. Sampling regimens should therefore be adapted accordingly by manufacturers, based on the likely bioburden of a given product. Susceptible products, such as tragacanth powder, should be examined at every delivery; less susceptible products may be sampled at defined intervals, for instance, 1 in 5 or 1 in 10 batches, others may not need to be sampled routinely. When testing indicates a decline in microbial quality, sampling should be switched from a normal level to a tightened regimen and remain at this level until there is evidence that the product quality has improved sufficiently.

Products of natural origin, whether plant, mineral, or animal, are likely to be heavily contaminated and may well contain pathogenic organisms, for example, talcum powder may contain *Clostridium* spp. and pancreatin extracts may contain *Salmonella* spp. and *E. coli*. Microbiological control of such products may rest with

---

TABLE 8.2

Comparison of BP (2005) and USP 27 (2004) Microbiological Test Requirements for Selected Raw Materials

Raw Material	BP Requires Absence of	USP requires Absence of
Acacia	*E. coli*[a]	Salmonellae
Agar	*E. coli*, salmonellae[a]	Salmonellae
Alginic acid	*E. coli*, salmonellae[a]	*E. coli*, salmonellae[b]
Aluminum hydroxide	*E. coli*, enterobacteria, certain Gram-negative bacteria[a]	*E. coli*[b]
Bentonite	[a]	*E. coli*
Cochineal	*E. coli*, salmonellae	
Digitalis, powdered		Salmonellae
Gelatin	*E. coli*, salmonellae[a]	*E. coli*, salmonellae[b]
Kaolin	[a]	*E. coli*
Lactose	*E. coli*[a]	*E. coli*[b]
Pancreatin	*E. coli*, salmonellae	Salmonellae, *E. coli*
Starch	*E. coli*[a]	*E. coli*, salmonellae
Sterculia	*E. coli*	
Talc, purified	[a]	[b]
Tragacanth	*E. coli*, salmonellae[a]	*E. coli*, salmonellae

[a] EP requirement in some member states for total viable count (TVC) (see individual monographs in BP).
[b] USP requirement for TVC (see individual monographs).

Source: Based on British Pharmacopoeia. (2005). *British Pharmacopoeia*. Vol. IV, Appendix XVIB. The Stationery Office, London; and United States Pharmacopeia. (2004). *United States Pharmacopeia 27— National Formulary 22*. U.S. Pharmacopeial Convention, Rockville, MD. With permission.

the supplier, who must provide materials of an agreed specification, detailed on an accompanying certificate of analysis. In contrast, synthetic materials are rarely contaminated and generally provide contaminants with poor growth opportunities. In other instances, the manufacturing process itself may have a profound effect on the microbiological quality of the raw material concerned. For example, extremes of pH employed in an extraction process will significantly reduce the natural bioburden. For other materials that do not meet the required microbiological standard, some form of in-house treatment may, however, be required.

8.4.2 Control of Sterile Products

8.4.2.1 Sterility Test

Much has been written about the deficiencies of the sterility test, as described in BP (2005) and USP 27 (2004) (see also Chapter 12). In order to make an objective assessment of its value, it is necessary to understand not only the microbiological problems involved in carrying out the test, but also the difficulties which may be encountered in obtaining representative samples from a given batch. Although essentially a qualitative test, the sterility test can be adapted to provide quantitative information on the numbers and types of contaminants present. Such data can be invaluable when investigating potential sources of contamination.

Statistical considerations—By its nature, the test is destructive and clearly cannot be applied to the entire batch. The test therefore relies heavily on a statistical method of withdrawing random samples; results are determined both by the number of samples taken and the incidence of contamination in the batch. In mathematical terms, this may be written as

$$\text{The probability of rejection } 1 - (1 - p)^n \qquad (8.1)$$

where p is the proportion of contaminated containers and n is the number of containers tested. The number of containers to be tested in a given batch is determined by the size of the batch and the type of product. Table 8.3 summarizes these requirements as stated in the BP (2005) and the *European Pharmacopoeia* (EP) (2004). Clearly, as the sample size increases, so the probability of approving a defective consignment decreases. Conversely, the probability of approving a defective consignment increases

---

TABLE 8.3
Sample Requirements of the BP (2005) and EP (2004) for Sterility Testing

Product	Batch Size	Minimum Number of Samples Required
Parenterals	Not more than 100	10% or 4, whichever is the greater
	101–500	10
	>500	2% or 20, whichever is the lesser
Ophthalmics	Not more than 200	5% or 2, whichever is the greater
	>200	10

TABLE 8.4
Probability of Rejecting a Batch of Parenteral Product as Nonsterile According to Sample Size (Based on One EP Sterility Test)

Batch Size	Sample Size	Probability of Rejection According to Frequency (%) of Contamination in Batch				
		0.1	1.0	5.0	10	50
40	4	0.004	0.039	0.185	0.344	0.937
101–500	10	0.010	0.096	0.401	0.653	0.999
1,000	20	0.020	0.180	0.640	0.878	0.999

Source: Adapted from Brown, M.R.W. and Gilbert, P. (1977). *J. Pharm. Pharmacol.,* 29, 517–523. With permission.

as the incidence of contamination decreases. Based on the result of a single test and taking no account of possible microbiological problems, the probability of rejecting single batches of different sizes and varying frequencies of contamination is shown in Table 8.4.

In the event of a positive sterility test result, the test may only be considered invalid if the following conditions have been fulfilled: a fault is shown in the sterility testing facility, test procedure materials, or technique used by the operator; or microbial growth is found in the negative controls. In such cases, a retest is permitted with the same number of units as in the original test. From a statistical point of view, the probability of approving a defective batch increases with a retest. In mathematical terms, the probability of passing a defective batch at a retest is

$$(1-p)^n \left[2-(1-p)^n\right] \tag{8.2}$$

The practice of retesting batches is therefore at best of questionable value. A further complication may be introduced when the sample size itself is reduced. BP (2005) states that the minimum quantity to be tested per container may vary according to the container size. For parenteral products of less than 1 ml volume, the entire contents must be sampled, whereas for those containing 1 ml or more, half the contents are required. On the other hand, the USP 27 (2004) requires the entire contents of parenteral products of less than 1 ml to be tested; for other volumes, a proportionate sample is allowed (1 to 40 ml, half the contents; >40 ml, 20 ml to be tested; >100 ml, 10% of contents, but not less than 20 ml). Clearly, such testing schemes can have a profound effect on the results of samples with low levels of contamination.

Microbiological considerations—There are a number of inherent problems in devising a test to demonstrate the sterility of a product. As potential contaminants are of unknown identity, a compromise on the choice of media is inevitably required. In practice, only two media, soya bean casein digest and fluid thioglycollate broths, are recommended in the most recent editions of the BP, EP, and USP. It is worth noting that fluid thioglycollate is known to be toxic to damaged cells. Compromises on incubation temperatures and times are also required.

The nature of the product under test may pose problems from a sampling point of view. In the case of oily products, such as eye ointments, microbial cells may well be embedded within the matrix of the product, thus requiring extraction with a suitable solvent, for example, sterile isopropyl myristate. Likewise the product itself or its components (including preservatives) may possess antimicrobial activity and these will need inactivating either by a dilution technique or by the addition of a specific inactivating agent to the culture medium (see Table 8.1). In some instances, specific inactivators may not be available, in which case, a multipurpose neutralizing agent such as Lubrol W or Tween 80 may be used. A third method involves the physical separation of microbial cells from the antimicrobial components of the product using a membrane filtration technique. A hydrophobic-edged membrane filter is used for such purposes and this is then washed with a sterile diluent, such as peptone saline or quarter-strength Ringer's solution. Antibiotic preparations have traditionally been sampled in this way.

Microbial cells exposed to the effects of such antimicrobial compounds are likely to be sublethally damaged, as are cells exposed to a heating process, as discussed earlier. It is noteworthy that the current sterility test method fails to incorporate a resuscitation step to recover such cells.

As far as the test itself is concerned, appropriate precautions should be taken to avoid accidental contamination of the sample under test. In practice, the test is carried out in a laminar airflow cabinet or positive pressure isolator using sterile equipment and sterile diluents. Personnel responsible for sterility testing should be appropriately clad, that is, wearing cleanroom clothing, and should be properly trained in aseptic manipulations. Rotational staffs are considered unsuitable for such work.

Accidental contamination of test material will inevitably occur from time to time. In some circumstances, the test then may be considered invalid and a retest submitted (see BP 2005, Appendix XVIA). As mentioned earlier, the practice of retesting actually increases the probability of passing a defective batch and must therefore be considered of dubious value.

Test methods—BP (2005) recommends the use of two test methods: membrane filtration and direct inoculation, the former being the method of choice. Two culture media are recommended: fluid thioglycollate medium, incubated at 30 to 35°C for not less than 14 days, is used primarily for isolating anaerobic bacteria, but will also sustain the growth of aerobic bacteria; soya bean casein digest medium, incubated at 20 to 25°C for not less than 14 days, is intended mainly to support the growth of fungi. Where a direct inoculation method replaces a membrane filtration technique, the incubation period is at least 14 days.

Quality control tests on all batches of culture media should include the following: a sterility test; a productivity test to demonstrate the growth of small numbers of organisms (approximately 10 to 100 cells), including where appropriate an aerobe such as *Staphylococcus aureus* and *Pseudomonas aeruginosa,* a spore-forming aerobe such as *Bacillus subtilis,* an anaerobe such as *Clostridium sporogenes* and a fungus such as *Candida albicans* and *Aspergillus niger;* and a similar test in the presence of the product to demonstrate effective neutralization of antimicrobial activity in the test sample.

1. Membrane filtration—This is the method of choice for aqueous and alcoholic preparations that can be filtered and for preparations miscible with or soluble in aqueous or oily solvents. Fluids are filtered through a sterile cellulose nitrate or cellulose acetate membrane filter, pore size not greater than 0.45 μm. Cells retained on the surface of the filter will then multiply when placed in appropriate growth media. In practice, the filter may be divided into two portions after filtration or the sample itself may be filtered through two separate filters. Suitable filtration equipment includes both traditional glass, metal or polycarbonate filter holders, and a number of dual-purpose, commercially available enclosed filter units. In the case of the latter, appropriate media may be introduced and incubated directly, thereby reducing the risk of operator contamination.
2. Direct inoculation—Using this technique, any antimicrobial agents must first be inactivated before the sample is added directly to the culture medium. The sample should be diluted by approximately tenfold. Where large volumes of liquid are concerned, the growth medium may be added directly to the sample in a concentrated form.

Overview—In summary, the time-honored sterility test, first described in the 1932 BP and the 1936 USP, has in recent years lost some of its former standing. Nowadays it is regarded by most industrialists as a rather poor indicator of product quality. In the case of terminally sterilized products, alternative methods are now available for assessing the probability of sterility. In such cases, confidence can be justifiably placed in the proper control of the manufacturing and sterilizing processes and in the associated use of biovalidation data (see Chapter 12). However, in the case of aseptic preparations, heat-labile and filter-sterilized products, the sterility test must still be regarded as an essential quality control test. It should not be forgotten, however, that the sterility test can only provide information on the sample tested under the conditions specified.

8.4.2.2 Testing for Pyrogens

The traditional way of detecting pyrogens has involved the injection of test material into rabbits and the measurement of any subsequent febrile response (see BP 2005, Appendix XIVD). The test originated from the observation that man and rabbits exhibit a comparable degree of sensitivity to pyrogenic material on a weight-for-weight basis. The test has a number of drawbacks, mainly relating to its biological nature. Over a number of years, the *Limulus* amoebocyte lysate (LAL) test has been used as a promising alternative in vitro method for the detection of Gram-negative bacterial lipopolysaccharides, known as endotoxins; the test will not, however, detect other pyrogens, and in such cases, the rabbit test remains the method of choice. The LAL test is based on the highly specific interaction of microbial endotoxin with amoebocyte lysate of the horseshoe crab *Limulus polyphemus,* producing turbidity, precipitation, or gelation of the mixture. This relatively simple, sensitive, and inexpensive test provides rapid results, the rate of reaction being dependent on the concentration of endotoxin, the pH, and the temperature. Some therapeutic products are known, however, to interfere with the test reaction and it must therefore be shown that a valid test can be carried out on the product under test, by demonstrating that any interfering factors have first been neutralized or removed.

The original LAL test for bacterial endotoxins was a limit test based on the presence or absence of a firm gel, visible to the naked eye. The product under test was therefore shown to have an endotoxin concentration above or below the defined threshold endotoxin concentration, known to bring about the gelation reaction. The BP (2005) lists six methods for determining endotoxin concentration in a product: the original gelation limit test; a semiquantitative gelation method; two kinetic methods, the turbidimetric kinetic method and the chromogenic peptide kinetic method, both of which utilize the linear regression of the log of the response on the log of the endotoxin concentration; a turbidimetric endpoint test; and a chromogenic peptide endpoint method that measures the intensity of dye color liberated from a suitable chromogenic peptide as it complexes with the endotoxin in the presence of lysate (Friberger 1987). These four quantitative methods, although requiring more instrumentation, provide for easier automation of regular testing of large sample numbers of product.

In recent times, considerable experience has been gained in both the performance of the LAL test through collaborative studies and also in refinement of the test to improve its sensitivity. The test continues to gain acceptance among the pharmacopoeial bodies, the licensing authorities, and individual manufacturers; it is now the preferred test method for bacterial endotoxin detection not only when screening raw materials and in monitoring production processes but also when releasing selected sterile products. It is also the specified test method for water for injections (EP). For a more detailed discussion of the test, the reader is referred to the companion *Handbook* (Baines 2000) and the BP (2005, Appendix XIVC).

8.4.3 MICROBIOLOGICAL TESTS ON NONSTERILE PRODUCTS

Microbiological control of pharmaceuticals has become increasingly important in the past three decades. Three factors in particular have contributed to this development: the occurrence of microbial contamination in medicines has been widely reported; the origins and avoidance of such contamination have been well documented; at the same time, there is now a better understanding of the health risks presented to patients from the inadvertent exposure to contaminated medicines. Microbiological quality-control tests will establish whether the finished product meets its release specification. Where appropriate, the pharmacopoeial standard is used (see current edition of BP, USP, and EP). In other cases, an in-house specification may be used alone or in combination with the pharmacopoeial standard. In-house specifications are generally more detailed, reflecting the known inherent microbiological problems associated with the product itself or its manufacture. Essentially, microbiological quality control tests comprise both quantitative techniques for counting microorganisms and qualitative tests to show the absence of specified organisms. Both sets of tests are recognized by the current pharmacopoeias, although the emphasis placed on such tests may differ. The tests may be carried out on raw materials, in process samples, or on finished products. Techniques commonly used are described below. For further details, the reader is referred to the companion *Handbook* (Millar 2000).

8.4.3.1   Enumeration Methods

Pour plate—In this method, the sample is mixed with molten agar at not more than 45°C in Petri dishes and allowed to set. Following incubation, colonies are counted and a total viable count per milliliter or gram can then be calculated. The method is more sensitive for counting small numbers of colonies than the spread plate technique, because a larger sample volume (up to 5 ml) can be examined. However, in preparing the plates, it is important not to exceed the holding temperature of the molten agar, as this may affect the count. Colonies held within the matrix of the agar will require an increased period of incubation (usually an extra 24 h) before they can be seen. Colonies should not be confused with undissolved particles of the test sample, nor with air bubbles, introduced through vigorous mixing of the plate contents.

Spread plate—Aliquots (usually 0.1 ml) of the prepared sample are spread in duplicate over the plate surface and incubated. Colonies are counted and results may then be converted to total viable counts using the appropriate dilution factor. Non-selective media, such as nutrient agar or tryptone soya agar, are usually employed for counting purposes.

Drop count—Less commonly used, the drop count method may be useful as a screening test when relatively high viable counts are suspected. The technique is a variation of the surface count method developed by Miles et al. (1938). Aliquots (0.002 ml) of the diluted sample are deposited from a special dropper onto previously dried plates. Following incubation, individual colonies are counted; the total viable count can then be calculated using the appropriate dilution factor.

Most probable number (MPN)—This technique uses liquid culture media and depends upon the observation of certain growth characteristics, such as turbidity, acid, or gas production. Decimal dilutions of the sample are first prepared from which aliquots are inoculated into the broth concerned. Following incubation, the number of tubes showing the required growth characteristics are counted. Through the use of reference tables (Meynell and Meynell 1970), the most probable number of organisms in the original sample can then be calculated. In using the tables it is assumed first that microorganisms are randomly distributed in the sample and second that a positive reaction is obtained if, and only if, the aliquot of sample tested contains one or more microorganisms.

Although the MPN technique has been widely used for counting coliforms in the water industry, it is regarded as somewhat cumbersome for routine pharmaceutical purposes. It has, however, been recommended for counting microorganisms in USP 27 (2004) and BP (2005) and is also considered to be useful when counting small numbers of microorganisms (van Doorne et al. 1981).

Filtration—Membrane filtration techniques, as described earlier, are commonly used to examine water and other liquid samples. Following filtration, membranes (0.45-μm pore size) are transferred to solid media for incubation. Resulting colonies are counted and a total viable count per sample volume may be calculated.

An enhanced recovery of microorganisms from water has been shown on a "low nutrient" medium when incubated for longer periods at a lower temperature. The use of the R2A medium incubated at 28°C for 5 days or 22°C for 7 days has provided a more accurate assessment of waterborne organisms, often stressed through starvation or as a result of processing, than standard examination methods (Reasoner and Geldreich 1985, Gibbs and Hayes 1988).

8.4.3.2 Tests for Specified Microorganisms

Individual methods vary according to the particular organism sought. In most instances, a broth enrichment step precedes the isolation stage on solid media which is then followed by confirmatory tests to establish the identity of isolates. The latter are generally based upon the growth characteristics of pure cultures on selective media and on their biochemical reactions to selected tests. A detailed discussion of identification schemes is outside the scope of this chapter, but appropriate reference sources should be consulted (e.g., Cowan 1993, Hodges 2000).

As discussed previously, those organisms considered to be hazardous from a health point of view are reflected in the various pharmacopoeial requirements. The BP (2005) details tests for the absence of Enterobacteria, *E. coli,* salmonellae, *P. aeruginosa, Staphylococcus aureus,* and clostridia. While recognizing why these particular organisms have been selected for pharmacopoeial standards, the presence of other organisms in pharmaceuticals cannot be overlooked. Nowadays it is well known that the presence of many other Gram-negative organisms, and particularly *Pseudomonas* and *Burkholderia* spp., cannot be condoned. Such contaminants often have simple nutritional requirements and are able to multiply to appreciable numbers within a product. Furthermore, the presence of such opportunist pathogens can be particularly hazardous for certain patients, for example, the very young, the elderly, immunosuppressed patients, or those with burns (see Chapter 2).

Nowadays many in-house specifications for nonsterile pharmaceuticals include a requirement to test for the presence of additional organisms other than those detailed in the various pharmacopoeias. An upper limit on the number of so-called undesirable organisms is often specified.

The BP (2005) lists methods for the isolation and identification of Enterobacteria *E. coli,* salmonellae, *P. aeruginosa, S. aureus,* and clostridia as summarized in Table 8.5. In all cases, the methods incorporate an initial recovery step, a broth enrichment stage (nonselective in the case of casein digest, but selective in the case of EEB-Mossel), followed by an isolation stage on solid media. Confirmation of isolates is then based upon suggested biochemical tests and observed growth characteristics on selective media. The reader is referred to the companion *Handbook* (Hodges, 2000) for a detailed discussion of these methods.

Although such tests will provide the required microbiological evidence that specified organisms are indeed absent, the methods may be considered to be time-consuming and somewhat cumbersome in practice, particularly when large numbers of samples must be examined. In-house specifications for both raw materials and nonsterile products are often therefore based on more generalized screening methods combining counting techniques with isolation techniques for specific microbial types.

TABLE 8.5
Isolation and Identification Tests for Specified Microorganisms

Organism	Enrichment	Primary Test	Secondary Test	Confirmation
Enterobacteriaceae	Lactose broth, 35–37°C for 2–5 h	EEB-Mossel, 35–37°C for 18–48 h	VRBGLA, 35–37°C for 18–24 h	Growth of Gram-negatives
E. coli	CSB, 35–37°C for 18–48 h	MacConkey broth, 43–45°C for 18–24 h	MacConkey agar, 35–37°C for 18–72 h	Indole at 43.5–44.5°C/biochemical
Salmonella	As above for 18–24 h	TBBG broth, 41–43°C for 18–24 h then subculture on DCA, XLDA, or BGA at 35–37°C for 18–72 h	TSI agar, 35–37°C for 18–72 h	Biochemical/serological
P. aeruginosa	Saline peptone, 35–37°C for 2–5 h	CSB, 35–37°C for 18–48 h	Cetrimide agar, 35–37°C for 18–72 h	Growth of Gram-negatives at 41–43°C for 18–24 h
S. aureus	As above	As above	Baird-Parker, 35–37°C for 18–72 h	Coagulase, DNase tests
Clostridia	RMC, 35–37°C for 48 h anaerobic	Columbia agar with gentamicin, 35–37°C for 48 h anaerobic	Columbia agar, 35–37°C for 48 h aerobic/anaerobic	Catalase test

Note: EEB-Mossel — Enterobacteriaceae enrichment broth-Mossel; VRBGLA—violet red bile agar with glucose and lactose; TBBG—tetrathionate bile brilliant green broth; DCA—deoxycholate citrate agar; XLDA—xylose lysine deoxycholate agar; BGA—brilliant green agar; TSI—triple sugar iron agar; CSB—casein soya bean digest broth; DNase—deoxyribonuclease test; RMC—reinforced medium for clostridia, anaerobic incubation.

Source: From British Pharmacopoeia. (2005). *British Pharmacopoeia*. Vol. IV, Appendix XVIB. The Stationery Office, London. With permission.

The latter have also been widely used in a number of surveys comparing the incidence of microbial contamination in different types of pharmaceutical products (Anonymous 1971, Baird 1985).

8.4.4 BIOBURDEN TESTING OF MEDICAL DEVICES

In the manufacture of sterile medical devices, the sterilization operation is itself recognized as a special process under the EN 46001/EN 46002 (1993) and EN 724 (1994) series of European standards, because sterility assurance cannot be verified by inspection and testing of the product; for this reason, sterilization processes need to be validated before use, the performance of each process must be monitored routinely, and the equipment must be properly maintained. In addition, it is also necessary to know the nature of the microbiological challenge to the sterilization process, not only in terms of numbers and identification of contaminants but also their characteristics.

The presterilization bioburden on a device comprises the viable microorganisms derived from the component raw materials, their subsequent storage, and the environment in which the product is manufactured, assembled, and packaged. For a more detailed discussion of how the microbial ecology of the manufacturing process can affect the microbial quality of a medical device, the reader is referred to Chapter 3, Chapter 4, and Chapter 5B. As the exact bioburden cannot be determined, a viable count using a defined technique is determined in practice and this is then related to a bioburden estimate by the application of a correction factor. Bioburden estimations can thus be used as part of the validation and revalidation of a sterilization process and also in the routine control of the manufacturing process. Furthermore, they may form part of the overall quality system: as an integral part of the environmental monitoring program; as an indicator of the efficacy of a cleaning process in removing contaminants; as a process monitor for products supplied as nonsterile, but for which the microbiological cleanliness is specified; and also for monitoring raw materials, components or packaging.

In estimating the bioburden of a medical device, four stages are involved: removal of microbial isolates from the device, their transfer to culture conditions, their subsequent enumeration and identification, and lastly, the application of the previously determined correction factor to the presterilization count in order to calculate the bioburden estimate (Anonymous 1996b, 1997a). Owing to the wide variety of materials used in the manufacture of medical devices, there is no single technique that can be universally applied; equally, the precise selection of culture conditions may be influenced by contaminants, expected on the basis of historical data. The number of contaminants can vary considerably as shown in Figure 8.2. Sampling of the device and selection of an appropriate technique for removing contaminants are clearly important issues. Microbial adhesion to some surfaces may cause problems, particularly where biofilms have been formed; a variety of removal techniques may be used including stomaching, ultrasonication, shaking, vortexing, flushing, blending, swabbing, agar overlaying, or contact plating. Account must also be taken of the physical and chemical nature of the product and the likelihood of inhibitory substances being released and how these might be counteracted. The resulting suspension of microorganisms can then be examined for the presence of viable microorganisms by a variety of methods, including membrane filtration, pour plates, spread plates, a MPN method using serial dilutions, or spiral plating. In selecting appropriate media and incubation conditions, some guidance is available in the references mentioned above; however, this may need to be adapted in the light of likely contaminants and their known characteristics.

Considerable emphasis is placed on the importance of validation of bioburden techniques (Hoxey 1993, Anonymous 1997b). First, the technique must be validated for its efficiency of removal of microorganisms: either a repetitive treatment method

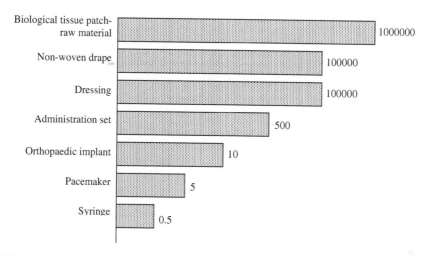

FIGURE 8.2 Magnitude of bioburden on devices, expressed as number of CFUs. (Reproduced by kind permission of E. Hoxey, personal communication.)

is used to demonstrate that the method can be repeatedly applied until there is no significant increase in the accumulated number of microorganisms recovered or the recovery efficiency can be established by inoculating the product with a known artificial bioburden. By establishing the recovery efficiency of the method concerned, the correction factor to be applied can be calculated. Second, the ability of the counting techniques and culture conditions to enumerate any removed microorganisms must be validated; the chosen culture conditions cannot be expected to detect all potential contaminants, hence, in practice, the bioburden inevitably will be underestimated. By comparing a preselected range of media and incubation conditions, the optimum recovery conditions can be assessed. Third, the ability of the chosen technique to recover low numbers of potentially fragile microorganisms requires validation; there should be no significant difference between the number artificially inoculated and those recovered. If inhibitory substances are thought to be present, these should be neutralized or filtered before testing proceeds.

8.4.5 Environmental Monitoring

Significant advances in cleanroom design and construction in recent years have had a considerable impact on the overall microbiological quality of manufacturing environments for both sterile and nonsterile products. In addition, the development of cleanroom standards (see Chapters 4, 6, and 7) have, in turn, identified the environmental conditions required for the manufacture of both sterile and nonsterile products.

Microbiological monitoring programs are carried out routinely in cleanrooms and other controlled environments to assess the effectiveness of cleaning and sanitization practices on bioburden levels in those environments. Such programs are not intended to enumerate or identify every microbial contaminant, but should provide sufficient information to confirm that the environment is operating within its required limits of control. Sampling should occur during normal operating conditions, that is, while production is in progress, and with the usual complement of operating staff.

The purpose of the monitoring program is, then, to collect representative data on the environmental bioburden that cannot only be evaluated and reviewed per se by trained staff, but also can be scrutinized for any trends over extended sampling periods. Any suggested deterioration in environmental quality over a period of time should initiate a thorough documentary review of, at least, the current status of area maintenance, environmental parameters (for example, temperature, relative humidity, differential air pressures, number of air changes, and status of HEPA filters), cleaning and sanitization, and operator training records, followed by a documented plan for corrective action.

In essence, a microbiological monitoring program sets out to establish the environmental bioburden contributed collectively by the air, the surfaces, equipment, and personnel, either shed directly or dispersed from their garments. The program should be tailored to specific facilities and conditions. Sampling sites should be identified and evaluated for their appropriateness at the time of commissioning but may need to be adjusted from time to time based on accumulated results.

8.4.5.1 Frequency and Extent of Sampling

The frequency of microbiological sampling will depend on the nature of the manufacturing environment and the work undertaken there. Clearly the potential for product contamination depends on the extent of normal intervention and personnel contact. Thus environmental monitoring has a higher profile in processes involving aseptic rather than terminally sterilized manufacture. BS 5295 (1989) and the most recent Medicines Control Agency [(MCA) 2002] guide to pharmaceutical manufacture give no firm indication of the frequency of sampling for aseptic products, although other published guidance (Parenteral Society 2002) suggests daily or batch-wise sampling for aseptic products and weekly sampling for terminally sterilized products. The most recent USP guidance (USP NF 22 2004) suggests that aseptic and adjacent supporting areas should be sampled at each operating shift, whereas other support areas should be sampled twice a week or once a week in the case of nonproduct contact areas.

8.4.5.2 Microbiological Limits

Recommended limits for microbial monitoring of clean areas in operation have been published (MCA 2002), as detailed in Table 8.6. The USP NF 22 (2004) has also published guidelines on the maximum acceptable number of colony forming units (CFUs) with regard to air cleanliness, surface cleanliness of equipment and facilities, and also with regard to operating personnel garments in controlled environments, as summarized in Table 8.7. The reader is referred to Chapter 7 for a detailed discussion of airborne contaminants and contamination risks in cleanrooms.

In practice, it is recommended that maximum microbial limits for individual manufacturing units are drawn up on the basis of what can be achieved under optimum operating conditions, based on historical data. Additionally, a definitive

TABLE 8.6
Recommended Limits for Microbiological Monitoring of Clean Areas in Operation

Environmental Classification	Air Sample CFU.m^{-3}	Settle Plates[b] (diam. 90 mm), CFU/4 h	Contact Plates (diam. 55 mm), CFU/plate	Glove Print 5 Fingers, CFU/glove
	Recommended Limits for Microbial Contamination[a]			
A	<1	<1	<1	<1
B	10	5	5	5
C	100	50	25	—
D	200	100	50	—

Note: Dashes indicate no recommended data.

[a] These are average values.

[b] Individual settle plates may be exposed for less than 4 h.

Source: Abstracted from U.S. Pharmacopeia. (2004). *National Formulary 22*. U.S. Pharmacopeial Convention, Rockville, MD. With permission.

TABLE 8.7
USP NF 22 (2004) Guidelines on Microbiological Cleanliness in Controlled Environments

Environmental Classification	CFU.m^{-3} of Air[a]	CFU/Surface Contact Plate	CFU/Clothing Contact Plate[b]
100	<3	3 (including floor)	3/5
10,000	<20	5	10/20
		10 floor	

[a] Using a slit-to-agar sampler or equivalent and taking a sufficient volume of air to detect excursions above the limits specified.

[b] The first figure is a limit for gloves, the second a limit for gowns.

Source: Abstracted from U.S. Pharmacopeia. (2004). *National Formulary 22*. U.S. Pharmacopeial Convention, Rockville, MD. With permission.

course of action should be established should such limits be exceeded. Many manufacturers adopt a three-class attribute plan for implementing environmental limits (see Section 8.2.2.2). Here environmental limits are defined as falling into one of three categories: an acceptable level; an "alert" level indicating a potential drift from normal operating conditions; or finally, an "action" level requiring immediate followup and potentially corrective action. From time to time these limits should be reviewed for their relevance and appropriateness; if necessary, they should be modified to reflect current conditions. As mentioned above, where analysis of environmental trends suggests deteriorating quality, the cause of this should be thoroughly investigated and appropriate corrective action taken.

8.4.5.3 Methodology

Methods for assessing the microbial bioburden in the air, on surfaces, on equipment, and on garments are discussed below. All methods should be validated for their appropriateness before being used routinely.

Air sampling—The traditional method of monitoring air quality has involved the use of settle plates. Petri dishes, usually containing a nonselective agar, are exposed to the atmosphere for defined periods during which time particles carrying microorganisms (>12 μm) will settle onto the agar surface. Following incubation, microbial colonies are counted and results from different exposure areas may then be compared. However, this "passive" method is of limited quantitative value because the results are affected by air currents in the immediate area, the particle settling rate, and the time of exposure (see Chapter 7). Moreover, the technique will not detect those organisms that remain suspended in air currents and hence fail to settle onto the agar surface.

Air quality can also be usefully monitored using "active" techniques, such as slit samplers, sieve impactors, surface air system (SAS) samplers, centrifugal samplers, and those using gelatin filtration. The essential features of these techniques are summarized in Table 8.8. All provide a quantitative evaluation of the number of organisms per volume of air sampled, although the estimate itself may be affected by the instrument; thus where alternative methods are used to those recommended, general equivalence must first be demonstrated (Ljungqvist and Reinmüller 1998). These devices should be regularly calibrated and used according to the manufacturer's instructions.

TABLE 8.8
Techniques for Monitoring Airborne Contamination

Sampler	Sampling Rate (L.min⁻¹)	Collection Method	Advantages	Limitations
Centrifugal	40–50	Centrifugation	Portable, speed, convenient, no subculture	Sampling rate may vary with particle size, detection limit
Gelatin filter	130	Impaction	High capture efficiency	Fragile filter, extra handling
Sieve impactor	28	Impaction	Separates particles into size range, no subculture	Unsuitable for high levels of contamination because particles may be superimposed
Slit sampler	50–700	Impaction	Time–concentration relationship gives high capture efficiency, no subculture	Unsuitable for high levels of contamination, cumbersome
Surface air sampler	180	Impaction	High capture efficiency, portable	Cell viability

Successful use of these samplers is dependent on the statistics of sampling. Sample volumes of air should therefore be adjusted according to expected contamination levels; for example, in a critical (filling) area, larger volumes of air will be required to provide statistically valid results than in a less clean environment where high levels of contamination might reasonably be expected.

Surface sampling of equipment and facilities—Surface sampling will indicate the effectiveness of cleaning and disinfection policies. Rodac or contact plates containing nonselective media can be pressed lightly onto flat surfaces. Following incubation, numbers of colonies can be counted, thus providing a quantitative estimate of contamination levels. If disinfectants or antibiotics have been used in the controlled area, the medium should include a suitable and appropriately validated inactivating agar. Other less accessible surfaces (drains, taps, ledges, etc.) are best sampled using a nonquantitative technique with moistened swabs. These can be cultured directly onto solid media or first incubated in nutrient broth and then subcultured onto solid media.

Surface sampling of garments—Clothing, including gloves, can be sampled using a contact plate technique. The technique can also be adapted for training purposes to demonstrate the effectiveness of various hand-washing techniques. Finger dabs, whereby the finger tips are pressed lightly on the surface of nutrient agar plates, can demonstrate the value of different hand-washing routines and are also widely used to monitor whether gloved hands have indeed offered the required level of product protection in practice.

Broth filling or process simulation—In an aseptic manufacturing process, the microbiological status of the process must be qualified through broth filling runs, whereby the filling process itself is simulated using nutrient media in place of product. In all other respects (including routine processing, planned interventions, and the normal complement of staff), the run should be identical to product filling. Any lapse in the aseptic manufacturing process would be indicated by visible growth (turbidity) in the nutrient broth following incubation. In validating a new aseptic process, at least three successful consecutive media fill runs are required and these should cover all product shifts for all line/product/container combinations (Anonymous 1987). Before establishing such a program, a number of issues must first be addressed: the filling method, fill volume, choice of media, incubation conditions, inspection procedures, documentation, interpretation of results, and possible corrective action (Anonymous 2004). Additional critical issues include the number of units to be filled. Traditionally a minimum of 3000 units have been required to demonstrate a contamination rate of 0.1%, but nowadays many pharmaceutical companies are filling well in excess of this in a single media–fill run and can demonstrate that contamination rates are well below this level. Smaller media fills are acceptable for pilot plant facilities used for manufacturing clinical trial material. For a more detailed discussion, the reader is referred to Parenteral Society (1993).

8.4.5.4 Environmental Monitoring in Advanced Aseptic Processing

The past decade has witnessed the emergence of a number of new techniques, including barrier, blow-fill-seal, and isolator systems, all of which incorporate the

JVERPOOL JOHN MOORES UNIVERSITY
LEARNING SERVICES

principle of removing human intervention, and hence contamination, from the critical manufacturing zones. It therefore follows that where personnel have been excluded from the critical zone, the necessity for microbial and particulate monitoring can be significantly reduced.

8.5 AUDITING MICROBIOLOGICAL ACTIVITIES

As in other areas of pharmaceutical production, the activities of the microbiology laboratory should be audited on a regular basis (see Chapter 21). Depending on the nature of the work undertaken there, this will require a critical appraisal of the following aspects: the use and suitability of existing premises; the use and mainte-nance of equipment; the type and range of activities in relation to the changing demands of production activities; staffing levels and staff ability, particularly with regard to individual training needs. For a more detailed discussion, the reader is referred to Chapter 21 and the companion *Handbook* (Lush 2000).

8.5.1 PREMISES

Sufficient working space should be available to segregate the various activities of the laboratory and to provide a streamlined direction of workflow. Separate areas should be designated for the following activities: sample receipt; sample preparation and testing, including a separate area for sterility testing with adjacent changing facilities; sample incubation; the preparation of culture media; decontamination and wash-up; the reading of results, documentation, and office work. The workflow through these areas should be such that the opportunity for crossover of samples is kept to a minimum.

8.5.2 EQUIPMENT

A wide range of equipment may be found in the microbiology laboratory, all of which requires regular maintenance. Separate log books should be kept for each item of equipment. Good laboratory practice requires that detailed records are kept on the operation of equipment including, for example, daily fluctuations of temper-ature within incubators and the temperature, pressure, and time readings monitored during autoclave cycles.

8.5.3 ACTIVITIES

The quality of the microbiological service provided by the laboratory is clearly an important issue. Details of accreditation for microbiological testing are available through the National Measurement Accreditation Service [(NAMAS) 1992]. The suitability, reproducibility, and precision of methods for isolating, counting, and identifying microbial contaminants should be reviewed from time to time and mod-ified if necessary. In the past, the quality control of culture media has been sadly neglected in pharmaceutical microbiology laboratories. All batches of media should meet release specifications, as discussed earlier. Technical skills of individual mem-bers of staff need to be examined periodically.

As in other areas of quality control, the microbiologist's involvement in laboratory-based activities, particularly in end-product analysis, has in recent years been overtaken by a more open approach to quality assurance activities generally. Nowadays, he may well be expected to advise on the microbiological aspects of building and equipment design, equipment cleaning, and sanitization methods, as well as cleaning and disinfection policies. Staff training may also form an important part of his job, involving both the teaching of basic microbiology and the education of production personnel in aspects of cleanliness and hygiene. In addition, the microbiology laboratory has an important function in monitoring whether acceptable conditions are being maintained, particularly in the testing of environmental and water samples. This involves not only the setting of appropriate standards with warning and action levels, but also close monitoring of test results against such limits.

Some laboratories will clearly be unable to meet all these requirements and some work may well be contracted out to a specialist microbiology laboratory. The work of such laboratories should be carried out to an agreed specification and audited on a regular basis. Certain aspects of the work, including environmental testing, are, however, best carried out in-house so that corrective action may be taken at an early stage, if required.

8.5.4 Staff

Demands on staff time change constantly as new problems arise within the laboratory. Sufficient trained staff should be available to cope with these demands, which may range from time-consuming challenge testing or stability testing of new formulations to the simple examination of in-process samples. Additionally, there should be sufficient flexibility in staff time to meet the increasing demands of the production department, whether such input is required in the form of training, or in an advisory or monitoring capacity.

REFERENCES

Anonymous. (1971). Microbial contamination of medicines administered to hospital patients: report by Public Health Laboratory Working Party. *Pharm. J.,* 207, 96–99.

Anonymous. (1982). *The Bacteriological Examination of Drinking Water Supplies.* Her Majesty's Stationery Office, London.

Anonymous. (1987). Guidelines on Sterile Drug Products Produced by Aseptic Processing. U.S. Food and Drug Administration, Rockville, MD.

Anonymous. (1996a). *Microbial Quality Management Limits and Guidelines.* Cosmetics, Toiletry and Perfumery Association, London.

Anonymous. (1996b). *Sterilization of Medical Devices—Estimation of the Population of Micro-organisms on Product Part 1. Requirements. BS EN 1174-1:1996.* British Standards Institute, London.

Anonymous. (1997a). *Sterilization of Medical Devices—Estimation of the Population of Micro-organisms on Product Part 2. Guidance. BS EN 1174-2:1997.* British Standards Institute, London.

Anonymous. (1997b). *Sterilization of Medical Devices—Estimation of the Population of Micro-organisms on Product Part 3. Guide to the Methods of Validation of Microbiological Techniques. BS EN 1174-3:1997.* British Standards Institute, London.

Anonymous. (2004). Guidance on Sterile Drug Products Produced by Aseptic Processing. U.S. Food and Drug Administration, Rockville, MD.

Baines, A. (2000). Endotoxin testing. In *Handbook of Microbiological Quality Control: Pharmaceuticals and Medical Devices*. Baird, R.M., Hodges, N.A., and Denyer, S.P., Eds., Taylor & Francis, London, pp. 144–167.

Baird, R.M. (1985). Microbial contamination of pharmaceutical products made in a hospital pharmacy: a nine year survey. *Pharm. J.*, 231, 54–55.

Baird, R.M. (2000a). Sampling: principles and practice. In *Handbook of Microbiological Quality Control: Pharmaceuticals and Medical Devices*. Baird, R.M., Hodges, N.A., and Denyer, S.P., Eds., Taylor & Francis, London, pp. 38–53.

Baird, R.M. (2000b). Culture media used in pharmaceutical microbiology. In *Handbook of Microbiological Quality Control: Pharmaceuticals and Medical Devices*. Baird, R.M., Hodges, N.A., and Denyer, S.P., Eds., Taylor & Francis, London, pp. 22–36.

Baird. R.M., Barnes, E.M., Corry, J.E.L., Curtis, G.D.W., and Mackey, B.M., Eds. (1985). Quality assurance and quality control of microbiological culture media. *Int. J. Food Microbiol.*, 2, 1–136.

Baird, R.M., Corry, J.E.L., and Curtis, G.D.W., Eds. (1987). Pharmacopoeia of culture media for food microbiology. *Int. J. Food Microbiol.*, 5, 187–300.

Barer, M.R., Gribbon, L.T., Harwood, C.R., and Nwoguh, C.E. (1993). The viable but non-culturable hypothesis and medical bacteriology. *Rev. Med. Microbiol.*, 4, 183–191.

Barer, M.R. (1997). Viable but non-culturable and dormant bacteria: time to resolve an oxymoron and a misnomer. *J. Med. Microbiol.*, 46, 629–631.

British Pharmacopoeia. (1932). *British Pharmacopoeia*. 6th ed., Appendix 16. Constable, London.

British Pharmacopoeia. (2005). *British Pharmacopoeia*. Vol. IV, Appendix XIVC, XIVD, XVIA, XVIB. The Stationery Office, London.

British Standards Institution. (1972). *British Standard 6000—The Use of BS 6001, Sampling Procedures and Tables for Inspection by Attributes*. British Standards Institution, London.

British Standards Institution. (1972). *British Standard 6001—Sampling Procedures and Tables for Inspection by Attributes*. British Standards Institution, London.

British Standards Institution. (1979). *British Standard 6002—Sampling Procedures and Charts for Inspection by Variables for Per Cent Defective*. British Standards Institution, London.

British Standards Institution. (1989). *British Standard 5295—Environmental Cleanliness in Enclosed Spaces*. British Standards Institution, London.

Brown, M.R.W. and Gilbert, P. (1977). Increasing the probability of sterility of medicinal products. *J. Pharm. Pharmacol.*, 29, 517–523.

Corry, J.E.L., Curtis, G.D.W., and Baird, R.M., Eds. (1995). *Culture Media for Food Microbiology*. Elsevier, Amsterdam.

Corry, J.E.L., Curtis, G.D.W., and Baird, R.M., Eds. (2003). *Handbook of Culture Media for Food Microbiology*. Elsevier, Amsterdam.

Cowan, S.T. (1993). *Cowan and Steel's Manual for the Identification of Medical Bacteria*. 3rd ed. Cambridge University Press, Cambridge.

European Committee for Standardization. (1993a). *EN 46001—Specification for Application of EN 29001 (BS 5750: Part 1) to the Manufacture of Medical Devices*. Comité Européen de Normalisation, Brussels.

European Committee for Standardization. (1993b). *EN 46002—Specification for Application of EN 29002 (BS 5750: Part 2) to the Manufacture of Medical Devices*. Comité Européen de Normalisation, Brussels.

European Committee for Standardization. (1994). *EN 724—Guidance on the Application of EN 29001 and EN 46001 and of EN 29002 and EN 46002 for Non-active Medical Devices.* Comité Européen de Normalisation, Brussels.

European Pharmacopoeia. (2004). *European Pharmacopoeia.* 5th ed. Council of Europe, Strasbourg.

Friberger, P. (1987). A new method of endotoxin determination. *ICPR,* July/August, 34–41.

Gibbs, R.A. and Hayes, C.R. (1988). The use of R2A medium and the spread plate method of enumeration of heterotrophic bacteria in drinking water. *Lett. Appl. Microbiol.,* 6, 19–21.

Hodges, N.A. (2000). Pharmacopoeial methods for the detection of specified micro-organisms. In *Handbook of Microbiological Quality Control: Pharmaceuticals and Medical Devices.* Baird, R.M., Hodges, N.A., and Denyer, S.P., Eds., Taylor & Francis, London, pp. 86–106.

Hoxey, E. (1993). Validation of methods for bioburden estimation. In *Sterilization of Medical Products,* Morrissey, R.F., Ed., PolyScience, Movin Heights, Canada.

Hurst, A. (1977). Bacterial injury: a review. *Canad. J. Microbiol.,* 23, 935–944.

Kallings, L.O., Ringertz, O., Silverstolpe, L., and Emerfeldt, F. (1966). Microbiological contamination of medicinal preparations *Acta Pharm. Suec.,* 3, 219–230.

Ljungqvist, B. and Reinmüller, B. (1998). Active sampling of airborne viable particles in controlled environments: a comparative study of common instruments. *Europ. J. Parent. Sci.,* 3, 59–62.

Lush, M. (2000). Microbiological hazard analysis and audit. In *Handbook of Microbiological Quality Control: Pharmaceuticals and Medical Devices.* Baird, R.M., Hodges, N.A., and Denyer, S.P., Eds., Taylor & Francis, London, pp. 144–167.

MCA (2002). Rules and Guidance for Pharmaceutical Manufacturers and Distributors. The Stationery Office, London.

Meynell, G.G. and Meynell, E., Eds. (1970). In *Theory and Practice in Experimental Bacteriology.* 2nd ed. Cambridge University Press, Cambridge.

Millar, R. (2000). Enumeration of micro-organisms: In *Handbook of Microbiological Quality Control: Pharmaceuticals and Medical Devices.* Baird, R.M., Hodges, N.A., and Denyer, S.P., Eds., Taylor & Francis, London, pp. 54–68.

Miles, A.A., Misra, S.S., and Irwin, J.O. (1938). The examination of the bacteriological power of blood. *J. Hyg. (Camb.),* 38, 732–749.

Mossel, D.A.A. (1982). *Microbiology of Foods. The Ecological Essentials of Assurance and Assessment of Safety and Quality.* 3rd ed. University of Utrecht, Utrecht, the Netherlands, pp. 89–97.

Mossel, D.A.A. and Corry, J.E.L. (1977). Detection and enumeration of sublethally injured pathogenic and index bacteria in foods and water processed for safety. *Alimenta,* 19–34.

Mossel, D.A.A., van Rossem, F., Koopmans, M., Henricks, M., Verouden, M., and Eelderink, I. (1980). Quality control of solid culture media: a comparison of the classical and the so-called ecographic technique. *J. Appl. Bact.,* 49, 439–454.

National Measurement Accreditation Service. (1992). *NIS 31—Accreditation for Microbiology Testing.* NAMAS Executive, London.

Newby, P.J. (2001). Viable but non-culturable (VBNC) microorganisms in the pharmaceutical industry—their significance and detection. *Eur. J. Parent. Sci.,* 6, 125–129.

Parenteral Society. (1993). *The Use of Process Simulation Tests in the Evaluation of Processes for the Manufacture of Sterile Products.* Technical monograph no. 4. The Parenteral Society, Swindon, U.K.

Parenteral Society. (2002). *Environmental Contamination Control Practice*. Technical monograph no. 2. The Parenteral Society, Swindon, U.K.

Ray, B. (1979). Methods to detect stressed micro-organisms. *J. Food Protect.*, 42, 346–355.

Reasoner, D.J. and Geldreich, E.E. (1985). A new medium for the enumeration and subculture of bacteria for potable water. *Appl. Env. Microbiol.*, 49, 1–7.

Russell, M. (1996). Microbiological control of raw materials. In *Microbial Quality Assurance in Cosmetics, Toiletries and Non-sterile Pharmaceuticals*. Baird, R.M. and Bloomfield, S.F., Eds. Taylor & Francis, London, pp. 31–47.

Russell, M.P., Purdie, R.M., Goldsmith, J.A., and Phillips, I. (1984). Computer-assisted evaluation of microbiological environmental control data. *J. Parent. Sci. Technol.*, 38, 98–102.

United States Pharmacopeia. (1936). *United States Pharmacopeia 11*. Mack Printing Co., Easton, PA.

United States Pharmacopeia. (2004). *United States Pharmacopeia 27—National Formulary 22*. U.S. Pharmacopeial Convention, Rockville, MD.

Van Doorne, H., Baird, R.M., Hendriksz, D.T., van der Kreek, D.M., and Pauwels, H.P. (1981). Liquid modification of Baird-Parker's medium for the selective enrichment of *Staph. aureus. Anton van Leeuwenhoek*, 47, 267–278.

9 Monitoring Microbiological Quality: Application of Rapid Microbiological Methods to Pharmaceuticals

Stephen P. Denyer

CONTENTS

9.1 INTRODUCTION

Traditional microbiological methods of detection, enumeration, and identification are generally time-consuming and labor-intensive. These practical considerations often limit the extent to which microbiological tests are routinely applied both at the formulation development stage (i.e., preservative screening) and for microbiological quality assurance (MQA). In the latter instance, the inevitable time delay associated with incubation often determines that MQA data are only of retrospective value. Modern pharmaceutical production and economic pressures can no longer accommodate this delay (Newby 2000). Considerable benefit would therefore be gained from the introduction of suitable, more rapid methods of microbiological analysis to the pharmaceutical sector. The typical requirements for such rapid methods are summarized in Table 9.1.

TABLE 9.1
Features of an Ideal Rapid Method for Application to Pharmaceutical Production

Rapid
Sensitive
Broad spectrum detection
Potential for specificity (identification)
Viability assessment
Simple
Potential for automation
Compatible with sample matrices

9.2 RAPID METHODS AVAILABLE

The development of rapid methods has been largely fed by the food, dairy, water, and medical diagnostics industries and has resulted in a diverse range of methods (Table 9.2) not all necessarily suited to pharmaceutical application. Their means of detection may be direct, in which individual microorganisms or populations of organisms are directly observed, or indirect, whereby microbial metabolism, metabolites, or components are monitored. Some methods may be highly developed with extensive equipment and information support, and others can still be considered to be at relatively early stages of research or currently developed for only a narrow application range. Only a few appear able to meet the challenges of pharmaceutical microbiology (Newby 2000; see also Section 9.3). It is also important to remember that the term "rapid" is variously applied to techniques of 5 min to 24 h duration, the definition often reflecting the expectations of the user. Furthermore, a method that may be deemed rapid in applications with high bioburden may require an extended enrichment period in situations of lower contamination.

Useful discussion and comparison of the principal methods can be found in the following works: general aspects (Jarvis and Easter 1987, Balows et al. 1989, Stannard et al. 1989, Blackburn 1993, Watling and Leech 1996, Stewart 1997, Geis 2006); adenosine triphosphate (ATP) bioluminescence (Jago et al. 1989, Stanley et al. 1989, Stewart et al. 1989, Stewart 1990, Stewart 1997); fluorescent labeling (Pettipher 1983, Hutcheson et al. 1988, Rodrigues and Kroll 1988, Rodrigues and Kroll 1990, Diaper and Edwards 1994, Nebe-von Caron et al. 1998, Newby 2000); electrical resistance (Baynes et al. 1983, Firstenberg-Eden and Eden 1984, Owens and Wacher-Viveros 1986, Silley and Forsythe 1996, Newby 2000); enzyme monitoring (Kroll and Rodrigues 1986, Watling and Leech 1996, Newby 2000); *Limulus* amoebocyte lysate (Jorgensen and Alexander 1981, Bussey and Tsuji 1984, Baines 2000); nucleic acid probes (Jordan 2000, Newby 2000, Dunsmoor et al. 2001); phage-interaction technology (Wolber and Green 1990, Turpin et al. 1993, Stewart et al. 1996, Stewart et al. 1998, Mole et al. 1999, Wu et al. 2001); and carbon dioxide radiometry (Cutler et al. 1989). In addition to the methods summarized in Table 9.2,

TABLE 9.2
A Selection of Rapid Microbiological Methods

Method	Detection Principle	Level of Development for Detection/ Identification
	Direct	
Fluorescent labeling (chemiluminescence)	Stain microorganisms using a viability-indicating fluorophore; direct enumeration, usually after filter capture, by light excitation (epifluorescent microscopy or laser scanning) and image analysis.	+/−
	Indirect	
ATP bioluminescence	Light emission from microbial ATP by luciferin/luciferase reaction. Amenable to amplification by intracellular adenylate kinase.	+/−
Carbon dioxide detection	Monitoring of microbial metabolism using ^{14}C-radiolabeled substrate to produce ^{14}C-labeled carbon dioxide. Infrared CO_2 detection offers a more acceptable substitute.	(+)/−
Chromatography and spectrophotometry	Detection of microbial metabolites and cellular components; gas chromatographic analysis of microbial fatty acid has been employed in identification.	−/(+)
Dye reduction	Monitoring microbial metabolism of specified substrates by color changes in redox dyes; can form the basis of identification profiles.	−/+
Electrical resistance	Measurement of electrical changes (conductance, impedance) in specialized media due to microbial growth; enumeration based on time to exceed a specified detection level.	−/+
Enzyme monitoring	Detection of microbial enzymes. By using appropriate substrates can form the basis of identification profiles.	(+)/+
Limulus amoebocyte lysate	Detection of (principally) Gram-negative bacterial lipopolysaccharide by gelation or colorimetric reaction	+/−
Nucleic acid probes	Labeled DNA or RNA probe hybridization to specific target sequences. Amplification of target by the polymerase chain reaction (PCR) increases sensitivity; competitive quantitative PCR offers enumeration.	(+)/(+)
Phage-interaction technology	Host-specific bacteriophage infects target cells leading to phage DNA replication. Detection by expression of new protein (using recombinant phage) or cell lysis.	(+)/(+)

+, highly developed; (+), moderate level of development; −, early stages of development or development limited to a narrow range of applications.

other techniques have been investigated but have received only modest development. These include electrochemical methods (Patchett et al. 1989); electronic particle counting (Kubitschek 1969); microcalorimetry (Forrest 1972, Beezer 1980, Watling and Leech 1996); and biophotometry (Thomas et al. 1985). For practical details of methods potentially applicable to the examination of microorganisms attached to medical devices, Denyer et al. (1993) should be consulted.

We now recognize that many different types of bacteria, while remaining physiologically active, can enter periods of nonculturability: in this form they are termed viable but nonculturable [(VNC) Colwell 1987, Kell et al. 1998, McDougald et al. 1998]. This may be an adaptive response to inimical environments; there is even evidence that this characteristic may be the dominant form in some environmental niches (Bloomfield et al. 1998). VNC organisms are theoretically capable of product spoilage and may be a potential infectious threat (Colwell et al. 1996, Rahman et al. 1996). It is perhaps reassuring therefore to discover that direct fluorescent staining (labeling) techniques offer a suitable approach to the detection of VNC organisms (Kawai et al. 1999). The *ChemScan*® process (laser scanning cytometry, section 9.3) routinely shows water bioburdens in excess of those determined by conventional culture, indicative of an otherwise undisclosed VNC population (e.g., Wallner et al. 1997).

9.3 USE OF RAPID METHODS IN PHARMACEUTICALS

From a pharmaceutical perspective, the principal areas in which rapid methods may find application are given in Table 9.3. A method may be required to provide quantitative or qualitative evidence of microbial presence (survival), some mechanism of contamination tracking, or to offer rapid confirmation of the absence of microorganisms. Few methods show complete promise in their range and relevance of reported applications (Table 9.4); some of the practical implications of their use are considered by Newby (2000) in the companion *Handbook*.

9.3.1 PRODUCT QUALITY ASSESSMENT

The pharmaceutical industry has tended to be conservative in its approach to rapid methods for assessment of product quality, largely because of the regulatory constraints

---

TABLE 9.3
Principal Areas of Application for Rapid Microbiological Methods in Pharmaceuticals

Area	Application
Product quality assessment	Microbial limit tests for raw materials and final nonsterile products (includes total viable count and detection of pathogens)
	Sterility tests
Process hygiene	In-process samples
	Site hygiene
	Air quality
Preservative efficacy	Screening potential preservatives
	Examining the influence of formulation on preservative behavior
	Challenge testing
Sterilizer testing	Biological indicators

TABLE 9.4
Some Examples of Rapid Methods Applied to the Detection of Microorganisms in Pharmaceuticals and Related Environments

Method	Sensitivity (CFU)	Limitations	Applications	References
ATP bioluminescence	>10^2 yeasts. >10^2–10^3 bacteria, reduced to 1–10 range with enrichment or an MPN-based approach.	Presence of high levels of nonmicrobial ATP. Interfering factors quenching light or adversely affecting luciferase reaction.	Cosmetics/toiletries Intravenous fluids Medical devices Packaging materials Sterility testing Surface hygiene Water	Neilsen and van Dellen (1989), Watling and Leech (1996), Anonymous (1996) Bopp and Wachsmith (1981), Anderson et al. (1986) Wassall et al. (1997) Senior et al. (1989) Bussey and Tsuji (1986) Blackburn et al. (1989), Anonymous (2001) Webster (1986), Woolridge (1989), Tanaka et al. (1997), Newby (2000)
Electrical resistance	Threshold for detection ~10^6 mL^{-1}.	Narrow spectrum of detectable organisms without careful media selection; may be overcome with indirect impedance method.	Cosmetics Preservative testing Sterility testing Toiletries Water	Kahn and Firstenberg-Eden (1984), Kaiserman et al. (1989) Connolly et al. (1983, 1994) Dal Maso (1998) Watling and Leech (1996) Wilkins et al. (1980)
Fluorescent labeling: DE(F)T	Air, >10^5; liquid, generally 10^3–10^4 mL^{-1} but down to 25 mL^{-1}; liquid (+enrichment), 6 organisms irrespective of sample volume.	Cannot be applied to highly viscous or particulate materials. Direct correlation with viable count not always possible.	Air Intravenous fluids Medical devices Preservative testing Surface hygiene Water	Palmgren et al. (1986) Denyer and Ward (1983), Denyer and Lynn (1987), Denyer et al. (1989) Ladd et al. (1985), Bridgett et al. (1993) Connolly et al. (1993) Holah et al. (1988) Mittelman et al. (1983, 1985), Newby (1991), Kawai et al. (1999)

(continued)

TABLE 9.4 (continued)
Some Examples of Rapid Methods Applied to the Detection of Microorganisms in Pharmaceuticals and Related Environments

Method	Sensitivity (CFU)	Limitations	Applications	References
Laser scanning cytometry	Liquid, to single organism level; in flow cytometry, ~50 mL^{-1}.	Spores must be germinated. Viable nonculturable (VNC) organisms may be detected requiring reappraisal of limits. Nonfilterable products need to be tested by flow cytometry.	Cosmetics/toiletries Sterility testing Water	Newby (2000) Anonymous (1995) Wallner et al. (1997), Gapp et al. (1999), Reynolds and Fricker (1999), Newby (2000)
Nucleic acid probes	Better than 0.1 using PCR amplification	Interference from formulation excipients. Nonviable organisms are also detected limiting utility. Current development focused on specific organisms.	Air Blood products Contamination tracing Water	Alvarez et al. (1994) Jordan (2000) Newby (2000) Atlas (1991), Maiwald et al. (1994)

Note: DE(F)T— direct epifluorescent (filtration) technique.

imposed upon these products. For this reason, much of the information accumulated in Table 9.4 is drawn from related industries, but using comparable products and environments. This table clearly demonstrates the current low probability that any single method will satisfy the requirements for all types of pharmaceutical application, although some manufacturers are now seeking a collection of related products to achieve this (Newby 2000). It is unlikely that any rapid method can be immediately applied in a wide range of situations without first undertaking extensive protocol development. The sensitivity of all methods can be enhanced by sample enrichment but this will lead to an inevitable increase in analysis time; additionally, contaminants grow at different rates and this may result in a substantially different microbial flora from the original sampled product. In sterility testing, where the bioburden (if any) is likely to be low, rapid methods generally require sample enrichment or extended incubation to reach the microbial levels required for detection.

9.3.2 PROCESS HYGIENE

In general, examination of in-process product samples can utilize the same methods of rapid analysis as raw materials and final product (see Table 9.4). Surface hygiene assessment, using appropriate swabbing or surface sampling techniques, may require an enrichment period if low counts are expected; similarly, large volumes of process water or air may need to be sampled and concentrated by filtration to ensure a sufficient microbial burden before examination.

9.3.3 PRESERVATIVE EFFICACY

The official preservative efficacy test methods (see Chapter 17) require challenge periods of up to 28 days and the introduction of rapid methods in this situation would confer no meaningful benefit. Where rapid methodology can have an important role to play is in the rapid examination of several candidate preservative systems (and their possible permutations of concentration and combination) for use in new or developing formulations. Here, kinetic data from D-value determinations (see Chapter 18) or estimation of growth-inhibitory concentrations can quickly provide a useful indication of preservative or formulation incompatibilities and can be used to compare the relative merits of potential preservative systems. Rapid methods have been applied to preservative evaluation (see, for example, Denyer 1990; Connolly et al. 1993, 1994). It is important to distinguish between those methods used primarily to explore bacteriostatic behavior and those able to examine bactericidal activity. In the latter instance, enrichment or extended incubation times may be necessary to detect low numbers of survivors, thereby extending the overall detection time.

In a product challenge designed to explore the capacity of a preservative system to withstand repeated microbial insults or to study the ability of spoilage organisms to survive and grow, kinetic information is of less importance and the detection methods summarized in Table 9.4 are potentially applicable.

9.3.4 STERILIZER TESTING

Sterilization protocols require regular microbiological validation; for some processes, continual efficacy monitoring with biological indicators is necessary (see

Chapter 10). To ensure that every reasonable opportunity is given for the recovery of stressed indicator spores, a long incubation period, often in excess of one week, is allowed before assurance of sterilizer efficacy can be given; this provides little opportunity for early detection of partial sterilizer failure. A commercial detection system in which a spore enzyme, α-glucosidase (reflective of spore viability), converts a nonfluorescent substrate into a fluorescent product within an hour offers one solution. Other approaches examined include ATP bioluminescence to detect spores surviving suboptimal sterilization processes (Webster et al. 1988) and *in vivo* bioluminescence as a reporter of recombinant spore viability (Stewart et al. 1989).

9.4   OVERVIEW

There are several well-developed rapid microbiological methods now becoming available that may have useful applications in pharmaceuticals; of these, ATP bioluminescence, fluorescent labeling, electrical resistance, and nucleic acid probes appear among the most promising (Table 9.5). Inevitably, no single method will satisfy easily all requirements, and further development will be needed to adapt them to the specific demands of the pharmaceutical situation. In this context, it is encouraging to see the developments in ATP bioluminescence offering increased sensitivity [adenylate kinase amplification (Corbitt et al. 2000)], potential specificity [phage lysins (Stewart 1997), phage lysis (Wu et al. 2001)], and internal calibration against excipient effects [caged ATP (Calvert et al. 2000)], while proposed developments using fluorescently labeled antibodies may offer specificity to laser scanning cytometry (Reynolds and Fricker 1999). It is particularly pleasing to see that in the reference

---

TABLE 9.5
Current Performance of Selected Rapid Methods against the Ideal Features Listed in Table 9.1

Assay Feature	Performance of Rapid Method			
	ATP	FL	ER	NAP
Rapidity	30 min–24 h	~ 2 h	1–24 h	4–5 h (with PCR)
Sensitivity (CFU)	100–1	1–200	10^6–10	0.1
Broad spectrum	+	+	+	−
Specificity	−	−	(+)	+
Viability assessment	+	+	+	−
Simplicity	+	+	+	(+)
Automation	+	(+)	+	(+)
Sample compatibility	(+)	(+)	+	(+)

Note: ATP—ATP bioluminescence; FL— fluorescent labeling; ER—electrical resistance; NAP—nucleic acid probe.

+, good; (+), moderate performance or potential; −, currently limited performance or potential.

of Wu et al. (2001), two technologies—phage interaction and ATP bioluminescence—come together in one application. The revolution in applied DNA technologies, particularly driven by medical diagnostics, offers major promise in the future for miniaturized nucleic acid–based detection systems; the recognition of gene families associated with particular microbial characteristics (Stewart 1997) offers a route to the detection of specific deteriogens and pathogens.

In choosing to employ rapid methods, the pharmaceutical microbiologist should examine their prospective performances against the specific requirements for that sector. Useful guidance is available in the PDA Technical Report 33 (PDA 2000) and the European Pharmacopoeia (2005). Some methods may require expensive equipment and offer full automation, and others represent only a small investment. The regulatory view of these methods is changing (Newby 2000), but it will still be up to the microbiologist to demonstrate that the method chosen is fit for the purpose intended.

REFERENCES

Alvarez, A.J., Buttner, M.P., Toranzos, G.A., Dvorsky, E.A., Toro, A., Heikes, T.B., Mertikas-Pifer, L.E., and Stetzenbach, L.D. (1994). Use of solid-phase PCR for enhanced detection of airborne microorganisms. *Appl. Env. Mic.,* 60, 374–376.

Anderson, R.L., Highsmith, A.K., and Holland, B.W. (1986). Comparison of standard pour plate procedure and the ATP and *Limulus* amoebocyte lysate procedures for the detection of microbial contamination in intravenous fluids. *J. Clin. Microbiol.,* 23, 465–468.

Anonymous. (1995). Multi-site evaluation of *ChemScan*® system for real-time water sterility testing. *ChemScan* product literature, progress report c541/95, Chemunex, Cambridge.

Anonymous. (1996). Using Bioluminescence for end-product release in the cosmetics industry. *Microbiology Europe,* 4, 22.

Anonymous. (2001). Pall Gelman Laboratory's Bioprobe luminometer for ATP testing of microbial contamination. PCT 2001 Event Guide, Coventry, February 27–28, pp. 26–27.

Atlas, R.M. (1991). Environmental applications of the polymerase chain reaction. *ASM News,* 57, 630–632.

Baines, A. (2000). Endotoxin testing. In *Handbook of Microbiological Quality Control: Pharmaceuticals and Medical Devices.* Baird, R.M., Hodges, N.A., and Denyer, S.P., Eds. Taylor & Francis, London, pp. 144–167.

Balows, A., Tilton, R.C., and Turano, A. (1989). *Rapid Methods and Automation in Microbiology and Immunology.* Brixia Academic Press, Brescia.

Baynes, N.C., Comrie, J., and Prain, J.H. (1983). Detection of bacterial growth by the Malthus conductance meter. *Med. Lab. Sci.,* 40, 149–158.

Beezer, A.E. (1980). *Biological Microcalorimetry.* Academic Press, London.

Blackburn, C. de W., Gibbs, P.A., Roller, S.D., and Johal, S. (1989). Use of ATP in microbial adhesion studies. In *ATP Luminescence: Rapid Methods in Microbiology.* Stanley, P.E., McCarthy, B.J., and Smither, R., Eds. SAB Technical Series 26. Blackwell Scientific Publications, Oxford, pp. 145–152.

Blackburn, C. de W. (1993). Rapid and alternative methods for the detection of salmonellas in foods. *J. Appl. Bact.,* 75, 199–214.

Bloomfield, S.F., Stewart, G.S.A.B., Dodd, C.E.R., Booth, I.R., and Power, E.G.M. (1998). The viable but non-culturable phenomenon explained? *Microbiology,* 144, 1–3.

Bridgett, M.J., Davies, M.C., Denyer, S.P., and Eldridge, P.R. (1993). *In vitro* assessment of bacterial adhesion to Hydromer-coated cerebrospinal fluid shunts. *Biomaterials,* 14, 184–188.

Bopp, C.A. and Wachsmith, I.K. (1981). Luciferase assay to detect bacterial contamination of intravenous fluids. *Am. J. Hosp. Pharm.,* 38, 1747–1750.

Bussey, D.M. and Tsuji, K. (1984). Optimization of chromogenic *Limulus* amebocyte lysate (LAL) assay for automated endotoxin detection. *J. Parent. Sci. Technol.,* 38, 228–233.

Bussey, D.M. and Tsuji, K. (1986). Bioluminescence of USP sterility testing of pharmaceutical suspension products. *Appl. Environ. Microbiol.,* 51, 349–355.

Calvert, R.M., Hopkins, H.C., Reilly, M.J., and Forsythe, S.J. (2000). Caged ATP—an internal calibration method for ATP bioluminescence assays. *Lett. Appl. Mic.,* 30, 223–227.

Colwell, R.R. (1987). From counts to clones. *J. Appl. Bact. Symp. Suppl.,* 69, 15–65.

Colwell, R.R., Brayton, P., Herrington, D., Tall, B., Huq, A., and Levine, M.M. (1996). Viable but non-culturable *Vibrio cholerae* 01 revert to a cultivable state in the human intestine. *World J. Microbiol. Biotechnol.,* 12, 28–31.

Connolly, P., Bloomfield, S.F., and Denyer, S.P. (1993). A study of the use of rapid methods for preservative efficacy testing of pharmaceuticals and cosmetics. *J. Appl. Bact.,* 75, 456–462.

Connolly, P., Bloomfield, S.F., and Denyer, S.P. (1994). The use of impedance for preservative efficacy testing of pharmaceutical and cosmetic products. *J. Appl. Bact.,* 76, 68–75.

Corbitt, A.J., Benman, N., and Forsythe, S.J. (2000). Adenylate kinase amplification of ATP bioluminescence for hygiene monitoring in the food and beverage industry. *Lett. Appl. Mic.,* 30, 443–447.

Cutler, R.R., Wilson, P., and Clarke, F.V. (1989). Evaluation of a radiometric method for studying bacterial activity in the presence of antimicrobial agents. *J. Appl. Bact.,* 66, 515–521.

Dal Maso, G. (1998). The use of the bioMerieux bactometer for sterility testing of pharmaceutical products. *Annali di Microbiologica ed Enzymologia,* 48, R7–R13.

Denyer, S.P. (1990). Monitoring microbiological quality: application of rapid microbiological methods to pharmaceuticals. In *Guide to Microbiological Control in Pharmaceuticals.* Denyer, S. and Baird, R., Eds. 1st ed. Ellis Horwood, Chichester, U.K., pp. 146–156.

Denyer, S.P. and Lynn, R. (1987). A sensitive method for the rapid detection of bacterial contaminants in intravenous fluids. *J. Parent. Sci. Technol.,* 41, 60–66.

Denyer, S.P. and Ward, K.H. (1983). A rapid method for the detection of bacterial contaminants in intravenous fluids using membrane filtration and epifluorescence microscopy. *J. Parent. Sci. Technol.,* 37, 156–158.

Denyer, S.P., Lynn, R.A.P., and Pover, P.S. (1989). Medical and pharmaceutical applications of the direct epifluorescent filter technique (DEFT). In *Rapid Microbiological Methods for Foods, Beverages and Pharmaceuticals.* Stannard, C.J., Petitt, S.B., and Skinner, F.A., Eds. SAB Technical Series 25. Blackwell Scientific Publications, Oxford, pp. 59–71.

Denyer, S.P., Gorman, S.P., and Sussman, M., Eds. (1993). *Microbial Biofilms: Formation and Control.* SAB Technical Series 30. Blackwell Scientific Publications, Oxford.

Diaper, J.P. and Edwards, C. (1994). The use of fluorogenic esters to detect viable bacteria by flow cytometry. *J. Appl. Bact.,* 77, 221–228.

Dunsmoor, C., Sanders, J., Ferrance, J., and Landers, J. (2001). Microchip electrophoresis: an emerging technology for molecular diagnostics. *PharmaGenomics,* August, 39–45.

European Pharmacopoeia (2005). 5th edition, supplement 5.5. General Text 5.1.6. Alternative methods for control of microbiological quality. In *European Pharmacopoeia*, Council of Europe, Strasbourg, pp. 4131–4142.

Firstenberg-Eden, R. and Eden, G. (1984). *Impedance Microbiology*. Research Studies Press, Letchworth.

Forrest, W.W. (1972). Microcalorimetry. In *Methods in Microbiology*. Norris, J.R. and Ribbons, D.W., Eds. Vol. 6B. Academic Press, London, pp. 285–318.

Gapp, G., Guyomard, S., Nabet, P., and Scouvart, J. (1999). Evaluation of the applications of a system for real-time microbial analysis of pharmaceutical water systems. *Eur. J. Parent. Sci.,* 4, 131–136.

Geis, P.A. (2006). Evolution of cosmetic microbiology beyond agar plating. In *Cosmetic and Drug Microbiology*. Orth, D.S., Kabara, J.J., Denyer, S.P., and Tan, S.K., Eds. Informa Healthcare, New York, pp. 327–343.

Holah, J.T., Betts, R.P., and Thorpe, R.H. (1988). The use of direct epifluorescent microscopy (DEM) and the direct epifluorescent filter technique (DEFT) to assess microbial populations on food contact surfaces. *J. Appl. Bacteriol.,* 65, 215–221.

Hutcheson, T.C., McKay, T., Farr, L., and Seddon, B. (1988). Evaluation of the stain Viablue for the rapid estimation of viable yeast cells. *Lett. Appl. Microbiol.,* 6, 85–88.

Jago, P.H., Simpson, W.J., Denyer, S.P., Evans, A.W., Griffiths, M.W., Hammond, J.R.M., Ingram, T.P., Lacey, R.F., Macey, N.W., McCarthy, B.J., Salusbury, T.T., Senior, S.S., Sidorowicz, S., Smither, R., Stanfield, G., and Stanley, P.E. (1989). An evaluation of the performance of ten commercial luminometers. *J. Biolumin. Chemilumin.,* 3, 131–145.

Jarvis, B. and Easter, M.C. (1987). Rapid methods in the assessment of microbiological quality: experiences and needs. *J. Appl. Bacteriol. Symp. Suppl.,* 63, 115S–126S.

Jordan, J.A. (2000). Real-time detection of PCR products and microbiology. *Trends in Microbiology: Special Issue. New Technologies for Life Sciences,* December, 61–66.

Jorgensen, J.H. and Alexander, G.A. (1981). Automation of the *Limulus* amoebocyte lysate test by using the Abbott MS-2 microbiology system. *Appl. Environ. Microbiol.,* 41, 1316–1320.

Kaiserman, J.M., Moral, J., and Wolf, B.A. (1989). A rapid impedimetric procedure to determine bacterial content in cosmetic formulations. *J. Soc. Cosmet. Chem.,* 40, 21–31.

Kawai, M., Yamaguchi, N., and Nasu, M. (1999). Rapid enumeration of physiologically active bacteria in purified water used in the pharmaceutical manufacturing process. *J. Appl. Mic.,* 86, 496–504.

Kell, D.B., Kaprelyants, A.S., Weichart, D.H., Harwood, C.R., and Barer, M.R. (1998). Viability and activity in readily culturable bacteria: a review and discussion of the practical issues. *Antonie van Leeuwenhoek,* 73, 169–187.

Khan, P. and Firstenberg-Eden, R. (1984). A new cosmetic sterility test. *Soap/Cosmet./Chem. Specialities,* 60, 46–48, 101.

Kroll, R.G. and Rodrigues, U.M. (1986). Prediction of the keeping quality of pasteurised milk by the detection of cytochrome c oxidase. *J. Appl. Bacteriol.,* 60, 21–27.

Kubitschek, H.E. (1969). Counting and sizing micro-organisms with the Coulter Counter. In *Methods in Microbiology*. Norris, J.R. and Ribbons, D.W., Eds. Vol. 1. Academic Press, London, pp. 593–610.

Ladd, T.I., Schmiel, D., Nickel, J.C., and Costerton, J.W. (1985). Rapid detection of adherent bacteria on Foley urinary catheters. *J. Clin. Mic.,* 21, 1004–1006.

Maiwald, M., Kissel, K., Srimuang, S., von Knebel Doeberitz, M., and Sonntag, H.-G. (1994). Comparison of polymerase chain reaction and conventional culture for the detection of legionellas in hospital water samples. *J. Appl. Bact.,* 76, 216–225.

McDougald, D., Rice, S.A., Weichart, D., and Kjelleberg, S. (1998). Nonculturability: adaptation or debilitation. *FEMS Microb. Ecol.,* 25, 1–9.

Mittelman, M.W., Geesey, G.G., and Hite, R.R. (1983). Epifluorescence microscopy: a rapid method for enumerating viable and non-viable bacteria in ultra-pure-water systems. *Microcontamination,* 1, 32–37, 52.

Mittelman, M.W., Geesey, G.G., and Platt, R.M. (1985). Rapid enumeration of bacteria in purified water systems. *Med. Dev. Diagnost. Indi.,* 7, 144–149.

Mole, R.J., Dhir, V.K., Denyer, S.P., and Stewart, G.S.A.B. (1999). Bacteriophage-based techniques for detection of food-borne pathogens. In *Encyclopaedia for Food Microbiology.* Patel, P., Ed. Academic Press, London, pp. 203–210.

Nebe-von Caron, G., Stephens, P., and Badtey, R.A. (1998). Assessment of bacterial viability status by flow cytometry and single cell sorting. *J. Appl. Mic.,* 84, 988–998.

Newby, P.J. (1991). Analysis of high-quality pharmaceutical-grade water by a direct epifluorescent filter technique microcolony method. *Lett. Appl. Mic.,* 13, 291–293.

Newby, P. (2000). Rapid methods for enumeration and identification in microbiology. In *Handbook of Microbiological Control: Pharmaceuticals and Medical Devices.* Baird, R.M., Hodges, N.A., and Denyer, S.P., Eds. Taylor & Francis, London, pp. 107–119.

Nielson, P. and van Dellen, E. (1989). Rapid bacteriological screening of cosmetic raw materials by using bioluminescence. *J. Assoc. Anal. Chem.,* 72, 708–711.

Owens, J.D. and Wacher-Viveros, M.C. (1986). Selection of pH buffers for use in conductimetric microbiological assays. *J. Appl. Bacteriol.,* 60, 395–400.

Palmgren, U., Ström, G., Blomquist, G., and Malmberg, P. (1986). Collection of airborne micro-organisms on Nucleopore filters, estimation and analysis—CAMNEA method. *J. Appl. Bacteriol.,* 61, 401–406.

Patchett, R.A., Kelly, A.F., and Kroll, R.G. (1989). Investigation of a simple amperometric electrode system to rapidly quantify and detect bacteria in foods. *J. Appl. Bacteriol.,* 66, 49–55.

PDA (2000). Technical Report No. 33. Evaluation, validation and implementation of new microbiological testing methods. *PDA J. Pharm. Sci. Technol.,* 54, 1–39.

Pettipher, G.L. (1983). *The Direct Epifluorescent Filtration Technique.* Research Studies Press, Letchworth.

Rahman, I., Shahamat, M., Chowdhury, M.A.R., and Colwell, R.R. (1996). Potential virulence of viable but non-culturable *Shigella dysenteriae* type 1. *Appl. Env. Microbiol.,* 62, 115–120.

Reynolds, D.T. and Fricker, C.R. (1999). Application of laser scanning for the rapid and automated detection of bacteria in water samples. *J. Appl. Mic.,* 86, 785–795.

Rodrigues, U.M. and Kroll, R.G. (1988). Rapid selective enumeration of bacteria in foods using a microcolony epifluorescence microscopy technique. *J. Appl. Bacteriol.,* 64, 65–78.

Rodrigues, U.M. and Kroll, R.G. (1990). Rapid detection of salmonellas in raw meats using a fluorescent antibody-microcolony technique. *J. Appl. Bacteriol.,* 64, 213–223.

Senior, P.S., Tyson, K.D., Parsons, B., White, R., and Wood, G.P. (1989). Bioluminescent assessment of microbial contamination on plastic packaging materials. In *ATP Luminescence: Rapid Methods in Microbiology.* Stanley, P.E., McCarthy, B.J., and Smither, R., Eds. SAB Technical Series 26. Blackwell Scientific Publications, Oxford, pp. 137–143.

Silley, P. and Forsythe, S. (1996). Impedance microbiology—a rapid change for microbiologists. *J. Appl. Bact.,* 80, 233–243.

Stanley, P.E., McCarthy, B.J., and Smither, R., Eds. (1989). *ATP Luminescence: Rapid Methods in Microbiology.* SAB Technical Series 26. Blackwell Scientific Publications, Oxford.

Stannard, C.J., Petitt, S.B., and Skinner, F.A., Eds. (1989). *Rapid Microbiological Methods for Foods, Beverages and Pharmaceuticals*. SAB Technical Series 25. Blackwell Scientific Publications, Oxford.

Stewart, G.S.A.B. (1990). *In vivo* bioluminescence: new potentials for microbiology. *Lett. Appl. Microbiol.,* 10, 1–8.

Stewart, G.S.A.B. (1997). Challenging food microbiology from a molecular perspective. *Microbiology,* 143, 2099–2108.

Stewart, G., Smith, T. and Denyer, S. (1989). Genetic engineering for bioluminescent bacteria: harnessing molecular genetics to provide revolutionary new methods for food microbiology. *Food Sci. Technol. Today,* 3, 19–22.

Stewart, G.S.A.B., Loessner, M.J., and Scherer, S. (1996). The bacterial *lux* gene bioluminescent biosensor revisited. *ASM News,* 62, 297–301.

Stewart, G.S.A.B., Jassim, S.A.A., Denyer, S.P., Newby, P., Linley, K., and Dhir, V.K. (1998). The specific and sensitive detection of bacterial pathogens within 4h using bacteriophage amplification. *J. Appl. Mic.,* 84, 777–783.

Tanaka, H., Shinji, T., Sawada, K. et al. (1997). Development and application of a bioluminescence ATP assay method for rapid detection of coliform bacteria. *Water Res.,* 31, 1918–1931.

Thomas, D.S., Henschke, P.A., Garland, B.A., and Tucknott, O.G. (1985). A microprocessor-controlled photometer for monitoring microbial growth in multi-welled plates *J. Appl. Bacteriol.,* 59, 337–346.

Turpin, P.E., Maycroft, K.A., Bedford, J., Rowlands, C.L., and Wellington, E.M.H. (1993). A rapid luminescent-phage based MPN method for the enumeration of *Salmonella typhimurium* in environmental samples. *Lett. Appl. Mic.,* 16, 24–27.

Wallner, G., Tillman, D., Harberer, K., Cornet, P., and Drocourt, J.L. (1997). The Chemscan system: a new method for rapid microbiological testing of water. *Eur. J. Parent. Sci.,* 2, 123–126.

Wassall, M.A. Santin, M., Isalberti, C., Cannas, M., and Denyer, S.P. (1997). Adhesion of bacteria to stainless steel and silver-coated orthopaedic fixation pins. *J. Biomed. Mater. Res.,* 36, 325–330.

Watling, E.M. and Leech, R. (1996). New methodology for microbiological quality assurance. In *Microbial Quality Assurance in Cosmetics, Toiletries and Non-Sterile Pharmaceuticals*. Baird, R.M. and Bloomfield, S.F., Eds. 2nd ed. Taylor & Francis, London, pp. 217–234.

Webster, W.A. (1986) cited in Olson, W.P. (1987). Sterility testing. In *Aseptic Pharmaceutical Manufacturing: Technology for the 1990's*. Olsen, W.P. and Groves, M.J., Eds. Interpharm Press, Englewood, CO, pp. 315–354.

Webster, J.J., Walker, B.G., Ford, S.R., and Leach, F.R. (1988). Determination of sterilization effectiveness by measuring bacterial growth in a biological indicator through firefly luciferase determination of ATP. *J. Biolumin. Chemilum.,* 2, 129–133.

Wilkins, J.R., Grana, D.C., and Fox, S.S. (1980). Combined membrane filtration-electrochemical microbial detection method. *Appl. Environ. Microbiol.,* 40, 852–853.

Wolber, P.K. and Green, R.L. (1990). Detection of bacteria by transduction of ice nucleation genes. *TIBTECH,* 8, 276–279.

Woolridge, C.A. (1989). ATP bioluminescence for microbial quality assurance of process water. In *ATP Luminescence: Rapid Methods in Microbiology*. Stanley, P.E., McCarthy, B.J., and Smither, R., Eds. SAB Technical Series 26. Blackwell Scientific Publications, Oxford, pp. 93–97.

Wu, Y., Brovko, L., and Griffiths, M.W. (2001). Influence of phage population on the phage-mediated biluminescent adenylate kinase (AK) assay for detection of bacteria. *Lett. Appl. Mic.,* 33, 311–315.

10 Principles of Sterilization

*Eamonn V. Hoxey, Nicolette Thomas,
and David J.G. Davies*

CONTENTS

10.1 INTRODUCTION

When the use of a medicinal product or a medical device breaches the body's natural defense against infection, it is necessary to ensure that the product is not contaminated by microorganisms; this is achieved by using a sterile product. Sterility is appropriate for medicinal products that are administered parenterally, for devices that are used to administer such products, and for invasive medical devices.

Sterility is defined as the absence of all viable life forms. The term is an absolute one and descriptions implying degrees of sterility are not only confusing, but erroneous. Sterilization is the process by which sterility is achieved, that is, the process of destroying or removing all viable life forms. In practice, this is achieved by either exposure to an inimical physical or chemical agent for a predetermined period or the physical removal of organisms. The inimical agents used are elevated temperature, ionizing radiation, or chemicals either as a gas, a gas plasma, or a liquid. Physical removal involves passing solutions or gases through a filter capable of retaining microorganisms. In all cases, it is necessary to prevent the product being recontaminated before it is used.

In order to discuss sterilization processes, it is necessary to define death as it applies to microorganisms. In practice, a microorganism is considered dead when it cannot be detected in culture media in which it previously has been shown to proliferate. Detection requires the production of a colony on the surface of solid medium or turbidity in liquid medium. A single organism must be able to proliferate through many generations to be detected and an organism that cannot reproduce or can only reproduce through a few generations would be classified as dead by this criterion. We do not have media capable of culturing all known organisms. Furthermore, organisms that have survived a potentially lethal process may have specific metabolic requirements, and if these are not known, it will not be possible to recover them in standard culture media. The absence of all viable life forms is therefore a negative state that can never be practically proved; the recognition of viable but nonculturable forms (see Chapter 9) adds further weight to the position.

A further problem in achieving sterility is that microorganisms exposed to a lethal agent do not all die at the same time. To a first approximation, the number of microorganisms decreases exponentially with the time of exposure; therefore, the absence of all viable organisms can only be guaranteed after infinite exposure to the agent. After a particular exposure time, even when viable organisms can no longer be detected, there will always be a finite probability of finding a viable organism. This probability decreases as the exposure time is increased, but never in real terms reaches zero. Sterility, the absence of all viable life forms, is therefore an absolute state, the achievement of which cannot be guaranteed absolutely. Nevertheless, the careful design of sterilization processes enables sterility in the absolute sense to be approached with a predicted probability of success.

The problems associated with the strict definition of sterility have led to the development of an approach that allows an acceptable sterilization process to be defined. In practical terms, sterility is expressed as a mathematical probability of a product item remaining contaminated with a microorganism after a defined sterilization process. This probability is often referred to as the sterility assurance level

(SAL). For medicinal products and medical devices, the designation "sterile" is generally applied to products that have been treated in such a manner that, on completion of the process, individual items have a probability of being contaminated with a single viable microorganism equal to or better than one in a million. This equates to having a SAL of 10^{-6}.

Inactivation of microorganisms by sterilizing agents involves the irreversible damage of essential molecules in the cell. Exposure to these agents is also likely to produce deleterious effects in the material to be sterilized. It is of particular importance that the sterilization process should not produce unacceptable changes in safety, quality, efficacy (for medicinal products), or performance (for medical devices). As a consequence, in some circumstances, it may be necessary to reach a compromise between achieving a desired SAL and not producing unacceptable effects upon the material to be sterilized.

10.2 MICROBIAL INACTIVATION

10.2.1 MICROBIAL INACTIVATION KINETICS

Our knowledge of microbial inactivation is derived from experiments in which pure cultures of microorganisms are exposed to lethal agents with the extent of treatment increasing incrementally. A population of a single microbial species is prepared and divided into a number of test pieces, called biological indicators, each containing an equal number of organisms. The biological indicator may be an ampoule containing a volume of microorganisms in suspension or a solid object onto which the microorganisms are deposited in suspension and then dried. One biological indicator remains unexposed to allow the initial number of microorganisms in each biological indicator to be determined. The remaining biological indicators are exposed to increasing extents of treatment with the lethal agent and, following treatment, the number of microorganisms surviving is determined. When plotted graphically, the data obtained from such an experiment typically take the form illustrated in Figure 10.1.

Figure 10.1 shows that not all the microorganisms die instantaneously and that the number of surviving microorganisms decreases with increasing exposure to the lethal agent. Furthermore, the curve is asymptotic to the x axis, making it impossible to predict the extent of treatment necessary to achieve sterility.

Alternatively, the data can be presented as a graph of either number of surviving microorganisms on a logarithmic scale or the logarithm of the number of surviving microorganisms against the extent of treatment on an arithmetic scale. This is illustrated in Figure 10.2. This type of "semilog" plot is referred to as a survivor curve.

The linear relationship of Figure 10.2 can be used to predict the number of survivors for any given treatment. When less than one surviving microorganism is predicted, this is considered as the region in which there is a defined probability of a microorganism surviving; thus, at the extent of treatment giving 10^{-1} survivors, a 1 in 10 chance of a microorganism surviving is predicted. Put another way, if 10 similar batches of organisms were treated to the same extent, then, on average, a single survivor in one batch would be found.

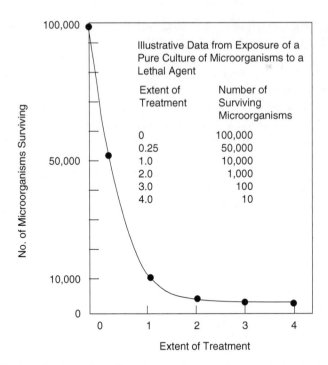

FIGURE 10.1 Graph of number of surviving microorganisms against extent of treatment with a sterilizing agent.

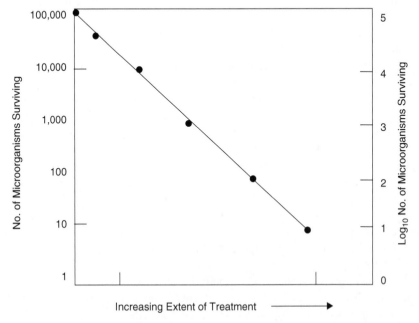

FIGURE 10.2 Graph of number of surviving microorganisms on a logarithmic scale against extent of treatment with a sterilizing agent.

The data represented in Figure 10.1 and Figure 10.2 may also be plotted as the surviving fraction of microorganisms at a given extent of treatment, that is, expressing the number of organisms surviving as a fraction of the initial number. This approach normalizes the data and allows separate experiments with different initial numbers of microorganisms to be directly compared. In all other respects, the survivor curves produced are identical.

The design of a sterilization process that will achieve a predetermined probability of occurrence of survivors depends upon a knowledge of the initial population of microorganisms in the product (the bioburden) and the kinetics of inactivation of such contaminating microorganisms when exposed to the lethal agent. As indicated earlier, in sterilization technology, the probability of a microorganism surviving a defined treatment has been termed the SAL and a value of 10^{-6} is generally accepted as the maximal value taken as the target for a sterilization process.*

A number of mathematical terms have been derived to describe microbial inactivation and to aid prediction of the SAL.

10.2.2 Microbial Inactivation Rate Constant (k)

Mathematically the inactivation process can be described in the same way as a first-order chemical reaction. A process where microbial inactivation is related to the time of exposure to a lethal agent can be described by the equation

$$N_t = N_0 e^{-kt} \qquad (10.1)$$

where N_t is the number of surviving organisms after time t, N_0 is the number of organisms at time zero, t is the exposure time, and k is the microbial inactivation rate constant. If the logarithm of the fraction of survivors (N_t/N_0) is plotted against exposure time, the resulting curve (survivor curve) will be linear with a negative slope (Figure 10.3). The slope of the line is $k/2.303$, from which the microbial inactivation rate constant k can be calculated.

The value taken by k is a measure of the resistance of an organism to the particular inactivation process; the larger the value of k, the more sensitive is the organism.

10.2.3 D Value

In sterilization microbiology, the D value is frequently used instead of k as a measure of the rate of microbial inactivation. The D value is the exposure time required for the number of survivors to change by a factor of 10 or the time taken for the microbial population to be reduced by 90%, in other words, the exposure time required to achieve a decrease of one logarithmic cycle in the survivor curve.

* Descriptors "higher" and "lower" are frequently used incorrectly in relation to SAL. Assurance of sterility (or sterility assurance) is a qualitative concept and one can discuss greater or lesser assurance of sterility. SAL, however, has a quantitative value and a SAL of 10^{-6} is a lower value than, for example, a SAL of 10^{-4}. Hence, one has a greater assurance of sterility associated with a lower SAL and a specified value for a SAL is taken to be a maximal value.

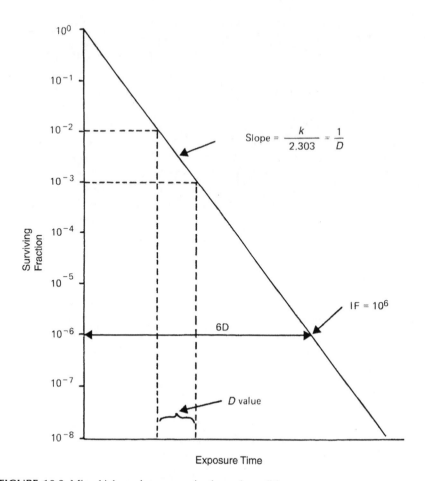

FIGURE 10.3 Microbial survivor curve: isothermal conditions.

The D value may be estimated graphically (Figure 10.3) or mathematically from the equation

$$D = \frac{t}{\log N_0 - \log N_t} \qquad (10.2)$$

The relationship between the D value and k for an exponential response is given by the equation

$$D = \frac{2.303}{k} \qquad (10.3)$$

The D value and k are specific for a particular inactivation process and when quoted, the conditions under which they are determined must be stipulated. Thus, with data for the heat inactivation of microorganisms, the temperature is shown as a subscript to the symbol, for example, D_{121} is the D value at 121°C. For radiation, the D value is stated in terms of absorbed dose (kGy). The D value is a measure of the resistance of an organism to the particular inactivation process; the smaller the D value, the more sensitive is the organism.

In the design of sterilization protocols, the microbial inactivation required of the system may be expressed in terms of the D value. An example is the use of the $12D$ endpoint, which defines that the sterilization process is required to produce 12 decimal reductions in the number of viable organisms or a decrease of 12 logarithmic cycles in the survivor curve.

10.2.4 INACTIVATION FACTOR

The total microbial inactivation of a sterilization process can also be described by the inactivation factor (IF), which is defined as the reduction in the number of viable organisms brought about by the process. The inactivation factor is expressed in terms of the D value as

$$IF = 10^{t/D} \qquad (10.4)$$

where t is the exposure time and D is the D value for the organism under the exposure conditions. A sterilization process that achieves $12D$ will therefore have an IF of 10^{12}.

10.2.5 NONLINEAR SURVIVOR CURVES

Although linear survivor curves have been determined experimentally, there are many examples in the literature of survivor curves that show significant deviations from linearity. Figure 10.4 illustrates the most common survivor curve shapes observed with the inactivation of microorganisms. Curve A is linear over the entire exposure time range, that is, it shows true logarithmic inactivation. Curve B is concave downward, having a low but increasing rate of inactivation during the initial exposure times, that is, an initial shoulder, followed by logarithmic inactivation at longer exposures. Curve C is concave upward, having a very high but decreasing rate of inactivation during the initial exposure times followed by logarithmic inactivation at longer exposures. A fourth type of survivor curve, Curve D, shows an initial sharp increase in the number of viable organisms prior to inactivation at a logarithmic rate. Survivor curves that are entirely curvilinear or sigmoid in shape also occur occasionally.

Type B curves have been explained as indicating the necessity for several critical events to occur in the organism before death occurs. Technical factors such as clumping of cells or a lag before the heat reaches the sensitive part of the organism could also result in this type of curve. It would also result if heat activation and inactivation occur concurrently, where the resulting number of viable organisms would never, unlike the type D curve, be greater than N_0. The initial shoulder of the type B curve may also be a measure of the capacity of the organism to repair heat-induced damage. The type C curve is usually attributed to differences in susceptibility of individual organisms, either as part of the natural distribution of resistance or as a result of cells being of different ages, or to the protective effect of the presence

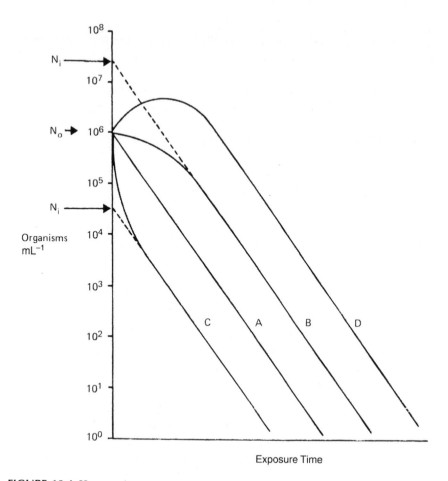

FIGURE 10.4 Heat survivor curve.

of dead cells. The nonlinearity of survivor curves may also be attributed to the effects of components in the environment during heating, a factor to be considered when organisms are heated in complex pharmaceutical dosage forms. Type D curves are frequently observed with exposure to moist heat of bacterial spores of low growth index; the initial sharp increase in the number of viable organisms is due to heat activation of dormant spores prior to heat inactivation at a logarithmic rate.

Deviations from linearity of survivor curves can present a particular problem if the D value is used as a measure of microbial resistance because the value is derived from logarithmic inactivation kinetics. It should be noted that D value determinations based on quantal data (see Section 10.3.1.2) take no account of survivor curve shape. If direct enumeration is used, corrections can be made if the D value is calculated using regression analysis on data from the logarithmic region of the survivor curve. The y intercept of the regression line fitted to the survivor data, $\log N_i$, or the intercept ratio (IR), $\log N_i / \log N_0$, can then be used in conjunction with the D value to describe the survivor curve.

A number of theoretical equations have been derived to describe nonlinear survivor curves but these are inevitably complex and have not found favor in sterilization microbiology. Historically, sterilization protocols were designed using a reference organism, usually a bacterial spore, which has known resistance to the lethal agent and shows reproducible logarithmic inactivation. The reference organism for sterilization by moist heat and low temperature steam and formaldehyde (LTSF) is spores of *Bacillus stearothermophilus* and for sterilization by dry heat and ethylene oxide spores of *Bacillus subtiltis* var. *niger* are used (see Chapter 12).

10.3 MEASURES OF MICROBIAL SENSITIVITY

10.3.1 EXPERIMENTAL DETERMINATION OF D VALUE

Because the D value is used extensively in the design of sterilization protocols, it is necessary to understand the methods that are used to determine it. These utilize direct enumeration and quantal data approaches. There may be considerable difference between a D value estimated from the regression line fitted to directly enumerated survivor data that extends over several logarithmic cycles and the D value estimated from quantal data where the value is calculated from N_0 and a point in the quantal region. For this reason, D values obtained by the different methods must not be used in combination, as, for example, in the calculation of z values (see Section 10.3.2.1).

10.3.1.1 Direct Enumeration

In this method, the experiment described in Section 10.2.1 is performed. Aliquots of a suspension of the surviving microorganisms are plated onto a suitable recovery medium and the number of colonies formed after incubation counted to determine the number of survivors. The data are used to generate a survivor curve and regression analysis can be used to determine the slope and its confidence limits. The D value can be calculated from the slope using Equation 10.3.

10.3.1.2 Quantal Data

In this approach, replicate test pieces containing a known population of microorganisms are exposed to different levels of inactivation. Each test piece is then recovered individually and incubated under suitable recovery conditions and each unit is assessed as showing growth or no growth. Exposure times or absorbed doses are selected so that at the first level all replicates show growth and at the last level all replicates show no growth. The intermediate levels of exposure then yield test pieces with zero to low numbers of surviving microorganisms, distributed about the mean according to a Poisson distribution. Inevitably, this will lead to some units showing no growth, yielding "quantal" or "fraction positive" (conversely, "fraction negative") data.

For a Poisson distribution, the probability (P) of occurrence of an integer (n) for a given mean (μ) is described by

$$P_{(n)} = \frac{\mu^n \cdot e^{-\mu}}{n!} \qquad (10.5)$$

The fraction of test pieces showing no growth establishes experimentally the probability (P) of zero survivors; hence the mean number of surviving microorganisms (μ) can be estimated from the fraction negative results by substituting in Equation 10.5 above, as follows

$$P_{(0)} = \frac{\mu^0 \cdot e^{-\mu}}{0!} \qquad (10.6)$$

or

$$\mu = -\log_e P_{(0)} \qquad (10.7)$$

If a total of n replicates are exposed and of these, r units show no growth, the probability of zero organisms occurring on a test piece is determined as

$$P_{(0)} = \frac{r}{n} \qquad (10.8)$$

Substituting in Equation 10.7, it follows that

$$\mu = -\log_e \frac{r}{n} \qquad (10.9)$$

or rearranging Equation 10.9, that

$$\mu = 2.303 \log \frac{n}{r} \qquad (10.10)$$

When 20 or more replicate test pieces are used, the quantal range extends from an average of 5 survivors per test piece to 0.01 survivors per test piece.

There are two methods that use this approach to calculate the D value: the Stumbo–Murphy–Cochran method and the Spearman–Karber method. Both use the principles outlined above to estimate the number of microorganisms surviving at a predetermined extent of treatment and thereby calculate the D value by substitution of the initial number of microorganisms, the calculated number of survivors, and the extent of treatment into the equations described in Sections 10.3.1.3 and 10.3.1.4.

10.3.1.3 The Stumbo–Murphy–Cochran Method

The Stumbo–Murphy–Cochran method uses Equation 10.10 to estimate the number of survivors at the respective inactivation level.

The D value is estimated from the equation

$$D = \frac{U}{\log N_0 - \log N_u} \tag{10.11}$$

where N_u is the number of surviving microorganisms (calculated from Equation 10.10) after an exposure time or absorbed dose U, and N_0 is the initial number of viable organisms per replicate unit.

10.3.1.4 The Spearman–Karber Method

The Spearman–Karber method calculates the mean exposure time or dose until no survivors are obtained, μ, using the equation

$$\mu = \sum_{i=1}^{k-1} \left(\frac{U_{i+1} + U_i}{2} \right) \left(\frac{r_{i+1}}{n_{i+1}} - \frac{r_i}{n_i} \right) \tag{10.12}$$

where U_i is the ith exposure time or absorbed dose, r_i is the number of replicate units showing no growth out of n_i after an exposure time or absorbed dose U_i, and k is the first exposure time or absorbed dose that results in all replicate units showing no growth. If the initial number of viable organisms per replicate unit is N_0, the D value can be calculated from the equation

$$D = \frac{\mu}{0.2507 + \log N_0} \tag{10.13}$$

The Spearman–Karber method is more resource intensive to perform than the Stumbo–Murphy–Cochran method but offers the advantage of permitting statistical analysis to be undertaken and allowing confidence intervals about the D value to be calculated.

10.3.2 EFFECT OF TEMPERATURE ON MICROBIAL RESISTANCE

Microbial resistance measurements and D value determinations are carried out under isothermal conditions. The resistance of the organism will change with alteration in temperature. The change in the rate of microbial inactivation with a change in temperature is the temperature coefficient for the lethal process. There are two measures of the effect of temperature on microbial resistance: z values are generally quoted for thermal processes and Q_{10} values for chemical processes.

10.3.2.1   z Value

If the logarithm of the D value is plotted against temperature a linear relationship results (Figure 10.5). The plot is called a thermal resistance curve. The negative reciprocal of the slope of this line is the z value and represents the increase in temperature required to reduce the D value of an organism by 90% or to produce a decrease of one logarithmic cycle in the thermal resistance plot. The units of the z value are degrees of temperature. The z value can be estimated graphically (Figure 10.5) or mathematically using the equation

$$z = \frac{T_2 - T_1}{\log D_1 - \log D_2} \qquad (10.14)$$

where D_1 and D_2 are the D values at temperatures T_1 and T_2, respectively.

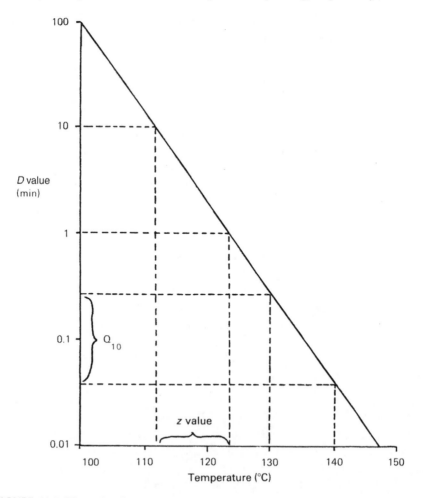

FIGURE 10.5 Thermal resistance curve.

The precision with which the z value can be estimated is dependent upon the precision with which the individual D values can be experimentally determined. If the data at the different temperatures are for different test conditions, they cannot be combined to yield a meaningful z value. Although the z value is a fundamental characteristic of an organism, it is not truly independent of temperature and is constant only for small temperature differences of the order of 20 to 25°C. However, because heat sterilization processes are usually carried out within a small temperature range, for example, 110 to 135°C for autoclaving, the z value is considered constant for all practical purposes.

10.3.2.2 Q_{10} Value

The Q_{10} value is also a measure of the change of microbial inactivation rate with temperature and is defined as the ratio of the microbial inactivation rate constants at two temperatures, 10°C apart. Thus

$$Q_{10} = \frac{k_1}{k_2} \qquad (10.15)$$

where k_1 and k_2 are the microbial inactivation rate constants (see Section 10.2.2) at temperatures T°C and $T + 10$°C, respectively. The relationship of the z value to Q_{10} may be expressed as

$$z = \frac{10}{Q_{10}} \qquad (10.16)$$

10.3.2.3 Activation Energy for Microbial Inactivation

In nearly every known case, the rate of a chemical reaction increases with temperature. This increase in rate is reflected in a rise in the reaction rate constant.

The same applies to microbial inactivation where an increase in temperature results in an increase in the microbial inactivation rate constant. The Arrhenius rate theory used in chemical reaction kinetics can also be utilized to estimate the activation energy for the microbial inactivation (E). The relationship between the microbial inactivation rate constant (k) and absolute temperature (T K) is stated by

$$k = Ae^{-E/RT} \qquad (10.17)$$

where A is termed the frequency factor, R is the universal gas constant and E is the activation energy for the microbial inactivation. When the logarithm of k is plotted against the reciprocal of absolute temperature, a linear relationship is obtained where the slope is equal to $E/2.303R$. The activation energy for microbial inactivation can also be calculated from the equation

$$\log \frac{k_2}{k_1} = \frac{E}{2.303R} \left(\frac{T_2 - T_1}{T_1 T_2} \right) \qquad (10.18)$$

where k_1 and k_2 are the microbial inactivation rate constants at temperatures T_1 K and T_2 K, respectively. The precision with which E can be estimated will depend upon the quality of the survivor curve data and the precision with which the values of k can be experimentally determined.

The relationship between the z value and the activation energy for microbial inactivation is represented by the equation

$$z = \left(\frac{2.303T_1T_2}{E} \right) R \qquad (10.19)$$

If microbial inactivation were a single chemical reaction or the result of a single event, the Arrhenius equation should hold for all temperature ranges. In practice, the plot of log k against $1/T$ over a wide temperature range often shows two straight portions of different slopes joined by a curved section, indicating a change from one lethal reaction to another as the temperature changes. In sterilization microbiology, this can have important significance if data obtained at low temperatures are used to predict the microbial inactivation at high process temperatures. However, over the temperature range normally used for moist heat sterilization (110 to 135°C), there is general obedience to the Arrhenius relationship, with bacterial spores having activation energies in the range 270 to 300 kJ per bacterium.

10.4 RESISTANCE OF MICROORGANISMS TO INACTIVATION

In sterilization processes, the agents that are used to inactivate microorganisms are elevated temperature, ionizing radiation, toxic gases and gas plasmas, and liquid sterilants. Although there is considerable variation in the resistance of microorganisms to inactivation, there is a general pattern in the degrees of resistance observed. Bacterial endospores are generally considered to be the most resistant forms, although there is a considerable range of resistance to individual agents within these spores. There is evidence that some of the causative agents of transmissible spongiform encephalopathies (TSEs), such as Creutzfeldt-Jakob disease (CJD), bovine spongiform encephalopathy (BSE), or scrapie, may be considerably more resistant to certain lethal agents than bacterial spores and may not be completely inactivated by traditional sterilization protocols. Nonsporing bacteria are the most sensitive to inactivation, along with the vegetative forms of yeasts and molds. The larger viruses show similar resistance to vegetative bacteria. Spores formed by molds and yeasts are generally more sensitive to inactivation than bacterial spores. In particular, the sexual spores of molds are usually only marginally more resistant than the vegetative cells from which they are derived. In contrast, asexual spores, for example, chlamydospores produced by molds, and the sexual ascospores formed by many yeasts, often show enhanced resistance, intermediate between that of vegetative cells and bacterial spores. Small viruses such as polio virus and hepatitis B virus also show enhanced resistance to inactivation.

Resistance to inactivation by a particular agent is a genetically determined characteristic of an organism and within the general pattern of microbial resistance there will often be considerable variation in the response of an organism to different agents. This is not surprising because the mechanisms by which each agent inactivates microorganisms may be different. It should be noted that there may be considerable genetic variation in the resistance to a specific agent between different strains of the same species, and the development and selection of resistant mutants may present a particular problem. For this reason, sterilization processes have traditionally been designed using, as the reference organism, isolates that are more resistant than the microorganisms that comprise the bioburden of the particular product. More recently, however, the determination of product-specific sterilization processes based on the resistance of the bioburden to the sterilizing agent has been applied more widely, particularly with radiation sterilization (see Chapter 12).

Microbial resistance to inactivation by a particular agent is affected by environmental influences during the formation and growth of the organism and during its exposure to the agent. Environmental influences during culture of the organism include its growth phase and age, the growth temperature and conditions, and the nutrient composition of the growth medium (see Chapter 1). These influences become especially apparent when resistant strains isolated from a product bioburden are cultivated in laboratory media, whereupon they invariably show reduced resistance to the inactivating agent. Environmental influences during exposure of the organism to the lethal agent include pH; ionic strength; carbohydrate, protein, and fat content of the menstruum; and the presence of soluble organic and inorganic compounds. Microbial water content and gaseous atmosphere may also be important, for example, with respect to inactivation by dry heat or radiation. In pharmaceutical dosage forms, the components may have considerable influence on the resistance of an organism to inactivation and it is necessary in the design of sterilization protocols to establish the resistance of the reference organism in the product to be sterilized or a menstruum similar to it.

The apparent response of an organism to a lethal agent will depend upon the success to which survivors can be recovered.

For these reasons, it is usual to design sterilization protocols using, as the reference organism, a well-characterized resistant laboratory strain, usually a bacterial spore, which is maintained, grown, and harvested under carefully controlled and defined conditions. Furthermore, the environmental conditions have to be carefully controlled, and standard, ideally optimal, recovery conditions, that is, temperature, duration of incubation, and medium composition are stipulated.

10.5 SELECTION OF A STERILIZATION METHOD

The selection of a method of sterilization requires balancing the advantages of the available methods against their disadvantages. There are no regulatory requirements that specify the method that is to be applied in a particular situation, but there are a number of basic principles that guide the selection of a sterilization method. These principles include:

- Terminal sterilization of the product in its final container is preferred to the use of aseptic manufacturing.
- Sterilization methods with process variables that can be directly controlled and monitored are preferred. Hence, the order of preference for the established technologies is moist or dry heat, irradiation, gaseous sterilants, and liquid sterilants.
- Sterilization methods that present the fewest hazards to operators or the environment and do not leave toxic residues within products are preferred.
- There are significant advantages in employing established sterilization methods with defined regulatory requirements. Gaining regulatory approval for novel processes and equipment can be expensive and time consuming.
- The choice of sterilization method has to be based on the characteristics of the product to be sterilized, not the limitations of a packaging material. If a product will withstand a sterilization method but the proposed package will not, the packaging material should be changed, not the sterilization method.
- Compatibility with established methods of sterilization can often be designed into products by the initial selection of appropriate materials. Incompatibility of materials with a sterilization method may be due to direct chemical or physical effects. Some incompatibilities may be due to the choice of combinations of materials; for example, two materials separately may withstand exposure to high temperatures but have different rates of expansion on heating, making the combination unable to withstand steam sterilization.
- The sterilizing agent has to come into contact with all parts of the product. Product configuration may present difficulties in the penetration of gaseous agents, including steam, or shield portions of the product from radiation. If temperature is a variable within the sterilization process, sufficient time needs to be allowed within the process for heat transfer.

Having selected a method of sterilization, the precise process specification needs to be established. There are two broad approaches to process development leading either to the use of established, compendial cycles (sometimes inappropriately referred to as the "overkill" approach) or to the development of product-specific processes.

Compendial cycles, such as moist heat at 121°C for 15 min, have generalized applications and were developed based on reference microorganisms considered to have high resistance to the sterilizing agent. In recent years, however, there has been a greater emphasis on "product-specific" processes. This is a rational approach based on a knowledge of the resistance to the sterilizing agent of the microorganisms associated with the product items. This approach may permit shorter sterilization processes to be employed, but can be significantly more demanding to validate and undertake. An extension to this position has been discussed by Gilbert and Allison (1996).

10.6   MECHANISMS AND CHARACTERISTICS OF MICROBIAL INACTIVATION

10.6.1   HEAT STERILIZATION

Heat is the most reliable and widely employed method of sterilization. The efficiency with which heat is able to inactivate microorganisms depends upon the degree of heat, the exposure time, and the presence of water. Heat comprises a basic form of energy produced by the vibratory motion or activity of molecules. It is nonquantized and will act on every molecule in an organism. It would be unlikely, therefore, that any one factor or event would be responsible for heat-induced microbial inactivation. Provided that enough heat is supplied, every site in the organism is likely to be damaged and death is probably due to an accumulation of irreparable damage to all metabolic functions.

The action of heat is due to the induction of lethal chemical events mediated through the action of water or oxygen. In the presence of water, much lower temperature–time exposures are needed to kill microorganisms than in its absence; coagulation of proteins also occurs at a lower temperature and in a shorter time in the presence of water. This observation has led to the hypothesis that, with moist heat, death results from denaturation and coagulation of essential proteinaceous sites, such as enzymes and structural proteins, within the organism. With dry heat, microbial inactivation is primarily the result of oxidative processes that require higher temperatures and longer exposure times.

Although sensitive vegetative organisms show significant inactivation by moist heat at 60 to 70°C, resistant bacterial spores require temperatures in excess of 100°C. Moist heat sterilization is therefore carried out using saturated steam under pressure and is the method of choice for aqueous preparations and surgical materials. Dry heat sterilization employs temperatures between 150 and 250°C and is used for nonaqueous preparations, powders, and glassware (see Chapter 11).

Because heat-induced microbial inactivation probably results from irreversible damage to metabolic functions, it would be expected that these reactions, like a number of other chemical processes, would follow apparent first-order kinetics and result in linear survivor curves when the logarithm of the fraction of surviving organisms is plotted against exposure time.

10.6.1.1   Measures of Heat Sterilization Efficiency

Compendial heat sterilization protocols have evolved from custom and practice and a long history of use rather than being carefully designed on the basis of scientific principle and equivalence of lethality (see Chapter 11). The modern approach to the design of sterilization protocols is based on a quantitative approach using reliable microbial inactivation data. To be effective, this requires a basis on which the equivalence of different sterilization processes can be established. The key terms for this are described below.

F **value**—The *F* value is a "unit of lethality" devised by the food industry and is a measure of the total lethality of a heat sterilization process with respect to a particular organism. It is defined as the equivalent in minutes at 250°F (121.1°C) of all heat considered with respect to its capacity to destroy spores or vegetative cells of a particular organism. Because the F value compares lethal effects at different temperatures, the z value of the organism should always be stated when an F value is quoted. For the reference temperature of 121°C and a z value of 10°C, the F value is referred to as F_0, the "reference unit of lethality," and can be determined from the equation

$$F_0 = D_{121}(\log N_0 - \log N) \tag{10.20}$$

where D_{121} is the D value at 121°C for the reference spores and N_0 and N are the initial and final numbers of viable spores, respectively. In pharmaceutical applications, B. stearothermophilus spores are usually used as the reference organism and z and D values of 10°C and 1.5 min, respectively, are generally assumed for aqueous solutions. For quantitative work, however, these values must be accurately determined in the particular product to be sterilized.

F_0 can also be expressed in terms of the inactivation factor as

$$F_0 = D_{121} \log \text{IF} \tag{10.21}$$

Any process that is stated to have an F_0 of x will have the same lethality on the reference spore as x minutes at 121°C; a worked example is given in Chapter 12. In practical terms, the F_0 value of a sterilization process is the lethality delivered by that process to the product in its final container. If used to describe the lethal effect upon reference spores at the coolest location in the sterilizer, the F_0 value represents the most conservative estimate of lethality and thus the safest conditions for determining cycle times.

The reader is referred to the companion Handbook (Hodges 2000) where the relationship between inactivation factor, sterility assurance level, and F_0 value are shown in a worked example.

Lethality factor (F_i)—This is the time at any other temperature equivalent, in terms of lethality to a particular organism, to 1 min at 121°C. This value will also be dependent on the z value of the organism. For a reference organism with a z value of 10°C, F_i is equal to the reciprocal of F_0.

Lethal rate (L)—The lethal rate L, at a temperature T, can be calculated as a fraction of that at a reference temperature T_{ref} from the equation

$$L = 10^{\left(\frac{T-T_{ref}}{z}\right)} \tag{10.22}$$

If the reference temperature is 121°C and the z value of the reference organism is 10°C

$$L = \frac{F_0}{F_T} \tag{10.23}$$

where F_T is the F value at temperature T. Tables of lethal rates for the reference temperature 121°C and a z value of 10°C are available (see, for example, Pflug 1973).

10.6.1.2 Integrated Lethality Determination

The compendial moist heat sterilization protocols specify a preferred combination of temperature and time, for example, 121°C for 15 min. The measurement of time begins when the temperature of the material being sterilized reaches the defined temperature. It ignores the heating-up and cooling-down stages of the cycle, which could be prolonged if the load has a large thermal capacity. The rate of heating and cooling of a product in a container is a function of container size, viscosity of the liquid, and the size of the load. The resultant heating and cooling curves are usually exponential with respect to time.

It is usually considered that moist heat has very little lethal effect on bacterial spores at temperatures below 90°C. However, the heat imparted to the load during the heating-up stage from 90°C to the holding temperature and during the cooling-down stage from the holding temperature to 90°C can make a considerable contribution to the total lethality of the sterilization cycle. Reduction in energy consumption, processing time, and product degradation can be achieved if this contribution is taken into account. The F_0 concept and the use of lethal rates facilitate integration of the lethality of the total heat process, including the heating and cooling stages. The requirement is for accurate time–temperature data during the whole of the process, from that part of the product where heating is slowest. With these data, the F_0 for the heating and cooling stages of the process can be determined and the holding time can then be reduced to that necessary to ensure that the predetermined F_0 for the process is attained. Two methods are commonly used to analyze time–temperature data.

Graphical method—This method involves plotting the time–temperature data on lethal rate or F reference paper, where the distance between each successive temperature line on the y (logarithmic) axis is a function of the lethality of that temperature compared to the lethality of the reference temperature. On the x axis, time is plotted on a linear scale. The area under the curve is a measure of the F value. This value can be calculated by multiplying the area by the appropriate scale factor. If the reference temperature is 121°C and the z value is 10°C, this gives the F_0 for the process.

An alternative method calculates the lethal rate for each temperature T as a function of that at the reference temperature T_{ref}, that is, minutes at T_{ref} per minute at T. The lethal rates are then plotted on a linear scale against time in minutes at T, and the area under the curve is the F value for the process.

Summation methods—These methods eliminate the plotting of lethal rate graphs and the measuring of the area under the curve. One such method is based on the trapezoidal rule for evaluating the area under the lethal rate curve. This is written as

$$area = F = \Delta t \left(\frac{L_{T_0}}{2} + L_{T_1} + L_{T_2} + ... + L_{T_{n-1}} + \frac{L_{T_{n...}}}{2} \right) \qquad (10.24)$$

where Δt is the constant time between measurements of T, and L is the lethal rate at temperature T. If the time–temperature data for the total process are analyzed, the data included in the analysis may be simplified so that the values of the first and last points, L_{T_0} and L_{T_n}, respectively, are zero. This simplifies the equation to

$$F = \Delta t \Sigma L \tag{10.25}$$

Interactive computer programs are available that evaluate the integral

$$F = \int_{t_1}^{t_2} L dt \tag{10.26}$$

according to the trapezoidal rule, directly from time–temperature data. These enable the F_0 value to be determined continuously and the process to be terminated when the predetermined F_0 for the process is attained. These form the basis of microprocessor controllers (temperature-integrating time controllers) for autoclaves.

10.6.1.3   Design of Optimum Autoclaving Protocols

Knowledge of the activation energy both for the killing of bacteria and for inducing damage to the product to be sterilized together with use of the F_0 concept can lead to the design of an optimum time-temperature combination for particular applications.

The *British Pharmacopoeia* (BP) (2005) prefers an autoclaving process for aqueous preparations consisting of a holding time of 15 min at 121°C together with the heating-up and cooling-down periods. It also accepts other time–temperature combinations and suggests an F_0 value of at least eight if that concept is used. The two processes are likely to differ in lethal efficiency by at least a factor of two and it is difficult to see the logic of recommending two such different processes. However, if, for example, a decision to use an F_0 of 12 is made, which is approximately halfway between the two BP methods, then the times necessary for achieving this at different temperatures can be calculated and are shown in Table 10.1.

TABLE 10.1
Different Time–Temperature Combinations Calculated to Give an F_0 of 12 for Reference B. stearothermophilus Spores with a D_{121} of 1.5 min and z Value of 10°C

Temperature (°C)	Time (min)
115	48
118	24
121	12
124	6
127	3

The activation energy for killing spores is of the order 270 to 300 kJ per bacterium, while for inducing chemical change (hydrolytic or oxidative) in aqueous solution it is generally of the order of 70 to 100 $kJ.mol^{-1}$ for drugs that pose a stability problem to manufacturers. When such solutions need to be sterilized, in order to minimize the amount of chemical change, it is always preferable to use a combination of the highest temperature possible with the shortest time, rather than to use a low temperature–longer time combination with the same lethal efficiency.

This is illustrated in Figure 10.6, which shows the Arrhenius plots for the killing of *B. stearothermophilus* spores, with an E of 285 kJ per bacterium and the hydrolysis of procaine hydrochloride at a pH of 6.2 with an E of 68 $kJ.mol^{-1}$. The times necessary to achieve an F_0 of 12 at different temperatures are first calculated from the inactivation data for *B. stearothermophilus* spores, resulting in the data shown in Table 10.1. By substituting these values for t together with values for k for procaine hydrochloride hydrolysis, obtained from Figure 10.6 for the $1/T$ values corresponding to the different required temperatures, into the equation

$$\log C_t = \log C_0 - \frac{k}{2.303}.t \qquad (10.27)$$

the percentage procaine hydrochloride remaining (C_t) when the initial concentration $C_0 = 100\%$ can be calculated for each temperature (inset in Figure 10.6).

10.6.2 RADIATION STERILIZATION

Radiation sterilization is a low-temperature method that can be used in certain situations where heat sterilization would produce an unacceptable amount of damage to the product. However, because of the difficulties and costs of producing the radiations, the method is only suitable for large-scale use. Radiations may be classified into two types: particulate, for example, alpha rays, beta rays, protons, and neutrons; and electromagnetic, such as, X-rays, gamma rays, and ultraviolet (UV) light. In general, particulate radiations have less penetrative power than electromagnetic radiations. Alternatively, radiations can be classified as ionizing and nonionizing; with ionizing radiations, the principal means of energy dissipation during their passage through matter is the ejection of an electron with production of a positively charged ion. In addition, ions produce free radicals and activated molecules in cells, some of which are lethal.

In practice, ionizing radiations are used for sterilization in the form of gamma rays or accelerated electrons (beta rays). Gamma radiation is the result of a transition of an atomic nucleus from an excited state to a ground state in radioactive materials such as cobalt-60 (^{60}Co) and caesium-137 (^{137}Ce); hence, it has the disadvantage of requiring a radioactive source. Accelerated electrons are generated from a machine source but have the disadvantage of a lower penetrative power. Advances in technology over the last few years, however, have led to an increase in the use of electron beams of high energy [around 10^7 eV (10 MeV)].

X-rays have the potential to be a practical form of radiation for sterilization. X-rays are emitted from an atom when there is a transition of an electron from an outer

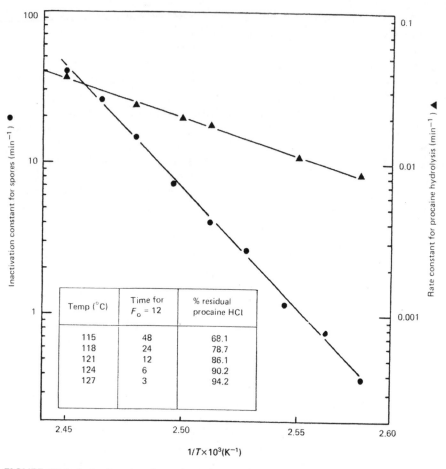

FIGURE 10.6 Arrhenius plots for moist heat inactivation of reference *B. stearothermophilus* spores (●) and hydrolysis of procaine hydrochloride solution (▲).

shell to a vacancy further within an inner shell. They are produced as secondary radiation by the bombardment of a heavy metal target with a beam of fast electrons in an accelerator. Power utilization in this process is low; X-ray production is consequently expensive and, to date, has only recently been employed on a commercial scale.

UV radiation has not been employed as a sterilizing agent but has found applications for disinfection of water and air. UV radiation comprises that portion of the electromagnetic spectrum with wavelengths from about 190 to 390 nm. The high-emission wavelength of mercury (254 nm) is most frequently used, being the wavelength closest to the absorption peak for DNA, the supposed chromophore and target. The quantum energies for UV are around 10^2 eV compared to values ranging from 10^6 to 10^9 eV for X-rays and gamma rays. As a consequence, UV has very low penetrating power and produces only increased excitation and not ionization of molecules that absorb it. Furthermore, most organisms possess enzymatic processes that are capable of repairing UV-induced damage, resulting in the recovery of UV-irradiated cells under certain conditions.

When a population of microorganisms is exposed to increasing doses of radiation, the plot of the logarithm of the surviving fraction against absorbed dose on a linear scale assumes one of three basic shapes (types A, B, and C as illustrated in Figure 10.4). The type A curve is represented by the equation

$$\frac{N_t}{N_0} = e^{-kD} \tag{10.28}$$

where N_t/N_0 is the surviving fraction after an absorbed dose D, and k is the microbial inactivation rate constant. This equation is comparable to that used to describe the linear survivor curve (Equation 10.1 and Section 10.2.2), but is derived on the basis that it represents the probability of a single randomly distributed radiation-induced lethal event or "hit" occurring in a single "target" within the cell, rather than on the basis of unimolecular chemical reaction kinetics. Although ionizing radiations can cause a wide variety of physical and biochemical effects in microorganisms, it is likely that DNA is the primary cell target and that unrepaired strand breakage or cross-linkage caused by ionization of a molecule in the DNA is the lethal event.

The equation for the type B curve is derived on the basis that a single hit is necessary on more than one target to cause cell death. The equation is

$$\frac{N_t}{N_0} = 1 - \left(1 - e^{kD}\right)^n \tag{10.29}$$

where n is the number of targets. Below a surviving fraction of 0.1 this closely approximates to the equation:

$$\frac{N_t}{N_0} = ne^{-kD} \tag{10.30}$$

which on log transformation is

$$\log \frac{N_t}{N_0} = \log n - \frac{kD}{2.303} \tag{10.31}$$

where n is the intercept with the log N_t/N_0 axis of the extrapolated linear portion of the survivor curve. Equations representing type B curves have also been derived on the basis that for cell death more than one hit is necessary on a single target. With radiation, as with heat, the type C curve cannot be readily explained and is often interpreted as being the result of nonhomogeneous resistance in the microbial population. Most radiation sterilization protocols have been derived on the assumption that the type C curve is rare and atypical.

As with resistance to heat, unicellular organisms are more resistant to radiation than multicellular organisms. Bacterial spores are the most resistant to radiation, vegetative bacteria, particularly Gram-negative rods, the most sensitive, and yeasts and fungi of intermediate resistance. There are exceptions; for example, the Gram-positive *Deinococcus radiodurans* and the Gram-negative *Moraxella osloensis* are both extremely resistant to radiation. In general, viruses, particularly small viruses, are more resistant to radiation than bacteria. Resistance to one agent does not necessarily imply resistance to another. Spores of *B. stearothermophilus and B. subtilis,* which exhibit considerable heat resistance, are not especially resistant to radiation and thus *B. pumilus* spores, with a *D* value of 3 kGy, were originally used as the reference organism for radiation sterilization.

Anoxic bacterial spores are more sensitive to radiation in the fully hydrated state than in the dry state due to the additional lethality induced by the radiolysis products of water. Resistance is also reduced when the organisms are irradiated in oxygen, particularly in the dry state. Radiation-induced degradation of materials is also greatest in the presence of water and this precludes the use of radiation as a method of sterilization for aqueous solutions of drugs. However, under dry anoxic conditions, radiation induces less damage than heat with the same lethal efficiency and is the method of choice for some pharmaceuticals in the dry state and for medical devices.

The presence of protective or sensitizing agents in the environment will have a considerable influence on the radiation resistance of microorganisms and ideally the resistance of organisms on and in the product to be sterilized should be taken into account when the radiation dose necessary for sterilization is being determined.

10.6.3 GASEOUS STERILIZATION

A wide variety of liquid chemicals are toxic to microorganisms and are useful as disinfectants, though very few are toxic to spores and are therefore of little use as sterilizing agents. This, together with the preference for terminal sterilization of items in their final packaging, often precludes the use of liquid antimicrobial chemicals for sterilization except in specialized applications such as the sterilization of particular medical devices containing tissues of animal origin.

Certain chemicals in their gaseous states and gas plasmas, however, are effective sterilants and are used for the sterilization of containers and delivery systems for parenteral medicines, for sterilization-in-place of equipment for aseptic manufacture, and for medical devices that cannot be subjected to heat or radiation sterilization. Such items are generally constructed of plastic materials.

Gaseous sterilizing agents can be categorized into alkylating agents and oxidizing agents based on their mechanism of microbicidal action.

10.6.3.1 Alkylating Agents

Alkylating agents such as beta-propiolactone, propylene oxide, and methyl bromide have been used for specialized sterilization purposes but only ethylene oxide and formaldehyde have been extensively employed in the gaseous sterilization of medical devices and pharmaceutical products.

Both ethylene oxide and formaldehyde are potent alkylating agents and are considered to exert their lethal effect on microorganisms by the alkylation of a number of cellular constituents. Possible reaction sites include amino, sulphydryl, and hydroxyl groups of proteins and purine nucleosides of nucleic acids. Formaldehyde reacts irreversibly with nucleic acids and this causes inhibition of spore germination. However, the revival of spores treated with aqueous formaldehyde and low temperature steam formaldehyde (LTSF) by postprocessing heat treatment, and of spores treated with glutaraldehyde solutions with postprocessing application of sodium hydroxide, suggests that the mechanism of action of alkylating agents is complex, and the precise mode of action remains unexplained.

Comparison of the sensitivities of various microorganisms to ethylene oxide and formaldehyde on the basis of published data is difficult due to the variations in exposure conditions and experimental methods employed. In general, bacterial spores show the greatest resistance to these sterilizing agents, being 5 to 10 times more resistant than vegetative bacteria. Molds and yeasts show intermediate resistance to gaseous alkylating agents. Both ethylene oxide and formaldehyde have marked antiviral activity. In contrast to many antimicrobial chemicals, genetically determined resistance to alkylating agents has not been a practical problem to date. Most antimicrobial chemicals induce nonlinear inactivation kinetics in microorganisms, resulting in sigmoid-shaped survivor curves, and ethylene oxide and formaldehyde are no exception. However, under carefully controlled conditions and in the high concentrations used in sterilization practice, the survivor curves approximate to type A curves, enabling the realistic estimation of SALs (see Section 10.2).

The activity of both gaseous ethylene oxide and gaseous formaldehyde is markedly influenced by the concentration of the gas, the temperature, the water content of the microorganisms, and the duration of exposure, and there is a complex interrelationship between these factors. Because the chemical must come into direct contact with the organism, the physical nature and penetrability of the material being treated are also of paramount importance when these chemicals are used as sterilizing agents. A particular problem with formaldehyde gas is its low penetrability necessitating preevacuation of the sterilizer chamber and operation of the LTSF cycles at pressures between 100 mbar (10 kPa) and 400 mbar (40 kPa).

Table 10.2 summarizes the effect of the temperature, concentration, and relative humidity (RH) on gaseous ethylene oxide and formaldehyde.

The interaction of all the above factors in defining an efficient gaseous sterilization process is described in Chapter 11.

10.6.3.2 Oxidizing Agents

Oxidizing agents used in the gaseous state include hydrogen peroxide, chlorine dioxide, ozone, and peracetic acid. In addition, gas plasmas have potent oxidizing action. Gas plasmas are considered as a fourth state of matter distinct from solids, liquids, and gases; they consist of a mixture of free radicals, electrons, ions, and excited radicals. Such gas plasmas occur naturally, such as in the Northern Lights, can be generated at low temperatures, such as in fluorescent lighting, and can be created in a sealed chamber under vacuum by a strong electric or magnetic field generated by radio frequency or microwave energy. Practical sterilizing systems use

TABLE 10.2
Factors Affecting the Microbiocidal Activity of Gaseous Sterilants

Alkylating Agents—Ethylene Oxide		
Temperature	Concentration	Relative Humidity (RH)
Activity increases with increase in temperature. Temperature coefficients (Q_{10} values) vary with concentration and with the temperature range. For the temperature range usually employed in sterilization cycles (40 to 60°C) and ethylene oxide concentrations above 400 mg.L^{-1}, the temperature coefficient for sporicidal activity is in the range of 2 to 3, that is, a temperature increase of 10°C approximately doubles the rate of inactivation and a decrease of 10°C will approximately halve the rate. Consequently, a spread of temperature of 10°C across a sterilization load will lead to a 100% difference in the rate of inactivation between the hottest and coldest locations.	Increasing the concentration increases the rate of microbial inactivation, reaching a plateau concentration above which further increases do not lead to greater effectiveness. The plateau concentration depends upon the temperature employed. At 30 to 50% RH, plateau concentrations have been reported as 800 mg.L^{-1} at 30°C and 500 mg.L^{-1} at 54°C. Doubling the gas concentration over the range of 50 to 800 mg.L^{-1} reduces the exposure time required to attain a given SAL by a factor of approximately 2, but increasing the concentration above 1000 mg.L^{-1} does not appreciably affect the exposure time.	Presence of water is critical for the alkylation reaction. Organisms in the dehydrated state are much more resistant than when preconditioned in an atmosphere of high RH. Similarly, organisms on hygroscopic carrier materials, such as paper or fabric, exhibit a marked decrease in resistance compared to organisms on nonhygroscopic materials such as metal or glass. RH levels of 20 to 40% are necessary for optimal sporicidal activity and reduced activity occurs at levels below 10% and above 65% RH. In practice, most processes operate between 40 and 80% RH because of the significant amount of moisture-absorbing material, such as packaging, included in a load. However, the presence of excess free water can lead to the formation of inactive ethylene glycol or ethylene oxide dissolving in the free water, reducing the gaseous ethylene oxide concentration.

Alkylating Agents—Formaldehyde		
Inactivation rate of microorganisms increases with increasing temperature over the range of 10 to 80°C but no clear temperature coefficient has been established. Investigations with gaseous formaldehyde over the range of 0 to 30°C showed little difference in inactivation rate with increasing temperature. With LTSF, increasing	With gaseous formaldehyde, activity increases with increasing concentration, with a significant increase in activity when increasing concentration from 0.1 to 1.3 mg.L^{-1}. A linear relationship has been shown between concentration and rate of inactivation between 0.04 and 0.31 mg.L^{-1} and for the concentration range 1.1 to 10.6 mg.L^{-1}. Concentrations of gas of	The microbiocidal effect of gaseous formaldehyde increases with increasing humidity up to 50% RH, with little further increase up to 95%. An RH in excess of 50% is essential and levels between 75 and 100% are generally considered to be desirable for rapid microbial inactivation. LTSF cycles effectively operate at 100% RH.

(continued)

TABLE 10.2 (continued)
Factors Affecting the Microbiocidal Activity of Gaseous Sterilants

temperature between 63 and 83°C only increased the rate of inactivation slightly.

the order of 3.5 mg.L^{-1} have been shown to be sporicidal at 25°C if the RH is above 50%. Concentrations of formaldehyde gas between 3.3 mg.L^{-1} and 100 mg.L^{-1} have been used in LTSF. Increasing concentration over the range 6 to 20 mg.L^{-1} increases the rate of inactivation, but a threshold around this concentration has been reported and further increases have little effect.

Oxidizing Agents

Sterilant	Concentration	Relative Humidity (RH)	Other
Chlorine dioxide	Sporicidal activity is concentration dependent. Activity against spores of *B. subtilis* var. *niger* at ambient relative humidity (20 to 40%) and at room temperature (23°C) is greater at a higher gas concentration.	Prehumidification is important for sporicidal activity and a RH of 70 to 75% greatly enhances the effectiveness.	
Ozone	No published data.	An increase in RH from 45 to 60 or 80% increases the biocidal effect. Desiccated spores and bacteria have been reported to be highly resistant to ozone.	Activity increases in combination with UV radiation, due to the formation of reactive hydroxyl radicals (see gas plasma below).
Gas plasma	Plasma produced from hydrogen peroxide has good sporicidal activity due to the lower energy required to create hydroxyl radicals and the fact that the species generated from hydrogen peroxide are among the most reactive. Increasing the concentration of precursor increases sporicidal activity.	Not applicable.	Sporicidal activity decreases as the distance from the generation source increases. Sporicidal activity increases with radio frequency power. Sporicidal activity increases with increasing temperature but no temperature coefficient has been quoted.

precursors such as hydrogen peroxide or peracetic acid to "seed" the gas plasma with reactive radicals to increase the sporicidal activity.

Reports of the microbiocidal properties of these oxidizing agents have been available for some time, but their use for sterilization in the pharmaceutical and medical device industries, and for the resterilization of reuseable medical devices in healthcare facilities, is relatively recent. There is a limited body of data on the effectiveness of these agents in the peer-reviewed literature.

Oxidizing agents used as gaseous sterilants are highly reactive and may interact with a number of cellular constituents, although as with alkylating agents, the precise mechanism of action has not been identified. Hydrogen peroxide in the liquid phase is probably the oxidizing agent that has been studied in greatest detail, although studies of its use in the vapor phase are less extensive. Its action has been reported as being due to the production of the hydroxyl radical, which reacts with membrane lipids, nucleic acids, and other cellular components. Bacterial destruction by ozone has been attributed to action at the cell surface, leading to disintegration of the bacterial cell wall. Ozone attacks glycoproteins, glycolipids, and certain amino acids within the bacterial membrane. Ozone also disrupts cellular enzymatic activity by reacting with sulfhydryl groups and acts on nucleic acids within the cell by modifying the purine and pyrimidine bases. Free radicals produced within gas plasma interact with essential cellular components, such as enzymes, nucleic acids, and the cell membrane, to inactivate microorganisms.

Table 10.2 summarizes those factors affecting the microbiocidal activity of chlorine dioxide, ozone, and gas plasmas.

10.6.4 FILTRATION STERILIZATION

Moist heat sterilization is the method of choice for liquid products such as parenteral and ophthalmic solutions. For thermolabile solutions, where even the use of high-temperature or short-exposure-time protocols imparts unacceptable levels of product degradation, filtration provides an alternative sterilization method. Filtration is also used to remove microorganisms and other particulate materials from gases as, for example, in the production of air with low particulate levels in aseptic production areas and operating theaters. This method differs from other sterilization methods in that microorganisms are physically removed from the product and not inactivated.

The requirement that the filter should not shed fibers or leach undesired materials into the solution being filtered has restricted the types of filter medium that can be used for sterile filtration of pharmaceutical products to those made of sintered glass, metal, or polymers. The polymeric membrane filters are almost universally used. Removal of particles by membrane filters is by a combination of sieving out of particles larger than the rated pore size and by adsorption within the filter matrix. Sieving is a function of the pore size of the surface pores of the membrane and adsorption is influenced by the thickness of the filter. Membrane filters most commonly used to effect sterilization are described as having a pore diameter of 0.2 to 0.22 μm. Although the filter will have either a "nominal" or an "absolute" pore size rating, the pores in the membrane consist of a range of sizes characterized by its pore size distribution. There is always the probability, although very remote, that a microorganism can pass through one of the few pores at the larger extreme of the pore size distribution. The absorption process in filtration also involves a degree of probability of retention. For these reasons, sterile filtration is a probability function and cannot be regarded as absolute. Aseptic precautions during the process are essential to avoid excessive bioburdens or recontamination and validation of sterile filtration must include stringent tests for sterility. With sterile filtration, as with all sterilization methods, attainment of a predetermined SAL depends upon good manufacturing practice and an initial low number of microorganisms in the product (see Chapter 12).

10.7 COMPARISON OF THE EFFICACY OF STERILIZATION METHODS

Because there is no microorganism that shows universal high resistance to all available sterilization methods, it is difficult to obtain a direct comparison of sterilization processes. An indication of the expected antimicrobial effectiveness of each process can be obtained by examining the effect of the process on the relevant biological indicator reference organism (see Chapter 11). In Table 10.3, the exposure time or absorbed dose required to achieve an inactivation factor of 10^{-8} is calculated for each sterilization method based on the highest acceptable D value for the reference organism. An IF of 10^8 would allow the attainment of a SAL of 10^{-6} or better with a bioburden of 100 microorganisms of resistance less than or equal to the reference microorganism. The time per absorbed dose recommended for sterilization by each method is also recorded. It should be noted that for moist heat, the D values for the reference B. stearothermophilus strain at 115, 126, and 134°C are calculated on the basis of a D_{121} of 1.5 min and a z value of 10°C, and for dry heat, the D values for the reference B. subtilis var. niger strain are calculated on the basis of a D_{160} of 10 min and a z value of 20°C.

It is apparent from Table 10.3 that, with the exception of moist heat at 115°C, the recommended sterilization protocols offer margins of safety if the process specification is an inactivation factor of 10^8. It is worthy of note, however, that an F_0 of 8 would only achieve an inactivation factor of $10^{5.3}$ against the reference organism for moist heat sterilization, and hence such a process could only be applied in the absence of organisms of resistance equal to or greater than the reference microorganism.

It must be remembered that these calculations are based on the response of carefully standardized organisms on selected carriers exposed to the sterilizing agent under ideal conditions. In practice, the contaminating microorganisms are not standardized and they are often exposed to the sterilizing agent under far from ideal conditions. To compensate for this, the exposure times or absorbed doses are often extended and this is particularly the case in gaseous sterilization where the situation is further complicated by the multiplicity of factors that are involved in the lethal process.

It seems clear, however, that even if the product to be sterilized is contaminated with an appreciable number of resistant spores, a high probability of achieving sterility can be ensured by the choice of an efficient sterilizing protocol. In reality, in industry and hospitals, far better levels of sterility assurance will be sought by

TABLE 10.3
Comparison of Microbial Inactivation Effectiveness of Sterilization Methods

Method	Biological Indicator Reference Organism	D Value	Time per Dose to Achieve an IF of 10^8	Recommended Sterilizing Time per Dose
Moist heat	*B. stearothermophilus*			
115°C	NCIMB 8157	6 min	48 min	30 min
121°C	(ATCC 7953, NCTC 10007)	1.5 min	12 min	15 min[a]
126°C		0.47 min	3.76 min	10 min
134°C	(z = 10°C)	0.075 min	0.6 min	3 min
Dry heat	*B. subtilis* var. *niger*			
160°C	NCIMB 8058 (ATCC 9372)	10 min	80 min	120 min
170°C		3.2 min	25.6 min	60 min
180°C	(z = 20°C)	1 min	8 min	30 min
Radiation	*B. pumilus* NCIMB 8982 (ATCC 14884, NCTC 8241)	3 kGy	24 kGy	25 kGy
Ethylene oxide 600 mg.L⁻¹, 54°C, 60% RH	*B. subtilis* var. *niger* NCIMB 8058 (ATCC 9372)	5.8 min	46.4 min	2 to 4 h+
Low temperature steam and formaldehyde 12 mg.L⁻¹, 73°C	*B. stearothermophilus* NCIMB 8224	5 min	40 min	1 to 2 h+

[a] 8 min if an F_0 of 8 is used.

Bacillus stearothermophilus renamed *Geobacillus stearothermophilus*.

applying good manufacturing practices. It is particularly important in this regard to recognize that the sterilization procedure is merely the last stage in a carefully planned and controlled manufacturing process and that all stages of manufacture must be controlled and monitored. It is particularly important that the microbial contamination levels be controlled at all stages, from the new materials at acquisition to the final product (see Chapters 3 and 4). The manufacturing environment must also be controlled and personnel adequately trained and their performance monitored (see Chapters 6 and 21). Only if all these aspects are carefully regulated can the health of the user be protected.

REFERENCES

British Pharmacopoeia. (2005). *British Pharmacopoeia*. The Stationery Office, London.

Gilbert, P. and Allison, D.G. (1996). Redefining the "sterility" of sterile products. *Eur. J. Parent. Sci.*, 1, 19–23.

Hodges, N.A. (2000). Relationship between inactivation factor, sterility assurance level and F_0 value — worked example. In *Handbook of Microbiological Quality Control: Pharmaceuticals and Medical Devices*. Baird, R.M. Hodges, N. A., and Denyer, S.P., Eds. Taylor & Francis, London, pp. 247–248.

Pflug, I.J. (1973). Heat sterilization. In *Industrial Sterilization: the Proceedings of the International Symposium, Amsterdam, 1972*. Briggs Phillips, G. and Miller, W.S. Eds. Duke University Press, Durham, NC, pp. 239–281.

FURTHER READING

Dring, G.J., Ellar, D.J., and Gould, G.W. (1985). *Fundamental and Applied Aspects of Bacterial Spores*. Academic Press, London.

Fraise, A.P., Lambert, P.A., and Maillard, J.-Y., Eds. (2004). *Russell, Hugo & Ayliffe's Principles and Practice of Disinfection, Preservation and Sterilization*. 4th ed. Blackwell Publishing, Oxford.

Gaughran, E.R.L. and Goudie, A.J. (1978). *Sterilization of Medical Products by Ionizing Radiation*. Multiscience Publications, Montreal.

Halls, N.A. (1994). *Achieving Sterility in Medical and Pharmaceutical Products*. Marcel Dekker, New York.

Hoskins, H.T. and Diffey, B.L. (1977). Tables for assessing the efficiency of autoclaves. *Pharm. J.*, 219, 218–219.

Hoxey, E.V. and Thomas N. (1999). Gaseous sterilization. In *Principles and Practice of Disinfection, Preservation and Sterilization*. Russell, A.D., Hugo, W.B. and Ayliffe, G.A.J., Eds. 3rd ed. Blackwell Scientific Publications, Oxford, pp. 703–732.

Johnson, E. and Brown, B. (1961). The Spearman estimator for serial dilution assays. *Biometrics*, 17, 79–88.

Meltzer, T.H. (1987). *Advances in Parenteral Sciences: 3. Filtration in the Pharmaceutical Industry*. Marcel Dekker, New York.

Morrissey, R.F. and Phillips, G.B. (1993). *Sterilization Technology*. Van Nostrand Reinhold, New York.

Pflug, I.J. (1988). *Selected Papers on the Microbiology and Engineering of Sterilization Processes*, 5th ed. Environmental Sterilization Laboratory, Minneapolis.

Pflug, I.J. and Holcomb, R.G. (1983). Principles of thermal destruction of microorganisms. In *Disinfection, Sterilization and Preservation*. Block, S.S., Ed. 3rd ed. Lea and Febiger, Philadelphia, pp. 751–810.

Richards, J.W. (1968). *Introduction to Industrial Sterilization*. Academic Press, London.

Stumbo. C.R. (1973). *Thermobacteriology in Food Processing*. 2nd ed. Academic Press, New York.

Stumbo, C.R., Murphy, J.R., and Cochran, J. (1950). Nature of thermal death time curves for PA3679 and *Clostridium botulinum*. *Food Technol.*, 4, 321–326.

11 Sterilization Methods

Eric L. Dewhurst and Eamonn V. Hoxey[*]

CONTENTS

[*] The opinions expressed in this chapter are those of the authors and do not necessarily represent their employers.

11.1   INTRODUCTION

Sterilization processes may be applied in the pharmaceutical and medical devices industries for three distinct activities:

1. Sterilization of products in their final containers.
2. Sterilization of components for subsequent aseptic assembly.
3. Sterilization of equipment to be used in aseptic manufacturing operations. Such sterilization may be carried out *in situ* or require disassembly, transfer to a sterilizer, and subsequent reassembly.

The effectiveness of sterilization cannot be demonstrated by testing a sample of product items drawn from a batch that has been exposed to the process. This is a consequence of the nature of microbial inactivation (see Chapter 10) and the limitations of the test for sterility (Brown and Gilbert 1977). Therefore, assurance of sterility is provided by (1) the initial validation of the process (see Chapter 12), (2) its periodic revalidation, and (3) the implementation of effective procedures for control and monitoring, demonstrating that the validated process has been delivered.

The validation and routine control of a sterilization process requires a number of interrelated activities to be carried out effectively. For this reason, sterilization processes are undertaken within a defined quality management system as defined by the requirements for good pharmaceutical manufacturing practice (GPMP) (Commission of the European Communities 2004) or in the international standards for quality management systems for medical devices [International Organisation for Standardisation (ISO) 2003a]. These quality management principles provide the structure for the effective control of general aspects such as:

- Calibration of instrumentation
- Planned, preventative maintenance of equipment
- Appointment of appropriately trained and qualified personnel to undertake designated activities
- Provision of appropriate, documented procedures together with the system of document control to ensure that these are available where required
- Safe retention of records, including batch records, calibration records, and maintenance records

In addition, GPMP (Commission of the European Communities 2004) and the quality management system requirements for medical devices (ISO 2003a) specify that, for the manufacture of sterile products, controls are established on:

- The manufacturing environment
- The health, cleanliness, and attire of personnel
- The raw materials, intermediate products, and manufacturing equipment

Specifying a sterilization process requires knowledge of the critical variables within the process, an understanding of the manner in which these variables interact, and an appreciation of the kinetics of microbial inactivation. For a given process, if the kinetics of microbial inactivation are understood and the critical variables can be measured directly, sterility can be ensured by monitoring the physical (and, if appropriate, chemical) conditions during sterilization. A system of this type is termed "parametric release" and has been defined as "release of sterile product based on process compliance to its specification" (Hoxey 1989, see also Chapter 12). For sterilization processes such as steam, dry heat, or irradiation, the physical conditions required are understood and can be directly monitored; therefore, a system of parametric release has been accepted (Hurrell 1986, British Pharmacopoeia 2001). Detailed requirements for the parametric release of terminally sterilized product are contained in Annex 17 to the European Union's *Guide to Good Manufacturing Practice* (Commission of the European Communities 2002). Standards for ethylene oxide sterilization (ISO 2006a) accepted the use of parametric release provided that a knowledge of the microbial inactivation kinetics was obtained during validation and all the process variables were directly monitored routinely. In general, however, for other gaseous sterilization processes and sterilization by liquid chemicals, parametric release is not yet accepted because the physical conditions cannot be readily and accurately measured. If parametric release is not accepted, biological indicators are incorporated into each sterilization batch. These indicators are removed after processing and cultured for surviving organisms, the growth of which is taken to indicate a failure to sterilize (see Chapter 12).

All sterilization methods have a finite capacity for destroying microorganisms (see Chapter 10). It is therefore necessary to control materials and processing prior to sterilization in order to reduce the presterilization microbial level (bioburden) and hence reduce the challenge to the process (ISO 2000a). Although determination of bioburden is outside the scope of this chapter (see Chapters 8 and 10), it should be emphasized that the frequency of bioburden determinations will vary for different processes and should be related to the capacity of the process for inactivation of microorganisms (ISO 2006b). The selection of a process with a low inactivation factor based on a low product bioburden will require extensive validation and monitoring of that bioburden (see Chapters 10 and 12). Such testing is invariably labor intensive and expensive. For a more detailed discussion on the practical issues, the reader is referred to the companion *Handbook* (Bill 2000).

For each method of sterilization, this chapter will consider the critical process variables and the manner in which these variables are routinely controlled and monitored, drawing on the established standards for the processes and equipment, and regulatory requirements. A number of references to standards is given in the text; standards are subject to continual review, amendment, and revision. Readers are advised to check for the latest version of any standard cited.

11.2 STERILIZATION METHODS: GENERAL ASPECTS

The properties of the ideal sterilization method would include:

- Absence of hazard to operators or to the environment
- High bactericidal, virucidal, and fungicidal activity producing a high assurance of sterility
- Fully understood physical conditions necessary for microbial inactivation that are easily controllable and measurable
- Permitting terminal sterilization in final packaging
- Compatibility with the range of materials to be sterilized
- Absence of toxic chemical residues in products
- Short processing time
- Low cost

The selection of a method of sterilization, therefore, depends on balancing the advantages of the available methods against their disadvantages. Table 11.1 summarizes these advantages and disadvantages and indicates the suitability or otherwise of these sterilization methods for a range of products and packaging materials.

The methods of sterilization have been presented in what is considered by the authors to be their order of preference. There are considerable advantages in utilizing

TABLE 11.1
Summary of Advantages and Disadvantages of the Various Sterilization Processes Together with Examples of Materials Suitable and Unsuitable for Processing

Method of Sterilization	Advantages	Materials Suitable for Processing	Materials Unsuitable for Processing	Disadvantages and Special Requirements
Steam Porous	Rapid cycle. Conditions required for sterilization known and documented. No requirement for biological indicators. Wrapped goods processed and suitable for storage. Automated equipment available.	Heat-stable materials, such as surgical instruments, dressing gowns and drapes, filters. Temperature-resistant plastics, such as high-density polypropylene. Packaging materials permeable to steam and air.	Temperature-, pressure-, or moisture-sensitive materials. Material combinations that will expand differentially and fracture on heating. Packaging materials impermeable to air or steam. Bottled fluids. Materials that are a barrier to air removal.	Will not depyrogenate.
Instrument and utensil	Rapid cycle. Conditions required for sterilization known and documented. Automated equipment available including small portable models. No requirement for biological indicators.	Heat-stable materials, such as unwrapped surgical instruments and utensils.	Wrapped goods. Bottled fluids. Temperature- or moisture-sensitive materials or combinations. Porous loads or devices with a narrow lumen from which air cannot be removed.	Products must be used immediately as they are unwrapped. Will not depyrogenate.
Bottled fluids	Conditions required for sterilization accepted and documented. Biological indicators not required. Automated equipment available.	Thermostable aqueous fluids. Packaging in rigid polymer, glass, or flexible polymer containers.	Nonaqueous fluids. Wrapped goods or porous loads. Temperature-sensitive products.	Cycle times may be long. Heat-up times require determination for each load and configuration. Explosion risk if containers overfilled. Will not depyrogenate. Need to control period between filling and sterilization to minimize microbial growth.

(continued)

TABLE 11.1 (continued)
Summary of Advantages and Disadvantages of the Various Sterilization Processes Together with Examples of Materials Suitable and Unsuitable for Processing

Method of Sterilization	Advantages	Materials Suitable for Processing	Materials Unsuitable for Processing	Disadvantages and Special Requirements
Dry Heat	Higher temperatures will depyrogenate. Sterilizers available with simple design and installation requirements. Sterilization conditions known and documented. No requirement for biological indicators.	Moisture-sensitive or steam-impermeable materials, for example, powders, nonaqueous fluids, some surgical instruments, such as sharps and powered drills. Aluminum foil preferred for packaging but glass or metal possible.	Aqueous fluids. Thermolabile materials.	Long cycle times. Heat-up times require determination for each load and configuration.
Irradiation	Product packaged in outer transit containers. No need for biological indicators. Gamma irradiation: highly penetrating, large-capacity sterilizers. Electron beam: no radioactive source, less degradation to plastic materials, rapid progress.	Range of plastic materials such as certain grades of polypropylene, styrene, acrylonitryl, polyethylene, and natural materials such as latex. Metallic products, subject to limits on density (e.g., orthopedic implants, scalpels).	Some plastics and glass undergo cross-linking leading to discoloration and embrittlement. Limited application to pharmaceutical products because of chemical alteration and breakdown products. Effect of radiation on packaging needs to be established.	High capital cost. Equipment requirements complex. Will not depyrogenate. Localized temperature increases. Gamma irradiators: control of radioactive material, slow process. Electron beam: low penetrating power, product thickness crucial, complex to control.

TABLE 11.1 (continued)
Summary of Advantages and Disadvantages of the Various Sterilization Processes Together with Examples of Materials Suitable and Unsuitable for Processing

Method of Sterilization	Advantages	Materials Suitable for Processing	Materials Unsuitable for Processing	Disadvantages and Special Requirements
Gaseous Sterilization Ethylene oxide	Cycle can be developed for particular products. Can process product packaged in transit containers. Automatic sterilizers available in variety of sizes from 1 to 30 m³ chamber capacity.	Polymeric materials such as low-density polypropylene, polyvinyl chloride, polymethyl-methacrylate, and polyurethane. Packaging materials must be permeable to air, water vapor, and gas, for example, sterilization-grade paper and spun-bonded polyolefin (Tyvek®, Dupont).	External surfaces of ampoules or vials because of ingress of gas via microfractures. Impermeable packaging, such as glass, metal. Products sensitive to high humidity present special problems. Products that are not clean because gas may not penetrate organic or inorganic soil.	Biological indicators required. No standard sterilization cycle available. Toxicity of gas. Residuals and explositivity in processed goods. Specialist expertise required. Prechamber humidification usually required. Will not depyrogenate.
Low temperature steam and formaldehyde (LTSF)	Automated sterilizers available. Can process some thermolabile materials.	Materials that will withstand temperatures up to 80°C. Packaging materials permeable to air, steam, and formaldehyde.	As ethylene oxide sterilization above.	Biological indicators required. Toxicity of gas. Residuals in processed goods. Specialist expertise required. Will not depyrogenate.
Gas plasma	Automated sterilizers available.	Thermolabile materials. Packaging materials permeable to air and gas.	As ethylene oxide sterilization above. Unsuitable for devices with a narrow lumen unless special adaptors used. No cellulosic products or packaging materials.	Expensive. Biological indicators required. Will not depyrogenate.

(continued)

TABLE 11.1 (continued)
Summary of Advantages and Disadvantages of the Various Sterilization Processes Together with Examples of Materials Suitable and Unsuitable for Processing

Method of Sterilization	Advantages	Materials Suitable for Processing	Materials Unsuitable for Processing	Disadvantages and Special Requirements
Filtration	Operates at ambient temperature. Large volumes on continuous basis. Also removes particles. Pyrogens not formed by processing. Choice of filter materials available which are compatible with even aggressive products.	Fluids that cannot be terminally sterilized.	Nonfluid products.	Aseptic processing after sterilization required. May not remove virus or mycoplasma. May not remove pyrogens from fluid stream. Sorption of some drugs, preservatives, and so on. Leaching of filter components. Integrity testing required.
Liquid Chemicals	Only method available for certain products.	Products that cannot be sterilized by other means, for example, materials of animal origin (heart valves, vascular prosthesis, tissue patches) and complex medical equipment, such as endoscopes.	Products that can be sterilized by other methods.	Aseptic processing usually required after sterilization. Control of production of sterilizing solution required, including filtration sterilization of sterilant. Solution must contact all parts of product. Biological monitoring and sterility testing required. Residual chemicals in product. Processing times extended. Will not depyrogenate. Regulatory requirements not established. Need to demonstrate effectiveness.
Emerging Sterilization Technologies	May be less hazardous to operators. May not leave residues in products. Processing times may be short.	May be compatible with items that cannot be processed in another way.	Will depend upon method.	

processes with established technology and regulatory requirements. It can be extremely expensive and time-consuming to perform research and development into novel processes and to break new ground with regulatory authorities.

11.2.1 STERILIZING EQUIPMENT

Each method of sterilization places constraints and requirements on the design of the equipment used for the process. The majority of these requirements are special to the particular process but one feature is common to all sterilization methods. Each sterilizer is designed to control sterilization conditions within established limits, and a recording of the actual conditions produced during the process is required. The control and system of recording must be independent of each other, using the output of separate sensors to confirm that the required conditions have been achieved and maintained.

11.2.2 PACKAGING FOR STERILE PRODUCTS

Sterile products are presented in sealed packages to maintain sterility up to the point of use. The principal requirements for such packaging have been presented in European and International standards (ISO 2006c; see also Chapter 16). The principal requirements are that the packaging:

- Is compatible with the product contained within it
- Is compatible with the sterilization system employed
- Is compatible with the labeling system used
- Provides physical, chemical, and microbiological protection to the product
- Meets the requirements of the user at the point of use, for example, by permitting aseptic presentation of the product

11.2.3 STERILIZATION SUBCONTRACTORS

Manufacturers may have sterilization processes carried out on their behalf by subcontractors. This may be for a number of reasons, including lack of appropriate facilities, shortage of capacity for sterilization, or limited expertise in sterilization. The most common sterilization processes to be performed by subcontractors are radiation and ethylene oxide. Specialist subcontractors for both these processes are located throughout the world.

When sterilization is undertaken by a subcontractor, the manufacturer of the product retains overall responsibility. There should be a formal contract defining the limits of responsibility of both parties and detailing the technical requirements of the processing. The subcontractor should be subject to formal audit (ISO 2003a, Commission of the European Communities 2004).

11.2.4 PRODUCT RELEASE FROM STERILIZATION

There are some essential requirements for product release following sterilization that are common to all the available sterilization methods. These general points include evidence that:

- The sterilization facility had been appropriately maintained, the process validated, and routine equipment tests performed in accordance with documented procedures.
- The documented procedures relating to product handling have been followed.
- The process conformed to specification.
- The automatic sterilization process performed satisfactorily.
- Sterilizer cycles were correctly and completely documented.
- The load and its packaging were not damaged during transport or processing.
- Any chemical indicators used to differentiate between processed and unprocessed products were satisfactory.

11.3 STEAM STERILIZATION

Steam is the method of choice for sterilization because it provides the majority of features of the ideal sterilization method. It is used either as a direct contact sterilant, such as in a porous load and "instrument and utensil" sterilizer, and for sterilization-in-place systems for manufacturing equipment, or as a heat transfer medium to raise the temperature of the contents of containers, for example, in a bottled fluids sterilizer. Standards for the validation and routine control of moist heat sterilization define requirements for use of steam in both applications (ISO 2006d). A standard for sterilization-in-place in aseptic processing is in preparation (ISO in preparation). Heat transfer is achieved in steam sterilizers by the release of the latent heat from steam as it condenses.

A series of time–temperature relationships are accepted as providing satisfactory sterilizing conditions [National Health Service (NHS) Estates Agency 1995a]. These conditions are illustrated in Table 11.2 but are not equipotent (see Table 11.4 and Chapter 10). The British Pharmacopoeia (2005) references a temperature of 121°C for 15 min and requires that any other time-temperature combination should result in a maximum sterility assurance level of 10^{-6}.

It is generally agreed that the highest temperature compatible with the product to be sterilized should be used as this combines the advantages of highest assurance

TABLE 11.2
Time–Temperature Relationships
for Steam Sterilization

Temperature Range (°C)	Minimum Holding Time at Temperature (min)
134–137	3
126–129	10
121–124	15
115–118	30

of sterility with shortest cycle time. Steam sterilization can be monitored by physical measurement of the conditions because the kinetics of microbial inactivation by steam has been extensively studied (reviewed by Russell 1982). The routine addition of biological indicators to steam sterilizers offers no advantages and does not increase the assurance of sterility (Anonymous 1959). Biological indicators may be of value in validation of sterilization of difficult loads such as filter cartridges. Some regulatory authorities require the use of biological indicators in validation studies.

11.3.1 POROUS LOAD STERILIZER

Steam sterilizers used to process wrapped goods and porous materials are commonly referred to as "porous load," "high vacuum," or "prevacuum" sterilizers. To ensure that the cycle achieves sterilizing conditions, it is important that all the air is removed, permitting contact between product and steam (Anonymous 1959). This can be achieved by a combination of evacuation and steam injection as a series of pulses that dilute and remove the air. Porous load sterilizers usually operate at the highest temperatures given in Table 11.2 (134 to 137°C) although sterilizers operating at 126 to 129°C and 121 to 124°C are in use.

11.3.1.1 Equipment

European standard EN 285 (CEN 1997a) details the requirements for porous load sterilizers and the major aspects are summarized in Table 11.3.

One of the most important factors for the correct operation of porous load sterilizers is the quality of the steam, the ideal quality being represented by dry saturated steam (NHS Estates Agency 1995a, 1997). If the steam is too dry, superheat may be created,

---

TABLE 11.3
Principal Features of Porous Load Steam Sterilizers

Feature	Rationale
Steam supply of suitable quality and quantity.	To ensure correct functioning of sterilizer.
Air removal system.	To ensure rapid and even steam penetration throughout chamber and load.
Air detector.	To ensure adequate air removal is achieved.
Automatic cycle controller.	For reproducible cycle control.
Time, temperature, and pressure recorder.	To provide batch or process record.
Sterilization hold time initiated by drain temperature using separate sensor from that supplying recorder.	To ensure coolest point of chamber (drain) reaches temperature prior to timing of hold period. Temperature record provides independent confirmation of attainment of correct conditions.
Cycle counter.	For maintenance and batch records.
Door interlock and noninterruptable cycle.	Operator safety and to prevent underprocessing.
Fault indication.	To indicate incorrect function of sterilizer.
Leak test gauge of suitable scale range.	For routine chamber integrity test when small changes in chamber pressure must be measured.

but overly wet steam can produce a wet load at the end of the sterilization cycle and wet packaging is not an effective microbial barrier (Placencia et al. 1986).

Noncondensable gases in the steam supply, caused, for example, by incorrect boiler feed water treatment, may affect sterilizer performance, influence effectiveness of the process, hinder steam penetration (see Section 11.3.1.3), cause chamber overheating, or result in inconsistent air detector performance (NHS Estates Agency 1995a).

Superheated steam is unsuitable for steam sterilization; it can be considered equivalent to a dry gas and therefore dry heat sterilization time–temperature combinations would be required (see Section 11.4). Superheat can be due to adiabatic expansion as a result of excessive reduction in pressure through a throttling device such as a reducing valve. The rehydration of exceptionally dry hygroscopic material is an exothermic reaction that can also cause superheating; this may persist for the entire sterilization hold time. For this reason, textiles, particularly those containing cotton, should be allowed to equilibrate in air after laundering before they are sterilized (NHS Estates Agency 1995c).

Wet steam supply to the sterilizer is one cause of wet loads at the end of the sterilization process. This may be due to inadequate design or maintenance of the steam main, such as inadequate sloping and draining of the main, poor thermal insulation, incorrect installation of separators and steam traps, or "priming" in the boiler leading to carryover of boiler water with the steam (NHS Estates Agency 1995a, 1997).

For these reasons, the importance of correct installation and maintenance of steam boilers and distribution systems, together with the determination of the quality of steam provided at the sterilizer location, cannot be overemphasized (NHS Estates Agency 1997).

Incomplete air removal can cause a failure to sterilize (Anonymous 1959). Air remaining in packages prior to the sterilization hold period will occur in random locations and may unpredictably delay or prevent steam from contacting surfaces over which air is present. Air removal is achieved by drawing a vacuum to between 40 and 60 mbar (4 to 6 kPa) absolute followed by dilution and removal of residual air by a series of steam injections and evacuations (see Figure 11.1). The correct function of the air removal portion of the sterilization cycle is monitored by the Bowie–Dick test on a daily basis and by the air detector for each sterilization cycle (see Section 11.3.1.3).

11.3.1.2 The Process

The essential features of a porous load steam sterilization cycle are illustrated in Figure 11.1. The variables that can affect the efficacy of the process are time, temperature, and pressure (because of its direct relationship with the temperature of dry saturated steam). Factors such as steam quality (see Section 11.3.1.1) and loading pattern will affect air removal and load dryness on cycle completion.

11.3.1.3 Routine Tests of Performance

The Bowie–Dick test (Bowie et al. 1963) is the standard test for steam penetration that should be performed on porous load sterilizers each day of operation. The test

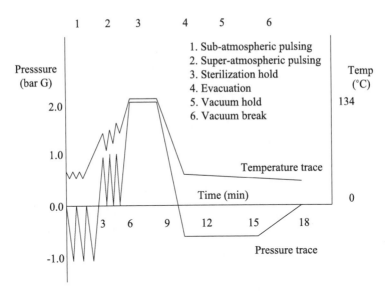

FIGURE 11.1 Schematic diagram of the stages of a steam sterilization cycle for porous loads.

consists of operating the cycle with a test pack containing a temperature-sensitive indicator as the only load. On completion of the cycle, the indicator is removed from the pack and examined for a uniform color change. The basis of the test is that inadequate air removal will leave a pocket of air in the center of the pack that will then act as a barrier to even steam penetration and result in an uneven color change in the indicator. Traditionally, this test was performed with a pack of "Huckaback" linen towels and a test sheet with a cross of autoclave indicator tape. However, other materials for this test have been developed and performance requirements for materials used in the Bowie–Dick test have been detailed in published specifications (ISO 2000b,c, ISO 2001).

The Bowie–Dick test is a test of air removal based on detecting rapid and even steam penetration into the pack. Steam penetration may be affected by the steam quality and, in particular, by superheat and the presence of noncondensable gas in the steam supply. Failed Bowie–Dick tests therefore require careful analysis of sterilizer performance.

Each porous load sterilization cycle should be monitored by an air detector (NHS Estates Agency 1995a). This device is part of the sterilizer control system and should be set to indicate a fault if there is sufficient air in the sterilizer to compromise the sterilization cycle (Pickerill et al. 1971). If the correct conditions are not established, this device should indicate a failure and the load be treated as nonsterile. The correct functioning of this device needs to be established at the commissioning of the sterilizer and should be routinely confirmed by an air detector performance test (NHS Estates Agency 1995b).

The integrity of the sterilizer chamber should be routinely checked by performing a leak test (NHS Estates Agency 1995b). A vacuum is drawn in the chamber and the subsequent pressure rise monitored over a timed period.

11.3.1.4 Product Release

The general points described earlier (see Section 11.2.4) for product release apply. A complete sterilization cycle record comprises time, temperature, and pressure.

11.3.2 "INSTRUMENT AND UTENSIL" STEAM STERILIZERS

"Instrument and utensil" steam sterilizers are also referred to as "bowl and instrument," "dropped instrument," "gravity displacement," or "downward displacement." The descriptions "gravity displacement" and "downward displacement" are, in fact, often incorrect as many sterilizers of this type generate steam within the sterilizer chamber and rely on upward displacement of air by steam. Like porous load sterilizers, their effectiveness relies on direct contact between steam and the surface to be sterilized. However, in these machines, air is directly displaced by steam, either admitted from a separate steam source or generated within the chamber; there is no evacuation either to aid air removal or to help dry the load at the end of the cycle. These machines are therefore unsuitable for sterilizing wrapped goods or porous loads (NHS Estates Agency 1994, Medical Devices Agency 1996a). Requirements for small sterilizers of this type are given in a European standard (CEN 2004).

Sterilizers of this type include the common portable machines that require a simple connection to an electrical power supply. They often operate at the highest temperatures given in Table 11.2, although machines operating over the range of conditions specified are available.

11.3.2.1 Equipment

Most machines are portable and are not equipped with temperature recorders. A monitoring sensor, separate from the control sensor, should be incorporated into the automatic control system to indicate a failed cycle should the required temperature not be maintained for the designated time. Temperature and pressure gauges should be provided to allow the operation of the equipment to be checked.

11.3.2.2 The Process

The essential features of an instrument and utensil sterilization cycle are air displacement, sterilization, and cooling. These are summarized in Figure 11.2.

11.3.2.3 Product Release

In addition to the general points described earlier (see Section 11.2.4), a complete sterilization cycle record of time and temperature is required when a temperature recorder is installed with the sterilizer. If a temperature recorder is not fitted, the operator should confirm the attainment of the correct conditions by checking the temperature and pressure indicated by the sterilizer instrumentation during the sterilization stage and entering these values in the sterilizer logbook.

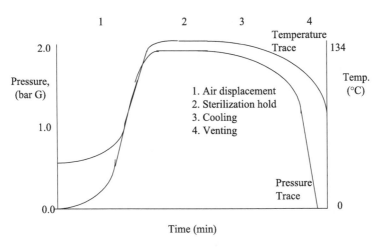

FIGURE 11.2 Schematic diagram of the stages of a steam sterilization cycle for instruments and utensils.

11.3.3 BOTTLED FLUIDS STERILIZER

Unlike porous load and instrument and utensil steam sterilizers, bottled fluids machines do not rely on contact between the steam and the product to effect sterilization. The steam is a transfer medium that heats the container; the actual sterilization occurs in the product fluid. Steam is used because it is an efficient heat transfer medium and the pressure associated with steam will help to counteract the pressure rise within a heated sealed container. It should be noted that the fluid to be sterilized must be aqueous. The presence of water is a critical factor in microbial inactivation leading to denaturation of proteinaceous materials. In nonaqueous fluids such as oils the microbial inactivation is by dry heat resulting in oxidative processes that require much higher temperatures and longer times to achieve sterilization (see Section 11.5 and Chapter 10).

The lower temperature treatments quoted in Table 11.2 are used for fluids loads because of the lower associated pressure employed, that is, 121 to 124°C for 15 min or 115 to 118°C for 30 min. The F_0 value may be used as an alternative to defining the sterilization process as a minimum hold time at a given temperature (see Chapter 12). The lethality of a process includes the effects of the heating and cooling periods as well as the sterilization hold time. In a large fluids load, the heat-up and cool-down times may be long and have a considerable microbiocidal effect. The F_0 value expresses the lethality of the whole process as an equivalent hold time at 121°C and its calculation has been extensively reviewed (Deindoerfer and Humphrey 1959, De Santis and Rudo 1986). It is generally accepted that an appropriate F_0 value should be delivered to every container in the load. For heat-labile products, F_0 values of less than 8 may be acceptable if the level and heat resistance of the product bioburden are known and are closely monitored; however, a sterility assurance level of 10^{-6} must be demonstrated (British Pharmacopoeia 2005; see also Chapter 12).

It should be noted that lower temperature processes impart significantly less lethal effect than those operating at higher temperatures. This is demonstrated in Table 11.4 where the lethal effects of the sterilization hold periods (as measured by the F_0 value) of accepted standard conditions are compared.

11.3.3.1   Equipment

The basic requirements for bottled fluids sterilizers are outlined in Table 11.5. The British Standard 3970 Part 2 (British Standards Institution 1991) refers to this type of equipment.

TABLE 11.4
F_0 of the Holding Period of Standard Sterilizing Cycles Previously Specified in Table 11.2

Nominal Temperature (°C)	Sterilization Holding Time (min)	F_0 Value
134	3	59
126	10	31
121	15	15
115	30	8

TABLE 11.5
Features of Bottled Fluids Steam Sterilizers

Feature	Rationale
Steam supply of suitable quality and quantity.	To ensure correct functioning of the sterilizer.
Automatic cycle controller.	For reproducible cycle control.
Time, temperature, and pressure recorder.	To provide batch or process records.
Sterilization hold period time, or F_0 integration, taken from load simulator using separate sensor from that supplying the recorder.	To ensure that coolest point of load controls the cycle and that the temperature record is an independent confirmation of the attainment of correct conditions.
Temperature control from a drain sensor during sterilization hold period.	More sensitive to temperature fluctuations in the chamber than a load simulator.
Cycle counter.	For maintenance and batch records.
If temperature is monitored or controlled from a load bottle this must not control door opening.	If the load bottle breaks, the probe will cool more rapidly than the load bottles and door opening may lead to explosion.
Chamber overpressure (air ballasting).	To prevent bottle breakage or distortion of flexible containers especially during cooling cycle.
Assisted cooling either by fan circulation of air or spraying load with recycled condensate.	To reduce the time for load to cool to below 80°C.
Door interlock and noninterruptable cycle. Door must not open until entire load is below 80°C.	Operator safety. Serious accidents have occurred due to load explosion when sterilizer door opened with load above 80°C.

As shown in Table 11.5, the design of a bottled fluids sterilizer has to consider many factors; certain special features merit detailed consideration.

Heating a sealed container—The heating of a sealed container of aqueous fluid will raise the pressure within that container. The steam pressure in the sterilizer chamber and the strength of the container itself will each counteract this pressure rise to some degree. However, the sterilizer chamber pressure is often increased with the addition of sterile air to above that associated with dry saturated steam at the sterilizing temperature; this prevents bottle breakage or deformation of polymeric containers and is known as air ballasting. Additional pressurization with air could introduce problems of temperature variations within the sterilizer chamber unless the air–steam mixture is homogeneous. Forced circulation of the air-steam mixture is usually employed to provide this homogeneity. Operation of forced circulation systems should be monitored during each cycle.

Cooling times—The cooling times for large loads may be extremely long. In order to decrease the total cycle time, a cooling system may be employed in which the load temperature is reduced by spraying with water. Contamination of the contents of containers with spray-cooling water has occurred and therefore water used for this purpose must be sterile. If the cooling water is derived from a water or steam service, it must itself be sterilized by exposure to sufficient heat to give an F_0 of 8. Where heat exchangers are used as part of the cooling system, they should be designed to fail-safe conditions and prevent cross-contamination from the primary circuit to the secondary circuit (NHS Estates Agency 1995).

Explosions—Serious explosions have resulted from the transfer of loads of hot fluids from the sterilizer into rooms at ambient temperature. Therefore, a door interlock must be installed to prevent removal of the load until the contents of all containers are below 80°C for glass or rigid polymer or 90°C for flexible polymer containers (NHS Estates Agency 1994). This interlock should operate through a load simulator, rather than a temperature sensor in an actual load container, in case the latter breaks during the process.

11.3.3.2 The Process

The essential features of a bottled fluids steam sterilization cycle are heat up combined with air displacement, sterilization, and cooling; these are illustrated in Figure 11.3.

The effectiveness of the process is determined by the time at sterilizing temperature. The size of the load and the size of the individual containers within the load will significantly affect heat-up and cool-down times and hence total cycle time.

11.3.3.3 Product Release

In addition to the general points described earlier (see Section 11.2.4), and to support product release following a bottled fluids steam sterilization cycle, the sterilization cycle temperature–time records or confirmation of attainment of required F_0 value are required. Where appropriate, a sterility test result will also be required.

11.3.4 STERILIZATION IN PLACE

Steam sterilization is becoming widespread for the sterilization of preassembled manufacturing and filling systems. Such sterilization-in-place (SIP) processes can

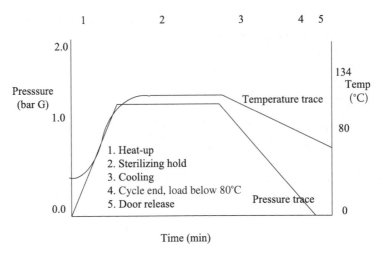

FIGURE 11.3 Schematic diagram of the stages of a steam sterilization cycle for fluids in sealed containers.

cover applications ranging from simple vessel-filter-filler systems up to complete biotechnology facilities.

The basic principles of porous load sterilization apply. For the steam to contact all surfaces, effective air removal and condensate removal are vital. The system must also be clean to ensure that organisms are not protected from steam contact by organic matter or product residues. A section on sterilization in place is in preparation as part of an International Standard on aseptic processing (ISO in preparation).

11.3.4.1 Equipment

It is critical that the system to be sterilized is designed and constructed to allow cleaning in place (CIP) (ISO 2003b) and SIP. Details of the engineering design criteria for CIP and SIP systems have been reviewed by Seiberling and Ratz (1995). The main points are summarized here.

The system design must withstand the pressures required for sterilization (dry saturated steam at 15 psi gives a temperature of 121°C). It must be adequately vented through microbial retentive filters to allow air removal, and condensate must be trapped off from all the low points.

Because the steam comes into contact with product-contact surfaces, it is generally necessary to carry out the sterilization using clean steam (NHS Estates Agency 1997). The steam should also meet the required criteria for dryness and noncondensable gases (see Section 11.3.1.1).

Many systems require the use of sterile air for cooling after sterilization and maintenance of overpressure to prevent recontamination of the system after sterilization. Temperature monitoring is required at multiple points within the system to demonstrate that sterilization temperatures have been achieved.

11.3.4.2 The Process

Although each process is unique, the following basic principles apply in most cases.

Following cleaning, steam is introduced into the system with all vent filter valves open and bypass valves on steam traps open. As air is displaced from the system, the temperature on the temperature sensors located at the traps rises and reaches a steady state equivalent to dry saturated steam at the pressure achieved in the system with the vents open. At this point, the vents are closed and the pressure is allowed to build to achieve sterilizing temperature. In systems with more than one steam entry point, great care must be taken to ensure that air is not trapped between two moving faces of steam.

Because sterilization of filters within the system occurs by passage of steam through the filter, care must be taken not to exceed the maximum pressure differential of the filter during introduction of the steam into the system.

11.3.4.3 Product Release

As with other steam sterilization applications, a complete cycle record comprising time, temperature, and pressure is required to support the release of the system for use (see also Section 11.2.4).

11.4 DRY HEAT STERILIZATION

Dry heat sterilization is the method of choice for heat-stable but moisture-sensitive items. The British Pharmacopoeia specifies a time–temperature combination of 160°C for 120 min (BP 2005). Other recognized time–temperature combinations for dry heat sterilization are 180°C for 30 min and 170°C for 60 min. It should be noted that the temperatures and times required for dry heat sterilization are significantly greater than for steam sterilization (see Section 11.3) and, as with time–temperature combinations for moist heat, the various dry heat combinations are not equipotent.

Dry heat at higher temperatures than those specified for sterilization may also be used for depyrogenation of some materials. The BP specifies a temperature of 220°C (BP 2005). The application of temperatures in excess of 250°C has been recommended (Halls 1994). An international standard covering the requirements for development, validation, and routine control of dry heat sterilization is in development (ISO in preparation).

11.4.1 EQUIPMENT

Details on the specification and installation of dry heat sterilizers are given in Health Technical Memorandum 2010 (NHS Estates Agency 1995a). An international standard describing safety requirements for dry heat sterilizers is available [International Electrotechnical Commission (IEC) 2005]. There are two main types of dry heat sterilizers: batch sterilizers that are basically hot air ovens, and tunnel sterilizers that can operate continuously as product is passed through a heated tunnel (Halls 1994). Table 11.6 illustrates the features of a hot air sterilizing oven.

TABLE 11.6
Features of Hot Air Sterilizing Ovens

Feature	Rationale
Open-mesh shelving and adequate circulating fan	To allow even heat distribution and maximum heating
Temperature recorder with no control function	To provide independent batch or process history
Overheat cutout	Safety and product protection
Thermally insulated chamber	
Cycle counter	Maintenance records (batch recording)
Forced cooling and filtered air	Increased turn-around or shortened cycle
Thermocouple inlet	Validation purposes
Door interlock and automatic noninterruptable cycle	Prevent inadvertent underprocessing
Sterilization hold period timer initiated by chamber temperature (separate sensor from chart recorder)	Timing of sterilization period
Fault indication	To indicate incorrect function of sterilizer

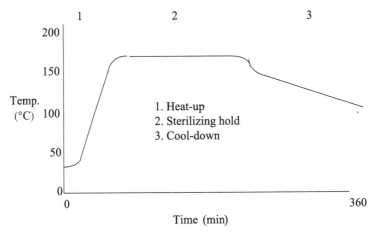

FIGURE 11.4 Schematic diagram of the stages of a dry heat sterilization process.

11.4.2 THE PROCESS

The essential features of a hot air sterilization cycle, heating, sterilizing, and cooling are illustrated in Figure 11.4.

The heating of the contents of a hot air sterilizer, batch, or tunnel process relies on sensible heat transfer. In a batch sterilizer, an even temperature distribution throughout the load is essential and design features are incorporated into the sterilizer to minimize temperature variations (see Table 11.6). However, the distribution of the load within the chamber can also dramatically affect the performance of the sterilizer. Variation from the established loading pattern may seriously affect the heat distribution and sterilization conditions may not be attained throughout the load (NHS Estates Agency 1995c). The loading pattern must therefore be controlled and hot air sterilizers must not be overfilled.

Tunnel sterilizers rely on passage of the load at a fixed speed on a conveyor through a heated tunnel. Failure of one or more heating elements within the tunnel or variation in conveyor speed can compromise the process. Where tunnels feed into classified areas, the effects of pressure differentials and the resulting airflows into the tunnel must be considered.

11.4.3 Product Release

Product release following a dry heat sterilization process in a batch sterilizer requires complete sterilization records of time and temperature, and evidence that any forced air circulation systems were fully operational throughout the cycle, in addition to the general points outlined earlier (see Section 11.2.4).

For tunnel sterilizers, records of temperature, belt speed, and processing times are required.

11.5 RADIATION STERILIZATION

The use of ionizing radiation is the method of choice for the sterilization of heat-labile materials that will withstand the high levels of radiation involved. The two methods currently available use either gamma rays from a radioactive source or accelerated electrons (beta particles) from an electron beam sterilizer (see Chapter 10). The use of machine-generated X-rays has been under investigation for sterilization (Saylor 1991) but is now becoming economical for sterilization on a commercial scale.

International and European standards are available that give the requirements for validation and routine control of sterilization by irradiation (ISO 2006e). These standards accept two approaches to establishing the dose of radiation that is required to achieve the preselected sterility assurance level (SAL): selection of a product-specific dose based on a knowledge of the numbers and resistance to radiation of the bioburden on product items, and use of a sterilizing dose of 25 kGy with evidence that this dose will achieve the preselected SAL (Kowalski and Tallentire 1999, 2003: Kowalski et al. 1999). Methods for establishing product-specific doses are described in the International Standard ISO 11137-2 (ISO 2006f).

11.5.1 Equipment

There are two types of plant used for gamma irradiation, batch or continuous; both have similar source arrangements. Within the irradiation cell an array of cobalt-60 rods (or caesium-137) is contained in a deep pit filled with water to a depth of about four m. In a continuous process, the cobalt source is raised from the water and product passes into the cell along a conveyor that tracks a zigzag path around the cell to ensure all sides of the load receive equal exposure. The dose delivered to the load is dependent on source strength and exposure time. A typical dwell time may be up to 12 h. In a batch process, the load is manually placed in a static position in the cell before the source is raised. In this batch approach, some variation will exist in the dose absorbed by product items in different locations within the load.

The basic principle of an electron beam sterilizer is that electrons from a hot filament or cathode source are emitted into a vacuum, accelerated by passage through high-voltage electrodes, and concentrated into a beam. The maximum energy for sterilizing is around 10 MeV although some machines operate at 3 to 4 MeV. The product passes on a conveyor under the beam, which scans from side to side across the product. The dose delivered to the product is determined by the energy of the electrons, the speed of the conveyor, the width of beam, and the speed at which it sweeps across the conveyor. As penetrating power is low, particularly at electron energies of 3 to 4 MeV, the dose received by the product at the bottom of the container is influenced by its thickness and density. In contrast to gamma irradiation, the dwell time of product in an electron beam sterilizer is usually only a few seconds or minutes.

11.5.2   THE PROCESS

For cobalt-60 and a fixed path through the irradiator, the only process variable is the source exposure time. For a load of given density and a source of given activity, the time of exposure to the source (dwell time) is adjusted to ensure the minimum required dose at the lowest dose point. The dose received at any point can be measured by the use of dosimeters. National standard traceable "red perspex" dosimeters such as those prepared by the U.K. Atomic Energy Authority at Harwell give a reproducible change in absorbance for a given dose of radiation. A calibration curve of change in absorbance at various irradiation doses is prepared for each batch of dosimeters. Thereafter the dose delivered to product containers can be determined after irradiation. Each batch of product should routinely contain at least two dosimeters in the first and last container in the batch located at a point which can be related to the minimum dose position by the validation data.

Routine control of a gamma irradiator is relatively straightforward. A chart record should show the position of the source (up or down) and the product dwell time.

Electron beam sterilizers are more difficult to control, as several factors affect the dose received by the product. These factors are electron voltage, beam current, width of the beam scan, conveyor speed, and product density. Dose to product is monitored with dosimeters, usually in the form of radiochromic dye film, and delivered dose is determined using a colorimeter. Routine controls on an electron beam sterilizer consist of continuous measurement of the characteristics of the beam and the conveyor speed (ISO 2006e).

The dose delivered to each batch of product must be determined by whichever method of irradiation is used. Color change monitors are attached to each container of product to indicate that it has been processed and products arriving at the irradiation facility are physically segregated from processed goods. For products sterilized by a contractor, all the above will be carried out by the contractor, who provides documentary evidence of the dose delivered to the product. The manufacturer may supply product boxes containing dosimeters to be returned with the load but such monitoring is problematical because it introduces a number of variables, particularly the temperature and times under which the dosimeters are stored and transported between irradiation and assessment. These will influence the dosimetric system; such monitoring, therefore, is of little value in practice (ISO 2006g).

11.5.3 PRODUCT RELEASE

Following an irradiation sterilization process, evidence that the required dose has been delivered to the product must be available in addition to the general points for product release discussed earlier (see Section 11.2.4). Evidence of dose delivered may be the actual dosimeter results or a certificate giving the dose delivered.

11.6 GASEOUS STERILIZATION

Many chemicals that can be generated in a gaseous phase have microbicidal activity including ethylene oxide (Phillips and Kaye 1949), formaldehyde (Nordgren 1939), propylene oxide (Bruch 1961), methyl bromide (Kolb and Schneiter 1950), betapropiolactone (Allen and Murphy 1960), peracetic acid (Portner and Hoffman 1968), chlorine dioxide (Knapp et al. 1986), and ozone (Hoffman 1971) (see Chapter 10). The use of all these agents as sterilants or disinfectants has been proposed at some time but ethylene oxide is the most widely used gaseous sterilant in the medical device and pharmaceutical industry (Hoxey 1989). Formaldehyde gas has been used for fumigation in a number of areas (Ackland et al. 1980), and gaseous formaldehyde has been combined with steam at subatmospheric pressure in the low temperature steam and formaldehyde (LTSF) sterilization process used in hospitals in Northern Europe for processing multiple-use, heat-sensitive items (Hoxey 1991). More recently, sterilization methods employing oxidizing agents have been available. The most widespread of these methods has been the use of gas plasma (see Chapter 10 and Section 11.6.3).

This section will consider only ethylene oxide, LTSF, and gas plasma. Other methods will be discussed in the later section on emerging sterilization technologies (see Section 11.9).

11.6.1 ETHYLENE OXIDE

Ethylene oxide gas is used to sterilize heat-sensitive materials, principally plastics, that will not withstand irradiation. There is no standard set of conditions for ethylene oxide sterilization. Each cycle in use is individually developed and microbiologically validated for the particular product to be sterilized (ISO 2006h,i). These standards are in the process of being revised. The effectiveness of the process is affected by a number of variables including time of exposure, temperature, humidity, gas concentration and pressure, gas penetration and distribution, and all must be carefully controlled (Hoxey 1989). The range of conditions in use is illustrated in Table 11.7.

Ethylene oxide sterilization cycles in current use vary enormously because each product presents distinct problems for cycle development. Although the interrelationship of factors that affect microbial inactivation by ethylene oxide have been extensively studied (reviewed by Hoffman 1971, Russell 1982), the product characteristics constrain the choice of conditions used such that an optimum cycle suitable for any product does not exist.

TABLE 11.7
The Range of Conditions Used for Ethylene Oxide Sterilization

Factor	Conditions
Concentration of ethylene oxide	250–1500 mg.L^{-1}
Temperature	30–65°C
Exposure time	1–30 h
Humidity	30–95%

Source: From Hoxey, E.V. (1989). In *Proceedings of the Eucomed Conference on Ethylene Oxide Sterilization.* April 21–22, 1989, Paris. Eucomed, Brussels, pp. 25–32. With permission.

Ethylene oxide is flammable and toxic and its use is controlled to protect the health and safety of workers. For example, the Health and Safety Executive has set the maximum exposure level to ethylene oxide in the United Kingdom at a time-weighted average over 8 h of 5 ppm (Health and Safety Executive 1996). The operation of an ethylene oxide sterilization process requires specialist knowledge and preferably microbiological facilities at the site of the operation.

11.6.1.1 Equipment

A European Standard specification for ethylene oxide sterilizers, EN 1422, has been published (CEN 1997b). A basic summary of equipment features for a sterilizer utilizing ethylene oxide is given in Table 11.8.

Ethylene oxide gas may be supplied to a sterilizer either as the pure gas or mixed with an inert diluent such as a hydrochlorofluorocarbon, carbon dioxide, or nitrogen. Chlorofluorocarbons (CFCs) were employed as a diluent for many years but have been removed because of their adverse effects on the environment (Jorkasky 1993). Hydrochlorofluorocarbons (HCFCs) have been introduced as a short-term replacement but these will have to be phased out in turn by 2020 (Commission of the European Communities 1994, UNEP 2002).

Supplies of ethylene oxide should be purchased to comply with a formal specification agreed with an appropriate supplier. Ethylene oxide should be treated as any other raw material and undergo documented incoming inspection. A certificate of analysis, which may be provided by the gas supplier, should be available for each batch of sterilant and this should demonstrate conformance with the purchasing specification.

11.6.1.2 The Process

The principal stages of an ethylene oxide sterilization cycle are illustrated in Figure 11.5.

The development of an ethylene oxide sterilization cycle allows selection of a suitable combination of factors including ethylene oxide gas supply (pure or mixed with an inert carrier gas), ethylene oxide concentration, temperature, pressure

TABLE 11.8
Features of Ethylene Oxide Sterilizers

Feature	Rationale
Prechamber humidification with temperature control and steam injection	To increase efficacy of process and reduce cycle time in the chamber
Automatically controlled sterilization process	To avoid variations between cycles inherent in manual control
Evacuation cycle	To remove air and ensure penetration of ethylene oxide
In-chamber humidification by steam injection	To replace moisture removed by evacuation stage of cycle
Jacketed chamber	To control cycle temperature
Time–temperature, pressure, and humidity recorders, separate to sterilizer controller	To provide independent batch process history
Over temperature cutout	To protect product
Heated gas vaporizer	To prevent admission of liquid ethylene oxide to sterilizer chamber
Cycle counter	Maintenance records (batch records)
Forced gas circulation in chamber	To minimize variations in conditions throughout sterilizer chamber
Thermocouple inlet	Validation purposes
Door lock and noninterruptable cycle	Product and operator protection
Ethylene oxide concentration in chamber controlled by pressure rise; independent monitor system such as weight loss from gas cylinder for in-chamber concentration measurement	Check of in-chamber concentration
Fault indication system	To indicate incorrect functioning of sterilizer
Degassing area with forced circulation and exhaust to safe place	To reduce sterilant residues in product and increase operator safety

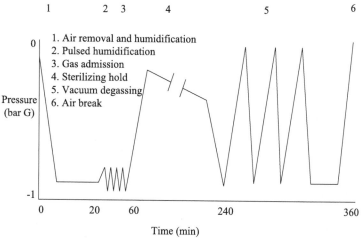

FIGURE 11.5 Schematic diagram of the stages of an ethylene oxide sterilization cycle, using 100% ethylene oxide.

(dependent on required gas concentration and type of gas mixture used), humidity, exposure time, preconditioning, and degassing conditions. Once the process has been selected and validated, the variables within the cycle that have to be controlled and monitored are temperature, pressure, relative humidity, gas concentration, exposure time, load configuration, and load density.

Each sterilization cycle may be monitored by including biological indicators in each sterilizer load (ISO 1996a; see also Chapter 12). These are generally spores of *Bacillus subtilis* var. *niger* deposited onto a suitable carrier material (Dadd et al. 1983). Specifications for these monitors are available (ISO 2006h,i). Biological indicators should be distributed throughout the sterilizer load prior to any prehumid-ification and removed as soon as possible after the sterilization process. Positioning of biological indicators should be determined on the basis of the validation studies and include those sites found most difficult to sterilize. The recommended number of biological indicators per sterilizer load is a minimum of ten indicators for steril-izers with a capacity up to 5000 l, with additional indicators added for larger chambers (ISO 2006b).

Alternatively, a system of parametric release is accepted provided that all the parameters of the process are monitored directly, and that the validation has generated knowledge of the microbial inactivation kinetics of the set of conditions being employed (CEN 1994a, ISO 1994b).

11.6.1.3 Product Release

In addition to the general points described earlier (Section 11.2.4), the following release data are required for a product sterilized by an ethylene oxide process:

- Preconditioning records (time, temperature, relative humidity)
- Time elapsed between preconditioning and sterilization
- Sterilization cycle records (time, temperature, pressure, relative humidity, measure of gas concentration, e.g., weight of gas used)
- Incoming inspection records of ethylene oxide supply
- Results of biological indicator incubation
- Degassing records (time, temperature)

11.6.2 Low Temperature Steam and Formaldehyde

The LTSF process generally operates in the temperature range between 70 and 80°C with a formaldehyde concentration of approximately 14 mg.L^{-1} of chamber volume. Like ethylene oxide, formaldehyde is a toxic gas and maximum exposure levels in air have been established in the United Kingdom by the Health and Safety Executive at a time-weighted average over 8 h of 2 ppm (Health and Safety Executive 1996).

11.6.2.1 Equipment

An LTSF sterilizer can, in some ways, be considered as similar to a porous load steam sterilizer but operating at subatmospheric pressure with injection of formaldehyde gas.

A European Standard specification (CEN 2003) and an international standard specifying safety requirements for sterilizers using toxic gases is also applicable to these sterilizers (IEC 2005). A steam supply of suitable quality is important to the satisfactory operation of a LTSF sterilizer for the same reasons as it is critical in porous load sterilizers (see Section 11.3.1.1).

11.6.2.2   The Process

The features of an LTSF sterilization cycle are illustrated in Figure 11.6. The cycle consists of a series of pulses in which formaldehyde is admitted to the evacuated chamber and allowed to diffuse through the load for approximately 2 min. This is followed by an injection of steam to the operating temperature and reevacuation. These stages are repeated to produce up to 20 pulses, which may be followed by a hold period (Hoxey 1991). The value of the hold period has been questioned as the chamber formaldehyde concentration decreases rapidly in static conditions due to polymerization and dissolution (Marcos and Wiseman 1979). Aeration prior to cycle completion is achieved by a series of pulses in which air admission through a bacteria-retentive filter is alternated with evacuation of the chamber.

The variables within an LTSF cycle are the same as the variables within an ethylene oxide sterilization cycle (Hoxey et al. 1984). Temperature, formaldehyde concentration and pressure, and sterilant penetration and distribution all affect the efficacy of the process. Humidity also affects the rate of microbial inactivation by gaseous formaldehyde (Nordgren 1939) but in the LTSF process a constant high humidity is provided by the steam injection.

The physical conditions in each sterilization cycle should be monitored and recorded. In addition, each sterilizer load should contain a biological indicator. To test gas penetration, this is placed in a Line-Pickerill helix (Line and Pickerill 1973): a stainless steel tube in a helical configuration attached to a small chamber into

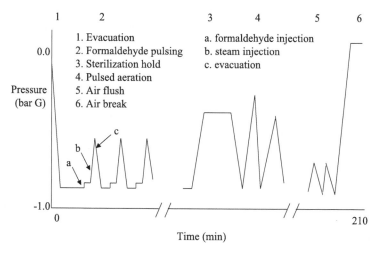

FIGURE 11.6   Schematic diagram of the stages of a low temperature steam and formaldehyde sterilization cycle.

which the biological indicator is placed. The length-bore ratio of the tube makes the helix a rigorous challenge to the process. The biological indicator generally comprises spores of *Bacillus stearothermophilus* deposited onto a suitable carrier material. These should comply with the general requirements for biological indicators (ISO 2006h,j).

11.6.2.3   Product Release

In addition to the general points described earlier (see Section 11.2.4), the release of product following LTSF sterilization requires the following:

- Sterilization cycle records (time, temperature, pressure)
- Results of biological indicator incubation
- Records to confirm that validated degassing procedures have been followed

11.6.3   Gas Plasma

Gas plasmas are a mixture of ions, free radicals, electrons, and neutral species (see Chapter 10). For sterilization processes, gas plasmas are created in a sealed chamber under vacuum using radio frequency or microwave energy to generate an electric or magnetic field to cause ionization and accelerate the resultant particles.

A gas plasma sterilization system was first patented for the sterilization of vials in 1968 (Menashi 1968). A later patent introduced the idea of "seeded plasma" employing a combination of gas plasma and aldehydes (Gut Boucher 1980). Two systems are available for the sterilization of instruments in a health-care setting; the more widely used system employs hydrogen peroxide to seed gas plasma in a chamber (Addy 1991). The other commercially available system uses secondary plasma seeded with peracetic acid that flows into an evacuated sterilization chamber (Caputo et al. 1993). Gas plasmas are suitable for sterilizing items that will not withstand elevated temperatures and are seen as alternatives to ethylene oxide, LTSF, and liquid chemical sterilants. The process is not compatible with cellulosic materials, including cellulose packaging materials.

11.6.3.1   Equipment

There are no standards available specifically for gas plasma sterilizers but there is a general standard specifying safety requirements for sterilizers using toxic gases that would be applicable to this type of sterilizer (IEC 2005).

11.6.3.2   The Process

Figure 11.7 illustrates a typical sterilization cycle for a gas plasma system seeded with hydrogen peroxide. The cycle comprises air removal, hydrogen peroxide injection, diffusion, gas plasma, and aeration.

The process operates at a deep vacuum, being evacuated to 0.3 mbar (30 Pa) initially. The initial evacuation can be prolonged, up to 20 min, and depends upon

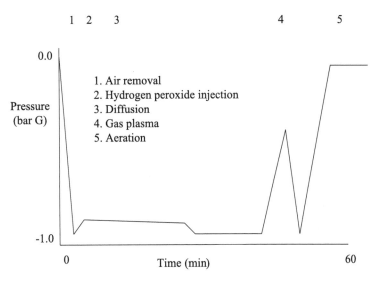

FIGURE 11.7 Schematic diagram of the stages of a hydrogen peroxide gas plasma sterilization cycle.

the level of moisture in the chamber and load. Excessive amounts of moisture can prevent the required vacuum being achieved and cause the cycle to abort. Following evacuation, hydrogen peroxide is admitted through a vaporizer into the sterilization chamber and allowed to diffuse throughout the chamber and load. Plasma is unable to penetrate long, narrow lumens and adaptors are available that inject the hydrogen peroxide directly into such lumens. Radio frequency energy is applied to generate the plasma within the chamber and load. Following the plasma stage, the chamber is flushed with air and returned to atmospheric pressure by the introduction of filtered air.

There are no standards for the validation or for routine control of gas plasma sterilization; therefore, a routine control protocol would have to be established for a particular application based on first principles. There is a general standard that can be applied to all processes for which particular standards are not available (ISO in preparation). The physical parameters of the process should be monitored: pressure, temperature, admission of the specified quantity of seed gas, plasma initiation, and the duration of each stage of the cycle. Additionally, monitoring of the process should include the use of biological indicators complying with the general requirements (ISO 1994d, CEN 1997j) but there is no particular standard laying down specific requirements for biological indicators for gas plasma sterilization.

11.6.3.3 Product Release

In addition to the general points described earlier (see Section 11.2.4), the release of product following gas plasma sterilization requires the following:

- Sterilization cycle records (time, temperature, pressure)
- Evidence of admission of seed gas

- Evidence of plasma initiation
- Results of biological indicator incubation

11.7 FILTRATION STERILIZATION

Filtration has a number of fundamental differences when compared to other sterilization processes. Unlike the other methods described in this chapter, filtration
physically removes organisms from the process stream rather than inactivating them
in situ. Thus the process cannot take place in the final container and is therefore not
a terminal sterilization step, meaning that further aseptic processing has to take place.
This can potentially compromise product sterility. Very high levels of microbial
retention can be demonstrated during filtration, but final product sterility is dependent
on subsequent aseptic processing, which is variable and difficult to quantify. Care
must be taken when using such terms as "sterility assurance level" for filter-sterilized
products (ISO 2003c, Chapters 10 and 12).

Filter sterilization remains the method of choice only for products that cannot
be terminally sterilized (British Pharmacopoeia 2004). The selection of an aseptic
filling process, when the solution can be terminally sterilized, is not readily acceptable to regulatory authorities [Committee for Proprietary Medicinal Products
(CPMP) 1996, 1998]. However, there has been much debate about consideration of
the whole product design. A solution that itself can withstand terminal sterilization
and be developed into a pack that confers significant clinical advantage may be a
case for filter sterilization and aseptic processing.

The requirements for aseptic filling of pharmaceuticals are beyond the scope
of this chapter, which will concentrate only on the use of filters to sterilize liquids
and gases. Much has been published on the mechanisms of filtration sterilization
and has been reviewed by Meltzer (1986) and Johnston (1992). Details of the
validation requirements for sterilization by filtration are covered in Errico (1986),
and Olson (1987).

It is currently generally accepted that filters for the sterilization of pharmaceutical
products should have a nominal rated pore size of 0.22 μm or less, or have at least
equivalent microorganism retaining properties (Commission of the European Communities 2004). It should be noted, however, that some organisms naturally occurring
in water supplies or cultured under specific conditions have been reported to pass
through 0.2-μm filters over extended time periods (Wallhaeusser 1979, Howard and
Duberstein 1980, Leo et al. 1997). Filters with pore size ratings of 0.1 μm are
commercially available but there is no universally recognized performance standard
for these filters (Lindenblatt et al. 2002).

Unlike sterilization by steam, dry heat, or irradiation, there are no specific standards for sterilization by filtration. The current universally adopted requirement for
a sterilizing filter is that a minimum challenge of 10^7 *Brevundimonas diminuta* (formerly *Pseudomonas diminuta*) per square centimeter of filter surface should be
retained with no passage into the effluent (Health Industry Manufacturers Association
1982, U.S. Food and Drug Administration 2004, British Pharmacopoeia 2005). A
challenge methodology is given in ASTM F838-83 (American Society for Testing
and Materials 1983). The Parenteral Drug Association (1998) has prepared a technical

TABLE 11.9
Features of Filter Media and Equipment used for Filtration Sterilization

Feature	Rationale
Filter Medium	
Acceptable flow rate	Product must flow through the filter at an acceptable rate within pressure drop limitations.
Biological capacity	Filter must retain a challenge of 10^7 *Brevundimonas diminuta* per square centimeter of filter surface area under simulated use conditions.
Strength	Filter must be capable of steam sterilization without exceeding the specified differential pressure.
Inertness	Filter must not shed particles or leach substances into the process flow.
Sorption	Filter must not sorb substances from the product.
Filtration Equipment	
Steam sterilizable in place	This avoids the need for aseptic assembly or presterilized equipment.
In situ integrity testing	Filters must be tested after sterilization and before use, and after use. This prevents expensive processing through incorrectly fitted or damaged filters and confirms that filter integrity was maintained throughout the process.
Equipment can be easily and effectively cleaned	As product contact occurs, all process equipment must be cleanable.

report on sterilizing filtration of liquids, which covers operation, selection, validation, bacterial retention, integrity testing, and sterilization of filters.

If the process stream has a low bioburden there is a very low probability that a filter will pass organisms because of the level of challenge. Nevertheless, current European Communities guidance (Commission of the European Communities 2004) is that the same filter should not be used for more than one working day unless such use has been validated, presumably due to the risk of microbial passage either by overchallenge or grow-through. Additionally this guidance suggests that a second filter, placed as close as possible to the point of fill, may be advisable to increase protection.

11.7.1 EQUIPMENT

A range of types of filter materials and configurations are available for liquids and gases (Denyer and Hodges 2004). Most of the major filter manufacturers offer technical services and prospective users are advised to discuss particular process requirements with them.

The essential features of the filter medium and filtration equipment are given in Table 11.9. Filtration equipment and downstream systems must be readily sterilizable, preferably in place, and must be designed so that they may be maintained in a clean condition.

11.7.2 THE PROCESS

The rate of filtration is dependent upon the viscosity of the fluid being filtered, the pore size of the filter medium, the surface area of the filter, and the pressure differential across the filter. It is therefore vitally important that these aspects are considered during process development and scale-up procedure.

In order to limit microbial growth prior to filtration, a maximum time should be specified between solution preparation and filtration. Ideally product should be filled into final containers and sealed immediately following filtration.

Filtration processes should be reproducible, and therefore, pressure differentials across the filter and the time taken to complete the filtration should be measured against values determined at validation. GPMP requires that these values be recorded in the batch record (Commission of the European Communities 2004).

The integrity of the filter should be checked before and after use. Postuse checking alone is not considered sufficient because, in theory, a filter that was originally passing organisms may become sufficiently blocked to pass a subsequent test. The filter system with the filter cartridge *in situ* must be tested and not the filter cartridge in isolation, as most problems occur with incorrect fitting in the housing. A number of different integrity testing methods are available including bubble point, diffusive flow, and pressure hold (Errico 1986, Meltzer 1986, Johnston 1992, Parenteral Drug Association 1998). Most major manufacturers produce automatic integrity testing equipment for their filters and provide technical support in validation of specific applications. This validation leads to an integrity test value specific to each product–filter combination. *In situ* integrity testing of hydrophobic filters for gas sterilization may be carried out by water intrusion testing (Jaenchen et al. 1997).

11.7.3 Product Release

In addition to the general requirements for compliance with documented procedures (see Section 11.2.4), the following are required for product release following sterilization by filtration:

- Identification of the filter used (filter manufacturer's batch and serial number must be traceable)
- Sterilization details for the filtration system
- Pre- and postfiltration integrity test details
- Pressure differential and filtration time records
- Prefiltration bioburden data
- Record of acceptable holding time for the bulk solution presterilization

These details refer only to the filtration process; the additional requirements for aseptic processing are outside the scope of this section (see Chapter 6).

11.8 LIQUID CHEMICAL STERILANTS

Liquid chemicals, predominantly alkylating agents such as glutaraldehyde or formaldehyde solutions, are used as a last resort to sterilize items that cannot be processed in any other way. For example, prostheses made from material of animal origin (heart valves, vascular prostheses and tissue patches) (Bruch 1991) and complex items of medical equipment such as flexible fiber-optic endoscopes may be sterilized in this way. For processing endoscopes, 2% glutaraldehyde solution is the most commonly used sterilant, but other chemicals such as formaldehyde or succinaldehyde alone or in combination with phenates or alcohols may also be employed (Medical Devices Agency 1996b). The effectiveness of the process will be affected by chemical concentration and temperature (Russell 1982) and these factors should be controlled.

Because these chemicals are toxic, their use must be carefully regulated. Maximum atmospheric exposure limits for formaldehyde have been set in the United Kingdom at a time-weighted average over 8 h of 2 ppm (Health and Safety Executive 1996). The approved occupational exposure standard for atmospheric glutaraldehyde over the same period is 0.2 ppm (Health and Safety Executive 1996). Incidents of allergic responses in personnel exposed to liquid chemical sterilants and their vapors have been reported (Ballantyne and Berman 1984, Corrado et al. 1986).

The use of liquid chemical sterilants for reprocessing medical equipment in clinical areas is not considered further in this chapter. For further information see, for example, Ayliffe and Babb (1999), Medical Devices Agency (1996b).

11.8.1 EQUIPMENT

The equipment requirements for liquid chemical sterilization are essentially simple: a closed container to hold the solution and product and a means to control and monitor the temperature of exposure.

11.8.2 THE PROCESS

The main variables in a liquid chemical sterilization process that must be controlled and monitored are time, concentration, and temperature. In addition, the pH of glutaraldehyde solutions affects their microbicidal efficacy (reviewed by Russell 1982). The sterilant solution must be carefully prepared and the concentration of active ingredient is generally determined by assay, as is the pH. Solutions are often sterilized by filtration prior to use to reduce the microbial challenge. Additionally, liquid chemical sterilization processes are generally monitored using biological indicators. However, these indicators are not exposed to the sterilant alongside the device being processed, but are used instead to test the solution after the device has been removed. Furthermore, a sterility test is often performed on samples of product or on some simulator that has been exposed to treatment identical to the product (ISO 1998).

11.8.3 PRODUCT RELEASE

In addition to the general requirements (see Section 11.2.4), the following information is required to release a product as sterile following a liquid sterilization process:

- Records of manufacture and test of sterilant solution (including records of preparation of the solution, assay results, pH, records of sterilant filtration)
- Records of sterilization conditions (temperature, time)

- Details of sterilization of packaging components
- Environmental monitoring data of area used for aseptic packaging
- Sterility test results

11.9 EMERGING STERILIZATION TECHNOLOGIES

Consideration of Table 11.1 shows that no single sterilization method satisfies all the criteria of an ideal process. Although steam sterilization is accepted as the method of choice because it comes closest to meeting the ideal criteria, investigations into alternatives continue, particularly alternatives to chemical sterilants (either gaseous or liquid). Current interest centers on materials that can be used in the vapor phase but which will break down to form nontoxic residues. Such developments are academically interesting but present many problems in routine use. The regulatory requirements for novel sterilization methods are not determined and use of established technologies remains the most straightforward route to product acceptance.

A potential user of a new method for sterilization would be expected to demonstrate the following (ISO 2000a):

- That established methods of sterilization are not practicable or that the new method demonstrates a clear advantage over established methods
- The effectiveness of the proposed method against a range of microorganisms
- The range and number of microorganisms likely to contaminate the product and activity of the proposed sterilization method against these organisms
- The critical variables of the process and their means of control
- Reproducibility of effectiveness for the proposed sterilization method
- The critical variables attained with physical records
- An appropriate standard organism for biological monitoring

11.9.1 CHLORINE DIOXIDE

The microbicidal properties of chlorine dioxide have long been recognized in use for the treatment of water supplies in Europe from the 1850s (Jeng and Woodworth 1990) but the sporicidal activity of chlorine dioxide in gaseous form was demonstrated only relatively recently (Orcutt et al. 1981).

A system for sterilizing medical devices using gaseous chlorine dioxide has been developed. Typically, it operates at ambient temperature (25 to 30°C) and a relative humidity of 70 to 90%. Chlorine dioxide is generated *in situ* from sodium chlorite and chlorine gas in a nitrogen carrier. Gaseous chlorine dioxide is then drawn into an evacuated chamber to the pressure required to achieve the required concentration of sterilant (in the range 10 to 50 mg.L^{-1}). The conditions are maintained for the required time (between 3 min and 2 h) with additions of further chlorine dioxide to maintain the concentration. The cycle is concluded by evacuation of the chamber and the exhaust gas is passed through a chemical column to absorb chlorine dioxide (Sintim-Damoa 1993).

There are no standards specifying requirements for chlorine dioxide sterilization at present but a general standard that can be applied to all processes has been prepared (ISO 2000a). Additionally, monitoring of the process should include the use of biological indicators complying with the general requirements for such indicators (ISO 2006h), but there is no particular standard laying down requirements specifically for biological indicators for chlorine dioxide sterilization (Allison 1998). Spores of *Bacillus subtilis* var. *niger* have been used as biological indicators for process monitoring.

11.9.2 Ozone

Like chlorine dioxide (see Section 11.9.1), ozone has been used extensively for the treatment of domestic water supplies (Symonds 1980). Ozone is used in the pharmaceutical industry for the treatment of deionized water systems and two types of sterilizers for use in health-care facilities have been reported (Karlson 1989, Stoddart 1989).

In the first system (Karlson 1989), ozone is generated from oxygen and used to displace air from the sterilizer chamber. Ozone is then passed continuously through the chamber for a defined period, after which the generation of ozone is terminated and oxygen admitted to flush the chamber.

The second system (Stoddart 1989) delivers ozone at a concentration of 10 to 12%. The sterilization cycle operates under vacuum and has a total cycle time of 30 to 60 min.

There are no standards specifying requirements for ozone sterilization at present but a general standard that can be applied to all processes has been prepared (ISO 2000a). Additionally, monitoring of the process should include the use of biological indicators complying with the general requirements for such indicators (ISO 2006h), but there is no particular standard laying down requirements specifically for biological indicators for ozone sterilization (Allison 1998).

11.9.3 Peracetic Acid

Peracetic acid is formed by the reaction between hydrogen peroxide and acetic acid and exists as an equilibrium mixture of peracetic acid, acetic acid, hydrogen peroxide, and water. Hence systems using peracetic acid inevitably also have hydrogen peroxide present as well. Peracetic acid has been applied as a sterilizing agent in solution (Malchesky 1993) and in the vapor phase, either as a seed gas for a gas plasma (see Section 11.6.3) or used to sterilize isolators and other enclosed spaces (Davenport 1989). The practical application of peracetic acid has been limited by its corrosive nature (Portner and Hoffman 1968, Malchesky 1993).

There are no standards specifying requirements for peracetic acid sterilization at present but a general standard that can be applied to all processes has been prepared (ISO 2000a). Additionally, monitoring of the process should include the use of biological indicators complying with the general requirements for such indicators (ISO 2006h), but there is no particular standard laying down requirements specifically for biological indicators for peracetic acid sterilization (Allison 1998). Spores of *Bacillus subtilis* var. *niger* have been used as biological indicators for process monitoring.

11.9.4   VAPOR PHASE HYDROGEN PEROXIDE

The microbicidal properties of hydrogen peroxide have long been recognized and it has been employed in the food industry since 1915 (Schlumb et al. 1955). Synergistic effects between hydrogen peroxide and UV light have been reported and attributed to the formation of reactive hydroxyl radicals (Bayliss and Waites 1979a, 1979b).

More recent applications for sterilization in the pharmaceutical and medical devices industries have been as a seed gas for gas plasma systems (see section 11.6.3) and in the vapor phase hydrogen peroxide (VPHP) process (Klapes and Vesley 1990).

VPHP has been developed in three modes: (1) a "deep vacuum" system using an evacuated chamber, (2) a "flow-through" system in which the hydrogen peroxide is delivered in a stream of filtered air acting as a carrier gas, and (3) a combination of these two for items of equipment that can act as their own chambers, such as freeze-driers and isolators.

The VPHP deep vacuum process employs a vacuum to pull a 30% solution of hydrogen peroxide through a heated vaporizer into the sterilizer chamber. The process operates at 55 to 60°C and the cycle time is approximately 90 min. Both the flow-through and combined approaches use a portable generator to vaporize a 30% hydrogen peroxide solution. In the flow-through approach, a continuous stream of hydrogen peroxide in a carrier gas, usually filtered air, is admitted into the enclosed space to be sterilized. In the combined approach, the equipment to be sterilized is evacuated, vaporized hydrogen peroxide admitted, followed by controlled admission of a small volume of air; further admissions of hydrogen peroxide and air are made at defined intervals.

VPHP has been proposed for the sterilization of medical devices, such as endoscopes, for decontaminating equipment for aseptic processing (e.g., isolators, pipework, freeze-driers) and for the decontamination of contaminated equipment, such as centrifuges and safety cabinets (Rickloff and Graham 1989). However, it cannot be used with cellulosic materials, including paper-based packaging, and some damage has been reported with nylon, certain anodized aluminum surfaces, and some epoxides (Rutala and Weber 1996).

There are no standards specifying requirements for VPHP sterilization at present but a general standard that can be applied to all processes has been prepared (ISO 2000a). Additionally, monitoring of the process should include the use of biological indicators complying with the general requirements for such indicators (ISO 2006h), but there is no particular standard laying down requirements specifically for biological indicators for VPHP sterilization (Allison 1998). Spores of *Bacillus stearothermophilus* have been used as biological indicators for process monitoring.

11.9.5   PULSED LIGHT

Broad spectrum white light in very high intensity short duration pulses is being developed as a means of sterilizing solutions in their final containers (Dunn et al.

1997). The technology is being developed specifically for the in-line sterilization of solutions in polyethylene containers produced by the blow-fill-seal (BFS) process.

The process generates pulses of light of nonionizing wavelengths from the ultraviolet through the visible to the infrared that are approximately 20,000 times more intense than sunlight. The antimicrobial properties are attributed to the high percentage of UV output (~20%) and the very high intensity. A small number of pulses, approximately 10 to 20, have been demonstrated to give high levels of kill against *Bacillus pumilus, B. stearothermophilus, B. subtilis* endospores, and *Aspergillus niger* conidiospores, when inoculated into water for injection in BFS containers (Dunn et al. 1997).

The advantage of this technology is that the temperature of the product is not significantly affected during sterilization. The pulsed light obviously requires access to the bioburden to be effective, and therefore, translucent containers and clear solutions are required although the technology may also be applicable to air, water, and surface sterilization.

There are no standards specifying requirements for pulsed light sterilization at present but a general standard that can be applied to all processes has been prepared (ISO 2000a). Additionally, monitoring of the process should include the use of biological indicators complying with the general requirements for such indicators (ISO 2006h), but there is no particular standard laying down requirements specifically for biological indicators for pulsed light sterilization (Allison 1998).

11.9.6 MICROWAVES

The use of microwave heating has been proposed for the sterilization of empty vials (Lohmann and Manique 1986) and hydrophilic contact lenses (Rohrer et al. 1986). The perceived advantage in microwave sterilization is that there may be a more intensive heating of organic materials such as microorganisms than the product to be sterilized, dependent upon the construction materials of the product (Lohmann and Manique 1986).

Microwaves have been reported as being used in equipment for in-line sterilization of solutions in glass ampoules (Sasaki et al. 1996). The process uses microwaves to generate heat within the solution. The effect of microwaves is to polarize the dielectric in the solution. When polarized, all the electron pairs align in the direction of the electric field produced by the microwave. By reversing the field, the alignment is also reversed and polar molecules can be made to oscillate. This molecular oscillation causes friction, thus generating heat. The major advantage of this process is that it can be used in-line rather than the traditional batch processing of steam sterilization and, because high temperatures are achieved extremely rapidly, this technology may also be applicable to heat-labile products where a short sterilization time could be employed.

It should be noted that in this process, each ampoule is sterilized separately. The validation approach that has been taken is therefore more akin to aseptic processing using a large number of media-filled ampoules inoculated with *B. stearothermophilus* spores. The ampoules are processed and incubated to demonstrate the effectiveness of the process (Sasaki et al. 1996).

There are no standards specifying requirements for microwave sterilization at present but a general standard that can be applied to all processes has been prepared (ISO 2000a). Additionally, monitoring of the process should include the use of biological indicators complying with the general requirements for such indicators (ISO 2006h), but there is no particular standard laying down requirements specifically for biological indicators for microwave sterilization (Allison 1998). Spores of *B. stearothermophilus* have been used as biological indicators for process monitoring.

11.10 CONCLUSION

During the 1970s and 1980s, the focus in sterilization technology was on the optimization of established processes and the preparation of agreed standards for validation and routine control of these established processes. In the 1990s, further sterilization technologies began to emerge. Developments have been driven by concerns over sterilant residues in products, environmental and operator safety legislation, the use of more complex materials, and the increasing application of aseptic processing, particularly BFS technology. These emerging technologies will require extensive investigation to establish confidence in their effectiveness and permit the drafting of standards for their validation and routine control.

REFERENCES

Ackland, N.R., Hinton, M.R., and Denmeade, K.R. (1980). Controlled formaldehyde fumigation system. *Appl. Env. Microbiol.,* 39, 480–487.

Addy, T.O. (1991). Low temperature plasma: a new sterilization technology for hospital applications. In *Sterilization of Medical Products. Vol V.* Morrissey, R.F. and Propopenko, Y.I., Eds. Polyscience Publications Inc., Morin Heights, Canada, pp. 80–95.

Allen, H.F. and Murphy, J.T. (1960). Sterilization of instruments and materials with beta-propiolactone. *JAMA,* 172, 1759–1763.

Allison, D.G. (1998). Commentary: validation of novel methods of sterilisation. *Eur. J. Parent. Sci.,* 3, 33–36.

American Society for Testing and Materials (ASTM). (1983). *ASTM F838-83—Standard Test Method for Determining Bacterial Retention of Membrane Filters Utilized for Liquid Filtration.* ASTM, Philadelphia, PA.

Anonymous (1959). Sterilization by steam under increased pressure. *Lancet,* i, 425–435.

Ayliffe, G.A.J. and Babb, J.R. (1999). Decontamination of the environment and medical equipment in hospitals. In *Principles and Practice of Disinfection, Preservation and Sterilization.* Russell, A.D., Hugo, W.B., and Ayliffe, G.A.J., Eds. 3rd ed. Blackwell Science, Oxford, pp. 395–415.

Ballantyne, B. and Berman, B. (1984). Dermal sensitizing potential of glutaraldehyde: a review and recent observations. *J. Toxicol. Cut. Ocular Toxicol.,* 3, 251–262.

Bayliss, C.E. and Waites, W.M. (1979a). The combined effect of hydrogen peroxide and ultraviolet irradiation on bacterial spores. *J. Appl. Bact.,* 47, 263–268.

Bayliss, C.E. and Waites, W.M. (1979b). The synergistic killing of spores of *Bacillus subtilis* by hydrogen peroxide and ultra-violet light irradiation. *FEMS Microbiol. Lett.,* 5, 331–333.

Bill, A. (2000). Microbiology laboratory methods in support of the sterility assurance system. In *Handbook of Microbiological Quality Control: Pharmaceuticals and Medical Devices*. Baird, R.M., Hodges, N.A. and Denyer, S.P., Eds. Taylor & Francis, London, pp. 120–143.

Bowie, J.H., Kelsey, J.C., and Thompson, R. (1963). The Bowie and Dick autoclave tape test. *Lancet, i*, 586–587.

British Pharmacopoeia. (2005). *British Pharmacopoeia*. The Stationery Office, London.

British Standards Institution (1991). *BS 3970-2—Sterilizing and Disinfecting Equipment for Medical Products. Part 2—Specification for Steam Sterilizers for Aqueous Fluids in Sealed Rigid Containers*. British Standards Institution, London.

Brown, M.R.W. and Gilbert, P. (1977). Increasing the probability of sterility of medicinal products. *J. Pharm. Pharmacol., 29*, 517–523.

Bruch, C.W. (1961). Gaseous sterilization. *Ann. Rev. Microbiol., 15*, 245–262.

Bruch, C.W. (1991). Role of glutaraldehyde and other liquid chemical sterilants in the processing of new medical devices. In *Sterilization of Medical Products. Vol V*. Morrissey, R.F. and Propopenko, Y.I., Eds. Polyscience Publications Inc., Morin Heights, Canada, pp. 376–396.

Caputo, R.A., Fisher, J., Jarzynski,V., and Martens, P.A. (1993). Validation testing of a gas plasma sterilization system. *Med. Dev. and Diag. Ind., 15*, 132–138.

CEN. (1997a). *EN 285—Sterilizers for Medical Purposes—Large Steam Sterilizer—Requirements and Test Methods*. CEN, Brussels.

CEN. (1997b). *EN 1422—Sterilizers for Medical Purposes—Ethylene Oxide Sterilizer—Requirements and Test Methods*. CEN, Brussels.

CEN. (2003). *EN 14180—Sterilizers for Medical Purposes—Low Temperature Steam and Formaldehyde Sterilizers—Requirements and Testing*. CEN, Brussels.

CEN. (2004). *EN 13060—Small Steam Sterilizers—Requirements and Test Methods*. CEN, Brussels.

Commission of the European Communities. (1994). Council Regulation (EC) no. 3093/94 of 15 December 1994 on substances that deplete the ozone layer. *Off. J. EC,* L333, 1–19.

Commission of the European Communities. (2002). *The Rules Governing Medicinal Products in the European Community. Vol IV—Good Manufacturing Practice for Medicinal Products. Annex 17—Parametric Release*. Office for Official Publications of the EC, Luxembourg.

Commission of the European Communities. (2004). *The Rules Governing Medicinal Products in the European Community. Vol IV—Good Manufacturing Practice for Medicinal Products*. Office for Official Publications of the EC, Luxembourg.

Committee for Proprietry Medicinal Products. (1996). CPMP/QWP/155/96. Notes for guidance on development pharmaceutics. EMEA, London.

Committee for Proprietry Medicinal Products. (1998). CPMP/QWP/054/98. Decision trees for the selection of sterilization methods—Annex to notes for guidance on development pharmaceutics. EMEA, London.

Corrado, O.J., Osman, J., and Davies, R.J. (1986). Asthma and rhinitis after exposure to lutaraldehyde in endoscopy units. *Human Toxicol., 5*, 325–327.

Dadd, A.H., Stewart, C.M., and Town, M.M. (1983). A standard monitor for the control of ethylene oxide sterilization cycles. *J. Hyg. Camb., 91*, 93–100.

Davenport, S.M. (1989). Design and use of a novel peracetic acid sterilizer for absolute barrier sterility testing chambers. *J. Parent. Sci. Tech., 43*, 158–166.

Deindoerfer, F.H. and Humphrey, A.E. (1959). Analytical method for calculating heat sterilization times. *Appl. Microbiol., 7*, 256–264.

Denyer, S.P. and Hodges N.A. (2004). Filtration sterilization. In *Russell, Hugo and Ayliffe's Principles and Practice of Disinfection, Preservation and Sterilization.* Fraise, A.P., Lambert, P.A., and Maillard, J.-Y., Eds. 4th ed. Blackwell Science, Oxford, pp. 436–472.

De Santis, P. and Rudo, V.S. (1986). Validation of steam sterilization in autoclaves. In *Validation of Aseptic Pharmaceutical Processes.* Carleton, F.J. and Agalloco, J.P., Eds. Marcel Dekker, New York, pp. 279–317.

Dunn, J., Burgess, D., and Leo, F. (1997). Investigations of pulsed light for terminal sterilization of water for injection filled blow/fill polyethylene seal containers. *J. Parent. Sci. and Tech.,* 51, 111–115.

Errico, J.J. (1986). Validation of aseptic processing filters. In *Validation of Aseptic Pharmaceutical Processes.* Carleton, F.J. and Agalloco, J.P., Eds. Marcel Dekker, New York, pp. 427–472.

Gut Boucher, R.M. (1980). Seeded Gas Plasma Sterilization Method. U.S. Patent 4,207,286.

Halls, N.A. (1994). *Achieving Sterility in Medical and Pharmaceutical Products.* Marcel Dekker, New York.

Health and Safety Executive. (1996). *Occupational Exposure Limits 1996. Guidance Note EH40.* Health and Safety Executive, Sheffield.

Health Industry Manufacturers Association (HIMA). (1982). Microbiological Evaluation of Filters for Sterilizing Liquids. HIMA Document no. 3, HIMA, Washington, D.C., p. 14.

Hoffman, R.K. (1971). Toxic gases. In *Inhibition and Destruction of the Microbial Cell.* Hugo, W.B., Ed. Academic Press, London, pp. 225–258.

Howard, B.A. and Duberstein, R. (1980). A case of penetration of 0.3 micron rated membrane filters by bacteria. *J. Parent. Drug Assoc.,* 34, 95–102.

Hoxey, E.V. (1989). The case for parametric release. In *Proceedings of the Eucomed Conference on Ethylene Oxide Sterilization.* April 21–22, 1989, Paris. Eucomed, Brussels, pp. 25–32.

Hoxey, E.V. (1991). Low temperature steam and formaldehyde sterilization. In *Sterilization of Medical Products. Vol. V.* Morrissey, R.F. and Propopenko, Y.I., Eds. Polyscience Publications Inc., Morin Heights, Canada, pp. 359–364.

Hoxey, E.V., Soper, C.J., and Davies, D.J.G. (1984). The effect of temperature and formaldehyde concentration on the inactivation of *Bacillus stearothermophilus* spores by LTSF. *J. Pharm. Pharmac.,* 36, 60P.

Hurrell, D.J. (1986). UK regulatory standards for monitoring LTSF and EO sterilization. *Proceedings of the Eucomed Workshop on Biological Monitoring of Sterilization.* April 21–23, Kerkrade, the Netherlands. Eucomed, Brussels, pp. 45–47.

IEC. (2005). *IEC 61010-2-040—Safety Requirements for Electrical Equipment for Measurement, Control, and Laboratory Use—Part 2-040: Particular Requirements for Sterilizers and Washer-Disinfectors Used to Treat Medical Materials.* IEC, Geneva.

ISO. (1998). *ISO 14160—Sterilization of Medical Devices—Validation and Routine Control of the by Liquid Chemical Sterilants of Single-Use Devices Incorporating Materials of Animal Origin.* ISO, Geneva.

ISO. (1998). *ISO 13408-1—Aseptic Processing of Health Care Products—Part 1: General Requirements.* ISO, Geneva

ISO. (2000a). *ISO 14937—Sterilization of Healthcare Products—General Requirements for Characterization of a Sterilizing Agent, and the Development, Validation and Routine Control of a Sterilization Process.* ISO, Geneva.

ISO. (2000b). *ISO 11140-3—Sterilization of Health Care Products—Chemical Indicators—Part 3: Class 2 Indicators for Steam Penetration Test Sheets.* ISO, Geneva.

ISO. (2000c). *ISO 11140-5—Sterilization of Health Care Products—Chemical Indicators—Part 5: Class 2 Indicators for Air Removal Test Sheets and Packs.* ISO, Geneva.

ISO. (2001). *ISO 11140-4—Sterilization of Health Care Products—Chemical Indicators—Part 4: Class 2 Indicators for Steam Penetration Test Packs.* ISO, Geneva.

ISO. (2003a). *ISO 13485—Medical Devices—Quality Management Systems—Requirements for Regulatory Purposes.* ISO, Geneva.

ISO. (2003b). *ISO 13408-4—Aseptic Processing of Health Care Products—Part 4: Clean-in-Place Technologies.* ISO, Geneva.

ISO. (2003c). *ISO 13408-2—Aseptic Processing of Health Care Products—Part 2: Filtration.* ISO, Geneva.

ISO. (2005). *ISO 11140-1—Sterilization of Health Care Products—Chemical Indicators—Part 1: General Requirements.* ISO, Geneva.

ISO. (2006a). *FDIS ISO 11135-1—Sterilization of Health Care Products—Ethylene Oxide—Part 1: Requirements for Development, Validation and Routine Control of a Sterilization Process for Medical Devices.* ISO, Geneva.

ISO. (2006b). *ISO 11737-1—Sterilization of Health Care Products—Microbiological Methods. Part 1: Estimation of the Population of Microorganisms on Product.* ISO, Geneva.

ISO. (2006c). *ISO 11607—Packaging for Terminally Sterilized Medical Devices—Part 1: Requirements for Materials, Sterile Barrier Systems and Packaging Systems.* ISO, Geneva.

ISO. (2006d). *ISO 17665—Sterilization of Health Care Products—Moist Heat—Part 1: Requirements for Development, Validation and Routine Control of a Sterilization Process for Medical Devices.* ISO, Geneva.

ISO. (2006e). *ISO 11137-1—Sterilization of Health Care Products—Radiation—Part 1: Requirements for Development, Validation and Routine Control of a Sterilization Process for Medical Devices ISO, Geneva.*

ISO. (2006f). *ISO 11137-2—Sterilization of Health Care Products—Radiation—Part 2: Establishing the Sterilization Dose.* ISO, Geneva.

ISO. (2006g). *ISO 11137-3—Sterilization of Health Care Products—Radiation—Part 3: Guidance on Dosimetric Aspects.* ISO, Geneva.

ISO. (2006h). *ISO 11138-1—Sterilization of Healthcare Products—Biological Indicators—Part 1: General Requirements.* ISO, Geneva.

ISO. (2006i). *ISO 11138-2—Sterilization of Healthcare Products—Biological Indicators —Part 2: Biological Indicators for Ethylene Oxide Sterilization.* ISO, Geneva.

ISO. (2006j). *ISO 11138-5—Sterilization of Health Care Products—Biological Indicators—Part 5: Biological Indicators for Low-Temperature Steam and Formaldehyde Sterilization Processes.* ISO, Geneva.

ISO. (in preparation). *ISO 20857-1—Sterilization of Health Care Products—Dry Heat—Part 1: Requirements for Development, Validation and Routine Control of a Sterilization Process for Medical Devices.* ISO, Geneva.

Jaenchen, R., Schubert, J., Jafari, S., and West, A. (1997). Studies on the theoretical basis of the water intrusion test (WIT). *Eur. J. Parenteral Sci.,* 2, 39–45.

Jeng, D.K. and Woodworth, A.G. (1990). Chlorine dioxide gas sterilization under square wave conditions. *Appl. Env. Microbiol.,* 56, 514–519.

Johnston, P.R. (1992). *Fluid Sterilization by Filtration.* Interpharm Press, Prairie View, IL.

Jorkasky, J.F. (1993). Special considerations for ethylene oxide: chlorofluorcarbons. In *Sterilization Technology.* Morrissey, R.F. and Phillips, G.B., Eds. Van Nostrand Reinhold, New York, pp. 391–401.

Karlson, E.L. (1989). Ozone sterilization. *J. Healthcare Mat. Man.,* 7, 42–45.

Klapes, N.A. and Vesley, D. (1990). Vapour-phase hydrogen peroxide as a surface decontaminant and sterilant. *Appl. Env. Microbiol.,* 56, 503–506.

Knapp, J.E., Rosenblatt, D.H., and Rosenblatt, A.A. (1986). Chlorine dioxide as a gaseous sterilant. *Med. Dev. Diagn. Ind.,* 8, 48–51.

Kolb, R.W. and Schneiter, R. (1950). The germicidal and sporicidal efficacy of methyl bromide on spores of *Bacillus anthracis. J. Bact.,* 59, 401–412.

Kowalski, J., Aoshuang, Y., and Tallentire, A. (1999). 2000 Radiation sterilization—Evaluation of a new method for substantiation of 25 kGy. *Radiat. Phys. Chem.,* 58, 77–86.

Kowalski, J., and Tallentire, A. (1999). Substantiation of 25 kGy as a sterilization dose: a rational approach to establishing verification dose. *Radiat. Phys. Chem.,* 54, 55–64.

Kowalski, J. and Tallentire, A. (2003). Aspects of putting into practice VD_{max}. *Radiat. Phys. Chem.,* 67, 137–141.

Leo, F., Auriemma, M., Ball, P., and Sundaram, S. (1997). Application of 0.1 micron filtration for enhanced sterility assurance in pharmaceutical filling operations. *Blow-Fill-Seal News,* (August), 15–24.

Lindenblatt, J., Jornitz, M., and Meltzer, T. (2002). Filter pore size versus process validation—a necessary debate? *Europ. J. Parent. Sci.,* 7(3), 61–71.

Line, S.J. and Pickerill, K.I. (1973). Testing a steam formaldehyde sterilizer for gas penetration efficiency. *J. Clin. Path.,* 26, 716–720.

Lohmann, S. and Manique, F. (1986). Microwave sterilization of vials. *J. Parent. Sci. Technol.,* 40, 25–30.

Malchesky, P.S. (1993). Peracetic acid and its application to medical instrument sterilization. *Art. Org.,* 17, 147–152.

Marcos, D. and Wiseman, D. (1979). Measurement of formaldehyde concentrations in a subatmospheric steam-formaldehyde autoclave. *J. Clin. Path.,* 32, 567–575.

Medical Devices Agency (1996a). Device Bulletin DB 9605—The purchase, operation and maintenance of benchtop steam sterilizers. Medical Devices Agency, London.

Medical Devices Agency (1996b). Device Bulletin DB 9607—Decontamination of endoscopes. Medical Devices Agency, London.

Meltzer, T.H. (1986). *Filtration in the Pharmaceutical Industry.* Marcel Dekker, New York.

Menashi, W.P. (1968). Treatment of Surfaces. U.S. Patent 3,383,163.

NHS Estates Agency. (1994). *Health Technical Memorandum no. 2010. Sterilization. Part 1—Management Policy.* Her Majesty's Stationery Office, London.

NHS Estates Agency. (1995a). *Health Technical Memorandum no. 2010. Sterilization. Part 2—Design Considerations.* Her Majesty's Stationery Office, London.

NHS Estates Agency. (1995b). *Health Technical Memorandum no. 2010. Sterilization. Part 3—Validation and Verification.* Her Majesty's Stationery Office, London.

NHS Estates Agency. (1995c). *Health Technical Memorandum no. 2010. Sterilization. Part 4—Operational Management.* Her Majesty's Stationery Office, London.

NHS Estates Agency. (1997). *Health Technical Memorandum no. 2031. Clean Steam for Sterilization.* Her Majesty's Stationery Office, London.

Nordgren, G. (1939). Investigations on the sterilization efficacy of gaseous formaldehyde. *Acta Pathol. Microbiol. Scand.,* 60, Suppl., 1–165.

Olson, W.P. (1987). Sterilization of small volume parenterals and therapeutic proteins by filtration. In *Aseptic Pharmaceutical Manufacturing Technology for the 1990s.* Olson, W.P. and Groves, M.J., Eds. Interpharm Press, Prairie View, IL, pp. 101–150.

Orcutt, R.P., Otis, A.P., and Alliger, H. (1981). Alcide TM: an alternative sterilant to peracetic acid. In *Recent Advances in Germ-Free Research. Proceedings of the VIIth International Symposium on Gnotobiology.* Sasaki, S., Ozawa, A., and Hashioto, K., Eds. Tokai University Press, Tokyo, pp. 79–81.

Parenteral Drug Association. (1998). Technical report 26, Sterilizing filtration of liquids. *PDA J. Pharm. Sci. and Technol.*, 52(3), Suppl.

Placencia, A.M., Oxborrow, G.S., and Peeler, J.T. (1986). Package integrity methodology for testing biobarrier properties of porous packaging. Part 1. Membrane agar plate strike-through method. *Med. Dev. Diagnost. Ind.*, 8, 61–65.

Phillips, C.R. and Kaye, S. (1949). The sterilizing action of gaseous ethylene oxide. Review. *Am. J. Hyg.*, 50, 270–279.

Pickerill, J.K., Perera, R., and Knox, R. (1971). Air detection in dressings steam sterilizers. *Lab. Pract.*, 20, 406–413.

Portner, D.M. and Hoffman, R.K. (1968). Sporicidal effect of peracetic acid vapour. *Appl. Microbiol.*, 16, 1782–1785.

Rickloff, J.R. and Graham, G.S. (1989). Vapour phase hydrogen peroxide sterilization. *J. Healthcare Mat. Man.*, 7, 45–49.

Rohrer, M.D., Terry, M.A., Bulard, R.A., Graves, D.C., and Taylor, E.M. (1986). Microwave sterilization of hydrophilic contact lenses. *Am. J. Ophthalmol.*, 101, 49–57.

Russell, A.D. (1982). *The Destruction of Bacterial Spores*. Academic Press, London.

Rutala, W.A. and Weber, D.J. (1996). Low temperature sterilization technologies: do we need to redefine sterilization? *Infect. Cont. Hosp. Epidem.*, 17, 87–91.

Sasaki, K., Honda, W., Shimizu, K., Iizima, K., Khara, T., Okuzzawa, K., and Miyake, Y. (1996). Microwave continuous sterilization of injection ampoules. *J. Pharm. Sci. Technol.*, 50, 172–179.

Saylor, M.C. (1991). Developments in radiation equipment including the application of machine-generated X-rays to medical product sterilization. In *Sterilization of Medical Products. Vol V.* Morrissey, R.F. and Propopenko, Y.I., Eds. Polyscience Publications Inc., Morin Heights, Canada, pp. 327–344.

Schlumb, W.C., Satterfield, C.N., and Wentworth, R.L. (1955). *Hydrogen Peroxide*. Reinhold, New York.

Seiberling, D.A. and Ratz, A.J. (1995). Engineering Considerations for CIP/SIP Systems. In *Sterile Pharmaceutical Products Process Engineering Applications.* Avis, K.E., Ed. Interpharm Press, Prairie View, IL.

Sintim-Damoa, K. (1993). Other gaseous methods. In *Sterilization Technology.* Morrissey, R.F. and Phillips, G.B., Eds. Van Nostrand Reinhold, New York, pp. 335–347.

Stoddart, G.M. (1989). Ozone as a sterilizing agent. *J. Healthcare Mat. Man.*, 7, 42–43.

Stumbo, C.R. (1973). *Thermobacteriology in Food Processing.* 2nd ed. Academic Press, New York.

Symonds, J.M. (1980). Ozone, chlorine dioxide and chloramines as alternatives to chlorine for disinfection of drinking water. In *Ozone and Chlorine Dioxide Technology for Disinfection of Drinking Water.* Katz, J., Ed. Noyes Data Corp., Park Ridge, NJ.

United Nations Environment Programme (UNEP). (2002). 2002 Report of the Aerosols, Sterilants, Miscellaneous Uses and Carbon Tetrachloride Technical Options Committee—2002 Assessment. UNEP/ Ozone Secretariat, Nairobi. http://www.teap.org.

U.S. Food and Drug Administration. (FDA) (2004). *Guidance for Industry. Sterile Drug Products Produced by Aseptic Processing—Current Good Manufacturing Practice.* FDA Center for Drug Evaluation and Research (CDER) Center for Biologics Evaluation and Research (CBER) and Office of Regulatory Affairs (ORA), Washington, D.C.

Wallhaeusser, K.H. (1979). Is the removal of micro-organisms by filtration really a sterilization method? *J. Parent. Sci. and Tech.*, 33, 156–170.

12 Assurance of Sterility by Process Validation

Klaus Haberer

CONTENTS

12.1 STERILITY AND ASSURANCE OF STERILITY

Sterility, the absence of any microorganism capable of reproduction, is an absolute requirement that cannot be compromised. For the purpose of sterility assurance, however, a maximum acceptable number of unsterile units is usually defined to describe the required safety of the sterilization process. For pharmaceuticals sterilized within their final container, the following definition is given by major pharmacopoeias, "The procedures and precautions employed are such as to give a sterility assurance level (SAL) of 10^{-6} or better" [European Pharmacopoeia (EP) 2006]. The SAL of a sterilizing process is the degree of assurance with which the process in

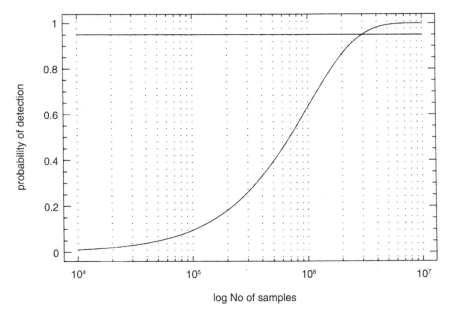

FIGURE 12.1 Theoretical number of samples required to detect one unsterile unit per 10^6 units purporting to be sterile. The figure is based on the expression $p = 1 - (1 - q)^n$, where p is the probability of detection, q is the relative frequency of contaminated units (10^{-6}), and n is the number of samples used in the test.

question renders a population of items sterile. "It is generally accepted that terminally sterilized injectables or critical devices purporting to be sterile, when processed in the autoclave, attain at least a 10^{-6} microbial survivor probability" [United States Pharmacopoeia (USP) 2004].

Such a level of sterility assurance can only be derived from theoretical calculations. The test for sterility cannot, for statistical reasons, contribute in any significant way to the assurance of sterility at the required safety level (Spicher and Peters 1975). Three million units of an article would theoretically have to be tested without any positive result to demonstrate sterility in 1×10^6 units with 95% confidence (Figure 12.1).

Furthermore, the test for sterility in itself constitutes an added source of error (see Chapter 8). For testing sterile articles, the protective sealed container has to be penetrated with the inherent risk of introducing contamination. Retesting procedures are no longer accepted as a means of determining whether a contaminant detected during a sterility test resulted from a true lack of sterility or was introduced during the test. Today, any growth in a sterility test must lead to a full investigation into the possible origin of the contaminant. The product can only be released if such investigations conclusively show that the contamination was introduced during the sterility testing procedure.

The reader is referred to the companion *Handbook* or Bill (2000) for a more detailed discussion on the practical implications of the sterility test in support of the sterility assurance system.

In consequence, sterility assurance must be based on validation of the process rather than on testing of the product. This was stated clearly in the EP (1997):

...the sterility of a product cannot be guaranteed by testing; it has to be assured by the application of a suitably validated production process; [...] when a fully validated terminal sterilisation method by steam, dry heat or ionising radiation is used, parametric release, that is the release of a batch of sterilised items based on process data rather than on the basis of submitting a sample of the items to sterility testing, may be carried out, subject to the approval of the competent authority.

A proposal on how to proceed with parametric release, in view of the minimal contribution of the sterility test to the safety of the released product, was given by the Parenteral Drug Association (PDA) (1999). This proposal concentrated mainly on the validation requirements for the sterilization process itself. The European authorities subsequently clarified their position in that approval of parametric release by the competent authority was based not only on the sterilization process, but on the complete manufacturing process [European Commission 2001, European Agency for the Evaluation of Medicinal Products (EMEA) 2001]. The necessary preapproval inspection was, therefore, to include not only the sterilization process, but also environmental monitoring and bioburden testing of the presterilization intermediate, and in addition, the general good manufacturing practice (GMP) procedures and the GMP attitude of the applying company [Pharmaceutical Inspection Cooperation Scheme (PIC/S) 2001a]. In other words, approval of parametric release in Europe is a procedure that involves the GMP inspectorates (European Commission 2001), as well as the authorities assessing pharmaceutical registration dossiers (EMEA 2001).

Although the European rules for parametric release have been clear now for several years, there have been only limited applications to the competent authorities. Additional GMP requirements, for example, on the manufacturing environment or on the control of the presterilization bioburden, are seen by many manufacturers as an unnecessary and irrational burden on the process of manufacturing sterile products that are terminally sterilized by an overkill procedure. It remains unclear on the other hand as to why a manufacturer should be permitted to produce a sterile product by a procedure that is deficient in GMP, only because the product is tested with the sterility test. For risk-based consideration of sterile product manufacture and parametric release, see Haberer (2004).

12.2 REGULATORY REQUIREMENTS

Validation of sterilization methods today is an absolute requirement in numerous official texts: "It is essential that the effect of the chosen sterilisation procedure on the product (including its final container or package) is investigated to ensure effectiveness and the integrity of the product and that the procedure is validated before being applied in practice" (EP 2006). The current view of the U.S. Food and Drug Adminstration (FDA) on sterilization process validation and the documentation thereof was published by the FDA (1994). Submission of data based on this document is mandatory in drug applications for the U.S. market.

The approach taken toward validation of sterilization processes is significantly different in Europe, compared with the United States. The EP relies on reference

TABLE 12.1
European Pharmacopoeia (2006) Sterilization Reference Conditions

Procedure	Minimum Reference Conditions
Steam sterilization	Saturated steam 121°C, 15 min
Dry heat sterilization	160°C, 2 h
Ionizing radiation sterilization	25 kGy

Note: For gaseous sterilization, no standard conditions are defined.

Source: Compiled from European Pharmacopoeia. (2006). *European Pharmacopoeia.* 5th ed. Council of Europe, Strasbourg.

overkill conditions as summarized in Table 12.1. Although it is accepted that reference conditions are effective sterilization conditions, the use of a reference method does not exempt a pharmaceutical manufacturer from validation of the process.

The expectation of the European authorities is clearly and comprehensively summarized in the British Pharmacopoeia (2005):

> For terminal sterilization it is essential to take into account the non-uniformity of the physical and, where relevant, chemical conditions within the sterilising chamber. The location within the sterilising chamber that is least accessible to the sterilising agent is determined for each loading configuration of each type and size of container or package (for example the coolest location in an autoclave). The minimum lethality delivered by the sterilising cycle and the reproducibility of the cycle are also determined in order to ensure that all loads will consistently receive the specified treatment.

> Having established a process, knowledge of its performance in routine use should, wherever possible, be gained by monitoring and suitably recording the physical and, where relevant, chemical conditions achieved within the load in the chamber throughout each sterilising cycle.

European authorities expect that the product should be designed wherever possible to withstand the application of pharmacopoeial standard sterilization conditions. This is specifically addressed in the Committee for Proprietary Medicinal Products (CPMP) note for guidance for development pharmaceutics: "wherever possible all such products should be terminally sterilised in their final container, using a fully validated terminal sterilisation method using steam, dry heat or ionising radiation as described in the European Pharmacopoeia" (EMEA 1998). In a decision tree published as an annex for this note for guidance (EMEA 1999), this same attitude was expressed even more strongly. However, with the increased introduction of sensitive biological pharmaceuticals into the market, this classical approach will not be so tenable in the future. Since 1997, the EP has included a chapter on the application of the F_0 concept, which allows more flexibility than the simplistic EMEA decision tree.

A different position is adopted in the United States. Priority is not only given to the sterilizing efficacy of a standardized process but also to the minimum destruction delivered by the sterilization procedure to the product. Hence, far greater weight is given to product-specific development of sterilization cycles and reference conditions are not a generally accepted approach. Typically, detailed studies on product compatibility and on the presterilization bioburden in the product are expected. The microbiological efficacy of the sterilization cycle must be demonstrated in relation to the bioburden found.

Even though the priorities are different, wherever validation requirements are given in regulatory texts, the same goals are to be met:

Suitability of the process
- The items must be designed to withstand the process and to maintain the desired quality.
- The process must be compatible with the items to be sterilized.
- The process must be effective in inactivating or removing all microorganisms to the required safety level.

Uniformity of the process
- All parts of the item to be sterilized must receive the specified minimum treatment.
- No part must receive a treatment destructive to the item.

Reproducibility of the process
- Measurements and recordings of suitable parameters must fully document the reproducibility of the process.
- Any conditions rendering the process ineffective must be recognized and procedures established to prevent the release of product so affected.

12.3 GENERAL PRINCIPLES OF STERILIZATION PROCESS VALIDATION

12.3.1 STERILIZATION CYCLE DEVELOPMENT

12.3.1.1 Sterilizing Principle

During the development of a new product intended to be sterile, it is critical that an optimized sterilization concept is chosen for the product. First it must be decided which sterilizing principle (steam, dry heat, gas, or ionizing radiation) is best tolerated by all components of the intended product, including active ingredients, excipients, and container closure or other primary packaging material. If it can be shown that none of these principles is tolerated by the product, sterilization by membrane filtration and subsequent aseptic processing or filling may be considered.

Once the sterilizing principle has been decided upon, the approach toward a sterilization cycle must be defined. The European authorities expect that first consideration should be given to application of a standard overkill cycle as defined in the EP (2006). Where feasible, products in development should be designed taking account of their ability to be sterilized. Where a newly developed sterile product formulation cannot tolerate a standard sterilization cycle, reformulation should be considered.

Standard overkill methods are designed to kill highly resistant test organisms by at least eight to ten orders of magnitude. For example, *Geobacillus stearothermophilus* (previously *Bacillus stearothermophilus*) spores in aqueous suspension are used as an indicator organism. A logarithmic reduction time (D value) of about 1.5 min at 121°C in saturated steam has been experimentally demonstrated. In this case, after application of standard autoclave conditions of 15 min at 121°C, a reduction of *G. stearothermophilus* spores of ten orders of magnitude should be achieved. Once it has been shown that such conditions are reached in every position of the load and that the lethality delivered within the product corresponds to the expected D value, no further biological validation would be required.

The U.S. definition of an overkill cycle was given in the PDA Technical Report no. 1 (1978) as follows: a cycle that provides more than a 12-log reduction of a resistant biological indicator with a known D value of not less than a minute. This approach ensures substantially greater than a 12-log reduction of the bioburden, and therefore, no further information on the bioburden is required. (PDA Technical Report No. 1 has been under revision for some time, and there will probably be more than one definition in the revised document.)

It is confusing that this definition is tied to bioindicators in general but not to a specific reference preparation. If the U.S. definition is applied, the European reference condition for steam sterilization would be considered an overkill procedure for a bioindicator with a D value of 1 min (15-log reduction), but not for a bioindicator preparation of *G. stearothermophilus* with a D value of 1.5 min (10-log reduction). This does not make sense, as neither presterilization bioburden nor cycle effectiveness is influenced in any way by the choice of bioindicator. *G. stearothermophilus* spores with a D_{121} value of more than 2.5 min are known to exist, and an overkill process for such a bioindicator would take at least 30 min at 121°C. As the bioburden of the product has nothing to do with the bioindicator chosen, it would be much more reasonable to define a standard D_{121}^0 value as a D_{121} value of 1 min and define an overkill cycle as any cycle equivalent to inactivate at least 12 logs of the theoretical D_{121}^0 bioindicator in or on the product to be sterilized. An alternative way of stating this is that an overkill sterilization process should have a F_0 of at least 12 min. In order to verify cycle effectiveness, any biological indicator can be used that has a sufficiently high D_{121} value to allow an equivalence calculation.

12.3.1.2 Equivalent Overkill Cycles

An alternative to reformulation of the product may be to choose other time–temperature combinations than the pharmacopoeial standard cycle. Short exposure to a higher temperature or longer exposure to a lower temperature may deliver equivalent lethality to microorganisms while being better tolerated by the product. Equivalent sterilization conditions can only be calculated for a given microorganism with known D and z values (defined in Chapter 10). Usually, Equation 12.1 is used for equivalence calculations:

$$F_T^z = \frac{F_{121}^z}{10^{(T-121/z)}} (\text{min}) \tag{12.1}$$

where F = effectiveness, T = process temperature, z = temperature difference leading to a tenfold change in the D value.

A range of D and z values have been described for resistant spores of *G. stearothermophilus* (e.g., Pflug 1995, Russell 1999). For resistant spores in aqueous solution, usually a z value of 10°C is assumed for equivalence calculations. For *G. stearothermophilus* with a z value of 10°C and an inactivation of 10-log orders within 15 min at 121°C (D value of 1.5 min), an equivalence calculation for 116°C would be as follows:

$$F_{116}^{10} = \frac{15}{10^{-5/10}} = 47.43 \,(\text{min}) \tag{12.2}$$

12.3.1.3 Bioburden-Based Cycles

Bioburden-based cycles can either be designated bioindicator cycles where the bioburden is represented by a specifically selected bioindicator, or straight bioburden-based cycles, where the effectiveness of the sterilization process has to be verified by actual bioburden determinations for each individual cycle.

12.3.1.3.1 Designated Bioindicator Cycles

For a designated bioindicator, the organism chosen should form endospores that are at least as resistant as the most resistant forms encountered in the bioburden of the product to be sterilized. This could be the most resistant organism ever isolated from the well-characterized bioburden of the product, or another organism that can be obtained from an official strain collection. During validation studies, it must be shown that appropriate bioindicator preparations of this organism are inactivated within the product at the positions most difficult to access by the sterilizing agent. By assuming that the bioburden found in the product comprises the most resistant variety, sterility assurance based on a maximum accepted number of bioburden organisms can be calculated as shown in Table 12.2.

The actual dwell time needed under validated sterilization conditions could be even shorter, if the effective sterilization time is calculated by summing the sterilizing effect of the temperatures achieved during heating and cooling of the product. This is especially important for heat-sensitive liquids sterilized in large volumes. The mathematical expression used for such calculations is given in Equation 12.3 (see also Chapter 10).

$$F_T^z = \int_{t=0}^{t=x} 10^{(T-121)/z} dt \tag{12.3}$$

where $t = 0$ is the beginning of the sterilization time, and $t = x$ is the end of the sterilization time.

TABLE 12.2
Sterilization Time for a Design Bioindicator Sterilization Cycle

Bioburden limit in presterilization solution	10^2 CFU.mL^{-1}
Volume per unit of finished product	10 mL
	(= 10^3 CFU per unit)
Logarithmic inactivation factor needed to achieve a sterility assurance level of 10^6 per unit	9
D value of design bioindicator under the validated sterilization conditions	0.8 min
Sterilization time needed under the validated sterilization conditions	$9D = 7.2$ min

Calculations can be based on temperature readouts taken every minute as described by Patashnik (1953) or else automatic F_0 calculators can be used to do the integration.

Where a designated bioindicator cycle is employed, assurance of sterility must rely on batch by batch confirmation that the bioburden limit is not exceeded. At scheduled intervals it should be confirmed that no organisms with higher resistance than the designated bioindicator are found in the bioburden.

12.3.1.3.2   Direct Bioburden-Based Cycles

If by design of the manufacturing process the presterilization bioburden is kept consistently and reliably at very low numbers, a direct bioburden-based cycle may be a possible approach. Prefiltration of a solution to be aseptically filled immediately prior to steam sterilization would be a typical example. In such a solution, the bioburden can be reliably kept well below 10 CFU in 100 mL, with a maximum number of heat-resistant spores well below 1 in 100 mL. For 10 mL units only five decimal reductions would be required to reach the necessary safety level.

For a direct bioburden-based cycle, the number of viable organisms and the absence of heat-resistant forms in a predefined volume must be determined for every batch as essential information for release of the product.

12.4   VALIDATION OF SPECIFIC STERILIZATION CYCLES

12.4.1   Sterilization by Moist Heat under Pressure

Sterilization in an autoclave is generally based upon highly efficient heat transfer from saturated steam to autoclave load, combined in some cases with a hydrating effect caused by the condensate that is formed during the process. Heat transfer is maximum if steam is kept along the phase separation line of the water–steam phase diagram. Neither unsaturated steam–air mixtures nor overheated steam exhibit the same favorable conditions (see Chapter 11).

For porous loads where the sterilization effect is achieved by direct steam contact, steam penetration into every part of the load must be ensured. Validation procedures to achieve this are described in detail in European Standards EN 554 (1994) and

EN 285 (1997). The validation in these standards is based upon hospital sterilization cycles, where stacks of garments have to be reliably sterilized.

It is important, however, to understand that the principle of saturated steam is deliberately set aside in many autoclaves used in the pharmaceutical industry to sterilize closed containers such as ampoules, vials, and even prefilled syringes. These containers usually contain a headspace filled with air or a protective gas. During sterilization, saturated steam conditions cannot always be reached within the containers; instead the steam pressure developing from the solution plus the headspace gas pressure will contribute to an increased internal container pressure. It may, therefore, be preferable to operate at suboptimal heat transfer capacity and instead to compensate for the increased container pressure by adding air to the autoclave (overpressure steam–air mixture) or even to fill the autoclave with water at sterilizing temperature (see Sections 12.4.2.3 and 12.4.2.4).

The important issue during validation of this type of autoclave cycle is the level of lethality delivered to resistant microorganisms within the containers. Neither the degree of steam saturation within the chamber, nor the pressure developed within the chamber, nor the heat transferred to the containers can give the complete picture during validation of the cycle. Bioindicator studies are needed to correlate the physically measured parameters to the delivered lethality. Once this correlation has been fully established, recording of the temperature transferred to containers in the established cold spot of the load would theoretically be sufficient to monitor routinely the correct (and effective) performance of the sterilizer. The actual monitoring requirements will, however, depend on the type of sterilization cycle chosen.

12.4.2 TYPES OF STEAM STERILIZATION CYCLES

Validation must generally be based on a thorough understanding of the process used. There are many types of autoclaves and to choose the correct cycle for the intended product is a critical step during validation. The experiments to be performed during process validation and the monitoring data required for routine process control will depend on the chosen type of cycle. If the wrong type of cycle is chosen, it may not be possible reliably to achieve sterility of the product.

12.4.2.1 Gravity Displacement Cycles

For autoclave loads with nonporous materials of simple geometry, the classical gravity displacement principle may be used. Steam enters at the top of the chamber and displaces the cold air that, due to its heavier specific weight, remains at the bottom of the chamber and exits through a trap.

The steam injection rate is critical in this type of process because turbulence may lead to trapped air pockets within the load. Conversely, if displacement is too slow, the air will be heated and, by losing specific weight, will then diffuse into the steam, leading to less efficient air–steam mixtures. Steam admission permits the gradual heating of product in advance of the required sterilization conditions being met.

For loads with complicated geometry or porous surfaces where air can easily be trapped, complete air displacement is very difficult or even impossible to achieve.

Validation must rely on a large number of individual measurements distributed within the load. Material and geometry for every type of autoclave load must be stated in written procedures and strictly adhered to in order to ensure reproducible conditions between validation runs and during routine operation.

Parameters to be routinely monitored during each cycle will depend on the type of load. Temperature in the chamber and at several positions in the load, chamber pressure, saturation of steam leaving the autoclave at the condensate outlet, and verification of steam penetration into porous loads are generally critical parameters.

12.4.2.2 Prevacuum Cycles

Penetration of the load by saturated steam can be more effectively secured by prevacuum or fractionated prevacuum cycles. Before saturated steam is injected into the chamber, air is removed by a vacuum pump. A vacuum as low as 55 mbar (5.5 kPa) is usually applied to remove air from the chamber (Wallhaeusser 1995). Fractionated vacuum cycles have therefore been developed to remove air more effectively by successive vacuum cycles and steam pulses.

Prevacuum cycles are especially suited for porous material and loads of complicated geometry, which are not sensitive to vacuum treatment (e.g., surgical dressings). Liquids (except when sealed in closed containers) may evaporate during the vacuum cycles and are therefore less suited or even unsuited for treatment in prevacuum cycles.

During validation, the suitability of the cycle to remove air effectively from the load must be established. This can be difficult for fabrics or porous materials, which release trapped gas slowly. For such critical loads, fractionated prevacuum cycles are most suited and typically used. The number of prevacuum cycles necessary and the depth of vacuum to be drawn must be established. Steam penetration and bioindicator studies are critical experiments during validation.

Another typical application for prevacuum cycles is the sterilization of equipment (e.g., filter assemblies) to be used for aseptic processing. Air may be rather difficult to remove from tubing and the housing of such equipment. Bioindicators must be placed within the equipment in positions that are least accessible to steam. The arrangement and setup of the equipment within an autoclave must be established, documented, and strictly adhered to in routine sterilization. These principles will also apply to certain medical devices where they may present similar complications.

For typical loads of pharmaceutical products in their final closed containers, removal of air presents less of a problem. Where prevacuum cycles are used, this generally serves to achieve homogeneous heat transfer conditions within the load.

Parameters to be routinely monitored during each cycle will depend on the type of load. Temperature should be measured in the chamber and at several positions in the load, and chamber pressure should be measured in all cases. Saturation of steam leaving the autoclave at the condensate outlet and verification of steam penetration are critical information for porous loads. For equipment, it should be decided case by case what data are needed to ensure that steam penetration and efficient sterilization were achieved.

12.4.2.3   Air–Steam Mixture Cycles

Whenever liquids sealed in their final container are to be sterilized, a situation quite different from other autoclave cycles arises. The headspace within every unit forms a gas pocket, which cannot be displaced or otherwise removed. If the autoclave is operated with saturated steam, a differential pressure between the outside and inside of the individual containers is generated with a risk of bursting containers. Therefore, for sterilizing a liquid in sealed containers, steam–air mixtures are often used to generate a higher chamber pressure, a technique known as air ballasting. For the purpose of sterilization, the autoclave chamber merely serves as a steam jacket.

For the validation of such sterilization cycles, the consideration of steam saturation is meaningless. Temperature recordings have to be taken from reference containers of the same size and configuration as the actual containers of the load. Such reference containers should be distributed throughout the load to check for temperature differences, which can easily arise if the air–steam mixture in the chamber is not sufficiently mixed. The cold spot of the load (coldest container) should be established in temperature mapping studies. Effectiveness of the sterilizing conditions at the cold spot should be established by use of bioindicator studies within the actual product.

For routine control, temperature profiles should be recorded within a sufficient number of reference containers of the same configuration as the product containers. One of the containers should represent the cold spot of the load. Alternatively, the correlation of the reference container to the cold spot must be established during validation, and this correlation is then used to extrapolate exposure at the cold spot.

12.4.2.4   Hot Water Shower and Hot Water Submerged Cycles

In order to achieve better temperature homogeneity for products in sealed containers, in some sterilizers, the load is showered with hot water. Alternatively, the chamber may be fully flooded with water, which is kept precisely at sterilizing temperature by circulation through a heat exchanger. Although temperature distribution is usually better controlled in such chambers, the questions to be asked during validation are the same as for steam–air mixture cycles.

For routine control, temperature recording in an adequate number of reference containers will be sufficient.

12.4.3   Sterilization by Dry Heat

Sterilization by dry heat must overcome the poor heat transfer capacity and conductivity of air. Therefore, higher temperatures or longer sterilization times are required as compared to the more efficient autoclave processes. For heat-resistant articles sensitive to or impenetrable by moisture, dry heat is nevertheless the method of choice. If the dry-heat sterilizer is operated at temperatures above 250°C, bacterial lipopolysaccharides (endotoxins) are destroyed under highly effective sterilization conditions (Wegel 1973). Dry heat is typically used to sterilize and simultaneously remove endotoxins from ampoules or glass vials.

Several types of sterilizers are used in the pharmaceutical industry; these require different considerations for validation.

12.4.3.1 Forced-Convection Batch Sterilizers

For batch sterilizers, homogeneous heat distribution within the chamber is difficult to achieve. Heated air should be blown through a high-efficiency particulate air (HEPA) filter to minimize particle contamination and distributed into the chamber via a baffle system. For validation studies, a large number of thermal sensors has to be distributed within the chamber and throughout the load to identify cold points. Usually several patterns of thermal sensor distribution should be used during validation to characterize heat distribution in a load as precisely as possible. For the sterilization of large items (e.g., parts of equipment), the positioning of sensors in contact with the surface or where feasible within the item is critical. For goods to be sterilized in containers, thermal sensors must be placed within each type of container to be sterilized to demonstrate sufficient heat penetration under all circumstances. Rearrangement of the load may alter convection patterns of the air considerably. Therefore exact loading patterns must be established, validated, and adhered to strictly.

For routine control, temperature profiles should be recorded at a sufficient number of positions, preferably in the established cold spots of the load. Alternatively, the correlation of the reference measuring positions to the cold spot must be established during validation, and this correlation is used to extrapolate exposure at the cold spot.

12.4.3.2 Dry Heat Sterilizing Tunnels

Dry heat sterilizing tunnels usually operate at temperatures well above 250°C. Items to be sterilized are placed on a conveyor belt. During passage through the heating zone, high temperatures of usually more than 300°C are reached for a few minutes in every item. Heat transfer is achieved by radiation in infrared tunnels or by forced convection in HEPA-filtered laminar-flow tunnels (Wegel 1973). Ventilation with HEPA-filtered air is employed in most sterilizers to cool the sterilized goods.

In addition to temperature probes, which should be placed all across the conveyor belt, reproducibility of belt speed and integrity of HEPA filters are considered critical parameters for validation.

For routine control, continuous temperature recording and recording of the belt speed are needed.

12.4.4 Gaseous Sterilization

The use of toxic gases to sterilize heat-sensitive goods is an attractive concept from a technological point of view, although there are concerns over their safety in use. In Europe, the use of ethylene oxide is no longer accepted for sterilization of pharmaceutical products. It is still widely used, however, for the sterilization of medical devices. The situation is different in the United States and Japan, where ethylene oxide is considered a viable alternative in spite of its toxic characteristics and the carcinogenic nature of its degradation products.

Gassing of rooms with formaldehyde is an effective disinfection method that can reach otherwise inaccessible positions. However, due to the carcinogenic and sensitizing properties of the gas, the associated health hazard to the personnel is considered too high and it is considered an obsolete practice today. In Europe, where formaldehyde gassing is still used as an exceptional method, for example, as a corrective measure after a major contamination event or where a basic state of cleanliness is to be achieved after construction work, a specifically trained responsible person for gassing must be registered with the authorities.

Surface sterilization by gaseous or vaporized hydrogen peroxide has not only gained widespread acceptance for isolator and room decontamination, but also as a means to surface sterilize tools, equipment, and other materials. For example, Sterrad® sterilizers, which use a cycle of hydrogen peroxide sterilization in conjunction with a plasma phase, are available on the market, but have been mainly used in the hospital environment and less so in the production of pharmaceuticals.

Gas sterilization is a complex process, where the concentration of the sterilant in the air, the condensation of the sterilant on surfaces, the sterilant distribution in the chamber, the temperature, relative humidity, and time of exposure are interacting in a complex way to achieve the sterilizing effect (Sigwart and Moirandat 2000). For example, sterilant concentration cannot be expected to remain constant over time in the gas or condensate phase. Development and characterization of such processes strongly depend on the use of suitable bioindicators, which must be of appropriate quality to give reliable indication of the gassing effect (Sigwart 2004).

During validation of gaseous sterilization, sterilizer loading patterns have to be established and strictly adhered to. Temperature and relative humidity should be measured at various geometrical positions within the load. For the validation of gas sterilizers, bioindicators are still the most effective and widely used method and their use is mandatory in the EP (2006) not only during cycle development and validation, but also for monitoring of the gas sterilization process.

Routine control measures comprise determination of gas concentration, temperature, and humidity, as well as the use of suitable bioindicator preparations.

12.4.5 Sterilization by Ionizing Radiation

Radiation is a highly effective means of sterilization and is used with increasing frequency. However, chemical changes are caused in many products by reaction with free radicals generated by ionizing radiation. In order to avoid such radiation damage, radiation sterilization can only be used if all parts of the product have been demonstrated to be compatible with this kind of treatment at the required energy level. The EP gives a standard sterilizing dose of 25 kGy (2.5 Mrad), which has been widely accepted in Europe.

For operational qualification, loading patterns have to be established and radiation penetration determined by the use of dosimeters distributed within the articles. Configuration of the load should be such that a minimal dose of 25 kGy (or another suitable dose established by bioburden studies) is delivered to every point in the load, and the maximum dose is still compatible with the product to be sterilized.

Bioindicators should only be used to determine inactivation characteristics in a product newly validated for radiation sterilization. For operational validation or revalidation, dosimetry is a much more reliable method.

12.4.6 STERILIZATION BY MEMBRANE FILTRATION

Sterilization by passage through a microorganism-retentive filter is a commonly used sterilization method for heat-sensitive fluids. In contrast to all other sterilization methods that inactivate living microorganisms within their environment, filtration is a physical process, removing contaminating particles from a fluid. Removal of particles is effected by sieving in combination with adsorptive properties of the filter surface. In most types of filters, the pores are in fact channels of irregular shape leading in a tortuous pathway through a spongiform matrix. The nominal pore size is a figure deduced by the filter manufacturers as a rating from the retentive capacity of the filter. It is clear that contaminating particles of smaller size than the normal pore size rating may pass through the filter. Hence, very small or pleomorphic forms of bacteria (e.g., mycoplasmas) or viruses are not fully retained by the usual sterilizing grade filters. Where such forms are to be expected in a solution, either a smaller nominal pore size has to be employed or else filtration may not be a feasible sterilization method for the product.

Validation of filtration is separated into several clearly distinct phases that should not be confused.

12.4.6.1 Filter Development

Filter development validation consists of basic filter characterization by the filter manufacturer. In this phase, chemical and physical compatibility of the filter and filter assembly is established. Toxicity of filter extractables in water is determined. The characteristics of gas flow across the filter membrane (diffusive flow and transition to capillary flow) and their correlation with the retentive capacity of the filter toward model organisms in aqueous suspension (ASTM 1988) are established as a basis for classification as a sterilizing filter and as a basis for physical integrity testing.

Quality assurance during filter production and assembly is the responsibility of the filter manufacturer. It must be ensured that the characteristics of each filter disk or assembly shipped to the filter user are fully in compliance with the requirements for the filter type as established during filter development validation.

12.4.6.2 Development of the Filtration Process

Development of the filtration process is the responsibility of the filter user. Retentive capacity of the filter, the accuracy of pore size during all conditions employed in the process, and integrity of the filter assembly are among the key factors with respect to sterilization (Wallhaeusser 1979, 1982). Other important factors are, for example, interactions between product to be filtered and the filter, the influence of product on the size and shape of microorganisms, the influence of product on filter retentivity, and parameters of the filtration process, that is, differential pressure across the filter, flow rate, processing time, and temperature (Jornitz and Meltzer 1998, Jornitz 2004). The considerations required of the filter user are compiled in ISO 13408-2 (2003).

Compatibility of the filter with the product to be filtered and with the filtration process to be applied has to be established. This includes physical and chemical compatibility studies and should comprise product-specific bacterial challenge testing to determine whether the product affects retentive properties of the filter (Carter and Levy 1998). As the product may also alter the properties of the microflora, microbial challenge testing should be performed where feasible on the product solution or under conditions simulating the influence of product on the challenge organisms. Considerations for the design of such studies are given in Technical Report no. 26 of the PDA (1998).

Frequently, such studies are offered by the filter manufacturers as a contract service to their customers. This makes sense, as the projection and extrapolation of filter behavior under process conditions is greatly facilitated by the in-depth knowledge of filter characteristics as established during filter development. Also, filter manufacturers have amassed a wealth of experience during collaboration with their clients, which can be used to facilitate the development of filtration processes.

A compilation of important activities required for membrane filtration validation is shown in Table 12.3.

12.5 METHODS USED FOR VALIDATION

In order to demonstrate the correct performance of sterilization processes, two concepts are used in combination, which yield different types of information:

- Observation of the effect exhibited by a chosen sterilization cycle on indicator organisms or indicator substances.
- Measurement and recording of the required physical or chemical conditions reached throughout the sterilization cycle within every part of the load.

Whereas the functional approach yields direct information on the effect of the process, the metrological approach yields data that are to be obtained and must be controlled during the sterilization process. The correlation between measured process data and the effect exhibited on the microorganisms to be eliminated should be established as the primary goal of validation.

This section reviews methods for characterization of the main pharmacopoeial sterilization methods and highlights those features that need to be considered in a validation program.

12.5.1 Physical Methods for Validation

12.5.1.1 Measuring Devices for Heat

For heat sterilizers, including autoclaves and dry heat sterilizers, determination and recording of heat distribution within the chamber are of prime importance. The most

TABLE 12.3
Validation of Membrane Filtration

Validation Aspects	Responsibility
Membrane Characterization	
Physicochemical characteristics:	Filter manufacturer
Particle shedding at various flow rates	
Extractables in standard solvents	
Chemical compatibility toward standard solvents	
Thermal stress resistance	
Hydraulic stress resistance	
Biological characteristic:	Filter manufacturer
Toxicity test	
Retention characteristics:	Filter manufacturer
Characterization of flow characteristics in water or standard solvent and correlation to microbial retentive characteristics (*Brevundimonas diminuta* in water)	
Establishment of physical integrity test conditions and test values	
Membrane manufacturing quality assurance	Filter manufacturer
Process Development	
Compatibility of product and filter material:	Filter user
Extractables in product solution	
Chemical compatibility toward product solution	
Thermal stress resistance in product solution (where resterilization is intended)	
Filter-process qualification:	Filter user
Required pore size rating	
Filter sizing and configuration	
Filtration rate and clogging	
Filtration system design	
Hydraulic stress resistance under process conditions	
Maximum time for filter use	
Microbial challenge testing:	Filter user
In product (where possible), in surrogate product, or on filters preconditioned with product; challenge organism *B. diminuta* or appropriate bioburden isolate	
Definition of maximum tolerable presterilization bioburden	
Integrity Test Procedure	
Qualification of integrity test equipment	Manufacturer
Establishment of integrity test procedure and parameters in product	Filter user
Routine Procedure	
Establishment of standard operating parameters	Filter user
Specification and individual batch determination of presterilization bioburden	
Validation of aseptic processing and filling	
Final product sterility testing (each batch)	
Documentation and release procedures	

commonly used equipment consists of resistance temperature detectors (RTDs) or thermocouple measuring systems.

RTDs are most commonly used as temperature standards for calibration studies, and thermocouple systems connected to multichannel electronic recording instruments are best suited for heat penetration studies. With thermocouple elements, care should be taken to use only high-grade thermocouple wire, if possible from the same production lot, to avoid errors due to noninterchangeable elements. Kemper (1986) has given a detailed description of thermocouple installation and calibration. The entire system should be calibrated before each run to ensure a measurement accuracy of at least ±0.1°C at 120°C, ±0.2°C at 200°C, and ±0.4°C at 300°C. Calibration should be repeated at the process temperature after each use to verify proper operation during the validation.

Data loggers are a modern alternative to thermocouples suitable for temperature distribution studies. These units consist of a temperature sensor and an integrated data recorder. They are designed to withstand autoclave temperature and pressure and can be distributed within the chamber of load. Temperature profiles are individually recorded in each such device and can be recalled via an interface to a personal computer after the run. This obviates the need to penetrate the autoclave chamber with wires through a validation port. The same calibration requirements apply as for thermocouples.

12.5.1.2 Measuring Devices for Pressure

Pressure sensors should be chosen to fit the purpose of the instrument. For autoclaves, precision requirements are specified; pressure gauges should show zero pressure at atmospheric pressure with a meter range between 1 and +5 bar (100 and +500 kPa), the scale being not wider than 0.2 bar (20 kPa). The precision of recording should be better than ±1.6% over the range of 0 to 5 bar according to ISO 11134 (1994).

12.5.1.3 Physical Methods for Determining Filter Integrity

Filter integrity is usually determined by suitable methods such as bubble point, pressure hold, or diffusion rate procedures, or for hydrophobic filters by a membrane intrusion test (Jornitz et al. 1998). Measurement of bubble point determines at which differential pressure the liquid contained in the weakest point (i.e., largest pore) of the matrix of a wetted filter is driven out. Besides the nominal pore size, the bubble point is also dependent on the filter material, the wetting fluid, and temperature. Diffusional flow measurements determine the flow of a test gas through a wetted filter at lower differential pressure than in the bubble point test. Results are dependent on the thickness and structure of the filter, the temperature, and interaction between test gas and test fluid (Meltzer 1987). Pressure decay testing can be considered a variation of diffusional flow testing with the advantage that the sterile system downstream of the filter is not broached during measurement (Denyer and Hodges 2004b).

For integrity testing, filters can either be wetted with a test fluid (e.g., water) or with the product solution to be filtered. If the filters are tested in product, correlation of the product-specific diffusive flow or bubble point data with the data given by the filter manufacturers for water must be established.

Integrity data should be obtained for every filter assembly before and after a filtration process is performed (Wallhaeusser 1982). If the results conform to the known characteristics of the filter type used, the filter assembly can be assumed to be integral and correctly assembled. Integrity testing is usually performed today with automated integrity test equipment. Care should be taken that these systems are appropriately validated (Docksey et al. 1999, Jornitz et al. 1998).

12.5.1.4   Dosimeters for Radiation Sterilization

Perspex (polymethacrylate) strips 1 to 3 mm thick are most frequently used as dosimeters. Perspex shows dose-dependent coloration when exposed to gamma radiation. The dose can then be deduced from photometric determination of the color.

Dosimetry for radiation processing was discussed in detail by McLaughlin et al. (1989).

12.5.2   Bioindicators for the Validation of Sterilizers

12.5.2.1   Bioindicator Preparations

Bioindicators are available as ready-to-use preparations of microorganisms inoculated into the product, adsorbed onto carriers (e.g., paper strips or glass beads), or suspended in liquid medium and sealed in ampoules for steam sterilization. The bacteria chosen are specifically selected for their high resistance against particular sterilization principles (Table 12.4). Whereas these marketed preparations are characterized by their

TABLE 12.4
Bioindicators for Sterilization Procedures

Procedure	Species	Strain	Proposed by
Steam	*Bacillus stearothermophilus*	ATCC 7953	USP 2004, EP 2006
		ATCC 12980	USP 2004
		NCTC 10007, NCIMB 8157, CIP 52.81	EP 2006
Dry heat	*B. subtilis* var. *niger*	ATCC 9372	USP 2004, EP 2006
		NCIMB 8058, CIP 77.18	EP 2006
Gas (EO)	*B. subtilis* var. *niger*	ATCC 9372	USP 2004, EP 2006
Radiation	*B. pumilus*	ATCC 27.124, NCTC 10327, NCIMB 10692, CIP 77.52	EP 2006
Membrane filtration	*Pseudomonas diminuta*	ATCC 19146	USP 2004, EP 2006
		NCIMB 11091, CIP 103020	EP 2006

Notes: ATCC—American Type Culture Collection, Rockville, MD; CIP—Collection de l'Institut Pasteur, Paris; NCTC—National Collection of Type Cultures, London; NCIMB—National Collection of Industrial and Marine Bacteria, Aberdeen, United Kingdom; EP 2006—European Pharmacopoeia (2006); USP 2004—United States Pharmacopeia (2004); EO—Ethylene oxide; *Bacillus stearothermophilus* renamed *Geobacillus stearothermophilus*; *Pseudomonas diminuta* renamed *Brevundimonas diminuta*.

manufacturer with regard to viable cell number and resistance properties, their suitability to reflect conditions delivered to microorganisms in a given product must be critically evaluated during process validation. For instance, microorganisms fixed on a paper strip may react differently from microorganisms in suspension, and bioindicators in ampoules are unsuitable where steam saturation is a critical factor.

Customized biological indicators can be prepared from suspensions of resistant microorganisms by direct inoculation into the product or onto the surface of components or equipment parts to be sterilized. In this way, it is possible to investigate the effects of a sterilization process much more accurately. It is also possible to inoculate specific remote positions in the product and to map the effect achieved during the sterilization process. The difficulty is, however, that these customized biological indicator preparations must be characterized under defined conditions to ensure that the results achieved during validation are indeed reliable. For such characterization experiments, biological evaluation resistometers (BIER) have been described (ANSI/ASTM 1992, Comité Européen de Normalisation 1997). Characterization of bioindicator preparations in the BIER is offered as a service by specialized laboratories.

If a specified high concentration of these organisms is inactivated or eliminated, the sterilization process has been demonstrated to be effective. Although this procedure initially may seem convincing, many questions arise if the approach is scrutinized in detail as discussed below.

12.5.2.2   Inactivation Kinetics of Bioindicators and Process Control

Inactivation of bioindicators is usually evaluated either by a fraction negative method as described by Stumbo et al. (1950) (most probable number approach) or by following the Spearmann–Karber (Holcomb and Pflug 1979) (mean time to survival) approach. Full evaluation of the inactivation kinetics by regression analysis is a more precise but tedious and time-consuming method.

In the United States, the efficacy evaluation of a sterilization process is often linked to the inactivation of bioindicators by a specified logarithmic factor, as evaluated by the fraction negative method. Bioindicators are, however, derived from cultivated strains of living organisms and as such are subject to change. Unless the D value of the bioindicators is always maintained at the same standardized value, this method leads to fluctuating estimation of the cycle effectiveness. An unusually resistant bioindicator preparation may even lead to the conclusion that the cycle is no longer valid. For this reason, cycle effectiveness should be evaluated during cycle development by use of a bioindicator with a well-characterized D value in the standard preparation and where possible in the product. The cycle effectiveness should then be expressed in terms of the standard inactivation conditions, such as those expressed as F_0 values.

12.5.2.3   Use of Biological Indicators in the Determination
of Homogeneous Conditions in an Autoclave

Inactivation of spores by heat is considered to follow first-order kinetics. Thus for a bioindicator containing 10^6 reference *G. stearothermophilus* spores ($D_{121} = 1.5$ min), after 9 min of treatment at 121°C, the average indicator would be assumed to contain $10^0 = 1$ surviving organisms. If a statistical distribution of survivors is assumed, according to Poisson's distribution (Equation 12.4), 37% of all indicators are expected to be sterile, 37% to contain just 1 survivor, 18% to contain 2 survivors, and 8% to contain more than 2 survivors. After 10.5 min of treatment, the respective expectations are 90% sterile and 10% nonsterile, containing 1 or more survivors (Lorenz 1988).

$$P_{(x=k)} = e^{-\mu} \frac{\mu^k}{k!} \tag{12.4}$$

where e = base of natural logarithm, k = expected value, μ = median value, and $P_{(x=k)}$ = probability of occurrence of units containing the expected number of $x = k$ viable cells.

The distribution of survivors within the autoclave is expected to be random if homogeneous conditions exist within the chamber. In order to demonstrate homogeneous conditions within the autoclave, at least ten indicators would have to be located at each measuring point to ensure that statistical distribution could be differentiated from true deviations.

12.5.2.4 Value of Bioindicator Use in Routine Sterilization Processes

In theory, for *G. stearothermophilus* with a z value of 6°C, a 1°C drop of temperature to 120°C would lead to a D value of 2.2 min (from 1.5 min). Thus 9 min of autoclaving of a bioindicator with a viable spore count of 10^6 would result in a median survivor probability of $10^{1.9}$ within every indicator. The probability of detecting a bioindicator without surviving spores would be highly unlikely, whereas 37% negatives would be expected for the theoretical process. At 10.5 min, $10^{0.8}$ survivors would be expected with a probability of still only about 2% to detect a negative one. These figures clearly show that even a small deviation in sterilization temperature could theoretically be detectable with bioindicators. However, it should be kept in mind that bioindicators are not precision instruments. Actual spore numbers and D values as well as z values vary with pretreatment or storage conditions of the indicators. For *G. stearothermophilus*, D_{120}/D_{121} values have been reported to vary between 1 and 5.8 (Russell 1999). As determination of these values is only possible by destructive treatment for any individual indicator, the uniformity of these critical parameters can only be determined by statistical studies to be conducted on every lot of indicators and for every validation experiment.

For these reasons, it is far easier and at least as reliable to determine the routine performance of a sterilizer by carefully designed and recorded physical measurements, compared with bioindicator studies. Bioindicators may be used to determine autoclave performance where physical measurements are not feasible, for example, in equipment not designed to allow access to thermal sensors. When bioindicators are employed in routine monitoring, data should not be overemphasized.

12.5.2.5 Bioindicators Used in Process Development

D values of bioindicators are strongly influenced by the environment in which they are sterilized. Valuable information for establishing a suitable sterilization cycle can be gained with bioindicators by comparing inactivation rates of the indicator organisms in a product with those in water. If the D value in the product is different from the D value found in water, the sterilization time would have to be adjusted. In this situation, either longer or shorter sterilization times may be required when compared to standard recommendations. Requalification of the efficacy of a standard sterilization procedure characterized in this way would only become necessary should the product formulation be subject to change.

In such studies, other factors that may influence the D value of spores must also be carefully considered, for example, preconditioning and storage of the bioindicator microorganisms or the type of suspending medium.

12.5.2.6 Bioindicators for Steam Sterilization

Recommended bioindicators for steam sterilization in the *European Pharmacopoeia* (2006) are spores of *G. stearothermophilus* with a D_{121} value of at least 1.5 min. It is, however, misleading that only a lower limit for the D value is given. Although for these spores, a decimal reduction time at 121°C (D_{121} value) of about 1.5 min is usually given in textbooks, commercially available bioindicator preparations frequently have a D_{121} value of 2 or even higher. Such bioindicators are in compliance with the *European Pharmacopoeia*, but their use in combination with the European standard sterilization cycle of 15 min at 121°C may lead to the recovery of positive units after sterilization. If biological indicators survive after a sterilization cycle where correct physical process parameters were achieved, investigation should always include verification of the D value of the bioindicator in addition to verification of the sterilization process.

The temperature difference leading to a tenfold change in the D value (z value) (Pflug 1973) is usually found to be between 6 and 10°C. Where bioindicators are used to develop or monitor sterilization processes at other than the standard temperature, the z value in the product would have to be evaluated in each individual case.[*]

12.5.2.7 Bioindicators for Dry Heat Sterilization

The use of bioindicators for validation of dry heat sterilizers is limited to forced convection batch sterilizers. Continuous dry heat tunnel sterilizers usually operate at such high temperatures that bacterial inactivation kinetics become practically impossible to follow due to the very fast inactivation of bioindicators. Also, these tunnels are generally used to sterilize primary packaging materials and at the same time to destroy bacterial lipopolysaccharides (bacterial endotoxins) that are inactivated by dry heat at temperatures above 200°C (Wegel 1973). The dry heat inactivation kinetics for endotoxins have been described by Tsuji and Harrison (1978).

---

[*] A range of D and z values have been described for *B. stearothermophilus* (*G. stearothermophilus*) (e.g., Russell 1999), the values being dependent upon the strain and status of the organism, its environment, and the temperature range employed.

Glassware can be impregnated with endotoxins and their inactivation rate used as a measure of the correct performance of a dry heat sterilizer. It is important, however, to standardize carefully both the amount of endotoxin adsorbed as well as the procedures for desorption and recovery (Jensch et al. 1987).

For forced-convection batch sterilizers operating at 160 to 200°C, bioindicators can be applied (Table 12.4). The same general factors as for steam sterilization should be taken into consideration.

12.5.2.8 Bioindicators for Gaseous Sterilization

For gaseous sterilization, physical measurements of gas concentrations and all other relevant parameters within a given load are difficult or impossible to perform. The use of bioindicators (Table 12.4) is a mandatory requirement for every cycle (EP 2004). The USP (2004) states that bioindicators may be employed in monitoring of routine runs. Their use should be regarded more favorably than in the more easily and precisely monitored heat sterilizers. It is clear, however, that the same statistical considerations apply as in all other cases. As the concentration of the sterilant does not normally remain constant over the sterilization time, and as condensation on or interaction of the sterilant with the surface of the materials to be sterilized may alter the sterilization effectiveness, evaluation of the suitability of the bioindicator for process monitoring is a mandatory part of sterilization process development. The quality of the spore preparation for each lot is also critical, because clumping of spores or embedding in a matrix of debris from the spore preparation may have a strong shielding effect, which can alter the sensitivity of the bioindicator. Mapping of the exact conditions in the sterilizer is difficult by means of bioindicator exposure, and hence, the most difficult to penetrate (worst case) positions in the load to be sterilized should be selected by theoretical considerations or from experience with previously validated gas sterilization processes.

12.5.2.9 Bioindicator Validation of Isolator Sterilization (Decontamination)

In the past few years, it has become widespread practice to use bioindicators for validation of gassing cycles for isolator cabinets. It remains controversial at this point whether sterilization in the sense of the pharmacopoeial definition [including a sterility assurance level (SAL) of 10^{-6}] is really needed to treat surfaces, which are not intended to come into contact with a product. Hence, it is internationally disputed what level of inactivation is required. Frequently, the term *decontamination* is used to indicate a less rigorous inactivation procedure applied in isolators. The FDA aseptic processing guide (FDA 2004) asks for a SAL of 10^{-6} on product contact surfaces (which would be equivalent to a surface sterilization), but concedes that a SAL of 10^{-4} to 10^{-6} can be justified on less critical surfaces. Bioindicators should be exposed in worst case positions, which are selected depending on the decontamination procedure. Where isolator decontamination by H_2O_2 is used at the same time to surface sterilize materials that are then used for sterility testing or processing in the closed chamber, the same considerations apply as for a gas sterilizer. Where decontamination is strictly a measure to control the surfaces of the enclosed manufacturing environment, the relevance of a position for maintenance of product sterility (e.g., proximity to critical product contact surfaces) would also be a relevant consideration besides gas penetration.

Internationally, H_2O_2 has become the agent of choice for isolator decontamination. Peracetic acid has practically disappeared as an alternative today. Consequently, *G. stearothermophilus* is used as the indicator organism of choice for gas sterilization processes.

12.5.2.10   Bioindicators for Radiation Sterilization

Radiation is easily and precisely monitored by exposure of dosimeters within the load. Bioindicator validation should only be used for initial characterization of inactivation rates within a given product.

12.5.2.11   Biological Indicators for Microorganism-Retentive Filters

Although the common definition of bioindicators is restricted to resistant spore preparations for characterization of other sterilization procedures, preparations of small microorganisms used to characterize membrane filters can also be considered a form of bioindicator. The microorganism-retentive capacity of a filter can only be determined by destructive testing. *Brevundimonas* (formerly *Pseudomonas*) *diminuta,* a Gram-negative rod-shaped bacterium, is commonly employed as a test organism for filters with a nominal pore size of 0.22 μm or less (Table 12.4). It has been chosen because of its relatively small dimensions of 0.5×1.0 to 4.0 μm (Palleroni 1984). The dimensions of bacteria are, however, strongly dependent on their growth conditions (Wallhaeusser 1979, 1982); fast-growing forms differ in appearance from slower-growing or starved forms. When validating filters, it is necessary to verify by microscopy or passage through a 0.45 μm rated filter that the challenge used predominantly consists of single cells with dimensions at the lower end of those given above. Less easily measured are the surface physiochemical properties of the organisms that might influence their sorption to filter media.

ASTM describes a detailed arrangement for a bacterial challenge test. In principle, filters are challenged with about 10^7 organisms per square centimeter of effective filter area. The filtrate is then analyzed for bacteria passing through the filter (Wallhaeusser 1982). Once a filter is loaded with 10^7 organisms per square centimeter of effective filter area, the test is usually considered complete. This level is considered a minimum requirement for sterilizing filters (FDA 2004) and is intended to exceed by a wide margin the challenge expected during the production process. This is based upon the observation of Bowman et al. (1967) that *B. diminuta* at that level can penetrate filters with a 0.45 μm nominal pore size rating.

The challenge organisms should be directly inoculated into the product to be filtered. This is seen as an assessment of any effects that the product may have on the filter matrix or on the challenge organisms (FDA 2004). Where this challenge method cannot be used (e.g., because the product has inherent antimicrobial properties that would lead to inactivation of the challenge organisms), the application of an alternative method must be sufficiently justified. Use of membranes that were retrieved from a simulated worst case production process and subsequently challenged by microorganisms suspended in a modified product without antimicrobial properties is an example of recommended alternative validation procedures (PDA 1998). In media fill simulations, the use of production filters to validate membrane filter sterilization as part of aseptic processing, as recommended by PIC/S (2001b), is not an acceptable procedure according to the U.S. authorities.

12.5.3 CHEMICAL METHODS FOR MONITORING STERILIZERS

Chemical monitoring of sterilization processes is based on the ability of heat, steam, sterilizing gases, and ionizing radiation to alter the chemical or physical characteristics of a variety of chemical substances (Denyer and Hodges 2004a). Ideally this change should not occur until the sterilization cycle has been satisfactorily completed, but in practice this condition is rarely met. Thus, with the exception of a radiation dosimeter, these devices should never be used as sole monitors of a sterilization process although they may be used in conjunction with other physical or biological methods. Again with the exception of radiation dosimeters, such devices would not be used for the validation of sterilizer efficacy because the changes recorded do not necessarily correspond to microbiocidal activity. Nevertheless, they do serve as useful indicators of the conditions prevailing at the coolest or most inaccessible points of a sterilizer or load and also offer a mechanism to differentiate between processed and unprocessed products.

For steam sterilization, thermochemical indicators that qualitatively show that the product has been exposed to a sterilization cycle are differentiated from thermochemical integrators, which yield a quantitative record of the sterilizing conditions achieved. In the United States, evaluation of thermochemical integrators is seen as a laboratory test to monitor sterilization processes. Such a test is deemed necessary to remain in compliance with current good manufacturing practice (cGMP) should the test for sterility be replaced by parametric release of products sterilized in their final container.

12.6 VALIDATION PLANS FOR STERILIZERS

For all sterilization procedures, a coherent validation plan should be developed. This plan should include:

Cycle development:
- Development of the product suitable for sterilization in the final container where possible
- Compatibility of the sterilization process with a given product
- Development of the sterilization cycle

Qualification of the sterilizer:
- Installation qualification
- Operational qualification
- Performance qualification

Process validation:
- Specification of data to be routinely collected and evaluated
- Revalidation, maintenance, and change control

12.6.1 Cycle Development

12.6.1.1 Development of the Product

In development of sterile products, the European authorities expect that preference should be given to formulations that can be sterilized in the final container by application of standard methods. Where this is not feasible, a scientific rationale should be given.

12.6.1.2 Compatibility of the Sterilization Process with a Given Product in Development

Once a sterilization method has been decided upon, compatibility of the product with the sterilizing principle should be investigated. It must be confirmed that all parts of the product including the container are equally resistant to sterilizing conditions. The maximum acceptable exposure of the product (e.g., maximum temperature or radiation dose) should be evaluated and defined. Product stability must also be taken into consideration during such studies. Where a pharmacopoeial standard is not applicable, alternative conditions should be evaluated, where the product may be compatible. For heat sterilization, not only lower temperatures but also higher temperatures, applied for a very short time, should be considered.

In radiation sterilization, possible radiolysis products caused by reactions with highly reactive free radicals have to be carefully investigated. Although some radiolysis products themselves may present a risk to the consumer, others, like those causing discoloration, may be merely aesthetic in nature, but may still lead to the decision that the sterilization method cannot be used.

For the highly reactive sterilizing gases, careful consideration should be given to possible damage to the product due to reactions with the sterilant. Another concern lies in the retention of the toxic gases or degradation products thereof within the product. Degassing procedures must be validated and carefully adhered to.

For membrane filtration, consideration should be given to possible effects of the filters on the product (e.g., retention of product on filters), but also to effects of the product on the filters (e.g., marginal compatibility of product solvent with the filter matrix or masking of filter surface charge where applicable).

12.6.1.3 Development of the Sterilization Procedure

Dependent on the product compatibility, the sterilization procedure should be developed. The most suitable sterilization approach for the product, including the type of autoclave cycle, should be chosen. These experiments should also include inactivation studies of bioindicators in the product itself to elucidate possible influences (positive or negative) of the product on inactivation rates. During this phase, the relevant criteria for process control should be established. Where a pharmacopoeial standard sterilization procedure cannot be applied, bioburden specifications should be established.

The main consideration with filtration processes will be one of practicality, with flow rates, filter fouling rates, and longevity being the main issues.

12.6.2 Qualification of Sterilizers

Before any sterilizer is taken into routine use in a production process, correct functioning of the equipment has to be verified.

12.6.2.1 Installation Qualification

Installation qualification (IQ) is performed to verify that the sterilizer has been correctly built and installed according to the specifications of the manufacturer or the specifications agreed upon when equipment was ordered. It also includes a check of completeness of all parts and documents supplied. This is usually done by the supplier in collaboration with the user of the equipment.

12.6.2.2 Operational Qualification

Operational qualification (OQ) serves to verify the reliable performance of the equipment. Calibration of all measuring instruments has to be performed. Qualification of computer systems and data gathering systems should be verified and documented. The functionality of all control systems and mechanical parts in operation in compliance with equipment specifications should be demonstrated in at least three consecutive runs. OQ is the responsibility of the user of the equipment, but it is frequently performed in collaboration with the supplier.

For autoclaves and other heat sterilizers, an important part of OQ consists of heat distribution studies with thermocouples placed at suitable positions within the empty chamber. The number and position of the thermocouples are determined by the size of the chamber as well as by the type of instrument and sterilization cycle used. Another part may consist of the qualification of the venting filter and venting filter integrity testing system.

For gas sterilizers, relative humidity and temperature should be measured by physical sensors distributed at suitable positions within the chamber.

Operational qualification has to be repeated, in part at least, whenever significant technical manipulations occur during maintenance or modification of the equipment.

12.6.2.3 Performance Qualification and Performance Validation

Performance qualification (PQ) consists of verification of the process-specific performance of the sterilizer. For any given type of load and sterilization cycle performed, it must be shown that the expected sterilizing activity is reached within every part of the load, and that conditions are always compatible with the article to be sterilized. As these studies are product-specific as well as equipment-specific, a sharp differentiation between performance qualification and validation cannot be drawn at this point. Studies should be done at least in triplicate and should use physical determinations as well as bioindicator studies as appropriate for the sterilization approach chosen. They are part of product-specific process development done by users of the equipment.

For heat sterilizers, loaded chamber heat distribution and heat penetration studies should be performed. Computerized multichannel recording systems are capable of handling a wealth of data and, using these systems, it is possible to map exactly the development of relevant conditions at various locations within the sterilizer. Usually thermocouples are placed at the positions within the load that are most difficult to heat up. Type and configuration of the load determine the number and position of the thermocouples as well as the type of instrument and the sterilization cycle used. Such validation systems must themselves be fully validated and carefully calibrated.

In porous loads, saturation of steam and steam penetration may be critical factors to determine in autoclaves, depending on the cycle used.

For gas sterilizers, relative humidity and temperature should be measured by physical sensors distributed at suitable positions within the load. Bioindicators distributed in the load must be used to verify sterilizing conditions at all positions within the load (EP 2006).

Penetration of ionizing radiation within a load is most reliably validated and monitored by the distribution of sufficient dosimeters within the load. Bioindicators may be used as an additional means of verification.

PQ of sterilizing filters is usually based on integrity testing, pressure differential, and flow rate measurements (Wallhaeusser 1982). It should be noted that fluids sterilized by passing through a membrane filter are usually exposed to the environment during further processing. In such cases, environmental control and proper validation of the aseptic handling area should be considered as an integral part of the sterile filtration process. Requirements for aseptic processing are given in the European GMP-Guide (European Commission 1996), the MCA Guide (Anonymous 2002), and also are described by ISO (1998).

12.6.3 PROCESS VALIDATION

12.6.3.1 Specification of Data to Be Routinely Collected and Evaluated

Validation of a sterilization cycle should not be seen as an isolated exercise but always as an integral part of a general validation program. Based on process development and PQ studies, a comprehensive plan should be established for each product, taking into account all parameters that contribute to the quality of the final product. Critical parameters and their limits to be met during the process should be noted. The documentation required should be clearly specified. A review and evaluation procedure should be established to ensure that documentation is adequately evaluated by the responsible person before the product is released.

A program, which is developed to the specific requirements of the product, is thus to be recommended as the basis of a validated production process. An example of such a program for a standard product (steam-sterilizable aqueous solution in 10-mL ampoules) is given in Table 12.5.

TABLE 12.5
General Validation Plan for a Sterilization Process

Item	Parameter	Requirements	Frequency
QUALIFICATION			
Raw Materials	Bioburden	<10^3 CFU.g^{-1} or CFU.mL^{-1} <10^2 CFU.g^{-1} or CFU.mL^{-1} (in some countries) Absence of specific pathogens	5 lots initially, then random samples
First cycle rinsing water	Bioburden	<10^2 CFU.mL^{-1} Absence of specific pathogens	Weekly
Process water and final cycle rinsing water	Bioburden	<10 CFU per 100 mL^{-1} Bacterial endotoxins <0.25 EU.mL^{-1}	Weekly
Production Area			
Setup	Air locks	Separate area for preparation and filling Entry of personnel only via air locks Entry of goods via air locks or sterilizers	
Air supply	Ventilation Filters	Enough air changes to provide short clean-up phase All air passed through HEPA filters Filters integrity tested	Every 6 months
Air quality at rest	Air particle count	>5 µm diameter, <2000 m^{-3} >0.5 µm diameter, <350,000 m^{-3}	Every 6 months
	Air viable microorganisms	<100 m^{-3}	
Maintenance	Maintenance plan	Maintenance plan containing scope, frequency of cleaning and disinfection, required measures, responsibilities Effectiveness of measures validated	Yearly

(continued)

TABLE 12.5 (continued)
General Validation Plan for a Sterilization Process

Item	Parameter	Requirements	Frequency
Construction	Floor	Even, smooth surface, resistant toward specified disinfectants, seams sealed, edges rounded and sealed	
	Walls	Smooth surface, washable, resistant toward specified disinfectants	
	Sewers, sinks, water taps	No open sewers, sinks, or water taps present	
Ampoule washer and sterilizer qualification	Calibration of measuring equipment	Correct values indicated with predetermined accuracy	Yearly
	Belt speed	Conform with specification	
	Temperature distribution	Required temperature reached for specified time at each point across the belt	
	Filter integrity	No leaks detected	
	Particle count	>5 μm diameter, <3500 particles	
	Endotoxin inactivation (only in high-temperature sterilizers >220°C)	>3 log-cycle inactivation of endotoxins adsorbed on 20-mL ampoules	
Autoclave qualification	Calibration of measuring equipment for temperature and pressure	Correct values indicated with predetermined accuracy, ± 2°C and ±10 kPa	2 to 4 times a year
	Temperature distribution, empty chamber	Correct temperature distribution with predetermined accuracy	After major modifications
	Cycle time	Correct cycle time with predetermined accuracy	Yearly
	Air filter	Passed integrity test	Twice a year
	Direct cooling water	<1 CFU.mL^{-1}	Monthly

Personnel	Hygiene training	Passed training program with detailed instruction in basic hygiene, correct behavior in clean room, correct wear of garments, correct use of disinfectants	Every 6 months
	Cleanroom garments	Hair and, where appropriate, beard cover; single- or two-piece trouser suit, gathered at wrists and with high neck; gloves permanently covering the sleeves, mask; goggles; appropriate shoes or overshoes	
VALIDATION			
Preparation of solution	Raw material	Integrity and orderly appearance of individual containers	Every container
	Process time	Specified process times, minimized waiting time between preparation, filling, and sterilization	Process development
	Presterilization bioburden	$<10^2$ CFU.mL^{-1} Bacterial endotoxins <15 EU.mL^{-1}	3 lots initially, then random samples
Production hygiene	Air viable microorganisms	<100 m^{-3} during work	Monthly active sampling
	Surface viable microorganisms	<40 per 100 cm^2	Monthly
	Gloves of personnel, viable microorganisms (contact plates)	<10 CFU per plate	Weekly
Sterilization cycle	Temperature sensitivity of product	No deterioration at 121°C for 15 min detectable	Cycle development
	Sterilization efficacy	Bioindicators as effectively inactivated as in water	Cycle development
	Autoclave loading pattern	Written description of autoclave loading pattern	Cycle development
	Temperature distribution within load	Determination of hot and cold points within the load Fixation of measuring points to measurement during routine run	Cycle development
Methylene blue bath	Bioburden	<1 CFU.mL^{-1}	Weekly
Test on final product	Sterility test area	Grade A (GMP) environment validated	
	Sterility test	Complies with pharmacopoeial requirement	Every lot unless parametrical release agreed by authorities

(continued)

TABLE 12.5 (continued)
General Validation Plan for a Sterilization Process

Item	Parameter	Requirements	Frequency
	Pyrogen test	Negative on 3 rabbits	Every lot unless endotoxin test permissible
	Endotoxin test	Bacterial endotoxins per hourly human dose <5 EU.kg^{-1}	Every lot
Documentation	Bioburden of raw materials, water intermediate	Trends recognizable	Continuously recorded
	Production hygiene	Trends recognizable	Continuously recorded
	Hygiene training	Frequency, subjects, and attending personnel documented	
	Sterilizers: temperature and pressure recordings	Available for every lot	Every lot
	Test for sterility	Performance: retest rate $<0.5\%$	Continuously monitored

Product: aqueous solution, ampoules 10 ml, sterile; sterilization: steam 121°C. CFU—colony-forming unit; USP—United States Pharmacopeia; EU—endotoxin unit; HEPA—high-efficiency particulate air; Pa—Pascal; FDA—U.S. Food and Drug Administration.

Source: This table is based on current requirements put forward by EEC authorities (European Commission, 1996, 2001), FDA (2004), the European Pharmacopoeia (2006), and United States Pharamacopeia (2004), also in part on industrial standards. The example applies only for the production of the stated product. Aseptic preparations would require different standards.

12.6.3.2 Revalidation, Maintenance, and Change Control

There should be a formal revalidation program, specifying frequency and types of qualification or validation studies to be repeated at scheduled intervals. Responsibilities should be clearly defined. A scheduled maintenance program for all equipment and premises should be established indicating responsibilities, routine calibration and maintenance measures, and the documentation required. There should be an action plan, linking marginal revalidation results to maintenance measures on occasion, as well as linking major maintenance measures to requalification or revalidation studies on occasion.

There should also be an action plan to specify the measures taken in case of outlying results in critical process parameters during routine production.

REFERENCES

Anonymous. (2002). *MCA Rules and Guidance for Pharmaceutical Manufacturers and Distributors*. The Stationery Office, London.

ANSI/ASTM. (1992). BIER/Steam vessels, American National Standard ST45−1992.

ASTM. (1988). Standard test methods for determining bacterial retention of membrane filters utilized for liquid filtration, *Annual Book of ASTM Standards,* American Society for Testing and Materials, Philadelphia, pp. 790–795.

Bill, A. (2000). Microbiology laboratory methods in support of the sterility assurance system. In *Handbook of Microbiological Quality Control: Pharmaceuticals and Medical Devices*. Baird, R.M., Hodges, N.A., and Denyer, S.P., Eds. Taylor & Francis, London, pp. 120–143.

Bowman, F., Calhoun, M.P., and White, M. (1967). Microbiological methods for quality control of membrane filters. *J. Pharm. Sci.,* 56, 222.

British Pharmacopoeia. (2005). *British Pharmacopoeia*. The Stationery Office, London.

Carter, J.R. and Levy, R.V. (1998). Microbiological retention testing in the validation of sterilizing filtration. In *Filtration in the Biopharmaceutical Industry*. Meltzer, T.H. and Jornitz, M.W., Eds. Marcel Dekker, New York, pp. 577–604.

Comite Europeen de Normalisation. (1997). *EN 866-3−Biological Systems for Testing Sterilizers and Sterilization Processes; Particular Systems for Use in Moist Heat Sterilizers. Annex 1*. CEN, Brussels.

Denyer, S.P. and Hodges, N.A. (2004a). Sterilization procedures and sterility assurance. In *Hugo and Russell's Pharmaceutical Microbiology*. Denyer, S.P., Hodges, N.A., and Gorman, S.P., Eds. 7th ed. Blackwell Publishing, Oxford, pp. 346–375.

Denyer, S.P. and Hodges, N.A. (2004b). Filtration sterilization. In *Russell, Hugo & Ayliffe's Principles and Practice of Disinfection, Preservation and Sterilization*. Fraise, A.P., Lambert, P.A., and Maillard, J.-Y., Eds. 4th ed. Blackwell Publishing, Oxford, pp. 436–472.

Docksey, S., Cappia, J.-M., and Rabine, D. (1999). A general approach to the validation of sterilising filtration used in aseptic processing. *Eur. J. Parent. Sci.,* 4, 73–77.

EMEA. (1998). CPMP−Note for Guidance for Development Pharmaceutics (CPMP/QWP/ 155/96).

EMEA. (1999). Decision Trees for the Selection of Sterilisation Methods (CPMP/QWP/ 054/98).

EMEA. (2001). CPMP−Note for Guidance on Parametric Release (CPMP/QWP/3015/99).

European Commission. (1996). Revision of the Annex I to the EU Guide to Good Manufacturing Practice for Medicinal Products, Manufacture of Sterile Medicinal Products.

European Commission. (2001). Annex 17 to the EU Guide to Good Manufacturing Practice, Parametric Release.

European Pharmacopoeia. (1997). *European Pharmacopoeia*. 3rd ed. Council of Europe, Strasbourg, France.

European Pharmacopoeia. (2006). *European Pharmacopoeia*. 5th ed. Council of Europe, Strasbourg, France (CD-ROM 5.5).

European Standards. (1994). EN 554—Validation and Routine Control of Steam Sterilisers.

European Standards. (1997). EN 285—Specification for Large Steam Sterilisers.

FDA. (1994). *Guidance for Industry for the Submission of Documentation for Sterilization Process Validation in Applications for Human and Veterinary Drug Products*. U.S. Food and Drug Administration, Rockville, MD.

FDA. (2004). *Guidance for Industry: Sterile Drug Products Produced by Aseptic Processing—Current Good Manufacturing Practice*. Centers for Drugs and Biologics, Office of Regulatory Affairs. U.S. Food and Drug Administration, Rockville, MD.

Haberer, K. (2004). Terminal sterilization and parametric release. In *Microbial Contamination Control in Parenteral Manufacturing*. Williams, K.L., Ed. Marcel Dekker, New York, pp. 419–447.

Holcomb, R.G. and Pflug, I.J. (1979). The Spearman–Karber method of analyzing quantal assay microbial destruction data. In *Microbiology and Engineering of Sterilization Processes*. 3rd ed. Environmental Sterilization Services, St. Paul, MN.

ISO. (1994). *ISO 11134—Sterilization of Healthcare Products—Validation and Routine Control—Industrial Moist Heath Sterilization Facilities*. ISO, Geneva.

ISO. (1998). *ISO 13408-1—Aseptic Processing of Health Care Products—Part 1: General Requirements*. ISO, Geneva.

ISO. (2003). *ISO 13408-2—Aseptic Processing of Health Care Products—Part 2: Membrane Filtration*. ISO, Geneva.

Jensch, U.-E., Gail, L., and Klavehn, M. (1987). Fixing and removing of bacterial endotoxins from glass surfaces for validation of dry heat sterilization. In *Detection of Bacterial Endotoxins with the Limulus Amoebocyte Lysate Test*. Watson, S.W., Levin, J., and Novitsky, T.J., Eds. Alan R. Liss, New York, pp. 273–281.

Jornitz, M.W. and Meltzer, T.H. (1998). Validation of filtrative sterilization. In *Filtration in the Biopharmaceutical Industry*. Meltzer, T.H. and Jornitz, M.W., Eds. Marcel Dekker, New York, pp. 897–924.

Jornitz, M.W. Trotter, A.M., and Meltzer, T.H. (1998). Integrity testing. In *Filtration in the Biopharmaceutical Industry*. Meltzer, T.H. and Jornitz, M.W., Eds. Marcel Dekker, New York, pp. 307–371.

Jornitz, M.W. (2004). Sterile filtration. In *Microbial Contamination Control in Parenteral Manufacturing*. Williams, K.L., Ed. Marcel Dekker, New York, pp. 283–339.

Kemper, C.A. (1986). Design, installation and calibration of thermocouple measuring systems. In *Validation of Aseptic Pharmaceutical Processes*. Carleton, F.J. and Agalloco, J.P., Eds. Marcel Dekker, New York, pp. 93–124.

Lorenz, R.J. (1988). *Grundbegriffe der Biometrie*. G. Fischer Verlag, Stuttgart, p. 207.

McLaughlin, W.L., Boyd, A.W., Chadwick, K.H., McDonald, J.C., and Miller, A. (1989). *Dosimetry for Radiation Processing*. Taylor & Francis, London, pp. 66–79.

Meltzer, T.H. (1987). The integrity tests. In *Filtration in the Pharmaceutical Industry*. Marcel Dekker, New York, pp. 219–295.

Palleroni, N.J. (1984). Pseudomonadaceae. In *Bergey's Manual of Systematic Bacteriology*. Krieg, N.R. and Holt, J.G., Eds. Vol. 1. Williams and Wilkins, Baltimore, pp. 140–199.

Patashnik, M. (1953). A simplified procedure for thermal process evaluation. *Food Technol.,* 7, 1–6.

PDA. (1978). Validation of Moist Heat Sterilization. Technical Report no. 1.

PDA. (1998). Sterilizing Filtration of Liquids. Technical Report no. 26. *PDA J. Pharm. Sci. Tech.,* 52, Suppl. S1.

PDA. (1999). Parametric Release of Pharmaceuticals Terminally Sterilized by Moist Heat. Technical Report no. 30. *PDA J. Pharm. Sci. Tech.,* 53, Suppl.

PIC/S. (2001a). Recommendation on the Guidance on Parametric Release. www.pic-scheme.org.

PIC/S. (2001b). Recommendation on the Validation of Aseptic Processes. www.pic-scheme.org.

Pflug, J.J. (1973). Heat sterilization. In *Industrial Sterilization.* Briggs Phillips, G. and Miller, W.S., Eds. Duke University Press, Durham, NC, pp. 239–282.

Pflug, I. (1995). In *Microbiology and Engineering of Sterilization Processes.* 8th ed. Environmental Sterilization Laboratory, Minneapolis, MN.

Russell, A.D. (1999). Destruction of bacterial spores by thermal methods. In *Principles and Practice of Disinfection, Preservation and Sterilization.* Russell, A.D., Hugo, W.B., and Ayliffe, G.A.J., Eds. 3rd ed. Blackwell Scientific Publications, Oxford, pp. 640–656.

Sigwarth, V. and Moirandat, C. (2000). Development and quantification of H_2O_2 decontamination cycles. *PDA J. Pharm. Sci. Tech.,* 54, 286–304.

Sigwarth, V. (2004). Process development of alternative sterilization methods. In *Microbial Contamination Control in Parenteral Manufacturing.* Williams, K.L., Ed. Marcel Dekker, New York, pp. 341–417.

Spicher, G. and Peters, J. (1975). Mathematische Grundlagen der Sterilitaetsprüfung. *Zbl. Bakt. Hyg. I. Abt. Orig. A,* 230, 112–138.

Stumbo, C.R., Murphy, J.R., and Cochran, J. (1950). Nature of thermal death time curves for P.A. 3679 and *Clostridium botulinum. Food Technol.,* 4, 321–326.

Tsuji, K. and Harrison, S.J. (1978). Dry heat destruction of lipopolysaccharide: dry heat destruction kinetics. *Appl. Environ. Microbiol.,* 36, 710–714.

United States Pharmacopeia. (2004). *United States Pharmacopeia 27.* U.S. Pharmacopeial Convention, Rockville, MD.

Wallhaeusser, K.-H. (1979). Is the removal of micro-organisms by filtration really a sterilization method? *Bull. Parenteral Drug Assoc.,* 33, 156–170.

Wallhaeusser, K.-H. (1982). Germ removal filtration. In *Advances in Pharmaceutical Sciences.* Bean, H.S., Beckett, A.H., and Cariess, J.E., Eds. Academic Press, London, pp. 1–116.

Wallhaeusser, K.-H. (1995). Sterilisation im Autoklaven. In *Praxis der Sterilisation—Desinfektion—Konservierung.* 5th ed. Thieme, Stuttgart, chap. 4.2.1.5.

Wegel, S. (1973). Kurzzeit-Sterilisationsverfahren nach dem Laminar-Flow-Prinzip. *Pharm. Ind.,* 35, 809–814.

13 Principles of Preservation

Donald S. Orth

CONTENTS

13.1 INTRODUCTION

Pharmaceutical products are subject to microbiological contamination and spoilage. Preservatives are defined in the sixth amendment to the Cosmetics Directive as substances added to products for the primary purpose of inhibiting microorganisms from growing (Cosmetics Directive 1993). Antimicrobial preservatives are used to reduce the likelihood of microbial growth in aqueous products and to reduce the chance of microbial survival in anhydrous products that may be contaminated or moistened during use. Sterile products, including parenterals, irrigating, and ophthalmic solutions, must remain sterile until they are used by the consumer. If products become contaminated before use, they are adulterated and are in violation of the EC Directive and U.S. Food, Drug and Cosmetic Act, as amended (Federal Food, Drug and Cosmetic Act 1976, Cosmetics Directive 1993).

Sterile drugs in multiple-dose containers must have a preservative system that is capable of self-sterilizing these products should contamination occur. Nonsterile aqueous products need preservative systems that are capable of reducing the microbial bioburden to an acceptable level in a reasonable time (Orth 1979, 1999a; Orth et al. 1998).

The objective of pharmaceutical product preservation is to ensure that the product is microbiologically safe and stable. Preservative efficacy testing is performed to determine the type and minimum effective concentration of preservatives required to preserve the product during manufacture and throughout its use by the consumer. This testing is an essential part of documenting the safety and stability of these products.

13.2   OBJECTIVES OF PRESERVATION

13.2.1   THE NEED FOR PRESERVATION OF PHARMACEUTICAL PRODUCTS

Bacteria, yeasts, and molds have diverse metabolic requirements and are able to grow in aqueous pharmaceutical products when nutrients are available and when environmental conditions are suitable. The regulation of microbial growth by physical and chemical agents was presented by Moat and Foster (1988). An understanding of how these factors control microbial growth is necessary to determine the most suitable preservative system needed in any product (refer to Chapter 1, where these factors are considered in detail).

Growth of bacteria, yeasts, or molds on, or in, products may make those products unsafe and unacceptable for use. The hazards of using contaminated products are due to the effect of microbial infections or harmful microbial by-products on human health. Several surveys conducted between 1969 and 1977 revealed contamination of cosmetic, toiletry, and pharmaceutical products (Dunnigan and Evans 1970, Bruch 1971, Baird 1977). McCarthy (1980) reported that similar patterns of contamination for nonsterile products were observed in both the pharmaceutical and cosmetic industries. Although most manufacturers have identified the critical control points in their processes and have implemented [validated] procedures to prevent microbial contamination, occasional contamination problems occur and have resulted in product recalls (Orth 1999b).

In the 1960s and 1970s, there were several reports of infections due to use of contaminated products (see Chapter 2). Hand lotions and creams were identified as sources of nosocomial infections that resulted in septicemia due to Gram-negative bacteria, particularly *Escherichia coli, Klebsiella pneumoniae, Enterobacter* spp., and *Serratia* spp. (Morse et al. 1967, Morse and Schonbeck 1968). Noble and Savin (1966) reported *Pseudomonas aeruginosa* contamination of a steroid cream preserved with chlorocresol, following modification of that cream by the addition of cetomacrogol emulsifying wax, paraffins, chlorocresol, and water. Although the final concentration of chlorocresol was 0.1% w/v, which should have been sufficient to inactivate contaminating *P. aeruginosa*, this microorganism persisted in the product. (Note: Noble and Savin reported that 0.01% w/v chlorocresol was sufficient to inhibit several strains of *P. aeruginosa* on nutrient agar.) The cause of contamination was traced to a decrease in preservative level in the aqueous phase of the product, due to partitioning of the preservative into the oil phase. Contamination of the product (possibly with adapted organisms) was facilitated by the practice of refilling used containers. It had been assumed that use of the same preservative system in the modified product would be satisfactory but, unfortunately, it was not. This demonstrates why a preservative system must be tailored to a specific product (Cowen and Steiger 1977).

Several studies have implicated inadequately preserved mascaras as the cause of eye injuries (Ahearn et al. 1974, Wilson et al. 1975, Wilson and Ahearn 1977). Although new mascaras were rarely contaminated, used mascaras often were contaminated with a variety of organisms (Ahearn et al. 1974). Bhadauria and Ahearn (1980) reported that the preservative systems in unused mascaras deteriorated with time. Similarly, Orth et al. (1987) followed the decrease in preservative efficacy in a protein-containing shampoo during stability testing at different temperatures. Preservatives have been reported to adsorb to packaging and to particulate materials in formulations (talc, iron oxides, and colors), and to be inactivated by surfactants (Orth 1997, and see Chapter 15).

The relative hazard created by microbiological contamination of cosmetic or pharmaceutical products may be related to the severity of infection or disease it causes. Dunnigan classified *Pseudomonas, Proteus, Staphylococcus, Serratia, Streptococcus, Penicillium, Aspergillus,* and *Candida* genera as health hazards (Dunnigan 1968). Bruch (1972) refined the classification of objectionable microorganisms according to product type. With input from Madden (personal communication 1990), Orth (1993) updated the list of objectionable microorganisms based on product type. The classification appearing in Table 13.1 has been further updated by inclusion of *Burkholderia* spp.

TABLE 13.1
Classification of Objectionable Microorganisms by Product Type

Sterile drugs	Any organism or pyrogen in a sterile product is **objectionable**.
Eye products	*Pseudomonas aeruginosa* is **always objectionable**. Other *Pseudomonas* spp., *Burkholderia* spp., *Staphylococcus aureus, Serratia marcescens,* and *S. liquifaciens* are **usually objectionable**.
Nonsterile oral products	Any enteric pathogen (i.e., *Salmonella* spp., *Yersinia* spp., *Campylobacter* spp.) and *Escherichia coli* are **always objectionable**. Other enteric organisms, such as *Enterobacter* spp., *Citrobacter* spp., *Pseudomonas* spp., *Burkholderia* spp., proteolytic *Clostridium* spp., enterotoxigenic *Staphylococcus aureus,* pathogenic yeasts (*Candida albicans*), and mycotoxin-producing fungi are **usually objectionable**.
Nonsterile topical products	*Pseudomonas aeruginosa, Klebsiella* spp., *Staphylococcus aureus, Serratia marcescens,* and *S. liquifaciens* are **always objectionable**; whereas *Pseudomonas putida, P. multivorans, Burkholderia cepacia, Clostridium perfringens, C. tetani,* and *C. novyi* are **usually objectionable**.
Genitourinary tract products	*Escherichia coli, Proteus* spp., *Serratia marcescens, Pseudomonas aeruginosa,* and *P. multivorans* are **always objectionable**; whereas *Klebsiella* spp., *Acinetobacter anitratus,* and *A. calcoaceticus* are **usually objectionable**.

Source: Adapted from Orth, D.S. (1993). *Handbook of Cosmetic Microbiology.* Marcel Dekker, New York. Used with permission.

Although the direct effects of microorganisms in infections and disease have been appreciated for many years, the insidious role they play in inflammation, immuno-modulation, and altering human physiology is only beginning to be appreciated.

Products intended for use on, or in, the body must be safe. Even though the aerobic plate count and total viable counts of the finished product may reveal the presence of <10 colony-forming units (CFU) mL^{-1}, residual microbial by-products may produce undesirable reactions. The problems created by microbial contamination can be minimized by use of raw materials that do not have a history of unacceptable microbial load (see Chapter 3), by adherence to validated manufacturing practices to reduce the risk of microbial contamination during processing (see Chapter 6), and by sterilizing products or using effective preservative systems in aqueous formulations.

13.2.2 PRESERVATION OF THE PRODUCT DURING USE

In 1970, Halleck published the recommendations of the Preservation Subcommittee of the Toiletry Goods Association (TGA) Microbiology Committee (Halleck 1970). These recommendations stated that preservation studies should consider product formulation, manufacturing conditions, packaging, product stability, and continued effectiveness of the preservative system during the intended use by the consumer.

In 1984, Eiermann noted that data obtained from surveys and during U.S. Food and Drug Administration (FDA) inspections of cosmetic manufacturers suggested that microbiological contamination of cosmetics during manufacturing was no longer a major regulatory issue (Eiermann 1984). He indicated that the question of whether these products remain uncontaminated when used by consumers had not been resolved. This is addressed in the tentative final order regulating over-the-counter (OTC) antimicrobial drug products, in which the FDA used the phrase "effectively preserved" to include preservation during use by the consumer (Eiermann 1984).

It is believed that normal use of some products by consumers repeatedly subjects these products to contamination. For example, hair care products (shampoos, con-ditioners, antidandruff products) are used while showering, which exposes these products to dilution with water and contamination with microorganisms. Repeated use of creams, which requires dipping a finger into a jar to obtain the product, may expose the cream to contamination and dilution with soil, microorganisms, and moisture on fingers. Also, adaptation may occur in product residues present on the threads of the cap or neck of the container if the residues become diluted with water or contaminated with body fluids (blood, urine, tissue fluids). These micro-organisms may become adapted to the product and may be introduced into that product when the cap is next removed, resulting in microbial contamination of the remaining product.

13.2.3 MICROBIAL ADAPTATION

Microorganisms have diverse metabolic capabilities and are able to utilize virtually any organic and some inorganic compounds as substrates for growth. The modulation of bacterial metabolism in response to substrates in the environment and the problem of microbial adaptation have been recognized by several works (Orth and Lutes 1985, Levy 1987, Orth 1993). Preservatives may be utilized as substrates. Close and Nielsen (1976) reported the isolation of a strain of *Burkholderia cepacia* from oil-in-water (O/W) emulsions preserved with methylparaben and propylparaben. This isolate could hydrolyze both paraben esters and it could use propylparaben as the sole source of carbon and energy in minimal media. Orth (1981) reported the growth of *Burk. cepacia* in an apparently adequately preserved hand and body lotion.

Yablonski (1978) reported the importance of equipment cleaning and sanitization in preventing microbial contamination in manufacturing plants. Improperly cleaned and sanitized equipment provides dilute product residues that enable microorganisms to adapt to the product and become "house organisms" (Orth 1981, Orth et al. 1996, Orth 1999a).

Levy (1987) stated that preservative efficacy testing should be performed using test organisms that exhibit the highest level of resistance toward a preserved product. Strains selected for such tests should be at least as difficult to inactivate as microorganisms that may contaminate the product, either during manufacturing or use by the consumer. The use of specifically adapted microorganisms is not, however, recommended for routine testing because it is believed that many organisms of importance to the cosmetic and pharmaceutical industries may be adapted to survive and grow in adequately preserved products (Orth 1981, 1984, 1999a). Thus, use of adapted organisms in preservative efficacy testing may make the tests impossible to perform because these organisms may not die when introduced into products to which they are adapted.

The proper solution to the problem of microbial adaptation does not rest with increasing the potency of preservative systems in all products, because they may lead to the use of preservative levels well in excess of those normally required. Instead, alternative types of packaging or product reformulation may be indicated to eliminate conditions that assist the adaptation process. Also, it may be possible to select a preservative that has an adaptation index (AI) close to unity (Orth and Lutes 1985) or a preservative with low dilution coefficient (η, also known as the concentration exponent) (Russell et al. 1979, Hugo and Denyer 1987, Hurwitz and McCarthy 1985, see also Chapter 14). This may help to minimize the risk of adaptative contamination arising through product abuse because preservatives of this type are less affected by dilution than are preservatives with larger AI or η values, and should, therefore, exercise more effective control against contaminating microorganisms.

Orth et al. (1996) observed that house organisms may contaminate a manufacturing plant because of inadequate preservative systems, inadequate cleaning and sanitization procedures (failure to comply with validated procedures or good manufacturing practices), contaminated raw materials, inadequate test methods, microbial limits set higher than <10 $CFU.g^{-1}$ for aqueous products, and lack of adequate follow-up when contamination is detected. These works (Orth 1993, Orth et al. 1996) reported that regrowth (i.e., the phoenix phenomenon) is an artifact due to the failure of test procedures to detect injured microorganisms. Orth (1999b) cautioned that procedures intended to detect viable microorganisms from product samples or stressful environments should not use conditions that are normally considered optimal for growth [rich culture media, optimum growth temperatures for mesophiles (e.g., 35 to 37°C), and shaking or aeration] because there is ample evidence to show that these conditions may result in decreased recoveries of microorganisms, irrespective of whether we call it the viable but nonculturable state or the phoenix phenomenon.

13.3 RECOGNITION OF THE PRESERVATIVE SYSTEM CONCEPT

The preservative action of a formulation often is considered to be due solely to the preservatives used. In practice, however, the preservative system of a product involves both specific preservative chemicals and the physicochemical constitution of the product (Orth et al. 1987, Orth et al. 1989). Preservative chemicals do not act independently of the product. Thus, factors such as pH, water activity (a_w), nutrient availability, surfactant concentration, sequestering agents, nonaqueous components, insoluble ingredients, and interfering materials (i.e., antibiotics, antioxidants) will influence the preservative action of any given formulation (see Chapter 15).

There are many formula components that may contribute to the preservative system of a product (Table 13.2). In some formulations, it is possible to use these factors, or hurdles to reduce or eliminate the use of preservatives (Kabara and Orth 1997, Orth and Kabara 1998). For example, it is possible to use hurdle technology (principles of preservation) to eliminate the use of preservatives in an alpha hydroxy acid lotion at pH 3.5 with sufficient glycerin to reduce the a_w to 0.85. Another example of a self-preserving product might be a hair conditioner with quaternary ammonium compounds at pH <4. In all cases, preservative efficacy testing is needed to demonstrate that the formulation is adequately preserved.

TABLE 13.2
Formula Components That May Contribute to the Preservative System of a Product

Preservatives, antibiotics
Acids, alkalis
Alcohols (e.g., ethyl, isopropyl, benzyl)
Cationic surfactants (e.g., cetyl pyridinium chloride)
Anionic surfactants (e.g., soap, sodium lauryl sulphate)
Esters (e.g., glyceryl monolaurate, glyceryl caprylate, sucrose hexadecanoate)
Humectants (e.g., glycerol, propylene glycol, butylene glycol, sorbitol)
Aqueous solutes (e.g., sugars, dextrins, salts)
Phenolic antioxidants [e.g., *t*-butylhydroxytoluene (BHT); *t*-butylhydroquinone (TBHQ)]
Chelating agents (e.g., tetrasodium ethylenediaminetetraacetic acid, citric acid)
Glycols (e.g., propylene glycol, butylene glycol, pentylene glycol)
Colors
Fragrances and flavors

Source: Adapted from Orth, D.S. and Milstein, S.R. (1989). *Cosmet. Toiletr.*, 104(10), 91, 92, 94–100, 102, 103. Used with permission.

The choice of preservative depends on the formulation (Cowen and Steiger 1977). Thus, preservatives may be unnecessary in an ointment base because of the absence of water. Also, the presence of antibiotics in a formula may make the use of specific preservatives unnecessary; however, the type and concentration of antibiotic will determine whether additional antimicrobial agents are required.

Microorganisms may be metabolically injured or "stressed" by exposure to various physical or chemical conditions, such as heating to sublethal temperatures, freezing, drying, hydrogen peroxide, acid pH, and disinfectants. Stressed microorganisms generally are more susceptible to secondary stresses created by adverse physicochemical conditions found in preservative systems than uninjured microorganisms (Denyer and Stewart 1998, Orth 1999a).

13.4 THE IDEAL PRESERVATIVE

Understanding the characteristics of an ideal preservative helps to provide the basis for rational selection of the most suitable agent(s) for a given formulation. The desired characteristics of an ideal preservative have been discussed by many authors and include the following:

- It should have a broad spectrum of activity. Ideally, a single preservative should be used as this will reduce costs and possibly may reduce the irritation or potential toxicity of the formula.
- It should be effective and stable over the range of pH values encountered in cosmetic and pharmaceutical products. Ideally, the preservative should be able to function effectively at any pH compatible with any product applied topically or taken internally. In addition, it should be chemically stable so that there is no loss of preservative efficacy during the expected shelf life of the product.
- It should be compatible with other ingredients in the formulation and with packaging materials. This attribute would prevent loss of preservative potency as a result of interactions with formula components or packaging materials. It should not alter the therapeutic properties of a drug (i.e., loss of potency of active ingredients or alteration in the pharmacokinetic behavior of the active ingredients), a phenomenon that may occur if the preservative reacted with formulation components.
- It should not affect the physical properties of the product (i.e., color, clarity, odor, flavor, viscosity, texture). Ideally, it should not produce any interactions with formulation components that may alter the appearance, texture, aroma, or performance of the formulation.
- It should have a suitable O/W partition coefficient to ensure an effective concentration of the preservative in the aqueous phase of the product. Biological reactions take place in aqueous systems or at the interface of O/W systems; consequently, it is necessary to have sufficient preservative in the water phase to ensure adequate preservation of the product.
- It should inactivate microorganisms quickly enough to prevent microbial adaptation to the preservative system. Preservatives are used in aqueous

products to make them bactericidal and fungicidal in a short enough time to meet acceptance criteria, prevent adaptation and growth (Orth 1997, 1999b), and to reduce the likelihood of microbial persistence in anhydrous products that may be contaminated and moistened during use. It is believed that contaminating microorganisms may be able to develop resistance to a product if the preservative system does not inactivate them quickly enough to prevent genetic or biochemical modifications (i.e., enzyme induction, modification of metabolic pathways, detoxification mediated by hydroperoxidases and oxygenases) that enable microorganisms to adapt to the product.

- It should be safe to use. Safety includes handling of pure or concentrated materials in the manufacturing plant as well as the effect of preservatives in the finished formulation on the consumer. Ideally, the product should be nontoxic by oral ingestion, nonirritating, and nonsensitizing.
- It should comply with governmental regulations and manufacturers of preservative chemicals should be registered. The preservative should be used in accordance with permissible levels, where applicable.
- It should be cost-effective to use. From a commercial perspective, an effective concentration should add little to the cost of the formulated product.

The parabens have been used more often than any other preservative in cosmetic products (Steinberg 2004); however, no single preservative meets all the above characteristics of the ideal preservative for all formulations. A more detailed discussion of some of these points can be found in Chapters 14, 15, 17, and 18.

13.5 SELECTION OF TYPE AND CONCENTRATION OF PRESERVATIVE

Having established candidate preservatives based on their individual characteristics and the nature of the formulation, the final selection and concentrations required for satisfactory preservation of products are ascertained by preservative efficacy testing (Chapter 17). The determination of preservative efficacy by the linear regression method (Orth 1979) is described in Chapter 18. The reader is directed to publications by Orth (1981, 1984, 1991, 1993, 1999a), Hodges and Denyer (1996), and Sutton et al. (1997) for comparisons of the linear regression method with official tests.

13.6 RATIONAL DEVELOPMENT OF A PRODUCT PRESERVATIVE SYSTEM

The steps required for the rational development of a product preservative have been discussed (Orth and Milstein 1989). The first step is to review the product formula and type to determine what are the most likely challenge organisms, then to decide which preservatives are indicated, and finally which preservative test method is the most appropriate. Samples of the product may then be prepared, with at least one sample containing an inadequate preservative system, one or two samples with the preservative level close to the expected target concentration, and at least one sample with excess preservative. This provides samples with a range of concentrations of the preservative under investigation and which may now be tested for preservative efficacy.

As mentioned earlier, several preservative efficacy test methods may be used (Lorenzetti 1984, Orth 1984, Parker 1984). The reader is referred to the companion *Handbook* for a discussion of the traditional methods of testing for preservative efficacy (Hodges and Hanlon 2000). However, it is important to employ a test method that is reliable and is capable of indicating the concentration of preservative required for the preservative system of the product to meet acceptance criteria. The linear regression method (Chapter 18) is recommended because it provides quantitative data on the kinetics of inactivation. Thus, the D value for each concentration of the preservative used may be determined with each test organism. If the preservative concentrations were selected correctly, a family of curves will be obtained. It is necessary to select the concentration of preservative required to achieve the desired rate of death so that microorganisms are killed too quickly to allow them to adapt and grow. Orth et al. (1998) reported that Gram-negative bacteria may survive and grow if initial rates of killing are too slow, when D values were about 30 h or greater. This suggests that formulations that kill Gram-negative bacteria at rates approaching the maximum allowable limits of the United States Pharmacopoeia and Cosmetic, Toiletry, and Fragrance Association methods should be used only with special manufacturing precautions (aseptic filling) or with packaging that prevents water or microbial intrusion into the product.

13.7 CROSS-RESISTANCE OF PRESERVATIVES WITH OTHER ANTIMICROBIAL AGENTS

Antibiotic resistance is increasing and is a serious global problem. Whereas antibiotics are known to have specific targets in microbial cells, biocides (including preservative agents) are believed to have multiple actions on the cell, including altering membrane permeability, inactivating enzymes, and interfering with nucleic acids. It has been assumed that bacteria cannot develop resistance to biocides because they do not have specific targets. This belief is being challenged by findings in recent years, and it has been demonstrated that bacteria have developed resistance to triclosan, chlorhexidine, quaternary ammonium compounds, and other biocides (Levy 1998, McMurry et al. 1998, Russell et al. 1998, McDonnell and Russell 1999). Increased resistance to antibiotics and disinfectants may be due to mutation or the acquisition of genetic material by horizontal gene transfer or plasmids (Orth 2000). Growth conditions may affect resistance to antimicrobial agents. Exposure to aerobic conditions enables cells to develop tolerance to oxidative stress (i.e., the SOS response, which includes production of enzymes to detoxify reactive oxygen species and repair DNA lesions). Exposure of cells to subinhibitory doses of hydrogen peroxide was reported to increase resistance of *Escherichia coli* and *Salmonella* spp.

LIVERPOOL JOHN MOORES UNIVERSITY
LEARNING SERVICES

to that agent (Demple and Halbrook 1983, Winquist et al. 1984). Starvation or low a_w resulted in slower rates of death (larger D values) for *P. aeruginosa* during preservative efficacy testing (Orth et al. 1998). Although these workers did not demonstrate the presence of different levels of heat shock proteins, their findings showed that exposure to one stress helps prepare a population to survive another type of stress.

Pine oil disinfectant, salicylate, and other weak acids (benzoate) may induce multiple antibiotic resistance in a number of organisms (Cohen et al. 1993, Lambert et al. 1997, Moken et al. 1997, Gustafson et al. 1999, McDonnell and Russell 1999). McDonnell and Russell (1999) reported increased cross-resistance to heat, ethanol, and hypochlorous acid. These reports suggest that exposure to sublethal concentrations of biocides may foster the development of microorganisms with increased tolerance to biocides and other antimicrobial agents (Orth 2000). At this time, we do not have sufficient data to state that development of antibiotic resistance caused by antimicrobials or biocides in laboratory studies occurs under actual product-use conditions. However, manufacturers of cosmetics and drugs should consider programs to ensure that their products will not encourage the development of drug-resistant microorganisms (Orth 2000).

13.8 OVERVIEW

The goal of a preservative system is to satisfactorily preserve a product against microbial challenge while it is in trade channels and in the hands of consumers. In order to successfully achieve this, full consideration must be given to all the factors that may influence preservative activity and to select for testing those preservative systems that come closest to the ideal. A preservative efficacy test may then be used to determine the type and minimum effective concentration of preservative(s) required to preserve a product satisfactorily to meet the recommended acceptance level.

In order to meet the objective of pharmaceutical product preservation, which is to ensure that the product is microbiologically safe and stable, it may be appropriate to perform a risk assessment. In this, it will be necessary to determine whether applying hurdle technology or increasing the potency of the preservative system (i.e., by use of higher concentrations of preservatives, by use of additional preservatives, or both) offers the best practical solution to the problem of product preservation, or whether reformulating the product or using alternative modes of packaging (i.e., unit-dose packages or contamination-resistant packages) should also be considered.

REFERENCES

Ahearn, D.G., Wilson, L.A., Julian, A.J., Reinhardt, D.J., and Ajello, G. (1974). Microbial growth in eye cosmetics: contamination during use. *Dev. Ind. Microbiol.,* 15, 211–216.

Baird, R.M. (1977). Microbial contamination of cosmetic products. *J. Soc. Cosmet. Chem.,* 28, 17–20.

Bhadauria, R. and Ahearn, D.G. (1980). Loss of effectiveness of preservative systems of mascaras with age. *Appl. Environ. Microbiol.,* 39, 665–667.

Bruch, C.W. (1971). Cosmetics: sterility vs. microbial control. *Am. Perfum. Cosmet.,* 86, 45–50.

Bruch, C.W. (1972). Objectionable micro-organisms in non-sterile drugs and cosmetics. *Drug Cosmet. Ind.,* 111(4), 51–54, 151–156.

Close, J. and Nielsen, P.A. (1976). Resistance of a strain of *Pseudomonas cepacia* to esters of *p*-hydroxybenzoic acid. *Appl. Environ. Microbiol.,* 31, 718–722.

Cohen, S.P., Levy, S.B., Foulds, J., and Rosner, J.L. (1993). Salicylate induction of antibiotic resistance in *Escherichia coli:* Activation of the *mar* operon and a *mar*-independent pathway. *J. Bacteriol.,* 175, 7856–7862.

Cosmetics Directive. (1993). 76/768/EEC, The European Commission, 6th Amendment.

Cowen, R.A. and Steiger, B. (1977). Why a preservative system must be tailored to a specific product. *Cosmet. Toiletr.,* 92(3), 15–16,18–20.

Demple, B. and Halbrook, J. (1983). Inducible repair of oxidative damage in *E. coli. Nature,* 304, 466–468.

Denyer, S.P. and Stewart, G.S.A.B. (1998). Mechanisms of action of disinfectants. *Internat. Biodet. & Biodeg.,* 41, 261–168.

Dunnigan, A.P. (1968). Microbiological control of cosmetic products. Proceedings of the Joint Conference. Cosmetic Science, Washington, DC, April 21–23, 1968. Cited in McCarthy, T.J. (1984). Formulated factors affecting the activity of preservatives. In *Cosmetic and Drug Preservation: Principles and Practice.* Kabara, J.J., Ed. Marcel Dekker, New York, pp. 359–388.

Dunnigan, A.P. and Evans, J.R. (1970). Report of a special survey: microbiological contamination of topical drugs and cosmetics. *TGA Cosmet. J.,* 2, 39–41.

Eiermann, H.J. (1984). Cosmetic product preservation: safety and regulatory issues. In *Cosmetic and Drug Preservation: Principles and Practice.* Kabara, J.J., Ed. Marcel Dekker, New York, pp. 559–569.

Federal Food, Drug and Cosmetic Act of 1938, as amended. (1976). Sections 601 and 602, 21 USC 361 and 362.

Gustafson, J.E., Candelaria, P.V., Fisher, S.A., Goodridge, J.P., Lichocik, T.M., McWilliams, T.M., Price, C.T.D., O'Brien, F.G., and Grubb, W.B. (1999). Growth in the presence of salicylate increases fluoroquinolone resistance in *Staphylococcus aureus. Antimicrob. Agents Chemother.,* 43, 990–992.

Halleck, F.E. (1970). A guideline for the determination of adequacy of preservation of cosmetics and toiletry formulations. *TGA Cosmet. J.,* 2, 20–23.

Hodges, N.A. and Hanlon, G. (2000). Antimicrobial preservative efficacy testing. In *Handbook of Microbiological Quality Control: Pharmaceuticals and Medical Devices.* Baird, R.M., Hodges, N.A., and Denyer, S.P., Eds. Taylor & Francis, London, pp.168–189.

Hodges, N.A. and Denyer, S.P. (1996). Preservative testing. In *Encyclopedia of Pharmaceutical Technology,* Swarbrick, J. and Boylan, J.C., Eds. Vol. 13. Marcel Dekker, New York, pp. 21–37.

Hugo, W.B. and Denyer, S.P. (1987). The concentration exponent of disinfectants and preservatives (biocides). In *Preservatives in the Food, Pharmaceutical and Environmental Industries.* Board, R.G., Allwood, M.C., and Banks, J.G., Eds. SAB Technical Series no. 22. Blackwell Scientific Publications, Oxford, pp. 281–291.

Hurwitz, S.J. and McCarthy, T.J. (1985). Dynamics of disinfection to selected preservatives against *Escherichia coli. J. Pharm. Sci.,* 74, 892–894.

Kabara, J.J. and Orth, D.S. (1997). Principles for product preservation. In *Preservative-Free and Self-Preserving Cosmetics and Drugs: Principles and Practice.* Kabara, J.J. and Orth, D.S., Eds. Marcel Dekker, New York, pp. 1–14.

Lambert, L.A., Abshire, K., Blankenhorn, D., and Slonczewski, J.L. (1997). Proteins induced in *Escherichia coli* by benzoic acid. *J. Bacteriol.,* 179, 7595–7599.

Levy, E. (1987). Insights into microbial adaptation to cosmetic and pharmaceutical products. *Cosmet. Toiletr.,* 102(12), 69–74.

Levy, S.B. (1998). The challenge of antibiotic resistance. *Sci. Am.,* 278(3), 46–53.

Lorenzetti, O.J. (1984). A preservative evaluation program for dermatological and cosmetic preparations. In *Cosmetic and Drug Preservation: Principles and Practice.* Kabara, J.J., Ed. Marcel Dekker, New York, pp. 441–463.

McCarthy, T.J. (1980). Microbiological control of cosmetic products. *Cosmet. Toiletr.,* 95(8), 23–27.

McDonnell, G. and Russell, A.D. (1999). Antiseptics and disinfectants: activity, action and resistance. *Clin. Microbiol. Rev.,* 12, 147–179.

McMurry, L.M., Oethinger, M., and Levy, S.B. (1998). Triclosan targets lipid synthesis. *Nature,* 394, 531.

Moat, A.G. and Foster, J.W. (1988). *Microbial Physiology.* 2nd ed. Wiley, New York, pp. 523–578.

Moken, M.C., McMurry, L.M., and Levy, S.B. (1997). Selection of multiple-antibiotic-resistant (Mar) mutants of *Escherichia coli* by using the disinfectant pine oil: roles of the *mar* and *acr*AB loci. *Antimicrob. Agents Chemother.,* 41, 2770–2772.

Morse, L.J. and Schonbeck, L.E. (1968). Hand lotions—a potential nosocomial hazard. *New Engl. J. Med.,* 278, 376–378.

Morse, L.J., Williams, H.L., Grenn, Jr, F.P., Eldridge, E.E., and Rotta, J.R. (1967). Septicemia due to *Klebsiella pneumoniae* originating from a hand-cream dispenser. *New Engl. J. Med.,* 277, 472–473.

Noble, W.C. and Savin, J.A. (1966). Steroid cream contaminated with *Pseudomonas aeruginosa. Lancet,* i, 347–349.

Orth, D.S. (1979). Linear regression method for rapid determination of cosmetic preservative efficacy. *J. Soc. Cosmet. Chem.,* 30, 321–332.

Orth, D.S. (1981). Principles of preservative efficacy testing. *Cosmet. Toiletr.,* 96(3), 43, 44, 48–52.

Orth, D.S. (1984). Evaluation of preservatives in cosmetic products. In *Cosmetic and Drug Preservation. Principles and Practice.* J.J. Kabara, Ed. Marcel Dekker, New York, pp. 403–421.

Orth, D.S. (1991). Standardizing preservative efficacy test data. *Cosmet. Toiletr.,* 106(3), 45–48, 51.

Orth, D.S. (1993). *Handbook of Cosmetic Microbiology.* Marcel Dekker, New York.

Orth, D.S. (1997). Inactivation of preservatives in surfactants. In *Surfactants in Cosmetics.* Rieger, M.M. and Rhein, L.D., Eds. 2nd ed. Marcel Dekker, New York, pp. 583–603.

Orth, D.S. (1999a). *An Introduction to Cosmetic Microbiology.* IFSCC Monograph Number 5. Micelle Press, Weymouth, U.K.

Orth, D.S. (1999b). Putting the Phoenix phenomenon into perspective. *Cosmet. Toiletr.,* 114(4), 61–66.

Orth, D.S. (2000). Cosmetics, toiletries and antibiotic-resistant "superbugs." *Cosmet. Toiletr.,* 115(3), 88.

Orth, D.S. and Lutes, C.M. (1985). Adaptation of bacteria to cosmetic preservatives. *Cosmet. Toiletr.,* 100(2), 57–59, 63–64.

Orth, D.S. and Milstein, S.R. (1989). Rational development of preservative systems for cosmetic products. *Cosmet. Toiletr.,* 104(10), 91, 92, 94–100, 102, 103.

Orth, D.S. and Kabara, J.J. (1998). Preservative-free and self-preserving cosmetics and drugs: application of hurdle technology. *Cosmet. Toiletr.,* 113(4), 51, 52, 54, 56–58.

Orth, D.S., Lutes, C.M., Milstein, S.R., and Allinger, J.J. (1987). Determination of shampoo preservative stability and apparent activation energies by the linear regression method of preservative efficacy testing. *J. Soc. Cosmet. Chem.,* 38, 307–319.

Orth, D.S., Anderson, C.M. Lutes, Smith, D.K., and Milstein, S.R. (1989). Synergy of preservative system components: use of the survival curve slope method to demonstrate anti-*Pseudomonas* synergy of methyl paraben and acrylic acid homopolymer/copolymers *in vitro. J. Soc. Cosmet. Chem.,* 40, 347–365.

Orth, D.S., Dumatol, C., and Zia, S. (1996). House organisms: dealing with the "bug in the plant." *Cosmet. Toiletr.,* 111(6), 59–66,68–70.

Orth, D.S., Delgadillo, K.S., and Dumatol, C. (1998). Maximum allowable D-values for Gram-negative bacteria: determining killing rates required in aqueous cosmetics. *Cosmet. Toiletr.,* 113(8), 53–56, 58, 59.

Parker, M.S. (1984). Design and assessment of preservative systems for cosmetics. In *Cosmetic and Drug Preservation: Principles and Practice.* Kabara, J.J., Ed. Marcel Dekker, New York, pp. 389–402.

Russell, A.D., Ahonkhai, I., and Rogers, D.T. (1979). A review. Microbiological applications of the inactivation of antibiotics and other antimicrobial agents. *J. Appl. Bacteriol.,* 46, 207–245.

Russell, A.D., Tattawasart, U., Maillard, J.-Y., and Furr, J.R. (1998). Possible link between bacterial resistance and use of antibiotics and biocides. *Antimicrob. Agents Chemother.,* 42, 2151.

Steinberg, D.C. (2004). Frequency of use of preservatives 2003. *Cosmet. Toiletr.,* 119(1), 55–58.

Sutton, S.V.W., Magee, M.A., and Brannan, D.K. (1997). Preservative efficacy, microbial content, and disinfectant testing. In *Cosmetic Microbiology. A Practical Handbook.* Brannan, D.K., Ed. CRC Press, Boca Raton, FL, pp. 95–126.

Wilson, L.A. and Ahearn, D.G. (1977). *Pseudomonas*-induced corneal ulcers associated with contaminated eye mascaras. *Am. J. Ophthamol.,* 84, 112–119.

Wilson, L.A., Julian, A.J., and Ahearn, D.G. (1975). The survival and growth of microorganisms in mascara during use. *Am. J. Ophthamol.,* 79, 596–601.

Winquist, L., Rannug, U., Rannug, A., and Ramel, C. (1984). Protection from toxic and mutagenic effects of hydrogen peroxide by catalase induction in *Salmonella typhimurium. Mutat. Res.,* 141, 145–147.

Yablonski, J.I. (1978). Microbiological aspects of sanitary cosmetic manufacturing. *Cosmet. Toiletr.,* 93(9), 37–50.

14 Antimicrobial Preservatives and Their Properties

Stephen P. Denyer

CONTENTS

14.1 INTRODUCTION

Choosing a suitable antimicrobial preservative agent for a pharmaceutical product requires careful consideration of the product type, its usage, formulation character- istics, and likely microbial challenge (see Chapter 13). Selection is then made largely from a common pool of agents used in food, cosmetic, toiletry, and pharmaceutical products, and experience has identified a limited range most generally suited to the pharmaceutical situation (Wallhaeusser 1974, Chapman 1987, Matthews 2003, English 2006; Table 14.1). This chapter focuses on this limited list of preservative agents in order to illustrate the varied properties and qualities of this diverse group of pharmaceutical excipients. On occasion, however, none of the agents described may be suited to a particular application, in which case, a wider range of monographs compiled, for example, by Wallhaeusser (1984, 1988) and Paulus (1993), is recom- mended for consultation. In the final analysis, the ultimate selection of a preservative agent may be a compromise between registration acceptability (Chapter 19), anti- microbial efficacy (Chapters 17 and 18), and product compatibility (Chapter 15).

14.2 FACTORS AFFECTING PRESERVATIVE ACTIVITY

Preservative efficacy is influenced by factors both intrinsic and extrinsic to the target organisms. Intrinsic factors include the nature, structure and composition, and con- dition of the microorganism together with its capacity to resist, degrade, or inactivate the preservative agent. Detailed treatment of these factors is given in Gilbert and Wright (1987), Gilbert (1988), Hugo (1988), Chopra (1990), Denyer and Maillard (2002), and Russell (2004).

To the formulator, the extrinsic factors that reflect the external environment in which the preservative acts are probably of the most immediate relevance and are undoubtedly of a more controllable nature. Changes in preservative concentration, product pH, storage temperature, and product composition can all significantly influence antimicrobial activity. To some extent, these influences are predictable provided the properties of preservative agents are fully recognized. The following sections consider the potential influence of such parameters on the performance of preservative agents.

TABLE 14.1

Principal Antimicrobially Active Preservative Agents Used in Pharmaceuticals and Their Major Potential Areas of Application

Preservative Agent	Pharmaceutical Products			
	Injectable	Ophthalmic	Topical	Oral
Benzalkonium chloride	+	+	+	
Benzethonium chloride	+	+	+	
Benzoic acid (+ salts)	+		+	+
Benzyl alcohol	+		+	+
Bronopol			+	
Butylated hydroxyanisole	+		+	+
Cetrimide		+	+	
Chlorbutanol	+	+		
Chlorhexidine		+	+	
Chlorocresol	+		+	
Cresol	+		+	
Diazolidinyl urea			+	
Edetic acid (+ salts)	+	+	+	+
Ethanol				+
Imidurea			+	
Parabens (methyl, ethyl, butyl, propyl, benzyl + salts)	(+)	(+)	+	+
Phenol	+		+	
Phenoxyethanol			+	
Phenylethanol		+	+	
Propionic acid (+ salts)			+	+
Propylene glycol			+	+
Sorbic acid (+ salts)			+	+
Sulphites, inorganic				+
Thiomersal	+	+	+	

Note: Regulatory approval may not exist for all applications in all countries. Current approved lists must be consulted to allow for changed status.

(+) Now generally regarded as unsuitable for these preparations.

14.2.1 CONCENTRATION

Investigations at the beginning of the last century (reviewed in Hugo and Denyer 1987) clearly demonstrated the exponential relationship between rate of microbial death and concentration of antimicrobial agent. This can be described by the equation:

$$C_1^\eta t_1 = C_2^\eta t_2 \tag{14.1}$$

where C_1 and C_2 represent two concentrations of the antimicrobial agent and t_1 and t_2, their respective times to achieve the same level of reduction in viable count. The exponent η is a measure of the effect of changes in concentration (or dilution level) on microbial death rate and is termed the concentration exponent or dilution coefficient (Russell and McDonnell 2000).

Information derived from such kinetic studies can be used to calculate the concentration exponent of a microbiocidal preservative agent from a mathematical rearrangement of Equation 14.1 (see Equation 14.2 below) or by a graphical method (Denyer and Wallhaeusser 1990).

$$\eta = \frac{\log t_2 - \log t_1}{\log C_1 - \log C_2} \tag{14.2}$$

Thus, any change in activity arising from a change in preservative concentration is exponential in accordance with the specific concentration exponent for that agent.

In practical terms, the activity of a compound with a high concentration exponent will be markedly decreased by dilution, whereas that of an agent with low η value will be less severely affected (Table 14.2). Conversely, an increase in concentration will be of much greater benefit to a compound of high η value than to a compound with a low concentration exponent. Example concentration exponents are given for a range of preservatives in Table 14.3.

It is likely that the concentration exponent in some way reflects the nature of the interaction between preservative agent and microbial target (Hugo and Denyer 1987). Indeed, variations in concentration exponent are reported and these appear to reflect differences in both experimental design (i.e., antimicrobial endpoint) and target organism. Some indication of this variation between organism type is seen in the average values of Table 14.3, and more detailed examples of this variation are given in the papers of Beveridge et al. (1980), Karabit et al. (1985, 1986, 1988), and Mackie et al. (1986). Clearly, experimentation is necessary if a precise value for the concentration exponent is required for a particular situation.

TABLE 14.2
The Influence of Concentration Exponent on Residual Antimicrobial Activity of a Preservative Following Its Depletion in a Product

| | Residual Antimicrobial Activity of Preservative (%) | |
Concentration Exponent	At Three-Quarters (75%) of Original Concentration	At One-Half (50%) of Original Concentration
1	75	50
2	56	25
3	42	12.5
4	32	6
5	24	3
6	18	1.6
7	13	0.8
8	10	0.4
9	7.5	0.2

TABLE 14.3
Preservative Characteristics

Preservative Agent	Concentration Exponent[a]	Optimal pH Range
Benzalkonium chloride	3.5, 1.8 (y), 9 (m)	Broad; 4–10
Benzethonium chloride	1–2[b]	Broad; 4–10
Benzoic acid (+ salts)	3.5 (y)	Acidic; 2–5
Benzyl alcohol	6.6, 4 (y), 2 (m)	Acidic; <5
Bronopol	0.9	Weakly acid; 5–7
Butylated hydroxyanisole	2–4[b]	Acidic
Cetrimide	1	Broad; 4–10
Chlorbutanol	2	Acidic; 4
Chlorhexidine	1.9	Neutral; 5–8
Chlorocresol	8.3	Acidic; <8.5
Cresol	8	Acidic; 9
Diazolidinyl urea	1–2[b]	Broad; 3–9
Edetic acid (+ salts)		Broad, but dependent on salt
Ethanol	4.5, 5.7 (y), 3 (m)	Acidic
Imidurea	1–2[b]	Broad; 3–9
Parabens (methyl, ethyl, butyl, propyl, benzyl + salts)	2.5	Broad; 3–9.5
Phenol	5.8, 4 (y), 4.3 (m)	Acidic; <9
Phenoxyethanol	9	Broad
Phenylethanol	5.6	Acidic; 7
Phenylmercuric salts	1	Neutral; 6–8
Propionic acid	2–4[b]	Acidic; S5
Propylene glycol	4–6[b]	Broad
Sorbic acid	3.1	Acidic; <6.5
Sulphites, inorganic	1.3, 1.6 (y), 1.8 (m)	Acidic; 4
Thiomersal	1	Neutral; 7–8

[a] Determined against bacteria unless indicated (m) for mold or (y) for yeast.
[b] Anticipated range.

In summary, the activity of a preservative system can be substantially influenced by changes in concentration and this may lead to its failure. Although this change may arise by simple dilution (i.e., in the mixing of two differently preserved creams) (Hugo et al. 1984) it will also occur by any mechanism that leads to a reduction in the active agent. This could include complexation of preservative with product ingredients (see Chapter 15), partitioning of antimicrobial agent into a nonaqueous phase (Orth 1997) or container material, or through pH influences on ionization and activity.

14.2.2 EFFECT OF PH

The antimicrobial activity of many preservative agents is strongly influenced by environmental pH, variously through changes in ionic status, altered interaction with target groups on the microbial cell, and variable partitioning between product and microorganism. In general, a negatively charged microbial cell will interact most strongly with a cationic agent at high pH while effectively repelling anionic agents. Furthermore, a compound that exerts its activity by partitioning into the target cell is likely to do so most efficiently in its unionized rather than its ionized form. For instance, the activity of weak acid preservatives such as benzoic and sorbic acids resides principally in the unionized (undissociated) form and greatest activity is therefore evident at pHs at or below their pK_a (4.2 and 4.76, respectively) where the fraction undissociated will be 50% or greater. With knowledge of preservative pK_a it is possible to determine the extent of ionization (dissociation) from Equation 14.3 and thereby predict the likely influence of pH on activity:

$$\text{Undissociated fraction of weak acid preservative} = \frac{1}{1 + \text{antilog}(\text{pH} - pK_a)} \tag{14.3}$$

The influence of pH may extend beyond the confines of activity alone, influencing preservative stability (Moore and Stretton 1981, Chapman 1987), interactions with pharmaceutical excipients (see Chapter 15), and partitioning behavior in multiple-phase systems (Section 14.2.4). Optimal pH ranges are given for selected preservative agents in Table 14.3.

14.2.3 EFFECT OF TEMPERATURE

As with many chemical reactions, the activity of preservative agents usually increases with an increase in temperature although the effects are often complicated by the temperature dependency of the target organism (Kostenbauder 1983). Over a narrow temperature range, which may extend from refrigeration temperatures to body heat, and in a limited concentration range, it is possible to describe the effect of temperature on preservative activity using Equation 14.4:

$$Q_{10}\text{(change in activity per 10°C change in temperature)} = \frac{t_{(T)}}{t_{(T+10)}} \tag{14.4}$$

where $t_{(T)}$ represents the death time at temperature T°C and $t_{(T+10)}$ the death time at $(T + 10)$°C. It is usual to report temperature sensitivity over a 10°C temperature difference (Q_{10} value), but a temperature coefficient (θ) for a 1°C temperature change can be similarly calculated; θ values are generally in the order of 1 to 1.5 as a consequence of the apparent geometric nature of the temperature-activity relationship (Berry and Michaels 1950).

Some temperature coefficients (Q_{10} values) for selected agents are given in Table 14.4. From these values, it can be seen that preservative agents respond differently to temperature variations and this may be important in extrapolating preservative efficacy test data from the usual room temperature to the recommended range of product storage temperatures (see also Chapter 15). In particular, this may have important implications in products which are recommended for refrigerated storage between periods of usage (Allwood 1982).

TABLE 14.4
Influence of Temperature on Preservative Activity

Preservative Agent	Temperature Coefficient (Q_{10})
Benzalkonium chloride	2.9–5.8
Benzyl alcohol	2.3–7.2
Bronopol	2.9
Chlorhexidine	3–16
Chlorocresol	3–5
Cresol	3–5
Ethanol	45
Phenol	5
Sorbic acid	2.3

14.2.4 EFFECT OF PARTITIONING IN MULTIPLE-PHASE SYSTEMS

It has long been recognized (Bean et al. 1962, 1965; Bean 1972) that antimicrobial activity primarily resides in the aqueous phase of a preserved two-phase system and is therefore dependent upon the equilibrium concentration of the preservative in this phase. Preservatives will partition between oil and water phases in accordance with their partition coefficients (Table 14.5) and the relative ratio of oil and water present in the system. This partitioning behavior and its influence on the aqueous concentration of preservative agent (C_w) can be described by Equation 14.5:

$$C_w = \frac{C(\theta+1)}{K_w^o\theta+1} \tag{14.5}$$

where C represents the overall concentration of preservative, K_w^o the oil-water partition coefficient, and θ the oil:water ratio. It follows that if K_w^o is high, then it becomes extremely difficult to maintain adequate preservative levels in the aqueous phase without an excessive total preservative concentration. This formula can be applied to follow the behavior of a preservative in a simple two-phase system and a worked example for chlorocresol is given by McCarthy (1984).

The oily phase can generally be considered to represent any water-immiscible material, but it must be remembered that the K_w^o of a preservative may vary considerably depending upon the type of oil present. Indeed, it is generally recognized that partitioning from water into vegetable oils is more efficient than into mineral oils (Bean 1972, McCarthy 1984); furthermore, water solubility is not necessarily a predictor of partitioning behavior (Table 14.5). Additionally, the pH of the system may substantially affect partitioning, as may also the droplet size, and hence surface interfacial area, of the oil phase during mixing.

The above mathematical method describes a simple system that would rarely be found in emulsion and cream formulations. In these situations, a stable formulation would usually be maintained by the addition of an emulgent, usually a nonionic

TABLE 14.5
Examples of Solubility and Partition Characteristics for Preservative Agents

Preservative Agent	Water Solubility (20–25°C)	Vegetable Oil: Water Partition (K_w^o)	Emulgent: Water Partition (R)
Benzalkonium chloride	High	<1.0	High
Benzethonium chloride	High	<1.0	High
Benzoic acid	Slight[a]	3.6, 1.9[e]	Medium
Benzyl alcohol	Moderate	1.3	Low
Bronopol	High	0.11	Low
Butylated hydroxyanisole	Poor	High[f]	
Cetrimide	Moderate	<1.0	High
Chlorbutanol	Slight	High[f]	Medium
Chlorhexidine	Slight to high[b]	0.04 (diacetate)	High
Chlorocresol	Slight	117	High
Cresol	Moderate	High[f]	Medium or low
Diazolidinyl urea	High	Poor[f]	
Edetic acid	Slight[a]	Poor[f]	
Ethanol	Miscible		
Imidurea	High	Poor[f]	
Parabens (methyl, ethyl, butyl, propyl, benzyl)	Slight to poor[a] (decrease with increasing chain length)	4 (methyl), 16 (ethyl), 52 (propyl), 280 (butyl)	High
Phenol	Moderate	High[f]	Low
Phenoxyethanol	Moderate	2.6	Low
Phenylethanol	Moderate	21.5[e]	Low
Phenylmercuric salts	Slight	0.4	Low
Propionic acid	Miscible[c]	0.33[e]	
Propylene glycol	Miscible		
Sorbic acid	Slight[d]	3.5	
Sulphites, inorganic	High	Poor[f]	
Thiomersal	High	Poor[f]	Low

Solubility limits: high, > 25%; moderate, > 1%; slight, > 0.05%; poor, < 0.05%.

[a] Sodium salts moderate to high solubility.
[b] Digluconate (high), diacetate (moderate), dihydrochloride (slight).
[c] Sodium salt (high).
[d] Potassium salt (high), calcium salt (moderate).
[e] Octanol: water partition.
[f] Oil solubility.

surface-active agent. In order to account for this third component, Bean et al. (1969) extended their equation to accommodate the possible partitioning of preservative into the micellized emulgent or their complexation:

$$C_w = \frac{C(\theta+1)}{K_w^o\theta + R} \tag{14.6}$$

In this equation, C, θ, and K_w^o have the same meanings as before, C_w is now the free concentration in the water, and a new term R, the ratio of total preservative in the aqueous phase to the free (unbound) preservative in that phase, is introduced. The relative order of magnitude of R for selected preservative agents is given in Table 14.5. In relatively simple systems, with a single nonionic emulgent and preservative, a clear linear relationship exists between R and surfactant concentration (Bean 1972, Kostenbauder 1983); relationships also dependent upon preservative concentration have been described (Dempsey 1996). More complex models have been developed along the above lines (Garrett 1966) to include multiple preservative agents or mixed emulgent systems (Kazmi and Mitchell 1976, 1978a, 1978b).

Within the literature there are examples of potentiation of antimicrobial activity associated with the presence of low concentrations of nonionic surfactants (Allwood 1973, Lehmann 1988; see also Chapter 15). This can be attributed to a surfactant-based permeabilization of the bacterial cell to the antimicrobial agent, but this phenomenon occurs only over a narrow concentration range of emulgent and above the critical micelle concentration preservative is displaced into the micellar system.

The effective preservation of multiple-phase systems is thus a complex task, and due consideration must be given to the partitioning properties of the preservative agent(s), the nature and amount of the oil phase, and the properties of the emulgent system. At the very best, partitioning of the preservative into the oil phase can be said to provide a reservoir of the active agent to replenish declining aqueous levels; more likely, unrecognized and unaccounted for partitioning will lead to loss of preservative capability.

14.2.5 PREDICTION OF PRESERVATIVE BEHAVIOR

The mathematical treatments described in earlier sections (14.2.1 through 14.2.3) can be employed to predict the consequences of formulation changes or other external influences on the antimicrobial activity of a preservative system. Indeed, Bean (1972) has shown, from a theoretical perspective, how changes in more than one factor can be accommodated in a combined mathematical approach. It is likely, however, that a more complex range of influences will arise in practice and the predictive capability of these formulae may then be only qualitative at best. Under these circumstances, experimentation will be required to discover the precise significance of every relevant variable. To this end, application of factorial design to the evaluation of preservative efficacy in pharmaceutical systems is highly relevant (Karabit et al. 1989).

14.3 ANTIMICROBIAL ACTIVITY

14.3.1 SPECTRUM OF ACTIVITY

In general, a preservative agent should offer a microbicidal capability although clearly the ability to hold contaminating organisms in stasis will be satisfactory in most instances. In this respect, the sporicidal activity of many preservatives at room temperature is often extremely limited, but good activity against the vegetative form of the organism should be sufficient. The capacity of a preservative system should also be sufficient to withstand the anticipated microbial challenge levels, especially in situations where multiple contamination episodes may arise.

An ideal preservative should possess broad-spectrum activity against both Gram-positive and Gram-negative bacteria, molds, and yeasts. In practice, few preservatives have this capability (Table 14.6) and this, combined with the constraints imposed by formulation design and product type, place further limitations on the spectrum that can be achieved by use of a single agent in a formulation. Great care must therefore be taken to ensure that a perceived gap in activity does not lead to the selection of insensitive organisms or the creation of conditions favorable for a less adaptable spoilage organism by the growth of the primary insensitive contaminant.

TABLE 14.6
Range of Antimicrobial Activities at Typical Use Concentrations

Preservative Agent	Bacteria		Yeasts	Molds
	Gram-Positive	Gram-Negative		
Benzalkonium chloride	++	(++)*	(++)	+
Benzethonium chloride	++	(++)*	(++)	+
Benzoic acid (+ salts)	++	+	+	+
Benzyl alcohol	++	+	+	+
Bronopol	(++)	++	+	+
Butylated hydroxyanisole	(++)	+*	(++)	(++)
Cetrimide	++	(++)*	(++)	+
Chlorbutanol	++	++	(++)	+
Chlorhexidine	++	(++)*	(++)	+
Chlorocresol	++	(++)	+	+
Cresol	(++)	+	+	+
Diazolidinyl urea	++	(++)	+	+
Edetic acid (+ salts)		+	+	+
Ethanol	++	++	(++)	(++)
Imidurea	(++)	(++)	+	+
Parabens (methyl, ethyl, propyl, butyl, benzyl + salts)	(++)	+*	(++)	(++)
Phenol	(++)	+	+	+
Phenoxyethanol	(++)	++	+	+
Phenylethanol	(++)	++	+	+
Phenylmercuric salts	++	++*	(++)	+
Propionic acid (+ salts)	+	+		++
Propylene glycol	++	++	(++)	+
Sorbic acid (+ salts)	+	+	++	(++)
Sulphites, inorganic	+	+	(++)	(++)
Thiomersal	++	(++)	(++)	+

++, Active; (++), moderately active; +, weakly active; *, poorly active against *Pseudomonas* spp.

In these respects, considerable concern has been raised regarding the recognized *Pseudomonas* gap of certain preservative agents (Croshaw 1977), because this genus represents a major, and potentially pathogenic, contaminant of many pharmaceutical preparations (see Chapter 2). This has led to the widespread use of preservative combinations, not necessarily to achieve potentiation or synergy (Section 14.4), but simply to combat the likely range of contaminants. Such combinations include mixtures of parabens in place of single esters (this has the principal benefit of increasing the total paraben level that can be held in solution), parabens with phe-noxyethanol, bronopol and parabens (particularly useful in alkaline preparations where the rapid "knock-down" effect of the alkali-labile bronopol can be supplemented by the long-term activity of the parabens), and benzalkonium chloride with chlorhexidine gluconate (Croshaw 1977).

14.3.2 RESISTANCE TO PRESERVATIVES

At in-use levels few preservatives show selectivity in their target site and activity is often associated with compound lesions (Denyer and Stewart 1998). Resistance, where it can be discovered, has normally been associated with cellular imperme-ability, efflux mechanisms and, rarely, metabolic inactivation (Denyer and Maillard 2002, Poole 2004, Russell 2004, Stickler 2004); occasionally these resistance pro-cesses are shown to be shared with antibiotic resistance mechanisms (Russell 1997, Poole 2004). Fueled by studies involving triclosan (Heath et al. 1998, Heath and Rock 2000), however, there is now an emerging concern that some preservatives or biocides may, at appropriate concentrations, share specific targets with antibiotics (McBain and Gilbert 2001). Although only poorly evidenced at present, the possi-bility of cross-resistance with antibiotics requires the formulator to review constantly the choice of preservative agents for pharmaceutical application.

14.4 POTENTIATION AND SYNERGY

An occasional benefit from the use of preservative agents in combination is an enhancement in activity which, if to a sufficiently marked degree, can be considered to represent true synergy. This phenomenon, although elusive, has been widely reported (see, for example, Lehmann 1988 and Denyer 1996). Synergy is rarely a property deliberately sought for in preservative combinations, and it should not be confused with the broadening of antimicrobial spectrum described in Section 14.3.1. If achieved, it is often by serendipity although the basis for its prediction has been described (Denyer et al. 1985; Denyer 1990, 1996).

The extent of synergy between preservative agents is dependent upon their relative ratios; this will be controlled not only by their initial concentrations within the formulation, but also by those factors that influence their availability in active form (Section 14.2). Thus, if attempts are made to select synergistic combinations of preservatives, due attention must be given to their individual physicochemical properties in order to predict their likely behavior in a formulated system. Complex preservative dose-response relationships can mislead in the interpretation of synergy (Gilliland et al. 1992) but this may be predicted from knowledge of concentration

---

TABLE 14.7
A Selection of Preservatives Whose Action Is
Potentiated by Edetic Acid

Benzalkonium chloride
Bronopol
Cetrimide
Chlorhexidine
Ethanol
Parabens
Phenol
Phenylethanol
Sorbic acid

---

exponents (Lambert et al. 2003). The extent of synergy can be tested in simple solution or in more complex formulations and some of the methodological approaches to this have been described (Denyer et al. 1985, Hodges and Hanlon 1990, McCarthy and Ferreira 1990, Denyer 1996, but see also Lambert and Lambert 2003). The laboratory examination of preservative combinations will also eliminate those with the potential for antagonism and the methods can be adapted to study excipient interactions (Pons et al. 1992).

An additional phenomenon to synergy, also known to arise in the presence of preservatives, is one of potentiation. Here the activity of an antimicrobial agent is enhanced by the inclusion of a microbiologically inactive or only weakly active component that presumably sensitizes the microbial target to preservative action. The classic example of potentiation is that involving edetic acid [ethylenediaminetetracetic acid (EDTA)], which has been clearly shown to enhance the activity of a range of preservative agents (Table 14.7), especially against *Pseudomonas aeruginosa* where its chelating power affects the stability of the protective outer membrane (Hart 1984, Lambert et al. 2004). Recently, a similar mechanism of potentiation has been attributed to the beneficial effect of polyacrylic acid polymers on paraben activity (Scalzo et al. 1996). A further example has already been considered in Section 14.2.4 where low concentrations of nonionic surfactants assist the antimicrobial action of preservative agents.

With the existing limitations in our current list of preservative agents, it is likely that the use of preservatives in combination will continue to increase. In order to maximize their potential, and to avoid unfortunate mismatches, their properties must be fully understood and exploited. In this respect, the potential benefits of synergy and the potentiating effects of formulation ingredients (see Chapter 15) as part of a "systems approach" (hurdle technology, Kabara 2006) should be exploited to advantage.

14.5 TOXICITY

While seeking to achieve maximum antimicrobial activity from preservative agents, we need also to minimize adverse reaction in the user. Thus preservative agents are required to be selectively toxic and acceptable in all aspects of toxicity including acute, subacute, and chronic exposure, local reactions, reproductive toxicity, mutagenicity, and carcinogenicity (Schneider 1984, Hayden 1996). These demands have resulted in a reduction in the range of preservatives considered acceptable for pharmaceutical use, limited the application of others, and imposed considerable restrictions upon novel biocidal agents (Allwood 1978, Bloomfield 1986, Matthews 2003). A useful classification scheme for toxigenic potential can be found in the Blue List (a collection of information on cosmetic ingredients) (Kemper et al. 2000). A further treatment of preservative toxicities can be found in Bronaugh and Maibach (1984), Lautier (1984), Wallhaeusser (1984, 1988), Rowe et al. (2004), and Goon et al. (2006).

Current opinion would suggest that a satisfactory balance between antimicrobial effect and toxicity may be achieved by careful selection of preservative concentration. In some instances, it may be acceptable to achieve only a growth inhibitory or weak killing effect in order to avoid the toxicological consequences that may arise from high levels of preservative agent. The need to exercise proper control over recalcitrant organisms is still, of course, a prerequisite to effective preservation.

One further approach taken in an attempt to minimize the toxicological potential of preservative systems is to use combinations where submaximal levels of the component agents can be employed; in this regard, the potential benefit of combining agents that demonstrate antimicrobial potentiation or synergy is obvious. It must be remembered, however, that the toxicological behavior of individual agents may also be modified in combination, as may also be the case when specific formulation ingredients are present.

14.6 PRESERVATIVE MONOGRAPHS

In this section, brief monographs are presented for the preservative agents summarized in Table 14.1 (see Section 14.1). The information given is designed to be read in conjunction with data given elsewhere in this chapter. The tolerance of challenge organisms to the inactivators suggested for use in the preservative efficacy test (PET) must be confirmed by preliminary experimentation and, if found unsuitable, membrane filtration should be considered as an alternative approach. Dilution will always be a useful adjunct to the use of specific inactivators, especially with those preservatives known to possess a high concentration exponent. Typical in-use preservative concentrations are given for guidance only, as the actual concentration employed will depend upon the product type. A further appreciation of preservative behavior in formulated products can be gained by consulting Chapter 13 and Chapter 15. Additional reference information is available in Wallhaeusser (1984, 1988), Paulus (1993), Martindale (2002), Rowe et al. (2004), and English (2006).

For a detailed discussion on the antimicrobial preservative efficacy test, the reader is referred to the companion *Handbook* (Hodges and Hanlon 2000).

14.6.1 BENZALKONIUM CHLORIDE

Synonym(s)—Mixture of alkyldimethylbenzylammonium chlorides (alkyldimethyl (phenylmethyl) ammonium chlorides); BKC

CAS number—800 1-54-5

Class of compound—Cationic quaternary ammonium

Empirical formula—General formula $[C_6H_5 \cdot CH_2 \cdot N(CH_3)_2 \cdot R]Cl$, where R represents a mixture of the alkyls from C_8H_{17} to $C_{18}H_{37}$ with the principal components represented by $C_{12}H_{25}, C_{14}H_{29}$, and $C_{16}H_{33}$ (average molecular weight 360)

Stability—Solutions are stable at room temperature for prolonged periods; stable to autoclaving conditions

Compatibility—Incompatible with anionic surfactants, soaps, citrates, nitrates, iodides, heavy metals (including silver salts), alkalis, some oxidants, some commercial rubber mixes, proteins; sorbed by some plastic and filter materials

PET inactivator—Lecithin and a nonionic surfactant such as Lubrol W or Tween 80 (Letheen broth)

Typical in-use concentration—0.002 to 0.02%

14.6.2 BENZETHONIUM CHLORIDE

Synonym(s)—N,N-Dimethyl-N-[2-[2-4-(1,1,3,3-tetramethyl-butyl) phenoxy] ethoxy]ethyl] benzene-methanaminium chloride; BZT

CAS number—121-54-0

Class of compound—Cationic quaternary ammonium

Empirical formula—$C_{27}H_{42} ClNO_2$, molecular weight 448.1

Stability—Good, aqueous solutions stable to autoclaving

Compatibility—Incompatible with soaps and anionic surfactants; may be precipitated in some salt solutions

PET inactivator—Lecithin and a nonionic surfactant such as Lubrol W or Tween 80 (Letheen broth)

Typical in-use concentration—0.01 to 0.5%

14.6.3 BENZOIC ACID (AND SALTS)

Synonym(s)—Benzene carboxylic acid; phenylcarboxylic acid; sodium benzoate salt

CAS number: 65-85-0

Class of compound—Organic weak acid (or salt)

Empirical formula—$C_6H_5 \cdot CO_2H$; molecular weight 122.1, sodium benzoate 162

Stability—Stable at low pH; stable to autoclaving

Compatibility—Incompatible with ferric salts and salts of heavy metals, nonionic surfactants, quaternary compounds and gelatin; loss of activity in presence of proteins, glycerol, and kaolin

PET inactivator—Dilution and nonionic surfactant such as Tween 80

Typical in-use concentration—0.1 to 0.5%

14.6.4 BENZYL ALCOHOL

Synonym(s)—Benzenemethanol; phenylcarbinol; phenylmethanol; α-hydroxytoluene

CAS number: 100-51-6

Class of compound—Alcohol

Empirical formula—$C_6H_5 \cdot CH_2OH$, molecular weight 108.1

Stability—Stable to autoclaving conditions; slowly oxidizes to benzaldehyde and benzoic acid in air, reduced by saturating solutions with nitrogen; dehydrates at low pH

Compatibility—Incompatible with oxidizing agents; activity reduced by nonionic surfactants; incompatible with some packaging materials and methyl cellulose

PET inactivator—Dilution and nonionic surfactant such as Tween 80

Typical in-use concentration—1 to 3%

14.6.5 BRONOPOL

Synonym(s)—2-bromo-2-nitropropane-1,3-diol; 2-bromo-2-nitro-1,3-propanediol; BNPD

CAS number—52-52-7

Class of compound—Aliphatic halogenonitro diol

Empirical formula—$C_3H_6BrNO_4$, molecular weight 200

Stability—Stable in aqueous solution at low pH (<5) and can be stored at room temperature for up to 2 years with no decomposition or for at least 1 month at 50°C; decomposition increases in the light, at elevated temperatures and in alkaline pH; decomposition accelerated in presence of iron and aluminum; unstable in anhydrous solutions of glycerol

Compatibility—Little or no inactivation by anionic, cationic, or nonionic surfactants, proteins, or serum; sulphydryl compounds (cysteine, thioglycollate), thiosulphate, and metabisulphite are markedly antagonistic

PET inactivator—Sulphydryl compounds such as cysteine or thioglycollate

Typical in-use concentration—0.01 to 0.1%

14.6.6 BUTYLATED HYDROXYANISOLE

Synonym(s)—2-tert-butyl-4-methoxyphenol; 1,1-dimethylethyl-4-methoxyphenol; BHA; E320

CAS number—25013-16-5

Class of compound—Phenolic

Empirical formula—$C_{11}H_{16} \cdot O_2$, molecular weight 180.3

Stability—Trace quantities of metals and exposure to light causes discoloration and loss of activity

Compatibility—Incompatible with oxidizing agents and ferric salts

PET inactivator—Dilution and nonionic surfactant such as Tween 80

Typical in-use concentration—0.005 to 0.02%

14.6.7 CETRIMIDE

Synonym(s)—Alkyltrimethylammonium bromide; tetradecyltrimethylammonium bromide; cetyltrimethylammonium bromide (CTAB); cetrimonium bromide and CTAB were formerly applied to cetrimide as the hexadecyltrimethylammonium bromide preparation

CAS number—Cetrimide, 8044-71-1; tetradecyltrimethylammonium bromide, 1119-97-71; hexadecylammonium bromide, 57-09-0

Class of compound—Cationic quaternary ammonium

Empirical formula—Chiefly tetradecyltrimethylammonium bromide with small amounts of dodecyl- and hexadecyltrimethylammonium bromides; contains not less than 96% of alkyltrimethylammonium bromides calculated as $C_{17}H_{38}BrN$, molecular weight 336.4; the main component was formerly $C_{19}H_{42}BrN$ (hexadecyltrimethylammonium bromide) (British Pharmacopoeia 1953)

Stability—Stable in solution and to autoclaving conditions

Compatibility—Incompatible with soaps and other anionic surfactants, high concentrations of nonionic surfactants, nitrates (including phenylmercuric nitrate), heavy metals, some oxidants, some alkalis, bentonite, some rubbers, and proteins

PET inactivator—Lecithin and a nonionic surfactant such as Lubrol W or Tween 80 (Letheen broth)

Typical in-use concentration—0.0 1 to 0.1%

14.6.8 CHLORBUTANOL

Synonym(s)—1,1,1-trichloro-2-methylpropan-2-ol; chlorbutol; chlorobutanol; acetonechloroform

CAS number: 57-15-8

Class of compound—Alcohol

Empirical formula—$C_4H_7Cl_3O$, molecular weight 177.5

Stability—Decomposes in aqueous solutions when heated, especially under alkaline conditions; losses from solution due to volatility

Compatibility—Incompatible with nonionic surfactants and alkalis; sorbed by polyethylene and polypropylene containers, bentonite, and magnesium trisilicate

PET inactivator—Dilution and nonionic surfactant such as Tween 80

Typical in-use concentration—0.3 to 0.5%

14.6.9 CHLORHEXIDINE

Synonym(s)—1,1-hexamethylene*bis*[5-(4-chlorophenyl) biguanide]; 1,6-*bis*(5-*p*-chlorophenylbiguanido)hexane, as (di)acetate, (di)gluconate, and (di)hydrochloride salts

CAS number: 55-56-1 (chlorhexidine base)

Class of compound—Cationic *bis*biguanide

Empirical formula—Base formula $C_{22}H_{30}Cl_2N_{10}$; diacetate ($2C_2H_4O_2$), molecular weight 625.6; digluconate ($2C_6H_{12}O_7$), molecular weight 897.9; dihydrochloride ($2HCl$), molecular weight 578.4

Stability—Generally unstable at high temperatures, decomposing to give trace amounts of 4-chloroaniline, but, under appropriate conditions, salts can be sterilized by autoclaving at 115°C for 30 min; alkaline pH promotes decomposition

Compatibility—Incompatible with soaps and other anionic agents, various gums and sodium alginate; forms insoluble salts with, for example, borates, bicarbonates, carbonates, chlorides, citrates, phosphates, and sulphates at concentrations of 0.05%

PET inactivator—Lecithin and a nonionic surfactant such as Lubrol W or Tween 80 (Letheen broth)

Typical in-use concentration—0.01 to 0.1%

14.6.10 CHLOROCRESOL

Synonym(s)—4-chloro-3-methylphenol; p-chloro-m-cresol; 2-chloro-5-hydroxytoluene; PCMC

CAS number—59-50-7

Class of compound—Halogenated phenolic

Empirical formula—C_7H_7ClO, molecular weight 142.6

Stability—Aqueous solutions turn yellow in light and air; solutions in water are stable to autoclaving and those in oil or glycerol to 160°C exposure for 1 h

Compatibility—Reduction in activity in the presence of nonionic surfactants; discoloration with iron salts; sorption to some plastics

PET inactivator—Dilution and nonionic surfactant such as Tween 80

Typical in-use concentration—0.1%

14.6.11 CRESOL

Synonym(s)—Methyl phenol; 3-cresol; m-cresol; cresylic acid; hydroxytoluene

CAS number—1319-77-3

Class of compound—Phenolic

Empirical formula—$CH_3 \cdot C_6H_4 \cdot OH$, a mixture of *ortho, meta* and *para* isomers but predominantly the *meta* isomer, molecular weight 108.1

Stability—Aqueous solutions turn yellow in light and air; not suitable for freeze-dried preparations

Compatibility—Reduced activity in the presence of nonionic surfactants

PET inactivator—Dilution and nonionic surfactant such as Tween 80

Typical in-use concentration—0.15 to 0.3%

14.6.12 DIAZOLIDINYL UREA

Synonym(s)—N-(hydroxymethyl)-N-(1,3-dihydroxymethyl-2,5-dioxo-4-imidazolidinyl)-N'-(hydroxymethyl) urea

CAS number: 78491-02-8

Class of compound—Heterocyclic urea derivative
Empirical formula—$C_8H_{14}N_4O_7$, molecular weight 278.2
Stability—Hygroscopic; very stable, with slow release of formaldehyde in water-based formulations
Compatibility—Compatible with many excipients including proteins, non-ionic surfactants and lecithin
PET inactivator—Dilution, possibly with lecithin–Tween 80–thioglycolate medium
Typical in-use concentration—0.1 to 0.5%

14.6.13 EDETIC ACID (AND SALTS)

Synonym(s)—*N,N*-1,2-ethanediylbis[*N*-(carboxymethyl)glycine]; ethylenediaminetetraacetic acid; EDTA; potassium, sodium, and calcium salts
CAS number—60-00-4
Class of compound—Polycarboxylic acid chelating agent
Empirical formula—$C_{10}H_{16}N_2O_8$, molecular weight 292.2
Stability—Aqueous solutions are stable to autoclaving
Compatibility—Incompatible with strong oxidizing agents, strong bases, and polyvalent metal ions; activity reduced in presence of divalent metal ions
PET inactivator—Dilution, with possible addition of calcium or magnesium salts
Typical in-use concentration—0.01 to 0.1%

14.6.14 ETHANOL

Synonym(s)—Ethyl alcohol; dehydrated alcohol; absolute alcohol; alcohol (95% ethanol)
CAS number—64-17-5
Class of compound—Alcohol
Empirical formula—C_2H_5OH, molecular weight 46.1
Stability—Aqueous solutions stable to autoclaving in closed containers, but special conditions are required
Compatibility—Used as a solvent or cosolvent in pharmaceutical preparations and as a basis for the production of tinctures; reduced antimicrobial activity in presence of nonionic surfactants
PET inactivator—Dilution
Typical in-use concentration—Rarely used solely for its preservative capability, which is only satisfactory at concentrations above 15 to 20% and is optimally bactericidal at 60 to 70%

14.6.15 IMIDUREA

Synonym(s)—*N,N*-methylenebis{*N*-(hydroxymethy)-2,5-dioxo-4-imidazolidinyl urea}; imidazolidinyl urea
CAS number—39236-46-9
Class of compound—Heterocyclic urea derivative

Empirical formula—$C_{11}H_{16}N_8O_8$, molecular weight 388.3

Stability—Hygroscopic; stable, but does release formaldehyde in water-based formulations

Compatibility—Compatible with many excipients including proteins, nonionic surfactants, and lecithin

PET inactivator: Dilution, possibly with lecithin–Tween 80–thioglycolate medium or glycine

Typical in-use concentration—0.03 to 0.5%

14.6.16 PARABENS

Synonym(s)—Esters of p-hydroxybenzoic (PHB) acid [methyl, ethyl, propyl, butyl, and benzyl (NIPA)], and their sodium salts; PHB esters; NIPA esters

CAS number—Methyl (99-76-3), ethyl (120-47-8), propyl (94-13-3), butyl (94-26-8), benzyl (94-18-8)

Class of compound—Benzoic acid ester

Empirical formula—$HO \cdot C_6H_5CO_2 \cdot R$, where R represents CH_3 (methyl ester, molecular weight 152.1), C_2H_5 (ethyl ester, molecular weight 166.2), C_3H_7 (propyl ester, molecular weight 180.2), C_4H_9 (butyl ester, molecular weight 194.2), or $CH_2 \cdot C_6H_5$ (benzyl ester, molecular weight 228.2); the general formula for the sodium salt is $Na \cdot O \cdot C_6H_5CO_2 \cdot R$ with a molecular weight increase of 22 over each respective ester

Stability—Stable; chemical stability decreases as pH increases, with significant hydrolysis taking place at strongly alkaline pH and elevated temperatures; acidic solutions can generally withstand autoclaving conditions; sensitive to excessive light exposure

Compatibility—Some reduction in activity seen with anionic agents, nonionic surfactants, methylcellulose, gelatin, povidone, and proteins; incompatible with alkalis and iron salts; some pigments adsorb parabens reducing their activity

PET inactivator—Dilution and nonionic surfactants such as Tween 80

Typical in-use concentration—Up to 0.4% single ester or 0.8% for a mixture of esters; generally 0.2% methyl paraben, 0.15% ethylparaben, 0.02% propyl- and butylparaben, and 0.006% benzylparaben

14.6.17 PHENOL

Synonym(s)—Carbolic acid; hydroxybenzene

CAS number—108-95-2

Class of compound—Phenolic

Empirical formula—$C_6H_5 \cdot OH$, molecular weight 94.1

Stability—Excessive light exposure in air will catalyze oxidation and lead to discoloration of solution; aqueous solutions are autoclavable, and oily solutions can be sterilized by 150°C dry heat treatment for 1 h

Compatibility—Incompatible with alkali salts, iron salts, and certain drugs; activity reduced by nonionic surfactants

PET inactivator—Dilution and nonionic surfactant such as Tween 80

Typical in-use concentration—0.25–0.5%

14.6.18 PHENOXYETHANOL

Synonym(s)—2-phenoxyethanol; phenoxetol; ethylene glycol monophenyl
  ether; β-phenoxyethyl alcohol
CAS number—122-99-6
Class of compound—Alcohol
Empirical formula—$C_8H_{10}O_2$, molecular weight 138.2
Stability—Stable; aqueous solutions sterilized by autoclaving
Compatibility—Compatible with anionic and cationic surfactants; reduced
  activity in presence of some nonionic agents
PET inactivator—Dilution and nonionic surfactant such as Tween 80
Typical in-use concentration—1.0%

14.6.19 PHENYLETHANOL

Synonym(s)—2-phenylethanol; β-phenylethyl alcohol; phenethyl alcohol;
  benzyl carbinol
CAS number—60-12-8
Class of compound—Alcohol
Empirical formula—$C_6H_5 \cdot CH_2 \cdot CH_2OH$, molecular weight 122.2
Stability—Poor stability with oxidants
Compatibility—Partially inactivated by nonionic surfactants
PET inactivator—Dilution and nonionic surfactant such as Tween 80
Typical in-use concentration—0.3 to 0.5%

14.6.20 PHENYLMERCURIC SALTS

Synonym(s)—As phenylmercuric acetate (PMA, acetoxyphenylmercury),
  phenylmercuric borate (PMB), and phenylmercuric nitrate (PMN) salts
CAS number—Acetate (62-38-4), borate (8017-88-7), nitrate (55-68-5)
Class of compound—Cationic organic mercurial
Empirical formula—$C_6H_5HgO \cdot CO \cdot CH_3$ (acetate salt, molecular weight 336.7),
  $C_6H_5Hg \cdot O \cdot B(OH)_2$ and $C_6H_5Hg \cdot OH$ (basic borate salt, molecular weight
  633.2), $C_6H_5 \cdot Hg \cdot NO_3$ and $C_6H_5 \cdot Hg \cdot OH$ (basic nitrate salt, molecular weight
  634.4)
Stability—Sensitive to excessive light and air exposure
Compatibility—Incompatible with halides, aluminium and other metals,
  ammonia and ammonium salts, sulphides, and thioglycollates; activity may
  be reduced in presence of anionic emulsifying and suspending agents; can
  adsorb to polyethylene and certain rubber components; compatible with
  some nonionic surfactants
PET inactivator—Sulphydryl compounds such as cysteine or thioglycollate
Typical in-use concentration—0.001 to 0.002%

14.6.21 PROPIONIC ACID (AND SALTS)

Synonym(s)—Ethylformic acid; methylacetic acid; propanoic acid

CAS number—79-09-4; 6700-17-0 (sodium salt)

Class of compound—Organic weak acid (or salt)

Empirical formula—$C_3H_6O_2$, molecular weight 74.1; $C_3H_5NaO_2 \cdot xH_2O$ (hydrated sodium salt), molecular weight of monohydrate 114.1

Stability—Deliquescent

Compatibility—Similar to benzoic acid. Incompatible with alkalis, ammonia, amines, and halogens; can be salted out from aqueous solution by addition of calcium chloride and some other salts

PET inactivator—Dilution and nonionic surfactant such as Tween 80

Typical in-use concentration—0.3%, up to 5% as an antifungal agent

14.6.22 PROPYLENE GLYCOL

Synonym(s)—1,2-dihydroxypropane; 2-hydroxypropanol; propane-1,2-diol; methyl glycol

CAS number—57-55-6

Class of compound—Alcohol

Empirical formula—$C_3H_8O_2$, molecular weight 76.1

Stability—Oxidizes in air at high temperatures; aqueous solutions may be sterilized by autoclaving; hygroscopic

Compatibility—Incompatible with oxidizing agents

PET inactivator—Dilution, possibly including a nonionic surfactant such as Tween 80

Typical in-use concentration—15 to 30%

14.6.23 SORBIC ACID

Synonym(s)—2,4-hexadienoic acid; 2-propenylacrylic acid; calcium or potassium sorbate salts

CAS number—22500-92-1

Class of compound—Organic weak acid

Empirical formula—$C_6H_8O_2$, molecular weight 112.1; potassium sorbate $C_6H_7KO_2$, molecular weight 150.2

Stability—Sensitive to light and air exposure; oxidation occurring most readily in aqueous solution; unstable in a variety of containers except when stored at refrigeration temperatures or in the presence of an antioxidant

Compatibility—Appears compatible with vegetable gums, proteins, and gelatin; only slightly incompatible with nonionic surfactants; incompatible with bases, oxidizing and reducing agents

PET inactivator—Dilution and nonionic surfactant such as Tween 80

Typical in-use concentration—0.05 to 0.2%

14.6.24 SULPHITES, INORGANIC

Synonym(s)—Inorganic sulphites, including sodium/potassium sulphite, and sodium-potassium metabisulphite (pyrosulphite)

CAS number: Sodium sulphite (7757-83-7), sodium metabisulphite (7681-57-4)

Class of compound—Inorganic acid

Empirical formula—Na_2SO_3 and K_2SO_3 (sodium and potassium sulphites, anhydrous molecular weights 126.1 and 158.3, respectively); NaS_2O_5 and $K_2S_2O_5$ (sodium and potassium metabisulphites, molecular weights 190.1 and 222.3, respectively)

Stability—Unstable in solution, decomposing in air especially on heating; autoclaving requires removal of air; dextrose reduces stability of the metabisulphite

Compatibility—Sodium metabisulphite usually employed in acid conditions where antimicrobial activity is encouraged by liberation of sulphur dioxide and sulphurous acid; sodium sulphite is incompatible with strong acids and is usually preferred for use in alkali preparations; inorganic sulphites are incompatible with oxidizing agents; sodium metabisulphite may interact with rubber

PET inactivator—Sulphydryl compounds such as cysteine and thioglycollate

Typical in-use concentration—0.1%

14.6.25 THIOMERSAL

Synonym(s)—Sodium (2-carboxyphenylthio)ethylmercury; sodium o-(ethylmercurithio)benzoate; sodium ethyl mercurithiosalicylate; thiomersalate; thimerosal; mercurothiolate

CAS number: 54-64-8

Class of compound—Anionic organic mercurial

Empirical formula—$C_9H_9HgNaO_2S$, molecular weight 404.8

Stability—Aqueous solutions are fairly stable to heat and can be sterilized by autoclaving, but are labile to light and less stable in alkaline conditions; traces of copper, iron, and zinc ions increase heat lability; stability reduced in sodium chloride solutions

Compatibility—Incompatible with strong acids and bases, iodine, heavy metal salts (including phenylmercuric compounds), quaternary ammonium compounds, lecithin, thioglycollate, and proteins; can adsorb to various types of rubbers and plastics, especially polyethylene; edetic acid and sodium metabisulphite may reduce preservative efficacy

PET inactivator—Sulphydryl compounds such as cysteine and thioglycollate or lecithin–Tween 80–thioglycollate medium

Typical in-use concentration—0.002 to 0.01%

REFERENCES

Allwood, M.C. (1973). Inhibition of *Staphylococcus aureus* by combinations of non-ionic surface-active agents and antibacterial substances. *Microbios,* 7, 209–214.

Allwood, M.C. (1978). Antimicrobial agents in single- and multi-dose injections. *J. Appl. Bateriol.,* 44, Svii–Sxvii.

Allwood, M.C. (1982). The effectiveness of preservatives in insulin injections. *Pharm. J.,* 229, 340.

Bean, H.S. (1972). Preservatives for pharmaceuticals. *J. Soc. Cosmet. Chem.,* 23, 703–720.

Bean, H.S., Richards, J.P., and Thomas, J. (1962). The bactericidal activity against *Escherichia coli* of phenol in oil-in-water dispersions. *Boll. Chim. Farm.,* 101, 339–346.

Bean, H.S., Heman-Ackah, S.M., and Thomas, J. (1965). The activity of antibacterials in two-phase systems. *J. Soc. Cosmet. Chem.,* 16, 15–30.

Bean, H.S., Konning, G.H., and Malcolm, S.A. (1969). A model for the influence of emulsion formulation on the activity of phenolic preservatives. *J. Pharm. Pharmacol.,* 21, 173S–180S.

Berry, H. and Michaels, I. (1950). The evaluation of the bactericidal activity of ethylene glycol and some of its monoalkyl ethers against *Bacterium coli. J. Pharm. Pharmacol.,* 2, 243–249.

Beveridge, E.G., Boyd, I., and Jessen, G.W. (1980). The action of 2-phenoxyethanol upon *Pseudomonas aeruginosa* NCTC 6749. *J. Pharm. Pharmacol.,* 32, 17P.

Bloomfield, S.F. (1996). Control of microbial contamination in cosmetics, toiletries and non-sterile pharmaceuticals. In *Microbial Quality Assurance in Cosmetics, Toiletries and Non-Sterile Pharmaceuticals.* Baird, R.M. and Bloomfield, S.F., Eds. 2nd ed. Taylor & Francis, London, pp. 3–8.

British Pharmacopoeia (1953). *British Pharmacopoeia.* Her Majesty's Stationery Office, London.

Bronaugh, R.L. and Maibach, H.I. (1984). Safety evaluation of cosmetic preservatives. In *Cosmetic and Drug Preservation: Principles and Practice.* Kabara, J.J., Ed. Cosmetic Science and Technology Series. Vol. 1. Marcel Dekker, New York, pp. 503–531.

Chapman, D.G. (1987). Preservatives available for use. In *Preservatives in the Food, Pharmaceutical and Environmental Industries.* Board, R.G., Allwood, M.C., and Banks, J.G., Eds. SAB Technical Series 22. Blackwell Scientific Publications, Oxford, pp. 177–195.

Chopra, I. (1990). Bacterial resistance to disinfectants, antiseptics and toxic metal ions. In *Mechanisms of Action of Chemical Biocides; Their Study and Exploitation.* Denyer, S.P. and Hugo, W.B., Eds. SAB Technical Series 27. Blackwell Scientific Publications, Oxford, pp. 45–64.

Croshaw, B. (1977). Preservatives for cosmetics and toiletries. *J. Soc. Cosmet. Chem.,* 28, 3–16.

Dempsey, G. (1996). The effect of container materials and multiple-phase formulation components on the activity of antimicrobial agents. In *Microbial Quality Assurance in Cosmetics, Toiletries and Non-Sterile Pharmaceuticals.* Baird, R.M. and Bloomfield, S.F., Eds. 2nd ed. Taylor & Francis, London, pp. 87–98.

Denyer, S.P. (1990). Mechanisms of action of biocides. *Int. Biodet.,* 26, 89–100.

Denyer, S.P. (1996). Development of preservative systems. In *Microbial Quality Assurance in Cosmetics, Toiletries and Non-Sterile Pharmaceuticals.* Baird, R.M. and Bloomfield, S.F., Eds. 2nd ed. Taylor & Francis, London, pp. 133–148.

Denyer, S.P. and Maillard, J.-Y. (2002). Cellular impermeability and uptake of biocides and antibiotics in Gram-negative bacteria. *J. Appl. Mic. Symp. Suppl.,* 92, 35S–45S.

Denyer, S.P. and Stewart, G.S.A.B. (1998). Mechanisms of action of disinfectants. *Int. Biodet. Biodeg.,* 41, 261–268.

Denyer, S.P., Hugo, W.B., and Harding, V.D. (1985). Synergy in preservative combinations. *Int. J Pharm.,* 25, 245–253.

Denyer, S.P. and Wallhaeusser, K.-H. (1990). Antimicrobial preservatives and their properties. In *Guide to Microbiological Control in Pharmaceuticals.* Denyer, S. and Baird, R., Eds. 1st ed. Ellis Horwood, Chichester, U.K., pp. 251–273.

English, D.J. (2006). Factors in selecting and testing preservatives in product formulations. In *Cosmetic and Drug Microbiology.* Orth, D.S., Kabara, J.J., Denyer, S.P., and Tan, S.K., Eds. Informa Healthcare, New York, pp. 57–108.

Garrett, E.R. (1966). A basic model for the evaluation and prediction of preservative action. *J. Pharm. Pharmacol.,* 18, 589–601.

Gilbert, P. (1988). Microbial resistance to preservative system. In *Microbial Quality Assurance in Pharmaceuticals, Cosmetics and Toiletries.* Bloomfield, S.F., Baird, R., Leak, R.E., and Leech, R., Eds. Ellis Horwood, Chichester, U.K., pp. 171–194.

Gilbert, P. and Wright, N. (1987). Non-plasmidic resistance towards preservatives of pharmaceutical products. In *Preservatives in the Food, Pharmaceutical and Environmental Industries,* Board, R.G., Allwood, M.C., and Banks, J.G.. Eds. SAB Technical Series 22. Blackwell Scientific Publications, Oxford, pp. 225–279.

Gilliland, D., Li Wan Po, A., and Scott, E. (1992). Kinetic evaluation of claimed synergistic paraben combinations using a factorial design. *J. Appl. Bact.,* 72, 258–261.

Goon, A., Leow, Y.H., and Goh, C.L. (2006). Safety and toxicological properties of preservatives. In *Cosmetic and Drug Microbiology.* Orth, D.S., Kabara, J.J., Denyer, S.P., and Tan, S.K., Eds. Informa Healthcare, New York, pp. 153–162.

Hart, J.R. (1984). Chelating agents as preservative potentiators. In *Cosmetic and Drug Preservation: Principles and Practice.* Karbara, J.J., Ed. Cosmetic Science and Technology Series. Vol. 1. Marcel Dekker, New York, pp. 323–337.

Hayden, J. (1996). Safety evaluation of preservatives. In *Microbial Quality Assurance in Cosmetics, Toiletries and Non-Sterile Pharmaceuticals.* Baird, R.M. and Bloomfield, S.F., Eds. 2nd ed. Taylor & Francis, London, pp. 175–183.

Heath, H.J. and Rock, C.O. (2000). A triclosan-resistant bacterial enzyme. *Nature,* 406, 145–146.

Heath, R.J., Yu, Y.T., Shapiro, M.A., Olson, E., and Rock, C.O. (1998). Broad spectrum antimicrobial biocides target the FabI component of fatty acid synthesis. *J. Biol. Chem.,* 273, 30316–30320.

Hodges, N.A. and Hanlon, W. (1990). Detection and measurement of combined biocide action. In *Mechanisms of Action of Chemical Biocides: Their Study and Exploitation.* Denyer, S.P. and Hugo, W.B., Eds. SAB Technical Series 27. Blackwell Scientific Publications, Oxford, pp. 297–310.

Hodges, N. and Hanlon, G. (2000). Antimicrobial preservative efficacy testing. In *Handbook of Microbiological Quality Control: Pharmaceuticals and Medical Devices.* Baird, R.M., Hodges, N.A., and Denyer, S.P., Eds. Taylor & Francis, London, pp. 168–189.

Hugo, W.B. (1988). The degradation of preservatives by microorganisms. In *Biodeterioration 7.* Houghton, D.R., Smith, R.N., and Eggins, H.O.W., Eds. Elsevier Applied Sciences, London, pp. 163–170.

Hugo, W.B. and Denyer, S.P. (1987). The concentration exponent of disinfectants and preservatives (biocides). In *Preservatives in the Food, Pharmaceutical and Environmental Industries.* Board, R.G., Allwood, M.C., and Banks, J.G., Eds. SAB Technical Series 22. Blackwell Scientific Publications, Oxford, pp. 281–291.

Hugo, W.B., Denyer, S.P., York, H.L., and Tucker, J.D. (1984). Preservative activity in diluted corticosteroid creams. *J. Hosp. Infect.,* 5, 329–333.

Kabara, J.J. (2006). Hurdle technology for cosmetic and drug preservation. In *Cosmetic and Drug Microbiology.* Orth, D.S., Kabara, J.J., Denyer, S.P., and Tan, S.K., Eds. Informa Healthcare, New York, pp. 163–184.

Karabit, M.S., Juneskans, O.T., and Lundgren, P. (1985). Studies on the evaluation of preservative efficacy. I. The determination of antimicrobial characteristics of phenol. *Acta Pharm. Suec.,* 22, 281–290.

Karabit, M.S., Juneskans, O.T., and Lundgren, P. (1986). Studies on the evaluation of preservative efficacy. II. The determination of antimicrobial characteristics of benzyl alcohol. *J. Clin. Hosp. Pharm.*, 11, 281–289.

Karabit, M.S., Juneskans, O.T., and Lundgren, P. (1988). Studies on the evaluation of preservative efficacy. III. The determination of antimicrobial characteristics of benzalkonium chloride. *Int. J. Pharm.*, 46, 141–147.

Karabit, M.S., Juneskans, O.T., and Lundgren, P. (1989). Factorial designs in the evaluation of preservative efficacy. *Int. J. Pharm.*, 56, 169–174.

Kazmi, S.J.A. and Mitchell, A.G. (1976). The interaction of preservative and non-ionic surfactant mixtures. *Can. J. Pharm. Sci.*, 11, 10–17.

Kazmi, S.J.A. and Mitchell, A.G. (1978a). Preservation of solubilized and emulsified systems. I. Correlation of mathematically predicted preservative availability with antimicrobial activity. *J. Pharm. Sci.*, 7, 1260–1265.

Kazmi, S.J.A. and Mitchell, A.G. (1978b). Preservation of solubilized and emulsified systems. II. Theoretical development of capacity and its role in antimicrobial activity of chlorocresol in cetamacrogol-stabilized systems. *J. Pharm. Sci.*, 67, 1266–1271.

Kemper, F.H., Luepke, N.P., and Umbach, W. (2000). *Blue List: Cosmetic Ingredients*. Editio Cantor, Aulendorf, Germany.

Kostenbauder, H.B. (1983). Physical factors influencing the activity of antimicrobial agents. In *Disinfection, Sterilization and Preservation*. Block, S.S., Ed. 3rd ed. Lea and Febiger, Philadelphia, pp. 811–828.

Lambert, R.J.W. and Lambert, R. (2003). A model for the efficacy of combined inhibitors. *J. Appl. Mic.*, 95, 734–743.

Lambert, R.J.W., Johnston, M.D., Hanlon, G.W., and Denyer, S.P. (2003). Theory of antimicrobial combinations: biocide mixtures-synergy or addition? *J. Appl. Mic.*, 94, 747–759.

Lambert, R.J.W., Hanlon, G.W., and Denyer, S.P. (2004). The synergistic effect of EDTA/antimicrobial combinations on *Pseudomonas aeruginosa*. *J. Appl. Mic.*, 96, 244–253.

Lautier, F. (1984). Dermal and ocular toxicity of antiseptics: methods for the appraisal of the safety of antiseptics. In *Cosmetic and Drug Preservation: Principles and Practice*. Karbara, J.J., Ed. Cosmetic Science and Technology Series. Vol. 1. Marcel Dekker, New York, pp. 483–501.

Lehmann, R.H. (1988). Synergism in disinfectant formulations. In *Industrial Biocides, Critical Reports on Applied Chemistry*. Payne, K.R., Ed. Vol. 23. Wiley, Chichester, U.K., pp. 68–90.

Mackie, M.A.L., Lyall, J., McBride, R.J., Murray, J.B., and Smith, G. (1986). Antimicrobial properties of some aromatic alcohols. *Pharm. Acta Helv.*, 61, 333–336.

Martindale. (2002). *The Extra Pharmacopoeia*. 33rd ed. Pharmaceutical Press, London.

Matthews, B.R. (2003). Preservation and preservative efficacy testing: European perspectives. *Eur. J. Parent. Pharm. Sci.*, 8, 99–107.

McBain, A.J. and Gilbert, P. (2001). Biocide tolerance and the harbingers of doom. *Int. Biodet. Biodeg.*, 47, 55–61.

McCarthy, T.J. (1984). Formulation factors affecting the activity of preservatives. In *Cosmetic and Drug Preservation: Principles and Practice*. Kabara, J.J., Ed. Cosmetic Science and Technology. Vol. 1. Marcel Dekker, New York, pp. 359–388.

McCarthy, T.J. and Ferreira, J.-H. (1990). Attempted measurement of the activity of selected preservative combinations. *J. Clin. Pharm. Therapeut.*, 15, 123–129.

Moore, K.E. and Stretton, R.J. (1981). The effect of pH, temperature and certain media constituents on the stability and activity of the preservative, Bronopol. *J. Appl. Bact.*, 51, 483–494.

Orth, D.S. (1997). Inactivation of preservatives by surfactants. In *Surfactants in Cosmetics.* Rieger, M.M. and Rhein, L.D., Eds. 2nd ed. Marcel Dekker, New York, pp. 583–603.

Paulus, W. (1993). *Microbicides for Protection of Materials.* Chapman & Hall, London.

Poole, K. (2004). Bacterial resistance: acquired resistance. In *Russell, Hugo & Ayliffe's Principles and Practice of Disinfection, Preservation and Sterilization.* Fraise, A.P., Lambert, P.A., and Maillard, J.-Y., Eds. 4th ed. Blackwell Publishing, Oxford, pp. 170–183.

Pons, J.-L., Bonnaveiro, N., Chevalier, J., and Crémieux, A. (1992). Evaluation of antimicorbial interactions between chlorhexidine, quaternary ammonium compounds, preservatives and excipients. *J. Appl. Bact.,* 73, 395–400.

Rowe, R.C., Sheskey, P.J., and Owen, S.C., Eds. (2004). *Pharmaceutical Excipients 2004.* Pharmaceutical Press and American Pharmaceutical Association, London and Washington, D.C.

Russell, A.D. (1997). Plasmids and bacterial resistance to biocides. *J. Appl. Mic.,* 83, 155–165.

Russell, A.D. (2004). Factors influencing the efficacy of antimicrobial agents: In *Russell, Hugo & Ayliffe's Principles and Practice of Disinfection, Preservation and Sterilisation.* Fraise, A.P., Lambert, P.A., and Maillard, J.-Y., Eds. 4th ed. Blackwell Publishing, Oxford, pp. 98–127.

Russell, A.D. and McDonnell, G. (2000). Concentration: a major factor in studying biocidal action. *J. Hosp. Infect.,* 44, 1–3.

Scalzo, M., Orlandi, C., Simonetti, N., and Cerreto, F. (1996). Study of interaction effects of polyacrylic acid polymers (carbopol 940) on antimicrobial activity of methyl parahydroxybenzoate against some Gram-negative, Gram-positive bacteria and yeast. *J. Pharm. Pharmac.,* 48, 1201–1205.

Schneider, F.H. (1984). Evaluation of chemical toxicology of chemicals. In *Cosmetic and Drug Preservation: Principles and Practice.* Kabara, J.J., Eds. Cosmetic Science and Technology. Vol. 1. Marcel Dekker, New York, pp. 533–558.

Stickler, D.J. (2004). Bacterial resistance: intrinsic resistance of Gram-negative bacteria. In *Russell, Hugo & Ayliffe's Principles and Practice of Disinfection, Preservation and Sterilization.* Fraise, A.P., Lambert, P.A., and Maillard, J.-Y., Eds. 4th ed. Blackwell Publishing, Oxford, pp. 154–169.

Wallhaeusser, K.-H. (1974). Antimicrobial preservatives in Europe: experience with preservatives used in pharmaceuticals and cosmetics. International symposium on preservatives in biological products. *Dev. Biol. Stand.,* 24, 9–28.

Wallhaeusser, K.-H. (1984). Appendix B. Antimicrobial preservatives used by the cosmetic industry. In *Cosmetic and Drug Preservation: Principles and Practice.* Kabara, J.J., Ed. Cosmetic Science and Technology Series. Vol. 1. Marcel Dekker, New York, pp. 605–745.

Wallhaeusser, K.-H. (1988). *Praxis der Sterilisation-Desinfektion-Konservierung.* 4th ed. Thieme, Stuttgart.

15 Preservative Stability and Efficacy: Formulation Influences

Hans van Doorne

CONTENTS

15.1 INTRODUCTION

Throughout their shelf life and use, pharmaceutical preparations must comply with compendial microbiological requirements with respect to microbiological quality and efficacy of antimicrobial preservation.

The conditions of manufacture and the microbiological quality of raw materials are responsible for the initial level of contamination (Fels et al. 1987). However, growth of contaminants may occur during storage, possibly leading to unacceptably high numbers of microorganisms. As water is an essential prerequisite for growth, water-containing products are especially susceptible to microbial multiplication.

Administration of drugs in the hospital is undertaken by health care professionals and it may be expected that hygienic precautions are observed. Nevertheless, microbial contamination has occurred, mainly due to poor hygienic practices (Patchell et al. 1994, Hamill et al. 1995, Reboli et al. 1996, Bach et al. 1997). For highly critical products, such as enteral nutrition solutions, observed levels of contamination up to 10^8 CFU.mL^{-1} have been ascribed to failing hygiene, lack of preservative capacity, and to unsuitable storage conditions (Fagerman et al. 1984, Patchell et al. 1994). At home, medicines are usually administered by untrained people, frequently the patient. The majority of the drugs will be nonsterile products, such as tablets and creams. However, there has been an increase in the use of parenteral medication at home; examples are epidural administration of opioids, intravenous antibiotic administration, and peritoneal dialysis. Here, too, there is always a risk of contamination, even of sterile products such as eye drops (Baird et al. 1979, Anders and Wiedemann 1985, Livingstone et al. 1998).

In-use contamination, either in the hospital or at home, can be significantly reduced by means of a properly designed primary package and through hygienic precautions during administration. New and better systems for drug administration have been introduced, such as tubes for creams, spray bottles for nasal preparations, and single-dose packages for ophthalmic preparations produced in form-fill-seal equipment. However, the mode of administration or the cost prevent the universal introduction of contamination-proof packaging for all nonsterile products, and the use of preservatives continues to play an important part in preventing high levels of contamination. New developments in pharmaceutical and medical technology may present new specific preservation challenges, including problems associated with the use of particular types of dosage forms. Examples are the relatively frequent infections with *Acanthamoeba* spp. observed in contact lens wearers (Bacon et al. 1993, Illingworth 1998, Schaumberg et al. 1998), the problems associated with the use of nebulizers (Barnes et al. 1987, Hamill et al. 1995, Reboli et al. 1996, Cobben et al. 1996), and long-term intrathecal infusion (Nitescu et al. 1992). New formulations such as liposomal products (Komatsu et al. 1986) or products with new excipients, such as cyclodextrins (see Section 15.2.6), may be difficult to preserve.

The choice of preservative and its concentration for a given pharmaceutical preparation are based on a number of factors. The type of preparation, the route of administration, and the required efficacy of the preservative system are the primary criteria on which such a choice is based (see also Chapter 13). Other aspects to be considered during formulation are:

- Allergic reactions and other unwanted side effects—Owing to the wide-spread use of preservatives in pharmaceuticals, foodstuffs, and cosmetics, such reactions are frequently observed

- Resistance—The emergence of microorganisms resistant to common preservatives is a serious problem to all involved in the formulation of pharmaceuticals

Despite many apparent drawbacks, the use of chemical preservatives remains for the time being the only efficient way to guarantee the microbiological quality of pharmaceutical preparations throughout shelf life and use. It is, therefore, important to understand the factors that affect the activity of antimicrobial agents (see also Chapter 14) and to know the possible interactions between preservatives and other components of pharmaceutical formulations. In this chapter, the most important interactions will be discussed.

15.2 PHYSICOCHEMICAL EFFECTS

15.2.1 The Effect of Concentration

The activity of a preservative depends on the free concentration of the active form of the molecule in the aqueous phase. Partitioning between water and oil phase, micellar solubilization, and dissociation may all decrease the efficacy of a preservative. Much work has been done on the preservation of emulsified systems and various mathematical approaches have been proposed to predict the total amount required to give a known concentration of free preservative in the aqueous phase. One frequently used formula is:

$$C_w = \frac{C(\theta+1)}{K_w^o \theta + R}$$

(15.1)

where C_w is the concentration of free preservative in the aqueous phase, C is the total concentration of preservative in the emulsion, θ is the volume ratio of oil to water, R is the ratio of total to free preservative in the aqueous phase (influenced by the emulsifier-preservative interaction), and K_w^o is the oil-water partition coefficient of the preservative. Equation 15.1, which was derived for simple emulsions containing only one oil and one emulsifier, was found also to be applicable to modern emulsions comprising more than one oil and surfactant (Kurup et al. 1991a).

The effect of concentration on the antimicrobial activity of a preservative is given by the following formula:

$$C^\eta t = k$$

(15.2)

where C is the concentration, η is the concentration exponent, t is the time to achieve a certain effect (e.g., one decimal reduction), and k is a constant.

The concentration exponent depends on the type of preservative and sometimes on the type of organism (Beveridge et al. 1980; Karabit et al. 1985, 1986, 1988; Mackie et al. 1986; Hugo and Denyer 1987). Observed values range from 1 for phenylmercuric salts and cetrimide to 2.5 for the parabens and up to 9 for phenoxyethanol (see also Chapter 14). Benzalkonium chloride is an example of a compound that shows different concentration exponents for bacteria (3.5), yeasts (1.8), and molds (9).

When discussing interactions between preservatives and other excipients, the importance of the concentration exponent should be acknowledged. A low concentration exponent of 1 means that the activity changes linearly with the concentration; however, a high concentration exponent of 9 means that a twofold decrease in available concentration (perhaps due to interaction with a different component of the product) will invoke an approximately 500-fold decrease in activity (see Chapter 14).

Numerous combinations of preservatives are claimed to be synergistic (see Section 15.2.7); however, the observed increase in activity sometimes may be attributable to an increase in the overall preservative concentration of compounds with similar modes of action having high concentration exponents (McCarthy and Ferreira 1990). Gilliland et al. (1992b) observed that combinations of methyl- and propylparaben were bactericidal, whereas only a bacteriostatic action was noted when the agents were used singly. This synergistic effect was attributed to a sigmoidal dose–response relationship.

15.2.2 The Effect of pH

The choice of pH of a preparation is based not only on solubility and stability of the therapeutically active components but also on the tolerance of body tissues that come into contact with the preparation. The pH of the formulation may affect the efficacy of the preservative system in a number of ways:

- pH affects the activity of preservatives, for example, sorbic acid is only active as the undissociated molecule (van Doorne and Dubois 1980, Eklund 1983).
- pH affects the oil–water partition coefficient of the preservative, the micellar solubilization, and the interaction with cyclodextrins, because the undissociated form is more hydrophobic than the dissociated form.
- Some preservative-component interactions may be pH-dependent because of ionization effects on components.

A more detailed discussion of the effect of pH on preservative activity is given in Chapter 14; it is evident that the free aqueous concentration of the preservative, and thereby the activity, may be affected by the pH.

15.2.3 Adsorption by Solids

Liquid or semisolid pharmaceutical preparations may contain varying amounts of solids, ranging from less than 1 to 50%. In addition to solid material, suspensions contain thickening agents such as natural clays (veegum, bentonite) or semisynthetic compounds such as modified celluloses, or polyvinylpyrrolidone (PVP). Solids in suspensions are used as therapeutically active components (e.g., suspensions of antibiotics or corticosteroids), suspending agents (e.g., bentonite and veegum), or coloring agents (e.g., calamine).

It is difficult to protect solid–liquid dispersions against microbial attack. A survey (Vanhaecke et al. 1987) has shown that 8 out of 12 marketed antacid suspensions failed to meet the *United States Pharmacopeia* (USP) *21* (1985) criteria for preservative activity. Although the main cause of the failures was the incompatibility of pH with the chosen preservative, interaction with the solid components could not be excluded. In a draft revision for the preservative efficacy test, the microbiology subcommittee of the USP suggested a special category for antacids with less stringent requirements than for other oral preparations (Anonymous 1995). Although the suggestion was never adopted, it illustrates the problems frequently encountered with this type of preparation.

The binding of preservatives by solid particles has been studied by many workers. Pioneer investigations were conducted by Batuyios and Brecht (1957), who studied the adsorption of quaternary ammonium compounds by talc and kaolin. In the years that followed, numerous papers were published (McCarthy 1969; Bean and Dempsey 1971; Yousef et al. 1973a, 1973b; Myburgh and McCarthy 1980a, 1980b). For a review of the early literature, the reader is referred to McCarthy and Myburgh (1977).

The available information clearly shows that binding almost invariably results in a decreased concentration of free preservative and hence a decrease in activity of the preservative. As a general rule, the ionic preservatives (e.g., quaternary ammonium compounds, chlorhexidine) are strongly bound, whereas the nonionic compounds such as the parabens are usually not.

15.2.4 INTERACTION WITH GEL-FORMING AND SUSPENDING AGENTS

Suspending agents are natural clays and hydrocolloids, such as bentonite, veegum, gelatine, and tragacanth, or semisynthetic thickeners, such as modified celluloses. Their main effect is to increase viscosity, thereby reducing settling velocities of particles.

McCarthy and Myburgh (1974) studied the interaction between tragacanth and eight different preservatives. Only chlorhexidine was found to be absorbed to any extent (33%). The interaction between benzalkonium chloride and some suspending agents including tragacanth, bentonite, veegum, and sodium alginate was studied by Yousef et al. (1973b). In general, a decrease in preservative activity was observed. According to the extent of the antagonism, they classified the materials into four categories:

1. Highly antagonistic (attapulgite, bentonite, veegum, and magnesium trisilicate)
2. Moderately antagonistic (kaolin, tragacanth, and sodium alginate)
3. Slightly antagonistic (talc, acacia, starch, calamine, sodium carboxymethylcellulose, and methylcellulose)
4. Nonantagonistic (polyethylene glycols)

Kurup et al. (1992) studied the interaction of methylparaben with varying concentrations of natural hydrocolloids, such as acacia, tragacanth, sodium alginate, guar gum, and carrageenan. Tragacanth and guar gum inhibited the activity to a greater extent than the other three excipients. Reduction in the availability of the preservative appeared to be the predominant mechanism by which tragacanth and guar gum reduced the activity of methylparaben. Acacia, sodium alginate, and carrageenan apparently acted by offering physical protection to the microbial cell. Miyawaki et al. (1959), using an equilibrium dialysis method, did not observe a significant interaction between methyl- or propylparaben and tragacanth.

The effect of starch on the efficacy of preservatives has also been investigated. Mansour and Guth (1968) studied the complexation of benzoic acid and sorbic acid by starch. They concluded that the amylose portion of the molecule, rather than the amylopectin part, was involved in complexation with the preservatives. McCarthy (1969) observed that sorbic acid was adsorbed only to a limited extent, whereas, chlorhexidine was almost completely (97%) adsorbed.

Semisynthetic thickeners, such as the modified celluloses, may also decrease the free aqueous concentration of preservatives. The nature of the interaction has not been studied in detail and conflicting results have been reported. Thus, DeLuca and Kostenbauder (1960) reported that benzalkonium chloride was not antagonized by methylcellulose, whereas Yousef et al. (1973a) observed a marked antagonism. Tromp et al. (1976) reported a 7% loss of benzalkonium chloride in the presence of hydroxypropylmethyl cellulose (HPMC) and Myburgh and McCarthy (1980b) observed a decreased activity of cetylpyridinium chloride in the presence of hypromellose. Richards (1976) pointed out that the reaction between HPMC and benzalkonium chloride occurred during sterilization. More recently, the activity of chlorocresol, methylparaben, and phenoxyethanol against *P. aeruginosa* in the presence of cellulose derivatives was studied by Kurup et al. (1995). Methylcellulose, sodium carboxymethylcellulose, and HPMC reduced the activities of the preservatives to varying degrees. Dependent on the preservative–cellulose derivative combination, interaction between preservative and cellulose derivative or physical shielding was the predominant mechanism for the reduction in activity. However, earlier Miyawaki et al. (1959) concluded from their studies that there was no evidence for a significant degree of interaction between parabens and carboxymethylcellulose. Scalzo et al. (1996) studied the interaction between polyacrylic acid polymers (Carbopol® 940) and methylparaben using four test organisms. For the Gram-negative bacteria *P. aeruginosa* and *Escherichia coli,* a synergistic effect was observed. This effect was attributed to the chelation of Ca^{2+} ions in the outer membrane of the cell. The synergy could be partially reversed by the addition of calcium chloride. For *Staphylococcus aureus* and *Candida albicans,* an antagonism was noted. This was attributed to an interaction between the preservative and the polymer and the absence of an outer membrane with Ca^{2+} ions in these organisms. These authors confirmed results obtained earlier by Orth et al. (1989), who reported a synergism between methylparaben and three other acrylic acid polymers.

15.2.5 Interaction with Emulsifiers

Many chemical preservatives are lipophilic compounds (e.g., parabens), or compounds containing a substantial lipophilic part in the molecule (e.g., quaternary ammonium compounds). Exceptions are bronopol and phenylmercuric compounds. The lipophilic nature of such compounds makes them liable to micellar solubilization by emulsifiers. Interaction between antimicrobials and nonionic emulsifiers was first reported by Bolle and Mirimanof (1950). The interaction between parahydroxybenzoic acid esters and Tween 80 was studied by Patel and Kostenbauder (1958) and Pisano and Kostenbauder (1959). These authors studied the binding of the parabens by Tween 80 by means of an equilibrium dialysis method and were able to correlate the binding data to preservative efficacy. Brown and Richards (1964) reported that the activity of benzalkonium chloride, and to a lesser extent of chlorhexidine, was increased by a low (0.02%) concentration of Tween 80, but that a higher concentration (0.5%) completely eliminated the inhibitory effect of both preservatives. Similar results were claimed by Kurup et al. (1991b), who studied the effect of different concentrations of Tween 80 on methylparaben, phenoxyethanol, and chlorocresol against *P. aeruginosa*. At concentrations below the critical micelle concentration, the bactericidal effect of these preservatives increased with decreasing surface tension. However, the effects were only small. Methylparaben can be solubilized by sodium lauryl sulphate when the latter compound is present in a concentration above the critical micelle concentration (0.2%) (van Doorne 1977). Sodium lauryl sulphate itself has an antimicrobial effect, particularly against Gram-positive microorganisms (Birkeland and Steinhause 1939), which might explain the potentiation of the activity of some preservatives (including methylparaben), particularly against *S. aureus,* observed by Jund and Carrère (1971).

Kazmi and Mitchell (1978a) investigated mathematical models for the distribution and antimicrobial activity of chlorocresol in solubilized and emulsified systems, stabilized with a nonionic emulsifier. The models could predict adequately the free aqueous concentration of preservative. It was confirmed that the antimicrobial effect was related to the free aqueous concentration, and that the preservative solubilized in the surfactant or partitioned into the oil phase did not contribute to the antimicrobial effect (see Section 15.2.1). However, the micelles and the oil containing the preservative can act as a reservoir for the antimicrobial agent when the aqueous concentration has decreased through chemical or physical instability (Kazmi and Mitchell 1987b).

In conclusion, the interaction between a preservative and emulsifying agents may cause an increase in activity at concentrations below the critical micelle concentration of the emulsifier. Above the critical micelle concentration, there is a decrease in the free aqueous concentration of the preservative and hence a decrease in antimicrobial activity.

15.2.6 INTERACTION WITH MISCELLANEOUS COMPOUNDS

Cyclodextrins, a group of compounds that were introduced recently as excipients for pharmaceutical preparations, interfere with the activity of preservatives by a well-understood mechanism. The doughnut-shaped molecules have a hydrophobic cavity, and thus can form inclusion complexes with numerous pharmaceutically relevant compounds. Complexation of preservatives and antibiotics results in a (partial) loss of antimicrobial activity (Loftsson et al. 1992, van Doorne 1993, Lehner et al. 1994).

Another compound that interacts with preservatives by means of a well-understood mechanism is ethylenediaminetetraacetic acid (EDTA). EDTA enhances the activity of preservatives such as chlorhexidine (Richards and McBride 1972) and quaternary ammonium compounds (Richards 1971, Clausen 1973). The Gram-negative bacterial outer membrane-peptidoglycan complex is stabilized by divalent action (Ca^{2+} and Mg^{2+}) bridges. Owing to the chelating action of EDTA, the divalent ions are removed from the complex, which causes destabilization of the outer membrane and loss of lipopolysaccharides. This results in an easier access to the inner part of the cell (e.g., cytoplasmic membrane) of molecules that are otherwise excluded (Gilbert and Das 1996). Combinations of EDTA with preservatives such as the quaternary ammonium compounds are frequently used in ophthalmic preparations and contact lens solutions.

Propylene glycol is particularly useful in creams. In addition to its synergistic activity (see Section 15.2.7), propylene glycol increases the aqueous solubility of lipophilic preservatives such as the parabens; as a consequence the oil–water coefficient is reduced, leading to an increased concentration in the aqueous phase (van Doorne and Dubois 1980).

15.2.7 INTERACTION BETWEEN PRESERVATIVES

When satisfactory preservation cannot be achieved by means of a single compound, a combination of two or more preservatives is necessary. If the effect of the combination is greater than the sum of the effects of the individual compounds, the combination is said to be synergistic. However, as explained in Section 15.2.1, if the two preservatives have similar modes of action and their concentration exponents are relatively high, the apparent increase in concentration may fully account for the observed increase in activity. Advantages of the use of synergistic combinations are the wider antimicrobial spectrum and fewer undesired side effects because the individual compounds may be used at lower concentrations. Denyer et al. (1985) listed 45 synergistic combinations, many of which have been used in commercial products.

15.3 THE EFFECT OF TEMPERATURE ON PRESERVATIVE EFFICACY AND AVAILABILITY

The microbiological stability of pharmaceuticals is influenced by temperature through at least three different, unrelated mechanisms:

1. Temperature affects the metabolic state of microorganisms.
2. An increase in temperature enhances the activity of a preservative.
3. Temperature affects the interaction of preservatives with other formulation ingredients.

15.3.1 METABOLIC STATE

Growth of microorganisms in any medium, including pharmaceutical or cosmetic preparations, depends on temperature. The optimum temperature of most pathogenic species is about 37°C. Incubation temperatures used for environmental monitoring and nonselective microbiological examination are usually lower (about 30°C) because at this temperature higher recoveries are obtained. Pharmaceutical preparations are usually stored at room temperature (about 20°C) or in a refrigerator (about 4°C). Only preparations containing extremely unstable compounds are stored in a freezer at 20°C. During their shelf life, therefore, most products are kept at temperatures that, although possibly selective toward certain types of microorganism, do not preclude microbial growth.

15.3.2 PRESERVATIVE ACTIVITY

The activity of a preservative generally increases with increasing temperature. The temperature dependence is frequently expressed by means of the Q_{10} value, which is defined as:

$$Q_{10} = \frac{\text{Activity at Temperature } (T + 10)}{\text{Activity at Temperature } T} \qquad (15.3)$$

The Q_{10} value gives an indication of the extent to which the activity of a preservative is changed by a 10°C change in temperature (see Chapter 14 for some examples). It is of limited predictive use, however, because the values are only constant over a limited range of temperatures (Karabit et al. 1985) and even for one compound they may be different for different microorganisms (Karabit et al. 1986). Gilliland et al. (1992b) demonstrated that the Arrhenius equation provided a good mathematical model to describe the relationship between temperature and the antimicrobial activity of methyl- and propylparaben.

15.3.3 INTERACTION WITH INGREDIENTS

All the interactions discussed in the previous sections are dependent on temperature, thus:

- Adsorption of preservatives by solid ingredients decreases with increasing temperature.
- Micellar solubilization is dependent on temperature. As compared to a true molecular dispersion, the solubilized system is thermodynamically less favorable. Therefore an increase in temperature will result in a decrease in the extent of solubilization.
- The oil–water partition coefficient, K_w^o, is dependent on temperature. Between 20°C and 70°C, an increase in the K_w^o of methylparaben between water and Cetiol V (a semisynthetic oil) was observed (van Doorne and Dubois 1980).

Owing to the complexity of these effects, it is impossible to predict how the activity of a preservative in a specific preparation will change with changing temperature. The test for preservative efficacy as described in the *European Pharmacopoeia*

(2005) is carried out at ambient temperature. For products that are to be stored under refrigeration conditions (such as some injection solutions or eye drops), it is therefore advisable to run additional tests under the intended storage conditions.

15.4 EFFECT OF WATER ON PRESERVATIVE EFFICACY AND AVAILABILITY

15.4.1 WATER AS AN INTRINSIC FACTOR

Lowering the water content has always been a convenient method of protecting foods from microbial spoilage. In the late 1930s, it was recognized that the water content per se was not the governing factor in this context. The concept of water activity allowed a more quantitative approach to the influence of water on microbial proliferation. The water activity (A_w value) is defined as the ratio between water vapor pressure of a preparation and the vapor pressure of pure water at the same temperature. Protection against microbial spoilage can thus be obtained by the addition of high concentrations of sugars or polyalcohols, which lower A_w values. Although microorganisms cannot develop resistance against reduced water activity (Curry 1985), its application as a means of preserving pharmaceutical preparations is only limited. Even syrups, which contain high concentrations of saccharose (>60%), are frequently preserved with methylparaben. The reason is that during production a layer of condensed water sometimes forms on the surface. In this film, growth of microorganisms can occur, and this is inhibited by methylparaben. Shihab et al. (1988) reported that the solubility of sorbic acid was reduced in the presence of high concentrations of different sugars. In the presence of 75% w/v sucrose, glucose, or sorbitol, the solubility of the preservative was reduced from 1.5 mg.mL^{-1} to about 0.5 mg.mL^{-1} but the effect on antimicrobial activity was not studied. The addition of 20% w/v sorbitol had no detectable effect on an oral caffeine formulation preserved with sorbic acid (Barnes et al. 1994).

15.4.2 WATER AS AN EXTRINSIC FACTOR

Although dry oral preparations do not rank high as microbial hazards, an outbreak of infection reported by Kallings et al. (1966) illustrated that tablets may contain dangerously high levels of bacteria. The presence of these organisms was ascribed to the use of contaminated raw materials, rather than to growth during storage. Raw materials, particularly those of natural origin, may contain high numbers of bacteria, including Enterobacteriaceae (see Chapter 3). In a recent survey on the microbiological quality of herbal products (Kabelitz 1997), it was reported that more than 95% of the investigated senna fruit samples contained more than 10^4 Enterobacteriaceae per gram. Even dry products may thus contain relatively high numbers of Gram-negative bacteria, which are generally considered to be sensitive to drying. As long as tablets are stored under dry conditions, spoilage due to growth is unlikely to occur (Flatan et al. 1996). However, in tropical regions with a hot and humid climate [31°C and 75 to 100% relative humidity (RH)], growth of microorganisms cannot be excluded, particularly when the products are stored and dispensed in a nonprotective packaging or even without any packaging. Only a few studies have been published on the microbiological stability of tablets (Blair et al. 1987). Parker (1984) suggested that the addition of preservatives to tablets could be effective. The effect of added preservatives on the microbiological stability of tablets stored under different conditions has been studied (Bos et al. 1989). No microbial growth was found on tablets stored at 75% RH, and the addition of preservatives (methylparaben or sorbic acid) yielded no measurable effect on the viability of the test organisms. However, when tablets were stored at 95% RH, visible growth of fungi was observed after four weeks of storage on tablets without preservative. Sorbic acid (1% w/w) and methylparaben (1% w/w) added to the tablet formulations were found to be fungicidal under these conditions of storage.

The need to add preservatives to ointments has been discussed (Anonymous 1996, Anonymous 1998). Even though USP 28 (2005) requires the antimicrobial effectiveness of all products including nonaqueous preparations to be tested, the *European Pharmacopoeia* (2005) considers such a test relevant for aqueous products only. The current test has not been developed for nonaqueous products and its application may lead to erratic results. Van Doorne et al. (1998) described a test to demonstrate whether a product blocked the penetration of microorganisms and was thus protected against microbial attack. Such a test could be used as an alternative to the classical preservative effectiveness test. The reader is referred to the companion *Handbook* (Hodges and Hanlon 2000) for a more detailed discussion of preservative efficacy testing.

15.5 INTRINSIC EFFECTS

A number of components of pharmaceutical preparations exert measurable antimicrobial effects, although they are used for reasons other than their antimicrobial activity. Some well-known examples are discussed briefly because they may form part of a complete preservative system (see Chapter 13).

15.5.1 REDUCING AGENTS

Lowering the redox potential reduces the growth rate of aerobic and facultatively anaerobic microorganisms. In pharmaceuticals, packaging under oxygen-free conditions or the addition of reducing compounds is routinely undertaken to protect the active compounds against oxidative degradation, but not to inhibit growth. Sodium metabisulphite, which is used in ophthalmic preparations and some injections, has a weak bactericidal effect on *P. aeruginosa* and *S. aureus* (Richards and Reary 1972). Butylated hydroxyanisole is added to fats in order to prevent rancidity. It has been shown to have some antimicrobial activity (Lamikanra 1982). Ascorbic acid, another well-known antioxidant, has only a very weak antimicrobial effect.

15.5.2 ESSENTIAL OILS

The essential oils of many plants such as lemon grass, thyme, marjoram, nutmeg, and bergamot have significant antimicrobial activity against bacteria, yeasts, and molds, and their possible use as preservatives has been reviewed by Kabara (1984). Essential oils are, however, unsuitable as preservatives for pharmaceutical products for the following reasons:

- They are mixtures of different compounds and their composition may change with the source of the oil.
- Many of the components are unsaturated compounds, which easily form peroxides. These peroxides may cause deterioration of other components of the product to be preserved.
- Essential oils are used in aromatherapy because of their alleged pharmacological effect (Lis-Balchin 1997); against this background, therefore, they *may* have undesired side effects.

15.5.3 CHELATING AGENTS

As mentioned in Section 15.2.6, EDTA is frequently employed in antimicrobial combinations to enhance the activity of preservatives such as the quaternary ammonium compounds. In addition to this potentiating effect, EDTA has weak antimicrobial activity itself.

15.5.4 ANTIBIOTICS

The presence of antibiotics and related compounds in pharmaceutical preparations will certainly affect the contaminant's ability to survive and grow. Antibiotics should not be used as preservatives, because of the danger of (cross-) resistance. Owing to their limited spectrum of activity, many antibiotic preparations such as eye drops contain a preservative. During formulation studies, it must be remembered that occasionally antagonism between an antibiotic and a preservative may be observed.

15.5.5 ALCOHOLS

Some alcohols, such as benzyl alcohol and phenylethanol, are used as preservatives. Alcohols such as ethanol and propylene glycol are also used as solvents or cosolvents. Homeopathic and phytotherapeutic preparations, in addition to many cosmetics, frequently contain high levels of ethanol. Any product containing more than 20% ethanol can be considered to be adequately preserved. Propylene glycol is an alcohol that is frequently used in creams and ointments. It is known to enhance the activity of the parabens (Prickett et al. 1961) by increasing their aqueous solubility, thereby reducing the oil–water partition coefficient; in a concentration over 15%, it has a significant antimicrobial activity (van Doorne and Dubois 1980).

15.5.6 MISCELLANEOUS COMPOUNDS

Systematic studies on the antimicrobial activity of pharmaceutical compounds have not been performed, and only some incidentally observed effects have been reported.

The antimicrobial activity of sodium lauryl sulphate has already been discussed in Section 15.2.5. Local anaesthetics such as tetracaine, benoxinate, and cocaine were found to be lethal for *C. albicans, P. aeruginosa,* and *S. epidermidis* (Kleinfeld and Ellis 1967). Unpreserved commercial preparations of methohexital sodium and sodium thiopental were found to be toxic for different strains of microorganisms (Highsmith et al. 1982). It is, however, uncertain whether the observed effects should be ascribed to toxic action of the compounds per se or whether the high pH (10.5) of the preparation was responsible. Chakrabarty et al. (1989) reported that the antihistamine compound promethazine possessed antimicrobial activity against many aerobic and anaerobic Gram-positive and Gram-negative bacteria.

Many antineoplastic drugs such as doxorubicin are, in fact, antibiotics, but are never used in antimicrobial therapy. Radiopharmaceuticals also have an antimicrobial effect. Stathis et al. (1983), studying the survival of *S. epidermidis* in solutions containing various levels of ^{99m}Tc, concluded than an inoculum of 100 CFU.mL^{-1} was killed after having received an estimated dose of about 10 Gy. The bactericidal effect could be ascribed to radiation damage and not to toxic effects of the pertechnate itself.

15.6 OVERVIEW

In the previous sections, many physicochemical effects and interactions between preservatives and components of pharmaceutical preparations have been discussed. The physicochemical (sorption, partition, or solubilization) or biological (complexation of Ca^{2+} in the outer membrane) basis of these effects is often quite well understood. These theoretical considerations, however, allow only a qualitative interpretation of observed phenomena and are usually not sufficient to predict whether any product will be adequately preserved. Moreover, occasionally unexpected and unpredictable phenomena may occur, because of a unique characteristic (resistance) of a certain microorganism. The current pharmacopoeial preservative efficacy tests are designed to establish preservative effectiveness specifications for the formulation at the development stage of the product (Cooper 1996). Even if such tests are performed under conditions that guarantee maximum reproducibility, a marked degree of variation in microbial inactivation may be observed (Hodges et al. 1996). Interlaboratory variations have not been published, but may well be significant. These limitations do not make such tests redundant. Their results must be used in combination with the results of microbiological examination of used preparations. Adequate preservation in combination with well-designed packaging and clear instructions for use will guarantee the microbiological purity of pharmaceutical preparations at the point of administration.

REFERENCES

Anders, B. and Wiedemann, B. (1985). Mikrobiologische Kontamination gebrauchter Augentropfen. *Pharmaz. Zeit.,* 130, 1648–1655.

Anonymous. (1995). Antimicrobial effectiveness testing. *Pharmacopeial Forum,* 21, 1040–1046.

Anonymous. (1996). Preservation methodology. Proceedings of international harmonisation of sterility tests and efficacy of antimicrobial preservation. Barcelona 1996. *Pharmeuropa,* Special Issue, October 1996.

Anonymous. (1998). *Proceedings of the USP Open Conference on Microbiology for the 21st Century,* May 3–5, New Orleans. United States Pharmacopeial Convention Inc., Rockville, MD, p. 49.

Bach, A., et al. (1997). In-use contamination of propofol. A clinical study. *Europ. J. Anaesthes.,* 14, 178–183.

Bacon, A.S. et al. (1993). A review of 72 consecutive cases of *Acanthamoeba keratitis,* 1984–1992, *Eye,* 7, 719–183.

Baird, R.M., Crowden, C.A., O'Farrell, S.M., and Shooter, R.A. (1979). Microbial contamination of pharmaceutical products in the home. *J. Hyg. Camb.,* 83, 277–283.

Barnes, A.R., Hebron, B.S., and Smith, J. (1994). Stability of caffeine oral formulations for neonatal use. *J. Clin. Pharm. Therapeut.,* 19, 391–396.

Barnes, K.L. et al. (1987). Bacterial contamination of home nebulizers. *Brit. Med. J.,* 295, 812.

Bean, H.S. and Dempsey, G. (1971). The effect of suspensions on the bactericidal activity of m-cresol and benzalkonium chloride. *J. Pharm. Pharmacol.,* 23, 699–704.

Batuyios, N.H. and Brecht, E.A. (1957). An investigation of the incompatibilities of quaternary ammonium germicides in compressed troches. I. The adsorption of cetylpyridinium chloride and benzalkonium chloride by talc and kaolin. *J. Am. Pharm. Assoc. Sci. Ed.,* 46, 490–492.

Beveridge, E.G., Boyd, I., and Jessen, G.W. (1980). The action of 2-phenoxyethanol upon *Pseudomonas aeruginosa* NCTC 6749. *J. Pharm. Pharmacol.,* 32, 17P.

Birkeland, J. and Steinhaus, E. (1939). Selective bacteriostatic action of sodium lauryl sulphate and of "Dreft." *Proc. Soc. Exp. Biol. Med.,* 40, 86–92.

Blair, T.C., Buckton, G., and Bloomfield, S.F. (1987). Water available to *Enterobacter cloacae* contaminating tablets stored at high relative humidities. *J. Pharm. Pharmacol.,* 39, 125P.

Bolle, A. and Mirimanoff, A. (1950). Antagonism between non-ionic detergents and antiseptics. *J. Pharm. Pharmacol.,* 2, 685–692.

Bos, C.E., van Doorne, H., and Lerk, C.F. (1989). Microbiological stability of tablets stored under tropical conditions. *Int. J. Pharm.,* 55, 175–182.

Brown, M.R.W. and Richards, R.M.E. (1964). Effect of polysorbate (Tween) 80 on the resistance of *Pseudomonas aeruginosa* to chemical inactivation. *J. Pharm. Pharmacol.,* 16, 51T–55T.

Chakrabarty, A.N., Acharya, D.P., Neogi, D., and Dastidar, S.G. (1989). Drug interaction of promethazine and other non-conventional antimicrobial chemotherapeutic agents. *Ind. J. Med. Res.,* 89, 233–237.

Clausen, O.G. (1973). An examination of the bactericidal and fungicidal effects of cetylpyridinium chloride, separately and in combinations with EDTA and benzyl alcohol. *Pharmaz. Ind.,* 35, 869–674.

Cobben, N.A et al. (1996). Outbreak of severe *Pseudomonas aeruginosa* respiratory infections due to contaminated nebulizers. *J. Hosp. Infect.,* 33, 63–70.

Cooper, M.S. (1996). Preservatives and preservative testing. *Microbiological Update,* 14, 1–4.

Curry, J. (1985). Water activity and preservation. *Cosmet. Toilet.,* 100, 53–55.

DeLuca, P.P. and Kostenbauder, H.B. (1960). Interaction of preservatives with macromolecules. IV. Binding of quaternary ammonium compounds by non-ionic agents. *J. Am. Pharm. Assoc. Sci. Ed.,* 49, 430–437.

Denyer, S.P., Hugo, W.B., and Harding, V.D. (1985). Synergy in preservative combinations. *Int. J. Pharm.,* 25, 245–253

Eklund, T. (1983). The antimicrobial effect of dissociated and undissociated sorbic acid at different pH levels. *J. Appl. Bact.,* 54, 383–389.

European Pharmacopoeia. (2005). *European Pharmacopoeia.* 5th ed. Council of Europe, Strasbourg, pp. 447–449.

Fagerman, K.E., Paauw, J.D., McCamish, M.A., and Dean, R.E. (1984). Effects of time, temperature, and preservative on bacterial growth in enteral nutrient solutions. *Am. J. Hosp. Pharm.,* 41, 631–637.

Fels, P., Gay, M., and Urban, S. (1987). Antimicrobial preservation. Manufacturers' experience with pharmaceuticals in the efficacy test and in practice. *Pharmaz. Ind.,* 49, 631–637.

Flatan, T.C., Bloomfield, S.F., and Buckton, G. (1996). Preservation of solid oral dosage forms. In *Microbial Quality Assurance in Cosmetics, Toiletries and Non-Sterile Pharmaceuticals.* Baird, R.M. and Bloomfield, S.F., Eds. 2nd ed. Taylor & Francis, London, pp. 113–132.

Gilbert, P. and Das, J.A. (1996). Microbial resistance to preservative systems. In *Microbial Quality Assurance in Cosmetics, Toiletries and Non-Sterile Pharmaceuticals.* Baird, R.M. and Bloomfield, S.F., Eds. 2nd ed. Taylor & Francis, London, pp. 148–174.

Gilliland, D., Li Wan Po, A., and Scott, E. (1992a). Kinetic evaluation of claimed synergistic paraben combination: isothermal and non-isothermal studies. *J. Appl. Bact.,* 72, 252–257.

Gilliland, D., Li Wan Po, A., and Scott, E. (1992b). Kinetic evaluation of claimed synergistic paraben combinations using a factorial design. *J. Appl. Bact.,* 72, 258–261.

Hamill, R.J. et al. (1995). An outbreak of *Burkholderia* (formerly *Pseudomonas*) *cepacia* respiratory tract colonization and infection associated with nebulized albuterol therapy. *Ann. Intern. Med.,* 122, 762–766.

Highsmith, A.K., Greenhood, G.P., and Allen, J.R. (1982). Growth of nosocomial pathogens in multiple-dose parenteral medication vials. *J. Clin. Microbiol.,* 15, 1024–1028.

Hodges, N.A., Denyer, S.P., Hanlon, G.W., and Reynolds J.P. (1996). Preservative efficacy tests in formulated nasal products: reproducibility and factors affecting preservative activity. *J. Pharm. Pharmac.,* 48, 1237–1242.

Hodges, N.A. and Hanlon, G. (2000). Antimicrobial preservative efficacy testing. In *Handbook of Microbiological Quality Control: Pharmaceuticals and Medical Devices.* Baird, R.M., Hodges, N.A., and Denyer, S.P., Eds. Taylor & Francis, London, pp. 168–189.

Hugo, W.B. and Denyer, S.P. (1987). The concentration exponent of disinfectants and preservatives (biocides). In *Preservatives in the Food, Pharmaceutical and Environmental Industries.* Board, R.G., Allwood, M.C., and Banks, J.G., Eds. SAB Technical Series 22. Blackwell Scientific Publications, Oxford, pp. 281–291.

Illingworth, C.D. (1998). *Acanthamoeba* keratitis. *Surv. Ophthalmol.,* 42, 493–508.

Jund, Y. and Carrère, C. (1971). Détermination de l'activité bactériostatique et fongistatique de quelques conservateurs en présence de différents types d'excipients modernes pour pommades. *Ann. Pharm. Franç.,* 29, 161–172.

Kabara, J.J. (1984). Aroma preservatives. In *Cosmetic and Drug Preservation.* Kabara, J.J., Ed. Marcel Dekker, New York, pp. 237–273.

Kabelitz, L. (1997). Are the current requirements regarding the microbiological purity of medicinal plant drugs practicable? *Pharmeuropa,* 9, 570–575.

Kallings, L.O., Ringertz, O., Silverstolpe, L., and Ernerfeldt, F. (1966). Microbiological contamination of medicinal preparations. *Acta Pharm. Suec.,* 3, 219–228.

Karabit, M.S., Juneskans, O.T., and Lundgren, P. (1985). Studies on the evaluation of preservative efficacy. I. The determination of the antimicrobial characteristics of phenol. *Acta Pharm. Suec.,* 22, 281–290.

Karabit, M.S., Juneskans, O.T., and Lundgren, P. (1986). Studies on the evaluation of preservative efficacy. II. The determination of antimicrobial characteristics of benzyl alcohol. *J. Clin. Hosp. Pharm.,* 11, 281–289.

Karabit, M.S., Juneskans, O.T., and Lundgren, P. (1988). Studies on the evaluation of preservative efficacy. III. The determination of the antimicrobial characteristics of benzalkonium chloride. *Int. J. Pharm.,* 46, 141–147.

Kazmi, S.J.A. and Mitchell, A.G. (1978a). Preservation of solubilized and emulsified systems I. Correlation of mathematically predicted preservative availability with antimicrobial activity. *J. Pharmaceut. Sci.,* 67, 1260–1265.

Kazmi, S.J.A. and Mitchell, A.G. (1978b). Preservation of solubilized and emulsified systems II. Theoretical development of capacity and its role in antimicrobial activity of chlorocresol in cetomacrogol-stabilized systems. *J. Pharm. Sci.,* 67, 1266–1271.

Kleinfield, J. and Ellis, P.P. (1967). Inhibition of micro-organisms by topical anesthetics. *Appl. Microbiol.,* 15, 1296–1298.

Komatsu, H. et al. (1986). Preservative activity and *in vivo* percutaneous penetration of butylparaben entrapped in liposomes. *Chem. Pharm. Bull.,* 34, 3415–3422.

Kurup, T.R.R., Wan, L.S.C., and Chan, L.W. (1991a). Availability and activity of preservatives in emulsified systems. *Pharm. Acta Helv.,* 66, 76–82.

Kurup, T.R.R., Wan, L.S.C., and Chan, L.W. (1991b). Effect of surfactants on the antibacterial activity of preservatives. *Pharm. Acta Helv.,* 66, 274–280.

Kurup, T.R.R., Wan, L.S.C., and Chan, L.W. (1992). Interaction of preservatives with macromolecules: Part 1. Natural hydrocolloids. *Pharm. Acta Helv.,* 67, 301–307.

Kurup, T.R.R., Wan, L.S.C., and Chan, L.W. (1995). Interaction of preservatives with macromolecules: Part II. Cellulose derivatives. *Pharm. Acta Helv.,* 70, 187–193.

Lamikanra, A. (1982). Effects of butyl hydroxyanisole (BHA) on the leakage of cytoplasmic materials from *Staphylococcus aureus* and *Escherichia coli. J. Appl. Bact.,* 53, xvi.

Lehner, S.J., Müller, B.W., and Sydel, J.K. (1994). Effect of hydroxypropyl-β-cyclodextrin on the antimicrobial action of preservatives. *J. Pharm. Pharmacol.,* 46, 186–191.

Lis-Balchin, M. (1997). Essential oils and "aromatherapy"; their modern role in healing. *J. Roy. Soc. Health,* 117, 324–329.

Livingstone, D.J., Hanlon, G.W., and Dyke, S. (1998). Evaluation of an extended period of use for preserved eye drops in hospital practice. *Br. J. Ophthalmol.,* B82, 473–475.

Loftsson, T., Stefansdottinr, O., Frioriksdottir, H., and Guomundsson, Ö. (1992). Interactions between preservatives and 2-hydroxypropyl-β-cyclodextrin. *Drug Dev. Indust. Pharm.,* 18, 1477–1484.

Mackie, M.A.L., Lyall, J., McBride, R.J., Murray, J.B., and Smith, G. (1986). Antimicrobial properties of some aromatic alcohols. *Pharmaceut. Acta Helvet.,* 61, 333–336.

Mansour, Z. and Guth, P. (1986). Complexing behaviour of starches with certain pharmaceuticals. *J. Pharm. Sci.,* 57, 404–411.

McCarthy, T.J. (1969). The influence of insoluble powders on preservatives in solution. *J. Mond. Pharm.,* 4, 321–329.

McCarthy, T.J. and Ferreira, J.-H. (1990). Attempted measurement of the activity of selected preservative combinations. *J. Clin. Pharm. Therapeut.,* 15, 123–129.

McCarthy, T.J. and Myburgh, J.A. (1974). The effect of tragacanth gel on preservative efficacy. *Pharm. Weekbl.,* 109, 265–268.

McCarthy, T.J. and Myburgh, J.A. (1977). Further studies on the influence of formulation on preservative activity. *Cosmet. Toilet.,* 92, 33–36.

Miyawaki, G.M., Patel, N.K., and Kostenbauder, H.B. (1959). Interaction of preservatives with macromolecules III. Parahydroxybenzoic acid esters in the presence of some hydrophilic polymers. *J. Am. Pharm. Assoc. Sci. Ed.,* 48, 315–318.

Myburgh, J.A. and McCarthy, T.J. (1980a). Inactivation of preservatives in the presence of particulate solids. *Pharm. Weekbl. Sci. Ed.,* 2, 137–142.

Myburgh, J.A. and McCarthy, T.J. (1980b). The influence of suspending agents on preservative activity in aqueous solid/liquid dispersions. *Pharm. Weekbl. Sci. Ed.,* 2, 1411–1416.

Nitescu, P. et al. (1992). Bacteriology, drug stability and exchange of percutaneous delivery systems and antibacterial filters in long term intrathecal infusions of opioid drugs and bupivacaine in refractory pain. *Clin. J. Pain*, 8, 324–337.

Orth, D.S., Lutes Anderson, C.M., Smith, D.K., and Milstein, S.R. (1989). Synergism of preservative components: use of the survival curve slope method to demonstrate anti-Pseudomonas synergy of methyl paraben and acrylic acid homopolymer/copolymers *in vitro*. *J. Soc. Cosmet. Chem.*, 40, 347–365.

Parker, M.S. (1984). The preservation of oral dosage forms. *Int. J. Pharm. Tech. Prod. Manuf.*, 5, 20–24.

Patchell, C.J. et al. (1994). Bacterial contamination of enteral feeds. *Arch. Dis. Child.*, 70, 327–330.

Patel, N.K. and Kostenbauder, H.B. (1958). Interaction of preservatives with macromolecules. I. Binding of parahydroxybenzoic acid esters by polyoxyethylene 20 sorbitan monooleate (Tween 80). *J. Am. Pharm. Assoc. Sci. Ed.*, 48, 289–293.

Pisano, F.D. and Kostenbauder, H.B. (1959). Interaction of preservatives with macromolecules. II. Correlation of binding data with required preservative concentration of p-hydroxybenzoates in the presence of Tween 80. *J. Am. Pharm. Assoc. Sci. Ed.*, 48, 310–314.

Prickett, P.S., Murray, H.L., and Mercer, N.H. (1961). Potentiation of preservatives (parabens) in pharmaceutical formulations by low concentrations of propylene glycol. *J. Pharm. Sci.*, 50, 316–320.

Reboli, A.C. et al. (1996). An outbreak of *Burkholderia cepacia* lower respiratory tract infection associated with contaminated albuterol nebulization solution. *Infect. Cont. Hosp. Epidemiol.*, 17, 741–743.

Richards, R.M.E. (1971). Inactivation of resistant *Pseudomonas aeruginosa* by antibacterial combinations. *J. Pharm. Pharmacol.*, 23, 141S–146S.

Richards, R.M.E. (1976). Effect of hypromellose on the antibacterial activity of benzalkonium chloride. *J. Pharm. Pharmacol.*, 28, 264.

Richards, R.M.E. and McBride, R.J. (1972). The preservatives of ophthalmic solutions with antibacterial combinations. *J. Pharm. Pharmacol.*, 24, 145–148.

Richards, R.M.E. and Reary, J.M.E. (1972). Changes in antibacterial activity of thiomersal and PMN on autoclaving with certain adjuvants. *J. Pharm. Pharmacol.*, 24, 84P–88P.

Scalzo, M., Orlandi, C., Simonetti, N., and Cerreto, F. (1996). Study of interaction effects of polyacrylic acid polymers (Carbopol 940) on antimicrobial activity of methyl parahydroxybenzoate against some Gram-negative, Gram-positive bacteria and yeast. *J. Pharm. Pharmacol.*, 48, 1201–1205.

Schaumberg, D.A., Snow, K.K., and Dana, M.R. (1998). The epidemic of *Acanthamoeba* keratitis: where do we stand? *Cornea*, 17, 3–10.

Shihab, F.A., Ezzeddeen, F.W., and Stohs, S.J. (1988). Effect of some syrup constituents on the solubility of sorbic acid. *J. Pharm. Sci.*, 77, 455–457.

Stathis, V.J., Miller, C.M., Doerr, G.F., Coffey, J.L., and Hladik, W.B. III (1983). Effect of Technetium Tc 99m pertechnate on bacterial survival in solution. *Am. J. Hosp. Pharm.*, 40, 634–637.

Tromp, Th.F.J., Dankert, J., De Rooy, S., and Huizinga, T. (1976). De conservering van oogdruppels. III. Een onderzoek naar de interactie van hydroxypropylmethylcellulose en benzalkonium chloride. *Pharm. Weekbl.*, 111, 561–569.

USP. (1985). *United States Pharmacopeia 21*. U.S. Pharmacopeial Convention, Rockville, MD, p. 1151.

USP. (2005). *United States Pharmacopeia 28*. U.S. Pharmacopeial Convention, Rockville, MD, pp. 2242–2243.

van Doorne, H. (1977). Interactions between Microorganisms and Some Components of Pharmaceutical Preparations. Ph.D. thesis, University of Leiden, Leiden, The Netherlands.

van Doorne, H. (1993). Interactions between cyclodextrins and ophthalmic drugs. *Europ. J. Pharm. Biopharm.,* 39, 133–139.

van Doorne, H. and Dubois, F.L. (1980). The preservation of lanette wax cream (FNA). *Pharm. Weekbl. Sci. Ed.,* 2, 19–24.

van Doorne, H., Colledge, J., and Calter, R.C.W. (1998). A simple method to establish the vulnerability of "hydrophobic" aqueous dermatological emulsions to potential microbial contamination. *Pharmeuropa,* 10, 480–483.

Vanhaecke, E., Remon, J.P., Pijck, J., Aerts, R., and Herman, J. (1987). A comparative study of the effectiveness of preservatives in twelve antacid suspensions. *Drug. Dev. Ind. Pharm.,* 13, 1429–1446.

Yousef, R.T., El-Nakeeb, M.A., and Salama, S. (1973a). Effect of some pharmaceutical materials on the bactericidal activities of preservatives. *Can. J. Pharm. Sci.,* 8, 54–58.

Yousef, R.T., El-Nakeeb, M.A., and Salama, S. (1973b). Effect of some pharmaceutical materials on the bacteriostatic and bactericidal activity of benzalkonium chloride. *Pharmaz. Ind.,* 35, 154–156.

16 Package Integrity Testing

Lee E. Kirsch

CONTENTS

16.1 INTRODUCTION

In the past, the packaging of pharmaceutical products often took little account of the microbiological aspects of maintaining product integrity. This was especially true of nonsterile products, where minimal consideration was given to microbiological risks during storage and patient use. Similarly, the lack of such considerations also applied to many traditional forms of packaging for sterile products. For instance, glass bottles with rubber closures were implicated in contamination episodes through a failure to maintain seal integrity during autoclaving and few were designed to avoid contamination during use (see Chapter 2). Consequently, in recent years there has been a rapid change to all-plastic containers with integral seals. This type of material offers major design and manufacturing benefits leading to improvements in systems to maintain the microbiological safety of products (Table 16.1).

Microbiological considerations are now an essential element of the product design brief. This is clearly evident in developments in both nonsterile and sterile

TABLE 16.1
Container Types, Construction Materials, and Their Applications

Pharmaceutical Product	Container System and Materials Employed	Potential Microbiological Problems in Use
	Nonsterile Product Packaging	
Liquids	Glass bottles largely superseded by plastic containers; bottle lip designed to minimize backflow and nonsurface wetting plastic to stop product accumulation	Exposure to airborne microorganisms during use; accumulation of product around container lip (minimized by lip design)
Creams/semisolid formulations	Wide-necked jars largely replaced by collapsible aluminum-lined or plastic tubes	Repeated exposure to user's skin, minimized by reduced contact area of tube nozzle
Solid oral dosage forms	Strip-foil unit-dose packaging now predominates over stock bottles	Minor risk of cross-contamination from handling doses
	Sterile Product Packaging	
Large-volume parenterals	Collapsible plastic (PVC) bags or blow-mold systems (polyethylene, polypropylene) have largely eliminated need for glass containers	Minor risk from manipulations in complex assemblies
Small-volume parenterals	Glass ampoules and (multidose) vials increasingly substituted by single-dose blow-molded (polyethylene) units; reconstitution systems employing diluent transfer devices associated with small (<100 ml) collapsible bags; multidose pen injectors with delivery mechanisms and replaceable prefilled cartridges	Touch contamination risks associated particularly with glass ampoules largely eliminated by plastic ampoules and purpose-built syringe connectors, transfer devices, and automatic injectors
Single-dose injections	Prefilled syringes, sometimes employing dual compartments, of glass or plastic construction	Largely eliminated
Eye-drop preparations	Traditional glass dropper bottles almost obsolete, replaced by multidose plastic eye-drop containers and increasingly single-dose (often polycarbonate) disposable units	Manipulation and exposure risks largely eliminated in plastic systems although the container materials may constrain the sterilization method used
Ophthalmic ointments	Limited size semirigid aluminum or plastic tubes with narrow nozzles	Virtually no risks except from contamination of container tip
Nebulizer solutions	Generally provided as single-dose, preservative-free solutions in blow-molded (polyethylene) ampoules	Minimal risk during transfer of contents to nebulizer
Aerosol sprays	Pump action or pressurized systems designed with one-way delivery of contents	Negligible risk of ingress through delivery valve

TABLE 16.2
Design Features of Containers That Influence Microbiological Security of the Product

The container system should:

Provide resistance to contamination through good design (e.g., cap seal efficiency, minimal product accumulation at container mouth)

Allow sterilization by the optimum method (in the case of sterile products); this requires features such as resistance to the sterilizing agents, maintenance of seals under stress, and complete access by sterilizing agent

Not interact with the formulation by sorption of active components or excipients, especially preservatives, so as not to compromise the preservative system

Not release materials into the product (e.g., alkali leach from glass, antioxidants from polyethylene, plasticizers from PVC) or alter product characteristics that may adversely affect preservative efficacy

Provide ready access for microbiological testing

Not offer opportunities for microbial adhesion

Provide light protection for photosensitive preservatives

Be size-limited to reduce in-use shelf life and minimize evaporative headspace

Source: Adapted from Allwood 1990.

pharmaceuticals where factors influencing product microbiological security during storage and use are well recognized. The major aspects relating to container design are summarized in Table 16.2.

Major steps have been taken in container design to improve functionality and minimize extrinsic microbiological contamination. Such microbiological protection can only be guaranteed, however, by appropriate design validation studies and suitable quality assurance processes. Package integrity testing is therefore an intrinsic part of any product development and quality assurance program.

16.2 PACKAGE INTEGRITY QUALITY ASSURANCE

Pharmaceutical quality reflects the ability of a product to perform its intended functions despite the stresses associated with its manufacturing, distribution, storage, and use. Quality assurance depends on the measurement of product attributes that relate to performance and the comparison of these measurements to meaningful criteria. For pharmaceutical packaging, performance may involve protection from light, oxygen, and moisture; prevention of microbial ingress; tamper evidence; poison prevention; and ease of use. Package integrity is typically associated with maintaining product stability or sterility by excluding microorganisms or reactive gases (e.g., water or oxygen), maintaining a vacuum usually to facilitate product use or to prevent mass loss or product leakage.

Package integrity quality assurance involves the selection of an appropriate method of measurement, qualification of the measurement method by demonstrating that it directly or indirectly measures the desired package performance attribute, and validation of the test method.

16.3   LEAKAGE THEORY

A leak is an unintended portal in a package capable of passing matter from one side of the enclosure to the other. Leaks relate to both egress and ingress. Leakage is the movement of mass (gas, liquid, or solid; viable or not) through the leak. Thus a leak may be present whether leakage occurs or not. Leaks are typically of unknown morphology; therefore, it is useful to define them, often in terms of mass or volume flow. Typical quantitative leakage rate units are pressure-volume per unit time (e.g., pascal-m^3.s^{-1}), mass per unit time (e.g., mol.s^{-1}) or volume per unit time at a reference pressure (e.g., standard cm^3.s^{-1}, where "standard cm^3" indicates the quantity of gas contained in one milliliter at sea level atmospheric pressure). The international standard (IS) unit is Pa-m^3.s^{-1}, which is equivalent to 0.001 bar-m^3.s^{-1} or 10 sccs (standard cubic centimeters per second) (McMaster 1982). Alternatively, leakage can be defined functionally, for example, the probability that a microorganism will penetrate a package under well-defined conditions or the extent of vacuum loss that will affect a product's performance. The ability to compare and evaluate methods of leak detection may center on the relationship between functional and quantitative descriptions of leakage rate.

How quickly mass moves through a leak depends on the dimensions of the leak, the properties of the mass, the conditions inside and outside the package, and the composition of the leaking material. The latter is sometimes important for liquids and suspended solids but is not usually critical for gases. Moreover, under low pressure conditions, ideal gas law behavior can be applied to leaking gases. Thus, leakage theory is most readily understandable through a discussion of the gas flow through a simple leak such as a rigid short circular tube present in the wall of a container.

For gases, mass transfer will occur by convection flux when the total pressure difference across the leak is sufficient to support bulk flow or by diffusion if there exists a concentration gradient across the leak. Both convective and diffusional flux can occur simultaneously, but convective flow will predominate when the pressure difference is large enough to support it. Diffusive flow will be predominant when there is no pressure gradient across the leak but a partial pressure gradient exists for specific gases, for example, the ingress of oxygen into a poorly sealed vial with a nitrogen headspace. However, the leakage rate determined under specific measurement conditions where convective flux predominates (e.g., under a nearly complete vacuum) will not be equivalent to leakage under ambient conditions where only diffusive flux occurs.

In the presence of a pressure differential across the leak, the leak size will affect the mode of gas flux. Molecular flow occurs when the mean free path length of the escaping gas is greater than the cross-sectional diameter of the leak. Viscous flow occurs if the diameter is about 100-fold larger than the mean free path length. Three types of viscous flow are typical: laminar, turbulent, or choked (sonic) flow. In laminar flow, a parabolic function describes the distribution of fluid velocity across the diameter of the leak tube, whereas the eddies and swirls associated with turbulent flow cause random fluid velocity fluctuations and a nonparabolic distribution function. The difference between turbulent and laminar flow frequently correlates to a critical value of the Reynold's number, which is the ratio of inertial to viscous force

(McMaster 1982). The relationship between leak size and leakage rates for various modes of mass flow are illustrated in the simulation depicted in Figure 16.1. Choked flow is a special case wherein the flow velocity approximates to the speed of sound in the gas. This is typically relevant only for large leaks.

The measured leakage rate depends on characteristics of the leak (e.g., size and length), the conditions of measurement (e.g., the total pressure difference across the package and the temperature), and the properties of the leaking gas (e.g., concentration-partial pressure and viscosity or molecular weight). When critical leakage rates that reportedly correspond to microbial barrier properties, product stability, or other product performance criteria are referenced, it is important to also refer to conditions of the leak rate measurements and to recognize that the critical value may vary with measurement methods and conditions.

The measured leakage rate may also reflect other sources of variation. Permeation is a diffusional process that describes the movement of mass through integral package material in response to a concentration gradient and the solubility of the escaping mass in the package material. The total measured leak rate will be the sum of permeation and leakage (Figure 16.1). Permeation may limit the sensitivity of a

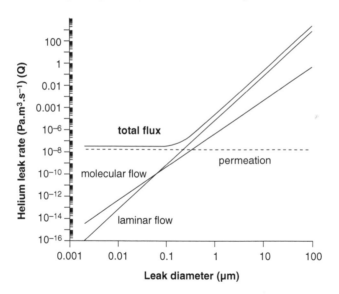

FIGURE 16.1 Simulated helium leak rates (mass flux, Q) as a function of leak diameter where the total leak rate is the sum of molecular and laminar flux through a 10-μm-long pore into a vacuum at 25°C and permeation through a 13-mm butyl rubber stopper. Flux equations used in the simulation were:

$$Q_{laminar} = \frac{\pi}{8\eta l} r^4 \left(P_1^2 - P_2^2 \right)$$

$$Q_{molecular} = \left(\frac{2\pi RT}{M} \right)^{1/2} \left(\frac{d^3}{6l} \right) (P_1 - P_2)$$

where η = 17.8x10^{-6} Pas), P_1 = 1 atmosphere, P_2 = 0, M = molecular weight, l = leak length.

leakage measurement method. For example, in Figure 16.1, leakage rates through holes with diameters less than approximately 0.1 μm are significantly less than the permeation rate through the rubber stopper under the conditions and assumptions used in this simulation. If the critical hole size is less than the minimum detectable hole size, then it cannot be measured (Kirsch et al. 1997a).

Another potentially important source of variation is mass adsorbed to the package surface, which may desorb during leak rate measurements. This is typically a time-dependent source of variation that is avoided by outgassing the package prior to measurement.

16.3.1 ESTABLISHING A CRITICAL LEAKAGE RATE

An important principle of leakage theory is that everything leaks. Thus quality assurance depends on quantitatively defining *acceptable* leakage rates in either a deterministic or probabilistic manner.

For some product performance characteristics, the establishment of quantitative, deterministic criteria is relatively straightforward. As an example, consider a product that contains a drug that is subject to oxygen-mediated instability. If the degradation rate is a function of the headspace oxygen content, then the total degradation over the product's shelf life can be predicted as a function of the rate of oxygen ingress. The simulations depicted in Figure 16.2A and Figure 16.2B illustrate the process of

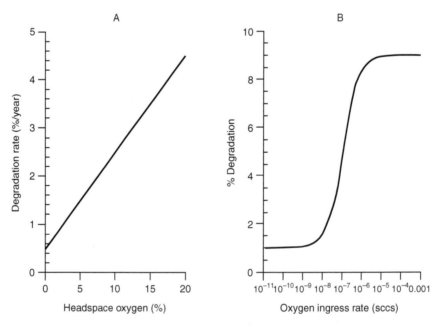

FIGURE 16.2 Simulated evaluation of the effect of oxygen ingress on product stability. (A) The relationship between oxygen headspace content and the degradation rate of a hypothetical, oxygen-sensitive product wherein degradation is assumed to occur in a zero-order fashion at ambient temperature. (B) The predicted total degradation in 2 years as a function of the diffusive flux of oxygen in the product headspace (5 mL).

estimating the total drug degradation in 2 years that would occur as a function of the increasing headspace oxygen content. In this simulation, at high ingress rates ($>10^{-5}$ sccs) the headspace quickly contains atmospheric levels of oxygen that correspond to the maximum degradation rate and therefore a maximum amount of degradation in 2 years. At low oxygen ingress rates ($<10^{-8}$ sccs), the concentration of headspace oxygen does not increase sufficiently in 2 years to cause significant oxidation, and the total degradation is minimal. Suppose the maximum allowable degradation in this product is <5%, then the oxygen ingress specification should be $\leq 10^{-7}$ sccs. Ideally, the package integrity specification should include a substantial margin of safety. For the measurement of gas leakage rates, the margin should be $\geq$10-fold, that is, if a product tolerable leakage rate is 10^{-7} sccs then the leakage rate specification should be 10^{-8} sccs.

Criteria for ensuring the microbial barrier properties of a package are probabilistic because microbial ingress involves a series of stochastic events. The invading microorganisms must find the leak. This depends on the external medium, the organisms' mobility, the concentration of organisms, how long the package is in the presence of the organisms, and environmental conditions. Then the microorganisms must be able to traverse the leak. This event depends on the organisms' mobility, the leak size and morphology, the composition of the leak (i.e., whether or not the organism adsorbs onto the leak walls), and the presence of a fluid path in the leak. Detection of microbial ingress may also depend on the ability of the microorganism to maintain its viability or to grow once it has entered the package (Kirsch et al. 1997b).

Historically, the ability of a pharmaceutical package to act as a microbial barrier has been demonstrated by microbial immersion or aerosol challenge testing (the details of which are discussed below). The practice of using routine sterility testing to demonstrate package integrity is not condoned by regulatory agencies [U.S. Food and Drug Administration (FDA) 1993, 1998; European Union (EU) 1996]. In the case of immersion testing, the presence of an airlock in the leak (that is, the absence of a fluid path through the leak by which microorganisms can travel into the package) will bias the results. In a study conducted using pin hole leaks (0.1 to 10 μm diameters) in standard pharmaceutical glass vials, it was found that no microbial ingress was observed in vials that failed to demonstrate liquid penetration (as measured by magnesium ion passage) regardless of leak size. In contrast, for vials that demonstrated liquid penetration, the probability of microbial ingress varied with leak size (Kirsch et al. 1997b).

The use of physical leak test methods in lieu of microbial challenges is a concept supported by many regulatory agencies (FDA 1993, 1998; EU 1996). The problem is that physical methods do not measure microbial ingress, but rather, they measure some physical property of the leak that may or may not relate to microbial failure. Thus the development of a physical method to characterize a microbial product quality issue requires the development of a relationship between microbial ingress and the physical method (FDA 1998).

A reasonable approach to establish a direct correlation between physical test method and microbial ingress is to (1) devise a representative test system containing minute leaks, (2) demonstrate the usefulness of the physical methods by measuring leakage rates as a function of leak size, (3) develop a reliable microbial challenge method, (4) establish the quantitative correlation between the probability of microbial failure and the measured leakage rate, and (5) demonstrate the utility of the physical method by comparing it to the microbial challenge testing on a significant scale. Ideally, the result of this process would be a quantitative relationship between the physical leakage rate method and the probability of microbial ingress. An example of this type of relationship is illustrated in Figure 16.3 wherein the frequency of microbial ingress is correlated to a helium leakage rate measurement. The test system was a standard 10-mL glass vial sealed with a butyl rubber stopper and aluminum crimp. It was modified to contain a leak composed of a borosilicate glass micropipette that was inserted through a 2-mm hole in the vial side wall and affixed there with an epoxy seal. For helium leakage rate measurements, helium gas was introduced into the vial headspace by enclosing the vials in a sealed helium-inflated glove bag. For microbial challenges, *Brevundimonas diminuta* and *Escherichia coli* were chosen as challenge organisms (Kirsch et al. 1997b).

The results of these studies indicated that the probability of microbial ingress increased dramatically for test units that contained leaks with helium leakage rates in the range $10^{-4.4}$ to $10^{-3.8}$ sccs (Figure 16.3). The microbial ingress cutoff was about $10^{-5.2}$ sccs.

The method of correlation will be test dependent. For example, microbial ingress into a single test unit is a discrete phenomenon; it either occurs or it does not. Helium leakage is a continuous measurement of leakage rate. Therefore, the probability of microbial ingress can be correlated to this quantitative measurement of leakage by

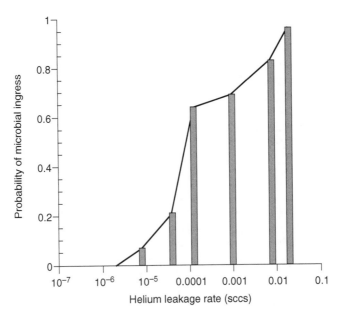

FIGURE 16.3 The direct correlation between the probability of microbial ingress and helium leakage rate for rubber-stoppered glass vials with glass micropipettes embedded in the vial side wall. Microbial challenge was conducted for 24 h at 37°C using 10^8 to 10^{10} organisms. Test units were incubated at 35°C for 13 days.

logistical regression methods (Kirsch et al. 1997b). Dye ingress is frequently used as a discrete measure of leakage: the presence or absence of detectable dye inside the container after an immersion challenge. In order for the dye test to be considered discriminatory in terms of microbial ingress, the correlation must demonstrate that dye ingress is present in all test units that show microbial ingress and preferably also in some test units in which microbial ingress was not possible. Most importantly, no test units that fail to show dye ingress can be the subject of microbial ingress (FDA 1998).

A validation study that demonstrates the rigor of the physical test method is also desirable. For example, a study was conducted to confirm the ability of the helium leak rate method to identify packages with defective microbial barrier properties (Kirsch et al. 1997c). In this study, a series of 11 batches of 90 broth-filled vials were seeded with 10% defective vials. Half of the defective vials contained leaks that were predicted to have a probability of microbial ingress greater than 0.1.

The broth-filled, sealed vials were charged with a tracer by placing them into a pressurized vessel filled with helium. The vials were removed and tested by determining whether the helium leakage rate was above a critical rate value based on the data presented in Figure 16.3. The vials were then subject to microbial immersion testing. For all batches, the helium leakage rate method was demonstrated to be more effective than microbial challenge testing in identifying defective test units. Thus the physical test method was shown to be as good or better than rigorous biological testing (Kirsch et al. 1997c).

Another reasonable approach for establishing a relationship between microbial barrier properties and a physical leakage rate method is to use an indirect correlation in which a physical test method is evaluated against an established standard. For example, a correlation between an alternative physical test method and the helium leakage rate method described above can be used to demonstrate the utility of the alternative method with respect to evaluating microbial barrier properties. As a case study, the vacuum decay test method has been evaluated (Nguyen and Kirsch 1998) and is outlined below.

The general approach involved development of a quantitative correlation between helium leakage rate and vacuum decay testing by subjecting a group of test vials, modified to contain glass micropipette leaks, to leak testing by both methods. The quantitative correlation between helium leakage rate and microbial ingress was used to relate the vacuum decay method to the probability of microbial ingress. The results are shown in Figure 16.4 and indicate that the maximum sensitivity for the vacuum decay method corresponded to a probability of microbial ingress > 0.5, whereas the helium leakage rate method was able to measure leakage rates corresponding to microbial ingress probabilities of less than 0.01. It can be reasonably concluded that the vacuum decay method was not suitable for microbial barrier quality assurance (Nguyen and Kirsch 1998).

16.4 INTEGRITY TESTING METHOD SELECTION

Package integrity testing method selection may depend on the desired product performance criteria, the type of product, and the type of package seal (FDA 1998).

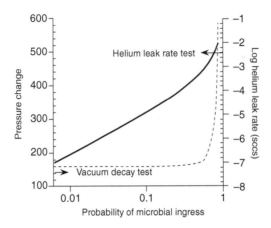

FIGURE 16.4 An indirect correlation between vacuum decay leak testing (increase in chamber pressure, Pa) and the probability of microbial ingress derived from the direct correlations between helium leakage rate versus the probability of microbial ingress and helium leakage rate versus vacuum decay testing. The left-hand vertical axis (pressure change in Pa) corresponds to the dashed line representing the vacuum decay method and demonstrates that the measurable leak detection range for this method, in terms of the probability of microbial ingress, is from 0.5 to 1. The right-hand vertical axis (log leakage rate in sccs) corresponds to the helium leakage rate method and shows that the useful range for this method, in terms of the probability of microbial ingress, is from <0.01 to 1.

Product performance criteria include microbial barrier characteristics in which the method needs to demonstrate that the packaging system is an effective barrier against the incursion of microorganisms. Appropriate test methods are either direct (microbial challenge) or indirect (physical methods correlated to microbial challenge; see Section 16.3.1). The maintenance of a vacuum, dry, or oxygen-free headspace may be the most rigorous product performance criteria. A variety of physical-chemical testing techniques have been used including chemical headspace analysis, vacuum or pressure decay, tracer gas or liquid analysis, residual gas ionization, and high-voltage leak detection.

The composition of product packaging or contents may affect the choice of an appropriate leak detection method. Some methods require specific physiochemical characteristics of the packaging materials or the contents. For example, high-voltage leak detection requires an electrical conductivity difference between the packaging materials and contents and that the more conductive product contents are present in or near the leak. Spark testing (residual gas ionization) is only applicable to packaging with headspace vacuums. The potential interaction between package contents and some liquid tracers (dyes) may significantly limit sensitivity. Gas permeability through packaging materials (e.g., rubber or plastic) may preclude the use of some gas tracer techniques.

Package seals include fusion seals in packages made of a single material that are typically formed by thermal treatment, e.g. plastic or glass ampoules, compression or friction fit closures such as rubber-stoppered glass vials, thermal or adhesive bonded closures made of different materials, and tortuous path closures such as screw threads. Tortuous path or porous seals preclude the use of package integrity methods that are based on liquid or gas tracer ingress or egress, although airborne particulate transmission or microbial aerosols have some usefulness (FDA 1998).

16.4.1 PHARMACEUTICAL PACKAGE INTEGRITY METHODS

16.4.1.1 Microbial Aerosol Challenge

Microbial aerosol challenge testing is accomplished by placing a nutrient medium-filled package into an aerosolization chamber that is charged with nebulized bacterial spores (usually *Bacillus subtilis*) under carefully controlled humidity, temperature, and air circulation conditions. The packaging may be subjected to pressure fluctuations to facilitate atmospheric ingress. Typically the packaging is exposed for a few hours and may be removed and incubated for a few weeks to allow invading organisms to multiply for detection. In addition to leak size and morphology, sources of test method variation occur including chamber temperature, humidity, and pressure; spore concentration, dimensions, and electrostatic condition; exposure time; and chamber air flow patterns. In general, this method is labor-intensive, expensive, and difficult to replicate (Korczynski 1987).

16.4.1.2 Microbial Immersion Challenge

Microbial immersion challenge testing involves immersing packages containing a sterile culture medium into a liquid suspension of viable microorganisms. An attempt is generally made to facilitate liquid penetration of the packaging. This is a critical aspect of the test because the absence of liquid in the leak path will preclude microbial ingress. Liquid penetration may be assisted by mild thermal stress of a liquid-filled package that is then allowed to cool to ambient conditions prior to inoculation of the challenge medium with viable microorganisms (Kirsch et al. 1997b); pressure fluctuations are occasionally used (FDA 1998).

After immersion for a specified duration (from a few minutes to 24 h), the packages are rinsed, incubated (1 to 2 weeks), and examined for microbial growth. Sources of variation include leak size, composition, and morphology; challenge medium formulation and surface tension; challenge organism dimensions, motility, concentration, and viability; package medium composition; and incubation and challenge duration and temperature.

Microbial immersion challenge testing is widely used and appears to be acceptable by regulatory agencies, but again it is labor-intensive, destructive, and difficult to replicate without careful design and the inclusion of large numbers of control test units (Korczynski 1987, Morton et al. 1989, Parenteral Society 1992). Especially difficult is the development of a meaningful positive control that will reliably demonstrate microbial ingress.

16.4.1.3 Liquid Tracer Immersion

Liquid tracer immersion challenges are frequently used in pharmaceutical package integrity testing (Morton et al. 1989, Parenteral Society 1992, Guazzo 1996, FDA

1998). The tracer may be a dye, metal ion, or other detectable solute. Migration of a tracer through a leak is determined visually or spectrophotometrically. The facilitation of liquid penetration is again critical in this method. If the immersion medium fails to wet the package surface due to packaging material porosity retaining air pockets, then a significant leak may be undetected; a vacuum is sometimes used to help degas package surfaces. Moreover, if diffusive rather than convective flux is relied upon for tracer ingress, the rate of ingress may be too slow for a sensitive test. Pressurization of the immersion bath may assist in creating a pressure gradient across the leak. Immersion medium should be particulate free to avoid clogging minute leaks.

The following are potentially critical sources of variation (FDA 1998): leak size, composition, and morphology; packaging surface conditions and immersion medium surface tension; duration of exposure; pressure differentials; stability of the tracer and its interaction with the product contents or packaging; and tracer concentration.

Visual detection of tracer ingress will be subject to sources of variation associated with visual inspection such as lighting, magnification, time of inspection, and inspector training and experience.

Liquid tracer methods can be quantitative when utilizing spectrophotometric detection techniques but are most frequently conducted in a qualitative, pass–fail mode.

16.4.1.4 High-Voltage Leak Detection

In high-voltage leak detection, critical packaging regions can be subjected to a high-frequency, high-voltage electrical current. If the packaging is nonconductive, and the product contents are conductive and are in or very near a leak, then a high current flow is detected. This method is very rapid and especially conducive to in-line, 100% package testing. It is widely used for blow-fill-seal packaging. It is a qualitative test. Sources of methodological variation include voltage setting and gain; probe positioning; product content composition; and package cleanliness, geometry, and conductivity (Guazzo 1996, Mull et al. 1998).

16.4.1.5 Vacuum Decay Testing

Vacuum decay testing is a nondestructive physical test method that involves the application of a differential pressure to a package and measurement of the relative change in vacuum of the test chamber containing the sealed package (Stauffer 1988, Chrai et al. 1994). The change in pressure is determined by an absolute pressure measurement. The method is fast and simple. Its sensitivity depends on the pressure differential applied across the leak and the ability of the vacuum gauge to detect the time-dependent chamber pressure change. Sensitivity is influenced by the test chamber geometry and volume, instrument sensor capabilities, differential pressure conditions, duration of test time, and environmental conditions such as temperature, relative humidity, and atmospheric pressure. Hence, with a smaller chamber volume and increasing length of test time, there is an increase in leak detection sensitivity (Nguyen and Kirsch 1998).

Vacuum decay leak detection is limited to pharmaceutical packages that do not rely on tortuous path or barrier seal types as the primary mechanism of container integrity. In addition, the package must tolerate the application of a differential pressure (maximum of 1 atm) to the seal region. A variation of this technique is based on the flash evaporation of liquid package contents that may be present in leaks in liquid-filled products. This results in a method sensitivity increase of greater than 10-fold.

16.4.1.6 Headspace Analysis

Changes in the chemical composition of packaging headspace can be used as a quantitative leakage rate measure (Wang et al. 1997, FDA 1998). Headspace sampling and analysis by gas chromatography or electrochemical detection have been used. The method typically requires sampling and is destructive. Sources of variation include analytical instrument sensitivity and sampling technique.

16.4.1.7 Spark Testing

Spark testing or residual gas ionization is used to determine the maintenance of a headspace vacuum (Guazzo 1996). This is performed by applying a high-voltage, high-frequency field to a sealed package. A conducting plate located near the package detects the ionization current. This is a fast, qualitative leak test method for evacuated packaging.

The measured leakage rate is a function of the size and morphology of the leak, the concentration and properties of the tracer, and the measurement conditions.

16.4.1.8 Helium Leakage Rate Testing

Helium leakage rate methods involve filling, or partially filling, the test units with a tracer gas and placing them into the leak detector sample chamber. The chamber is directly connected to a series of vacuum pumps. The chamber is evacuated and as headspace gases leak from the test unit, the tracer is separated from the other gases, ionized, and measured by mass spectrometry (Guazzo 1996).

Helium as a tracer gas is a good choice for leak detection applications because it is inert, inexpensive, and nontoxic (Varian 1995). Additionally it has a low atmospheric background and permeates easily through very tiny leaks. Unfortunately it also readily diffuses through plastics and rubbers and therefore may be less than ideal for some pharmaceutical applications. A tracer can be introduced into packaging by sealing in a helium-enriched environment or by charging. The latter involves placing a sealed container in a helium-pressurized chamber and allowing the tracer to enter the package through minute leaks (Kirsch et al. 1997c). Leakage rate measurements are relatively fast (<1 min) but sample charging can take hours.

One type of detector that is used is a counterflow instrument in which the separation of helium and other headspace gases is facilitated by a diffusion pump. There are three pumps in the detector: a roughing pump that quickly evacuates the sample chamber, a fore pump that continues to pull vacuum, and a diffusion pump that acts as a sample gas filter. At the diffusion pump outlet, helium diffuses into the diffusion pump against its flow. Other heavier gases are held back due to their higher compression ratio.

Sensitivity is not compromised, and the diffusion pump protects the spectrometer tube from contamination by water vapor and silicone (Kirsch et al. 1997a).

The spectrometer measures helium by ionizing the tracer (and any other gases that come through the diffusion pump) and accelerating the ionized gas into a magnetic field wherein the trajectory of the ions depends on their mass to charge ratio. The detector is positioned such that only helium ions are collected. The leakage rate is typically calculated by the use of an internal standardized leak (Kirsch et al. 1997a).

Ideally, the critical source of leakage rate variation is the leak size. For rubber-stoppered glass vials modified to contain a glass micropipette in the vial side wall, the logarithm of the "absolute" leakage rate increased with the squared leak diameter in the size range 0.5 to 10 μm. The "absolute leak rate" is the measured leakage rate when the headspace contained only helium (Kirsch et al. 1997a).

Other potential sources of measurement variation include the pressure differential across the container barrier, tracer concentration, and extraneous tracer sources such as permeation and surface adsorbed tracer (Morton 1987). Theoretically, the method is sensitive to leakage rates as low as 10^{-11} sccs, but permeation tends to determine the maximum sensitivity (Figure 16.1). Thus for helium-filled, rubber-stoppered glass vials, the minimal baseline leakage rate was found to be 2×10^{-7} sccs, which represents permeation of helium through the butyl rubber stoppers.

16.5 CONCLUSION

There is growing interest in packaging design and its influence on the safe use of medicines. Although primarily designed to offer significant safety advantages, this is often accomplished through increased complexity in the pack design, especially for injectables. This may introduce hidden microbiological risks. In particular, dead spaces may be introduced into packs in which product may become trapped, allowing microbial survival and proliferation. Further, unit-dose packs may require delivery devices, such as aerosol generators, which may become a reservoir for contamination and are difficult to clean adequately; moisture traps in nebulizer delivery equipment and noncleanable dead ends in delivery pumps for liquids must be avoided. The problems introduced by novel packaging design for parenterals require special thought. A major reason for introducing such packs is to reduce or eliminate chances of extrinsic contamination during reconstitution or administration. This is a laudable aim, but it is imperative to ensure that the pack itself does not compromise sterility by the presence of compartments not accessible to the sterilizing agent. Examples of packs that require such careful consideration are the prefilled syringe and double-chamber infusion system.

Careful scrutiny of novel packaging by the microbiologist and pharmacist is very necessary to ensure that potential microbial risks are identified and overcome. Suitable tests are available to integrity test new package designs and to quality assure microbiological integrity in manufactured products.

REFERENCES

Allwood, M.C. (1990). Package design and product integrity. In *Guide to Microbiological Control in Pharmaceuticals*. Denyer, S. and Baird, R. Eds. Ellis Horwood, London, pp. 342–365.

Chrai, S., Heffernan, G., and Myers, T. (1994). Glass vial container-closure integrity testing—an overview. *Pharm. Tech.*, 18(9), 162–173.

EU. (1996). *Manufacture of Sterile Medicinal Products, EU-CNT 5808/94*. Commission of the European Communities, Working Party on Control of Medicines, Brussels.

FDA. (1993). Guidance for industry for the submission of documentation for sterilization process validation in applications for human and veterinary drug products. *Federal Register*, 58, 231.

FDA. (1998). Guidance for industry container and closure integrity testing in lieu of sterility testing as a component of the stability protocol for sterile products. *Federal Register*, 63, 4272.

Guazzo, D. (1996). Current approaches in leak testing pharmaceutical packages, *PDA J. Pharm. Sci. Technol.*, 50(6), 378–385.

Kirsch, L., Nguyen, L., and Moeckly, C. (1997a). Pharmaceutical container/closure integrity. I: Mass spectrometry-based helium leak rate detection for rubber-stoppered glass vials. *PDA J. Pharm. Sci. Technol.*, 51(5), 187–194.

Kirsch, L., Nguyen, L., Gerth, R., and Moeckly, C. (1997b). Pharmaceutical container/closure integrity. II: The relationship between microbial ingress and helium leak rates in rubber-stoppered glass vials. *PDA J. Pharm. Sci. Technol.*, 51(5), 195–202.

Kirsch, L., Nguyen, L., and Gerth, R. (1997c). Pharmaceutical container/closure integrity. III: Validation of the helium leak rate method for rigid pharmaceutical containers. *PDA J. Pharm. Sci. Technol.*, 51(5), 203–207.

Korczynski, M. (1987). Evaluation of closure integrity. In *Aseptic Pharmaceutical Manufacturing*. Olsen, W. and Groves, M., Eds. Interpharm, Prairie View, IL.

McMaster, R.C., Ed. (1982). *Nondestructive Testing Handbook*. Vol. 1. American Society for Nondestructive Testing, Columbus, OH.

Morton, D. (1987). Container/closure integrity of parenteral vials, *J. Parent. Sci. Technol.*, 41, 145–158.

Morton, D., Lordi, N., Troutman, L., and Ambrosio, T. (1989). Quantitative and mechanistic measurements of container/closure integrity: bubble, liquid, and microbial leakage tests. *J. Parent. Sci. Technol.*, 43(3), 104–108.

Mull, F., Doyle, D., Haerer, M., and Guazzo, D. (1998). Validation of a high voltage leak detector for use with pharmaceutical blow-fill-seal containers—a practical approach. *PDA J. Pharm. Sci. Technol.*, 52(5), 215–227.

Nguyen, L. and Kirsch, L. (1998). Establishing the microbial barrier properties of pharmaceutical packaging by physical leak rate measurements. Poster presentation at the 1998 Annual American Association of Pharmaceutical Scientists, November.

Parenteral Society. (1992). The Prevention and Detection of Leaks in Ampoules, Vials and Other Parenteral Containers. Technical Monograph no. 3. The Parenteral Society, Swindon, U.K.

Stauffer, T. (1988). Non-destructive in-line detection of leaks in food and beverage packages—an analysis of methods. *J. Packaging Technol.*, 2(4), 147–149.

Varian. (1995). Introduction of Helium Mass Spectrometer Leak Detection. Varian Associates, Palo Alto, CA.

Wang, Y., Chen, H., Busch, M., and Baldwin, P. (1997). Headspace analysis for parenteral products: oxygen permeation and integrity test. *Pharm. Technol.*, 21(3), 108–122.

17 Official Methods of Preservative Evaluation and Testing

Michael J. Akers and Veda K. Walcott

CONTENTS

17.1   INTRODUCTION

Antimicrobial preservative agents are formulated in pharmaceutical products to assist in protecting the product from adventitious microbial contamination during manufacture, storage, and use. Such agents are chemicals that themselves are subject to the same environmental stress factors leading to degradation or inactivation as those experienced by the active ingredients they are attempting to protect. To assure the producer, regulatory assessor, and user of preserved products that antimicrobial preservative agents are active, valid scientific appraisal of antimicrobial activity in the finished product under a range of conditions must be carried out. This chapter focuses on testing procedures used to evaluate antimicrobial preservative effectiveness in pharmaceutical dosage forms. Such tests can be applied both to preserved sterile and nonsterile products and can be used to test formulations at the beginning, end, or anytime during their shelf life.

From a compendial standpoint, preservative efficacy testing was not a requirement of finished dosage forms until 1970 [United States Pharmacopeia (USP) 18, Chapter 51]. The USP method was derived from the work of several industrial scientists, including Eisman et al. (1963) and Kenney et al. (1964). The USP preservative efficacy test procedures have evolved little over the past 32 years, as seen in Table 17.1 comparing the procedures of the USP 18 (1970) with those of USP 29 (2006). The preservative efficacy test requires that inocula of given species of test microorganisms should be individually introduced into samples of the preserved product. This challenge should include representatives of the Gram-positive and Gram-negative bacterial species, molds, and yeasts and should be inoculated in sufficient numbers to enable kinetic information to be obtained. Compendial tests have been designed to ensure reproducible results across all laboratories. Even though the preservative efficacy test has survived scrutiny and criticism from many scientists over the years (Hodges and Denyer 1996), it still has several limitations that will be addressed in this chapter. The reader is also referred to the companion *Handbook* (Hodges and Hanlon 2000).

17.2   MATHEMATICAL APPROACH TO PRESERVATIVE EVALUATION

There are several references available explaining the growth and death kinetics of microorganisms (Han et al. 1976, Akers 1979, Davis et al. 1980, Avis and Akers 1986). In conducting a preservative efficacy test, growth of organisms is not as relevant as their death due to the antimicrobial activity of the formulation. Death kinetics generally are logarithmic and so the initial concentration of challenge microorganisms should be sufficient to evaluate microbial loss over several log-cycles. In this respect, therefore, it is far easier to measure a 10^3 or greater reduction in a microbial population if the initial population is high (10^5 to 10^6) than if it is low (10^3 to 10^4).

If microbial death were both logarithmic and linear, then simple decimal reduction time (D value) calculations probably would be acceptable for measuring preservative

TABLE 17.1
Basic Comparison of the Preservative Efficacy Test of USP 18 and USP 29

Criteria	USP 18 (1970)	USP 29 (2006)
Test organisms	C. albicans (ATCC 10231) A. niger (ATCC 16404) E. coli (ATCC 4352) Pseudomonas aeruginosa (ATCC 9027) S. aureus (ATCC 6538)	C. albicans (ATCC 10231) A. niger (ATCC 16404) E. coli (ATCC 8739) P. aeruginosa (ATCC 9027) S. aureus (ATCC 6538)
Medium	Suggest soybean casein digest agar	Soybean casein digest or Sabouraud dextrose agar medium
Inoculum growth, preparation, and final cell density	Bacteria: 37°C, 18–24 h C. albicans: 25°C, 48 h A. niger: 25°C, 1 week All at 5×10^7 CFU.mL^{-1}	Bacteria: 30–35°C, 18–24 h C. albicans: 20–25°C, 44–52 h A. niger: 20–25°C, 6–10 days All at 1×10^8 CFU.mL^{-1}
Test procedures	Add 20 ml of test product to each of 5 sterile tubes Each tube inoculated with $1.25–5 \times 10^5$ org.mL^{-1}; incubate tubes at 30–32°C	Conduct test, if possible, in 5 original product containers; otherwise, transfer 20 ml to each of 5 sterile tubes Each container or tube inoculated to achieve a concentration of $1–10 \times 10^5$ org.mL^{-1}; incubate at 20–25°C
Required pass result	No increase in C. albicans or A. niger 0.1% of initial bacteria left and remains below that level for a 7-day period within the 28-day test period	No increase in C. albicans or A. niger Not less than 3.0 log reduction by the 14th day, and no increase from 14 days count to 28 days

efficacy. However, in many cases, the death of microorganisms is nonlinear; such death curves are a result either of multiple bacterial cell sites that must be inactivated or the selection of more resistant survivors that may, in fact, cause "regrowth" ("grow-back") of the microorganism in the product (see Chapter 13). In the preservative efficacy test, *Pseudomonas aeruginosa* particularly demonstrates a propensity to regrow in pharmaceutical solutions.

Some manufacturers and researchers have advocated the use of *D* value calculations as rapid procedures for determining effectiveness of antimicrobial preservatives in pharmaceutical and cosmetic products (see Chapter 18). Such rapid methods are acceptable to screen and estimate antimicrobial activity of preservative agents alone and in combination (Moore 1978, Akers et al. 1984). However, as pointed out by Cooper (1989), these methods are limited in their ability to predict activity primarily due to the loss of exponential (logarithmic) activity for some preservatives over a 28-day period.

17.3 GENERAL PRINCIPLES OF TEST METHODS AND RECOVERY PROCEDURES

A preservative efficacy test must be designed to provide conditions that will permit microbial survival, and possibly growth, in the product under test should that product be poorly preserved. Further, the test method must ensure satisfactory and efficient recovery of survivors. Assurance that such conditions are met validates the scientific credibility of the test results with respect to the presence or absence of acceptable antimicrobial activity on the part of the pharmaceutical formulation. This section will briefly consider the type of challenge organism used in compendial preservative efficacy tests, and how the test conditions maximize survival and recovery potential for these organisms. For a more detailed discussion, the reader is referred to Hodges and Hanlon (2000).

17.3.1 CHOICE OF CHALLENGE ORGANISMS

Test microorganisms are chosen to represent potential contaminants in the environment in which preparations are manufactured, stored, and used. A common set of challenge organisms is generally employed (Table 17.2) representing Gram-positive and Gram-negative bacteria (with *Pseudomonas aeruginosa* especially selected because of its recognized resistance to many antimicrobial agents), molds, and yeasts. This choice is inevitably a compromise between the need to include representative challenges while seeking to avoid excessively lengthy testing programs. Additional organisms may be included for particular formulations. For example, in preparations with high sucrose content, growth of osmophilic yeasts is encouraged and these products should be tested against those organisms. Oral liquid preparations should be challenged with a suitable strain of *Escherichia coli* as specified by the particular pharmacopoeia. Brief details of the specific requirements and properties of these organisms are given in Chapter 1.

It is always recommended that any organism likely to be a particular contaminant during manufacture and use of the product also be tested in a preservative efficacy test. In addition, mixed challenge testing is an acceptable option in developmental phases of product formulation.

TABLE 17.2
Specified Test Organisms

	Organism	ATCC Reference	Other Reference
Bacteria	*S. aureus*	6538	NCIMB 9518, NCTC 10788 CIP4.83
	P. aeruginosa	9027	NCIMB 8626, CIP82.118
	E. coli	8739	NCIMB 8545, CIP 53.126
Molds	*A. niger*	16404	IMI 149007, IP 1431.83
Yeasts	*C. albicans*	10231	NCPF 3179, IP 48.72
	Z. rouxii	None	NCYC 381, IP 2021.92

17.3.2 Basic Preservative Efficacy Test Protocol

Preservative efficacy tests call for products to be inoculated with most or all of the above challenge microorganisms at approximately 10^5 to 10^6 viable cells per milliliter or gram. These organisms are generally cultured on soybean casein digest agar medium (soya tryptone agar, bacteria) or Sabouraud dextrose agar medium (Sabouraud agar, fungi), and the inoculated product stored at temperatures between 20 and 25°C. The inoculated product is incubated over 28 days and examined visually and by plate count procedures to determine the number of viable microorganisms remaining at each time interval specified in the particular compendium.

17.3.3 Preparation of the Test Inoculum

The test inocula are grown in "stock cultures" on specified nutrient media. *S. aureus, E. coli* and *P. aeruginosa* are grown on tryptone soya agar slopes, and *A. niger, Candida albicans,* and *Zygosaccharomyces rouxii* (an osmotolerant yeast, Chapter 1) are grown on Sabouraud agar slopes. The organisms are subcultured regularly, usually every 4 weeks. When required for the test, the bacteria are subcultured 1 day before use, whereas the fungus, *Asp. niger,* may require between 3 and 7 days for the spores to become established for sampling. The cells are harvested using a sterile diluent such as sterile 0.9% saline (EP, USP) or sterile peptone water 0.1% (JP). Adjuvants are added as required; in the EP, 0.1% peptone is added for the harvesting of bacteria, and all three compendia use 0.05% polysorbate 80 as an aid, if needed, to harvest microorganisms, for example, in the case of *A. niger.* Sterile glass beads can be used to remove mechanically the cells from the agar if necessary. Experience will usually indicate the volume of harvesting diluent to use to obtain a working inoculum of circa 10^8 CFU.mL^{-1}, but reference can be made to calibration graphs of optical density or total count against viable count (see Chapter 1 for methods) to determine the inoculum size. Ultimately, confirmation that the correct inoculum was employed is given in the preservative efficacy test controls. Inoculation of the test preparations is made with a known volume of the freshly prepared microorganisms to achieve a final concentration of approximately 10^6 organisms per milliliter or gram. All procedures are conducted under aseptic conditions.

17.3.4 Temperature Recommendations in the Preservative Efficacy Test

Most microorganisms can grow over a wide (30°C) temperature range, but optimal growth occurs within a more narrow range (usually 30 to 40°C for bacteria and yeasts, 20 to 25°C for molds). In practice, the 20 to 25°C range is a suitable compromise between a temperature that will permit growth of all challenge organisms used in the preservative efficacy test (should the preservative fail drastically) and yet also represents a likely product storage temperature.

17.3.5   OTHER BASIC PRINCIPLES OF PRESERVATIVE EFFICACY TESTING

17.3.5.1   Maintenance of Aseptic Environment and Aseptic Manipulations during the Test

This is of obvious importance to avoid or minimize false positives due to adventitious contamination. Appropriate and adequate training in aseptic techniques and continued environmental monitoring of the laminar airflow work area must take place for preservative efficacy test procedures to be valid. These aspects are discussed in Chapter 6 and Chapter 21; see also Hodges and Hanlon (2000).

17.3.5.2   Testing in the Original Product Container

Where the product container can be entered aseptically, the preservative efficacy test should be conducted in that original container. Potential problems, such as loss of antimicrobial preservative or adventitious contamination, can occur if the product is transferred from its original container. In addition, removal of the formulation to another container may result in preservative–container interactions (see Chapter 15) that are unrepresentative of the final product and which may lead to changes in preservative efficacy. However, in some instances, for example, glass-sealed ampoules, where the product cannot be entered and maintained aseptically, samples from original containers are removed, pooled, and 10 to 20 ml of product are placed in each of five sterile rubber-stoppered vials. Such vials should be validated to ensure that they do not allow the antimicrobial preservative to permeate through or be sorbed by the rubber closure.

On occasion, the preservative efficacy test may need to be performed on deliberately broached containers to demonstrate the maintenance of preservative efficacy during the in-use lifetime of a multidose product. Here the broaching procedure should be designed to mimic the circumstances of actual use.

17.3.5.3   Regrowth

Occasionally, an organism will be observed to "grow back" in the inoculated product. Microbial levels initially will decrease until, for example, when tested on the 14th day, then that level will increase when tested on the 21st or 28th day. This, of course, is extremely disconcerting as it raises more questions than usually can be answered. Grow-back can occur for a variety of reasons, including loss of preservative stability or activity during the test, assay error, or mutational changes in the microbial cell(s).

17.3.5.4   Methods of Sampling from Products

The sampling method from products is continually revalidated by the use of controls. Sterile diluent is inoculated under identical conditions to the test product and is sampled at the same time. The sampling intervals are determined by the guidelines laid down under each preservative efficacy test. The sample of inoculated product is then serially diluted before incubating with the appropriate nutrients and under optimum temperature conditions.

17.3.5.5   Inactivation of Antimicrobial Agents

When a product contains an active or excipient that is an antimicrobial agent, its effect must be neutralized before proceeding with the test. There are three main methods of inactivation (Russell et al. 1979):

1. Inactivation by chemical means, for example, by an enzyme—In the case of antibiotics, such as cephalosporin, inactivation is achieved by an enzyme such as β-lactamase. The amount of the enzyme, inactivation time, and temperature conditions should be predetermined. The enzyme is added after the first serial dilution of the product. The test procedure is then continued with the inactivated sample.
2. Dilution to subinhibitory level—Often the dilution of the product as part of the test is sufficient to inactivate the antimicrobial effect of excipients, such as, alcohols, organic acids and esters, and dyes (acridine). The inactivation of the preservative agent(s) may require the use of a specific neutralizer-inactivator solution as diluent (see Chapters 8 and 14).
3. Membrane filtration—If the product possesses antimicrobial activity that is not inactivated by either of the above approaches, membrane filtration is necessary. The sample is filtered through a 0.45-μm pore size filter after the first serial dilution has been completed. The membrane is rinsed with an appropriate diluent (e.g., 0.9% sodium chloride) and then aseptically transferred to a sterile solution of diluent. The microorganisms are dislodged from the membrane into the diluent mechanically.

17.4   COMPARISON OF PRESERVATIVE EFFICACY TESTING METHODS CURRENTLY IN USE

Preservative efficacy testing is covered in the three major pharmacopoeias: EP, USP, and JP. The latest is the *European Pharmacopoeia*, which has led to the harmonization of testing methods from the individual pharmacopoeias of many countries in Europe. The British, German, and Italian pharmacopoeias updated their testing requirements to be in line with the European protocol. With the re-opening of the Eastern European markets, the individual regulatory authorities have moved generally to a preference to follow the requirements of the EP and the USP as accepted references. The individual national pharmacopoeias are used in conjunction with the EP or USP. A list of countries that recognize the EP as a reference can be found at the front of that publication.

17.4.1   Choice of Microorganisms

There is a general consensus between the pharmacopoeias on the minimum range of challenge organisms to be employed and these are summarized in Table 17.3. The background to this selection has been given in Section 17.3.1.

17.4.2 Inoculation Conditions

The preparation of the inoculum uses fresh cultures of each microorganism, and the growth media, incubation conditions, and recovery procedures are detailed in the various pharmacopoeias. All tests require the formulation to be challenged with 10^5 to 10^6 microorganisms per milliliter or gram of product. The total volume of the inoculum is expressed as a percentage of the total volume of the formulation. The recommended storage conditions range between 20 and 25°C. The JP recommends a reduced concentration of viable cells for the testing of antacids to 10^3 to 10^4 microorganisms per milliliter or gram of product. Tests involve a single inoculum of each organism. It has been suggested that repeated inocula of the same organism into the same product would be more realistic for multiuse preparations such as creams and ointments, but this has not been verified scientifically (Cowen and Steiger 1976, Orth 1979). The formulator should decide on the appropriate challenge tests for a product to show satisfactory resistance to contamination based on the intended application.

TABLE 17.3
Criteria for the Interpretation of Preservative Efficacy Test Results

Type of Product	Organism	USP Time	USP Viable Count Reduced by	EP (criteria A) Time	EP (criteria A) Viable Count Reduced by	JP Time	JP Viable Count Reduced by
Parenterals (multidose and ophthalmics)	Bacteria	7 d	10^1	6 h	10^2	14 d	10^3
		14 d	10^3	24 h	10^3	28 d	NI
		28 d	NI	28 d	NR		
	Fungi	7 d	NI	7 d	10^2	14 d	NI
		14 d	NI	28 d	NI	28 d	NI
		28 d	NI				
Oral (liquid products only)	Bacteria	14 d	10^1	14 d	10^3	14 d	10^1
		28 d	NI	28 d	NI	28 d	NI
	Fungi	14 d	NI	14 d	10^1	14 d	NI
		28 d	NI	28 d	NI	28 d	NI
Topical preparations	Bacteria	14 d	10^2	2 d	10^2	14 d	10^2
		28 d	NI	7 d	10^3	28 d	NI
				28 d	NI		
	Fungi	14 d	NI	14 d	10^2	14 d	NI
		28 d	NI	28 d	NI	28 d	NI

NR—no recovery; NI—no increase.

17.4.3 INTERPRETATION OF RESULTS

The interpretation of results is based on the requirements of the regulatory bodies of each country. Table 17.3 summarizes the specifications of the USP, EP, and JP by product group. Since the introduction of the EP preservative efficacy test in 1997, there has been a general overall agreement between the three major players: Europe, United States, and Japan. Within Europe, the individual pharmacopoeias have continued to include additional criteria not covered by the EP. For example, the British Pharmacopoeia (2004) continued to publish its own criteria for the testing of ear preparations (but these were discontinued in the 2005 BP). The EP recommends the efficacy to be achieved on two levels: criteria A and the lesser criteria B (Table 17.4). Criteria B can only be used when there is a justifiable cause for the product not meeting the higher recommended levels of efficacy, for example, for reasons of an increased risk of adverse reaction. The validity of allowing a more relaxed set of criteria (as in criteria B in the EP) has been challenged (Spooner and Davison 1993). The pharmacopoeial tests are issued as guidelines for the pharmaceutical industry in response to the increasing expectations on product quality from regulatory agencies. As a consequence of this, they are open to the individual interpretation of criteria. The link between the evolution of the preservative efficacy test and the regulatory requirements of the European Union was presented in a review article by Irwin (1995).

TABLE 17.4
Application of Criteria A and B in the EP (2004)

| | | European Pharmacopoeia | | | |
| | | Criteria A | | Criteria B | |
Type of Product	Organism	Time	Viable Count Reduced by	Time	Viable Count Reduced by
Parenterals	Bacteria	6 h	10^2	24 h	10^1
(multidose and ophthalmics)		24 h	10^3	7 d	10^3
		28 d	NR	28 d	NI
	Fungi	7 d	10^2	14 d	10^1
		28 d	NI	28 d	NI
Oral	Bacteria	14 d	10^3	N/A	
(liquid products only)		28 d	NI		
	Fungi	14 d	10^1	N/A	
		28 d	NI		
Topical preparations	Bacteria	2 d	10^2	14 d	10^3
		7 d	10^3	28 d	NI
		28 d	NI		
	Fungi	14 d	10^2	14 d	10^1
		28 d	NI	28 d	NI

NR—no recovery; NI—no increase; N/A—no recommendation applies.

The test is applicable throughout a product's shelf life, and formulations are generally tested at the beginning and end of this period. Depending on the product and regulatory situation, once the product's preservation efficacy has been established initially, chemical assays of the antimicrobial preservative showing sufficient potency (e.g., greater than 90% of label) throughout its shelf life may suffice, that is, repeat preservative efficacy testing is not always required.

17.5 COSMETIC, FOOD AND TOILETRY ASSOCIATION GUIDELINES FOR PRESERVATIVE EFFICACY TESTING

The Preservation Subcommittee of the Toilet Goods Association Microbiological Committee has issued guidelines for testing the adequacy of preservation in cosmetics and toiletry formulations (Halleck 1970). To the authors' knowledge, these have not been updated since 1970. The recommended test organisms are primarily the same as for the pharmacopoeial tests (Table 17.2). There are two additional organisms in the test: *Penicillium luteum* (ATCC 9644), a second filamentous fungi, and *Bacillus cereus* or *Bacillus subtilis* var. *globigii*. Both are regarded as common contaminants by the committee. The inclusion of other microorganisms is left to the discretion of the investigator. The test conditions are summarized in Table 17.5.

It is recommended that the initial microbial load of the product is assessed prior to the start of the test. A high initial level of microbial contamination in the product may affect the interpretation of the test results. As a control, an unpreserved sample of the test product is challenged at the same time. A pure culture challenge on the product is preferred over a mixed challenge but a more severe test can be carried out by rechallenging the product after 28 days.

The preparation of the inoculum, sample preparation, sample dilution, and recovery procedures are similar to the pharmacopoeial methods. The full diluent and media details are to be found in the guidelines (Halleck 1970). The results are interpreted after the minimum 28-day time period. The committee has issued no definitive levels of contamination that are acceptable. The individual investigator is responsible for the final product decision.

TABLE 17.5
Test Conditions for Preservative Efficacy Testing of Cosmetics and Toiletries

Variable	Test Condition
Inoculum size (CFU per ml or g)	Not less than 10^6
Sample size (product)	Not less than 20 ml or g
Product storage	Room temperature or temperature for maximum growth of organism
Test intervals	0, between 1 and 2, 7, 14, and 28 days

Source: From Halleck, F.E. (1970). *TGA J.*, Winter, 20–23. With permission.

As the guidelines for the cosmetic industry are considered relatively unsatisfactory, it was believed that the Cosmetic Toiletry and Fragrance Association (CTFA), the Association of Official Analytical Chemists (AOAC), together with the FDA were working to produce a statistically based preservative efficacy testing protocol that could be used consistently across the industry. However, to the authors' knowledge this has never been accomplished. Such a protocol would probably not have addressed the key problem of consumer contamination of products and the correlation of preservative efficacy data with consumer use (Brannon 1996). However, a survey by the CTFA referred to in this review (Brannon 1996) would suggest that some cosmetic companies, at least, have derived their own correlation data.

17.6 NONPHARMACOPOEIAL TESTING OF NONSTERILE PRODUCTS FOR PRESERVATIVE EFFICACY

In the last 15 to 20 years, concern has increased over microbial contamination of nonsterile pharmaceutical formulations. A product being supplied to a country that does not have its own pharmacopoeial test has to comply with guidelines set down by the Committee of Official Laboratories and Drug Control Services and Section of Industrial Pharmacists (FIP) in its second report (Anonymous 1976).

17.7 PRACTICAL CONSIDERATIONS OF PRESERVATIVE EFFICACY TESTING IN COMPLEX FORMULATIONS

The preservative efficacy test for aqueous solutions and water-soluble solids is relatively easy to carry out and interpret. However, for products such as oily solutions, ointments, and other anhydrous items, the preservative efficacy test becomes more complex. This is because the microbial inocula are aqueous suspensions and thus depend on the water–oil partitioning properties of both the microbial cells and the antimicrobial preservative for interaction of preservative and organism to occur. It cannot be ensured that efficient water–oil partitioning occurs, and therefore, the preservative efficacy test for anhydrous-type products has its limitations. For example, a positive test result where bacterial levels do not decline is always questionable for anhydrous products because it cannot be assumed that the antimicrobial preservative in the oily product had adequate surface contact with the aqueous environment of the bacterial suspension. It could also be argued that by addition of an aqueous suspension of microorganisms to a nonaqueous system, the physicochemical characteristics of that system have so changed that a preservative efficacy test result would not reflect the true formulation behavior.

A common technique for improving interfacial interaction of antimicrobial preservatives in anhydrous products with microorganisms in aqueous suspensions is to use surface-active agents. Polysorbate 80 (Tween 80), Arlacel 80, and Tween 20 are commonly used surfactants in preservative efficacy testing of anhydrous products. A common procedure is to transfer 1 ml or 1 g of the anhydrous sample containing the microbial inoculum to 10 ml of a sterile mixture of equal volumes of Tween 20 and Arlacel 80, mix well, then add sterile phosphate buffer to make 100 ml, and mix well again to produce an emulsion. This emulsion (1:100 dilution) is then diluted further according to the usual compendial plate-counting procedure. Obviously, it is critical that thorough mixing of the dilutions takes place to maximize antimicrobial preservative interaction with the microbial inoculum. Once again, addition of such a surface-active mix will alter the physicochemical characteristics of the system and this may lead to interactions between the preservative and emulgent, or may alter the partitioning behavior (see Chapter 14 and Chapter 15).

In view of these difficulties, and because oil-based formulations are not usually considered suitable environments to support microbial growth, the question may be legitimately asked as to whether a preservative agent is really necessary. In such circumstances, the demonstration of low contamination at the time of manufacture may be sufficient to infer microbiological acceptability.

The pharmacopoeial testing of nonsterile formulations has been evaluated by several investigators. One particular area of interest has been in problems encountered when recovering microorganisms from cream and emulsion-type formulations. The British Pharmacopoeia (2005) recommends the addition of up to 10 g.L^{-1} Polysorbate 80 to aid in the dispersion of the product. Wide variation in the recovery of microorganisms by this method has been reported (Allwood and Hambleton 1972, 1973; Brown et al. 1986).

Another factor in interpreting efficacy test results is the possible potentiation of preservative action during the test. In this respect, the choice of dispersants and cosolvents in dilution and recovery fluids must be made with care. Certainly, ethanol in concentrations as low as 5% can have a profound potentiating effect on the activity of antimicrobial agents (McCarthy et al. 1988). The use of alcohol as a diluting fluid in tests must be discouraged. Other potentiators include salts, antioxidants, sugars, and polyols such as glycerol (Zeelie and McCarthy 1983, Griffith 1986).

REFERENCES

Akers, M.J. (1979). Dynamics of microbial growth and death in parenteral products. *J. Parenter. Drug. Assoc.,* 33, 372–388.

Akers, M.J., Boand, A.V., and Binkley, D.A. (1984). Preformulation method for parenteral preservative efficacy evaluation. *J. Pharm. Sci.,* 73, 903–905.

Allwood, M.C. and Hambleton, R. (1972). The recovery of *Bacillus megaterium* spores from WSP. *J. Pharm. Pharmacol.,* 24, 671–672.

Allwood, M.C. and Hambleton, R. (1973). The recovery of bacteria from white soft paraffin. *J. Pharm. Pharmacol.,* 25, 559–562.

Anonymous. (1976). Second joint report of the Committee of Official Laboratories and Drug Control Services and the Section of Industrial Pharmacists, FIP, July, 1975, *Pharm. Acta Helv.,* 51, 33–40.

Avis, K.E. and Akers, M.J. (1986). Sterilization. In *The Theory and Practice of Industrial Pharmacy.* Lachman, L., Lieberman, H.A., and Kanig, J.L., Eds. 3rd ed. Lea and Febiger, Philadelphia, pp. 620–622.

Brannon, D.K. (1996). Cosmetic Preservation, *Cosmetics and Toiletries Magazine,* 3, 69–83.

British Pharmacopoeia. (1999). *British Pharmacopoeia.* Appendix XVIC. The Stationery Office, London.

British Pharmacopoeia. (2004). *British Pharmacopoeia.* Appendix XVIC. The Stationery Office, London.

British Pharmacopoeia. (2005). *British Pharmacopoeia.* Appendix XVIC. The Stationery Office, London.

Brown, M.W., Evans. C.P., Ford. J.L., and Pilling, M. (1986). A note on the recovery of microorganisms from an oil-in-water cream. *J. Clin. Hosp. Pharm.,* 11, 117–123.

Cooper, M.S. (1989). *The Microbiological Update,* 7(1) (April).

Cowan, R.A. and Steiger, B. (1976). Antimicrobial activity: a critical review of test methods of preservative efficacy. *J. Soc. Cosmet. Chem.,* 27, 467–481.

Davis, B.D., Dulbecco, R., Eisen, H.N., and Ginsberg, H.S., Eds. (1980). *Microbiology.* 3rd ed. Harper & Row, Hagerstown, MD.

Eisman, P.C., Jaconia, D., and Lazarus, J. (1963). A proposed microbiological method for studying the stability of preservatives in parenteral solutions. *Bull. Parenter. Drug Assoc.,* 17, 10–17.

European Pharmacopoeia. (2004). *European Pharmacopoeia.* 5th ed. Council of Europe, Strasbourg, France, pp. 447–449.

German Pharmacopoeia. (1986). *German Pharmacopoeia.* Part VIII, No. 1. Deutsche Apothekerverlag, Stuttgart, Govi-Verlag GmbH, Frankfurt, pp. 369–370.

German Pharmacopoeia. (1989). *German Pharmacopoeia.* Supplement. Deutsche Apotheker-verlag, Stuttgart, Govi-Verlag GmbH, Frankfurt, pp. 71–72.

Griffith, I.P. (1986). Preservation of non-sterile pharmaceuticals: Part 2. *Aust. J. Hosp. Pharm.,* 16, 259–264.

Halleck, F.E. (1970). A guideline for the determination of adequacy of preservation of cosmetics and toiletry formulations. *TGA J.,* (Winter), 20–23.

Han, Y.W., Zhang, H.I., and Krochta, J.M. (1976). Death rates of bacterial spores: mathematical models. *Can. J. Microbiol.,* 22, 295–300.

Hodges, N.A. and Denyer, S.P. (1996). Preservative testing. In *Encyclopedia of Pharmaceutical Technology.* Swarbrick, J. and Boylan, J.C., Eds., Vol. 13. Marcel Dekker, New York, pp. 21–37.

Hodges, N. and Hanlon, G. (2000). Antimicrobial preservative efficacy testing. In *Handbook of Microbiological Quality Control: Pharmaceuticals and Medical Devices.* Baird, R.M., Hodges, N.A., and Denyer, S.P., Eds. Taylor & Francis, London, pp. 168–189.

Irwin, V.P. (1995). Regulatory requirements of the European Union concerning the inclusion and testing of antimicrobial preservatives in medicinal products. *Ir. Pharm. J.,* 73, 115–117.

Italian Pharmacopoeia. (1985). *Italian Pharmacopoeia.* 9th ed. Istituto Poligrafico e Zecca Dello StartoLibreria Dello Starto, Rome, pp. 509–512.

Japanese Pharmacopoeia. (2001). *Japanese Pharmacopoeia.* 14th ed. The Society of Japanese Pharmacopoeia, Tokyo, pp. 1321–1323.

Kenney, D.S., Grundy, W.E., and Otto, R.H. (1964). Spoilage and preservative tests as applied to pharmaceuticals. *Bull. Parenter. Drug Assoc.,* 18, 10–19.

McCarthy, T.J., Van Eeden, A., Stephenson, N., and Newman, C. (1988). Interaction between ethanol and selected antimicrobial agents. *S. Afr. J. Sci.,* 84, 128–131.

Moore, K.E. (1978). Evaluating preservative efficacy by challenge testing during the development stage of pharmaceutical products. *J. Appl. Bact.,* 44, Sxliii–Slv.

Orth, D.S. (1979). Linear regression method for rapid determination of cosmetic preservative efficacy. *J. Soc. Cosmet. Chem.,* 30, 321–332.

Russell, A.D., Ljeoma, A., and Rogers, D.T. (1979). A review: microbiological applications of the inactivation of antibiotics and other antimicrobial agents. *J. Appl. Bact.,* 46, 207–245.

Spooner, D.F. and Davison, A.L. (1993). The validity of the criteria for pharmacopoeial antimicrobial preservative efficacy tests. *Pharm. J.,* 251, 602–605.

United States Pharmacopeia. (1970). *United States Pharmacopeia 18.* Mack Publishing Company, Easton, PA.

United States Pharmacopeia. (2006). *United States Pharmacopeia 29.* U.S. Pharmacopeial Convention, Rockville, MD, pp. 2499–2500, General Chapter 51.

Zeelie, J.J. and McCarthy, T.J. (1983). Antioxidants: multifunctional preservatives for cosmetic and toiletry formulations. *Cosmet. Toilet.,* 98, 51–52.

18 Preservative Evaluation and Testing: The Linear Regression Method

Donald S. Orth

CONTENTS

18.1 INTRODUCTION

Official methods of preservative efficacy testing generally require a test period of up to 28 days, with incubation of yeast and mold plates for 5 to 7 days, giving a total of about 35 days (see Chapter 17). This is usually an unacceptable timescale for development pharmaceutics, and a more rapid method for predicting preservative performance is often required (Leak et al. 1996). One approach that has been used successfully for more than two decades is the linear regression method (Orth 1979), with modifications to include a rapid screening method for estimation of D values (Orth and Enigl 1993) and preservative efficacy testing without counting colonies (Orth and Eck 2001).

18.2 LINEAR REGRESSION METHOD

In the idealized microbiological growth curve, the four stages (phases) of growth are the lag phase, logarithmic phase, stationary phase, and death phase (see Chapter 1). Cellular death refers to the loss of viability of cells, as indicated by their ability to grow and form colonies on plating media. The death phase represents the dynamic response of the population of microorganisms in a given system when the rate of cell death exceeds the rate of new cell formation.

In preservative efficacy testing, products containing various concentrations of preservatives are challenged with test organisms to determine whether the preservative system inactivates the test organisms quickly enough to meet acceptance criteria. The survivor curve is obtained by plotting the logarithm of the aerobic plate count as a function of the time after inoculation. The slope of the survivor curve gives the rate of death of the population of test organisms in the test samples. The negative reciprocal of the slope gives the decimal reduction time (D value), which is the time required for a 1 log-cycle (90%) reduction in the population when exposed to a constant lethal treatment (Orth 1979).

The linear regression method provides a means of determining the D values for test organisms in samples of product and has been found to be a reliable method for determining the type and concentration of preservatives required in aqueous multiple-use products (Orth 1979, 1984, 1999; Orth and Brueggen 1982). Testing is performed by inoculating 0.1 ml of a saline suspension of each test organism into approximately 50 ml of test sample in screw-capped bottles. After inoculation, the bottles are shaken and the aerobic plate counts (APCs) are determined immediately (for the zero-time determination) and at various times thereafter, typically at 2, 4, and 24 h for bacteria, and at 4, 8, and 24 h for molds. In addition, APCs of the saline inocula of each test organism are determined as a check against the zero-time value. Additional samples are taken and APCs are performed at 2, 5, or 7 days after inoculation, unless the previous APC was <10 CFU.mL^{-1}, indicating that the preservative system had killed the test organisms introduced into the product.

A survivor curve for each test organism is constructed. Orth (1991, 1993) observed that the minimum acceptable rates of killing bacteria used in preservative efficacy testing may be determined from the slopes of the survivor curves and that the rates were represented by D values of ≤112, ≤56, ≤24, ≤4, and ≤28 h, respectively, for the then-current United States Pharmacopeia (USP), Cosmetic Toiletry and Fragrance Association (CTFA) guidelines (1973), British Pharmacopeia (BP) (1993), pathogens test (linear regression method), and nonpathogens test (linear regression method). Since the time of this publication by Orth in 1993, the USP (2004) criteria have been relaxed to require initial killing rates at D values ≤168 h. From these maximum acceptable D values, it is apparent that the USP (2004) and CTFA (2001) acceptance criteria are not as stringent as those required by the

European Pharmacopoeia (EP) (2005) and the linear regression method (see also Section 18.3).

Different organisms have different physiological and metabolic characteristics; consequently, they may exhibit differences in the rates of death when exposed to any given lethal treatment. The rationale for use of the linear regression method is that every organism has a characteristic rate of death when subjected to any lethal treatment (Orth 1979). Thus, the D value provides a quantitative expression of the rate of death of the population of each test organism in the test sample. The survivor curve obtained when performing preservative efficacy tests by this method is functionally equivalent to the death phase of the idealized bacteriological growth curve (Orth and Milstein 1989). Failure to kill the test organisms fast enough may result in deviation from linear rates of death and enable test organisms to adapt and survive, indicating that the product is not adequately preserved (Orth et al. 1998).

18.2.1 GRAPHICAL DETERMINATION OF D VALUES

The D value for each test organism in a test sample is determined from its survivor curve (Orth 1984). In order to illustrate this, let us assume that S. *aureus* was inoculated into a lotion and that APCs of 10^6, 4×10^5, 10^4, and <10 CFU.mL^{-1} were obtained at 0, 2, 4, and 24 h, respectively. The APCs and the times at which they are determined are plotted using seven-cycle semilogarithmic graph paper. Alternately, the logarithm of the APCs may be calculated and plotted as a function of the time on standard graph paper (values of <10 CFU.mL^{-1} are plotted as 0). The survivor curve is constructed by drawing a "best fit" straight line through the points, extrapolated to the x axis, as shown in Figure 18.1.

The x intercept of the survivor curve represents the time for complete inactivation of the test population in the test sample. Here, the x intercept is 24 h, which is the time required for complete inactivation of S. *aureus*. The D value is the time required for the preservative system to decrease the S. *aureus* population by 1 log-cycle and is equal to the negative reciprocal of the slope of the survivor curve. The slope in this example is 0.25 log reduction per hour; consequently, the D value is 4 h.

18.2.2 MATHEMATICAL DETERMINATION OF D VALUES

D values may be determined by use of least-squares linear regression analysis (Orth 1984) assuming a linear relationship of the form of Equation 18.1.

$$Y = mx + c \qquad (18.1)$$

where y = log number of S. *aureus* surviving per milliliter, x = time, m = slope, and c = y intercept.

This enables determination of the theoretical APC at the time of inoculation (y intercept), the slope of the survivor curve from which the D value can be calculated, and the correlation coefficient r. These values may be obtained directly by use of a calculator or software capable of least-squares linear regression analysis. In this approach, the APC values and the times at which they were determined are entered,

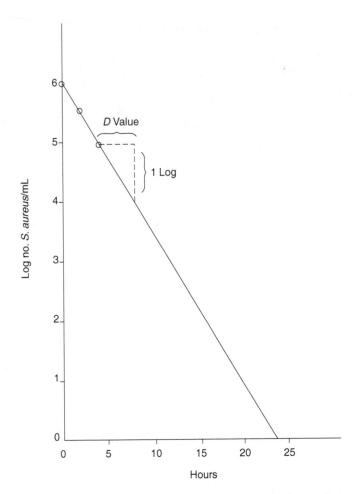

FIGURE 18.1 Survivor curve for *S. aureus* showing a decrease in the population from 10^6 organisms.mL^{-1} to <10 CFU.mL^{-1} after 24 h. The *D* value in this example is 4 h and is shown to be the time required for a 1 log-cycle reduction in the aerobic plate count. (From Orth, D.S. (1984). In *Cosmetic and Drug Preservation. Principles and Practice.* Kabara, J.J., Ed. Marcel Dekker, New York, p. 407. With permission.)

and the linear regression is produced. By using the data presented in Figure 18.1 for the inactivation of *S. aureus*, the following results were obtained:

> *y* intercept (c) = 6.041 (initial inoculum 1.1×10^6 CFU.mL^{-1})
>> Slope (*m*) = −0.25 h^{-1}
>>> *D* value = 4 h (*Note:* The *D* value is the negative reciprocal of the slope)
>>> *r* = −0.99
>> *x* intercept = 24.1 h [*Note:* Survivors are below the detection level
>>> (10 CFU.mL^{-1}) for APC (i.e., *y* = 0) at *x* intercept.]

These calculations demonstrate that the predicted x intercept was slightly greater than 24 h. The reason for this is that the data do not fit the regression perfectly, because the r value = 0.99. (*Note*: An r value of 1.00 indicates a perfect fit of the data to the regression, but, because of experimental variation, this is rarely achieved.) It is recommended that D values greater than 10 h should be reported to the nearest decimal point.

Although mathematical determination of D values was originally preferred (Orth 1984), software now permits easy graphical construction. The mathematical approach to linear regression analysis offers a means of statistical quality control for each assessment (see Section 18.2.4).

18.2.3 RELIABILITY OF THE LINEAR REGRESSION METHOD

Although there are reports indicating that survivor curves may deviate from linearity in some cases (Akers et al. 1984, Orth 1984, Geis and Hennessy 1997, Levy 1987), this has not been our normal experience when the preservative system inactivates test organisms quickly enough to meet acceptance criteria. Populations of spore-forming bacilli, however, appear to be inactivated in a biphasic manner with an initial, more rapid rate of inactivation of the vegetative cells being followed by a slower, sometimes negligible rate of inactivation of the spores (unpublished work). If the two portions of this biphasic curve are considered to be made up of two survivor curves, one for the population of vegetative cells and one for the population of spores, then D values may be determined for both vegetative cells and for spores in test samples.

The reliability of the linear regression method was evaluated by repetitive analysis of the same sample on the same and different days (Orth and Brueggen 1982). It was found that the D values obtained on each day were within 0.5 h^{-1} for four different bacteria, and the mean D values for each set of three samples examined on different days were within 1.1 h when D values were ≤ 2.9 h. An analysis of variance revealed that the standard deviation values for the triplicate means ranged from 0.06 to 0.59. It is believed that the variation between replicate analyses of the same sample may increase somewhat as the observed D values increase. In general, there has been an excellent agreement of the time predicted for the APC to be <10 $CFU.mL^{-1}$ and the actual time by performing APC determinations.

18.2.4 STATISTICAL QUALITY CONTROL OF THE LINEAR REGRESSION METHOD

Use of the linear regression method of preservative efficacy testing offers a statistical approach to examining assay reliability. The r value provides a measure of the precision of the assay and may be used to demonstrate preciseness of fit of the data to the regression (Orth and Brueggen 1982, Orth 1984). The data presented in Table 18.1 are the r values for a number of assays taken from a page in a laboratory notebook. All of the r values are negative for survivor curves in which the test organisms are dying. The closer the r value is to -1.00, the better the APC values fit the regression, as shown in the following example.

TABLE 18.1
Correlation Coefficients (*r* Values) Determined during
Calculation of *D* Values by Use of the Linear Regression Method

Assay No.	*r* Value
1	–0.9599
2	–0.9996
3	–0.9800
4	–0.9943
5	–0.9603
6	–0.9999
7	–0.9540
8	–0.9937
9	–0.9802
10	–0.9987
11	–0.9953
12	–0.8799
13	–0.9800

Note: See also Figure 18.2. *r* Values to four decimal places used for calculations.

To set up the statistical quality control procedures, one may take the mean value $\bar{x}$ for a number of analyses (a minimum of 10 is recommended). When this is done using the *r* values for assays 1 to 10 in Table 18.1, a $\bar{x}$ value of –0.98 is obtained. The standard deviation (*s*) about this mean is also calculated. In this example, the *s* value is 0.02, and the value for 2*s* is 0.04. One may set the 95% confidence limits about the mean by use of ±2*s*. Thus, any assay performed in which the *r* value lies within the 95% confidence limits [–0.98 ± 0.04, or between –0.94 and –1.00 (not –1.02 because –1.00 is the maximum *r* value)] will be considered to be "in control" (i.e., performing satisfactorily). This is illustrated in Figure 18.2. It is evident that the *r* values for the first 11 assays are within the 95% confidence limits. Assay 12 has a *r* value of –0.88, which is outside the 95% confidence limits. The failure of the *r* value to fall within the 95% confidence limits suggests that something went wrong during this assay or in performing the calculations. Failure of the data to fit the regression suggests that the survivor curve was not linear. This may be due to a mixed population of organisms, contamination in the test, lack of homogeneity in the sample or in the distribution of test organisms in the test samples, sampling errors, adaptation of the test organisms to the product, or other factors. It would be necessary therefore first to check the duplicate APC values to see that the numbers and means were recorded correctly and second to repeat the calculations for the linear regression to determine whether an error had been made there. If no error can be found, it should be concluded that the assay was outside the control limits, and it should be repeated. Note that assay 13 was in control because the *r* value for this assay fell within the 95% confidence limits.

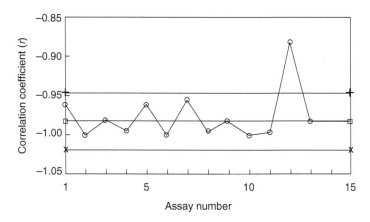

FIGURE 18.2 Statistical quality control of the linear regression method showing the $\bar{x}$ value ($\square - \square$) obtained in the first 10 assays and the 95% confidence limits about this $\bar{x}$ value (x – x, upper limit + – + lower limit).

Other criteria for statistical quality control may be used, at the discretion of the laboratory. For example, one may use the running average of the mean D value for a particular product, or $\pm 3s$ about $\bar{x}$ (for 99% confidence limits).

18.2.5 ESTIMATION OF D VALUES

Orth and Enigl (1993) published a rapid screening method to provide estimated D values (*ED* values). These tests are performed using saline suspensions from surface growth of each test organism, with incubation of bacteria and *C. albicans* at 30 to 32°C and molds at 25°C. The screening tests may be made using pooled inocula of groups of organisms with similar maximum acceptable D values (i.e., *Pseudomonas aeruginosa* and *S. aureus; Escherichia coli* and *Burkholderia cepacia*) and/or recovery media (i.e., *C. albicans* and *Aspergillus niger*). If viable organisms are recovered at 24 h or 7 d, D values can be calculated from APCs at time 0 and at 24 h or 7 d. If no organisms are recovered at these time points, the APCs are <10 CFU.mL^{-1}, which means that the test organisms died before testing at these time points. In this case, D values must be estimated. *ED* values are determined using APCs of test organisms immediately after inoculation into test samples and at 24 h for pathogenic microorganisms or at 7 d for nonpathogenic bacteria, yeasts, and molds. The most conservative D value would be the largest, which represents the slowest rate of death for the organism, and this is the *ED* value that is calculated using the APC at time 0 and <10 CFU.mL^{-1} at 24 h or 7 d. Products are judged to be adequately preserved if the *ED* values meet the acceptance criteria of the linear regression method for all test organisms (Orth 1999).

18.2.6 PRESERVATIVE EFFICACY TESTING WITHOUT COUNTING COLONIES

Orth and Eck (2001) reported a modification of the linear regression method that enables testing to be done without counting colonies. The system used broth enrichment in 96-well microtiter plates followed by streaking on agar or use of a fluorescent dye system to reveal growth. The reciprocal of the highest dilution showing growth was used as the log CFU.mL^{-1} bacteria at each time point, and D values were calculated. D values for *S. aureus, P. aeruginosa, Burkholderia cepacia,* and *E. coli* in aqueous cosmetic and drug products determined by use of the miniaturized system were comparable to those obtained by plating on agar and counting colonies. The miniaturized system increased laboratory efficiency by decreasing the time required for preparation of media, counting colonies, and sterilizing materials prior to disposal.

18.3 APPLICATION OF THE LINEAR REGRESSION METHOD TO DEVELOPMENT OF A PRODUCT PRESERVATIVE SYSTEM

The preservative system of a product involves both specific preservative chemicals and the physicochemical constitution of the formulation. Factors such as pH, water activity (a_w), sequestering agents, surfactants, the ability of formula components to serve as nutrients for contaminating organisms, and other factors determine the extent to which preservative action is manifest in a specific formulation (Orth and Milstein 1989).

18.3.1 RATIONAL DEVELOPMENT OF A PRESERVATIVE SYSTEM

The reliable and reproducible nature of the linear regression method enables it to be used to advantage during the early stages in the development of a preservative system. In providing quantitative data on the kinetics of inactivation, it allows D values to be determined for each test organism so that the preservative behavior can be predicted and compared. This method, or the rapid screening method (see Section 18.2.5), enables the preservative requirements of preformulations and developmental formulae to be determined in approximately 2 weeks, minimizing delays in product development time lines.

The first steps in the rational development of a product preservative system are to review the formula to determine which preservatives may be suitable and to select the test method to be used in evaluation of preservative efficacy. Samples of the test product with different concentrations of the preservative(s) are prepared and tested, and D values for each test organism at each preservative concentration are determined. Then, the amount of preservative required to meet acceptance criteria is calculated from the preservative death-time curve. For example, let us assume that preservative testing with *S. aureus* in a lotion containing 0, 0.1, 0.2, and 0.3% w/v methylparaben gave D values of >30, 10, 5, and 0.5 h, respectively (Orth 1984). The family of curves for *S. aureus* in this lotion is shown in Figure 18.3.

From these data, the preservative death-time curve may be constructed by plotting D values as a function of the concentration of preservative used to obtain these D values. The preservative death-time curve for *S. aureus* in this example is shown in Figure 18.4. This curve is used to determine the concentration of preservative needed to meet acceptance criteria. Here, it is seen that a concentration of 0.2% w/v methylparaben is required to give a D value of 4 h for *S. aureus*. This approach is

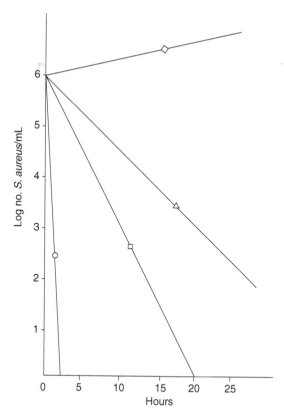

FIGURE 18.3 Series of survivor curves for *S. aureus* in lotion. *D* values of >30, 10, 5, and 0.5 h were obtained in test samples containing 0% (◊), 0.1% (△), 0.2% (□), and 0.3% (○) methylparaben, respectively. (From Orth, D.S. (1984). In *Cosmetic and Drug Preservation. Principles and Practice*. Kabara, J.J., Ed. Marcel Dekker, New York, p. 409. With permission.)

repeated for all test organisms, and the concentration of preservative needed for the product to meet the acceptance criteria with all the test organisms is then selected as the probable concentration required to preserve the formula satisfactorily. This is confirmed using this concentration in a new batch of product and retesting it to verify the predicted antimicrobial efficacy. Adequacy of preservation must be maintained during use of the product. It is recommended that manufacturers consider home use tests of the product in its final packaging to determine whether the product remains uncontaminated during use. Maintenance of preservative efficacy (as indicated by testing samples of product returned after a period of use) and an APC of <10 CFU.g⁻¹ (for aqueous products) indicates that the product preservative system is satisfactory in the packaging employed and in the manner used.

Samples of each preservative system being considered for use in the product are placed on an informal stability test for a minimum of 1 to 3 months so preservative system performance can be monitored after incubation for specific times at specific temperatures. The final preservative system is selected. Then pilot and production batches are made and placed on formal stability test to verify that the preservative

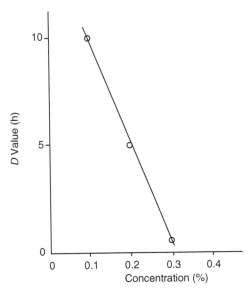

FIGURE 18.4 Preservative death time curve for *S. aureus* in lotion test samples used in Figure 18.3. The *D* value obtained in the absence of methylparaben (e.g., >30 h) was not used in constructing this preservative death time curve because the bacteria were not being inactivated. (From Orth, D.S. (1984). In *Cosmetic and Drug Preservation. Principles and Practice*. Kabara, J.J., Ed. Marcel Dekker, New York, p. 410. With permission.)

system continues to meet acceptance criteria throughout the shelf life of the product (i.e., up to 3 years).

18.3.2 Maximum Allowable *D* Values for Gram-Negative Bacteria

Pharmaceutical product manufacturers are aware that various products kill microorganisms at different rates. Although some preparations do not become contaminated with microorganisms, others do. This issue has been difficult to understand because product contamination may be sporadic and may occur in products that meet USP, EP, or CTFA acceptance criteria. A simple pharmacodynamic model was reported to describe the kinetics of death and regrowth when bacteria were exposed to concentrations of antibiotics that did not kill the population quickly enough (Li et al. 1994).

The kinetics of death of Gram-negative bacteria that may be used in preservative efficacy testing (*P. aeruginosa*, *Burkholderia cepacia*, and *E. coli*) following addition to diluted shampoos were determined to learn how fast these organisms must be killed to prevent adaptation or persistence. In the higher dilutions of shampoo, the APCs decreased 1 to 2 logs within 48 h and then appeared to stabilize or increase a little. It was found that these bacteria were capable of survival and growth when initial rates of killing were too slow, that is, when *D* values were about 30 h or greater in the product tested (Orth et al. 1998). Although all bacteria tested in this work started to die at a rate faster than required by CTFA preservative efficacy test acceptance criteria (*D* value ≤56 h), this work demonstrated that survival and growth was possible if rates of death were too slow. Gram-negative bacteria were used as the test organisms because they have been more troublesome for cosmetic and drug products than Gram-positive bacteria. Adequately preserved products kill test organisms at rates that may be described by linear survivor curves; however, it is believed that inadequately preserved products may not kill microorganisms at constant rates due to changes in the population (adaptation) that result in nonlinear rates of death. Although target acceptance criteria of the EP (2005) and linear regression methods require initial kill rates of <30 h for bacteria, the acceptance criteria of the USP (2004) and CTFA (2001) methods do not. The USP and CTFA criteria should be used with caution. Formulae that kill Gram-negative bacteria at rates greater than 30 h should be used only with special manufacturing precautions (aseptic filling) or with packaging that prevents water or microbial intrusion into the product.

18.3.3 THE REQUIRED D VALUE

Packaging and consumer use (abuse) play significant roles in determining whether a product becomes contaminated during use (see Chapter 16). The required D value (RDV) concept was introduced to provide a means for determining the adequacy of preservation of a product, bearing in mind the risks of contamination during use (Orth et al. 1992). Three variables that determine whether the product will become contaminated are:

1. The preservative system of the formulation, which may be expressed as the maximum D value for specific test organisms (i.e., target organisms) present in the formulation.
2. The packaging factor, which is based on a risk assessment of the extent to which the packaging allows contamination of the product during use.
3. A consumer use (abuse) factor, which is based on a risk assessment of the potential for microbial contamination of the product while in the hands of the consumer.

The RDV is the maximum permissible D value for specific organisms in a product. It relates formula, packaging, and consumer use factors by the following expression (Equation 18.2):

$$\text{RDV} = \frac{\text{Maximum acceptable } D \text{ value for target organisms}}{(\text{Packaging Factor}) \cdot (\text{Consumer Use/Abuse Factor})} \qquad (18.2)$$

In arriving at the RDV, the first step is determining adequacy of preservation. Target criteria for adequacy of preservation established by the linear regression method are:

- Products should have a D value of ≤ 4 h for pathogens.
- Products should have a D value of ≤ 28 h for nonpathogenic vegetative bacteria, yeasts, and molds.
- Products should be bacteriostatic or bactericidal for *Bacillus* spores.

The next step is to conduct a risk assessment of the packaging to determine whether it poses an unacceptable risk for a specific product. Generally, use of a bottle with a removable screw cap poses greater risk of product contamination than use of a pump dispenser or tube for the same product. The degree of risk of various packaging or closure options must be considered before final selection of the packaging is made. A risk assessment may reveal low risk (i.e., risk factor 1 with a single-use product, unit-dose container), moderate risk (i.e., risk factor 5 with a multiple-use container where contact with water or wet or soiled fingers is unlikely, such as a lotion in a pump dispenser), or high risk (i.e., risk factor 10 with a multiple-use container intended for use where repeated contact with water or wet fingers may occur, such as a shampoo in a pump dispenser). Use of a 1 to 10 scale is suitable for this risk assessment. Finally, a consumer use (abuse) risk assessment should be undertaken to predict the likelihood of microbial contamination that may occur during normal use of the product. Again use of a 1 to 10 scale is suitable.

Using the RDV concept, it is possible to establish the *required D* value for the preservative system against the target organism to ensure an *actual* level of protection equivalent to the target criteria. It draws attention to the multiple factors affecting contamination risk and encourages packaging improvements. An example can be found in Orth et al. (1992).

18.3.4 INTERFERENCES: INACTIVATION OF PRESERVATIVES BY SURFACTANTS

A number of physical and chemical parameters influence efficacy and stability of preservatives used in pharmaceutical preparations. The antimicrobial activity of a preservative system may decrease if preservatives are sequestered by solubilization or complex formation, if ionizable preservatives are made less effective as a result of pH effects, or if the concentration of preservatives in the aqueous phase is decreased as a result of partitioning into the oil phase of the product (Orth 1997; see also Chapters 14 and 15).

Incompatibility of preservatives with other formula components generally is detected during initial preservative efficacy testing. More subtle changes may occur slowly and be detected during stability testing. This results when preservative efficacy testing detects that the preservative system becomes inadequate because it fails to inactivate one or more of the test organisms quickly enough to meet acceptance criteria. Many preservatives are neutralized by nonionic surfactants and lecithin; cationic preservatives such as alkyltrimethylammonium bromide, chlorhexidine gluconate, and benzalkonium chloride are not compatible with soaps and anionic detergents. Formaldehyde and formaldehyde donors cannot withstand excessive heat treatments, and this is why they are generally added during the cooling phase of emulsion formation. Light is reported to inactivate chlorhexidine gluconate, and parabens are reported to be inactivated by pigments, talc, and titanium dioxide (Orth 1999).

Nonionic emulsifiers have enabled formulation chemists to develop elegant oil-in-water creams and lotions that have high water content and fairly neutral pH values (pH 5 to 9). In some instances, this has created difficulties for preserving these formulations because bacteria, yeasts, and molds frequently exhibit optimal growth under neutral pH conditions. Formulations that contain nonionic emulsifiers generally require higher concentration of lipophilic preservatives (e.g., paraben esters) for adequate preservation than are required when other emulsifiers are used. Explanations for this loss in preservative potency include complex formation and solubilization in surfactant micelles.

Although cationic, anionic, and nonionic surfactants may interfere with specific preservatives, the nonionics interfere with more classes of preservative than either cationic or anionics. The primary reason for this is the low critical micelle concentration value for nonionic surfactants; however, factors such as pH, water activity, and solubility may also be involved. Although physiochemical methods are useful for predicting the inactivation of preservatives by surfactants, preservative efficacy testing provides definitive information about the effect of a product preservative system on the rates of death of test organisms. Formulation chemists and microbiologists must be aware of the potential for interaction of preservatives with formula components to guide selection of systems that will be both effective and stable.

18.3.5 PRESERVATIVE EFFICACY TESTING DURING STABILITY TESTING

Preservative efficacy testing is performed to determine the minimum effective concentration of antimicrobial preservatives required for adequate preservation of pharmaceutical products. Products must be adequately preserved from the time they are manufactured until they are used up by consumers. Testing during product stability assessment is necessary to verify that the preservative system remains satisfactory during the expected shelf life of the product. Samples of the product (formulation in final packaging) are subjected to stability testing by storing samples at different temperatures (e.g., 3, 25, 40, and 50°C) for varying lengths of time before conducting preservative efficacy tests to ascertain whether the preservative system continues to meet acceptance criteria and is stable (Orth et al. 1987). This testing should be performed with all appropriate test organisms in samples of the product prepared with each preservative being considered for use in the product. Although many laboratories use "in-house organisms" in preservative efficacy testing (Orth et al. 1996), experience has shown that such organisms are not necessary when products meet the requirements of the linear regression method.

18.3.6 ACCEPTANCE CRITERIA

The acceptance criteria used to judge the adequacy of a preservative system in pharmaceuticals and cosmetics depend on the method of testing and the regulatory requirements for that type of product. The quantitative rates of death provided by the linear regression method enable quantitative acceptance criteria to be based on D values. For instance, pathogens should be killed within 24 h in all multiple-use cosmetic products that are capable of supporting microbial growth (Orth 1979). Thus, products should have a preservative system that will give D values of ≤ 4 h for pathogens so that 10^6 pathogens per gram will be inactivated completely within 24 h. The 24-h period thus provides time for a contaminated product to self-sterilize before use 1 day later. Less stringent criteria are considered to be appropriate for nonpathogens, giving D values of ≤ 28 h for nonpathogenic vegetative bacteria, yeasts, and molds. Again, this period was selected to enable a product contaminated with 10^6 organisms per gram to self-sterilize within 1 week (Orth 1979). Nonsterile aqueous cosmetic and pharmaceutical products should be bacteriostatic or slowly bactericidal for *Bacillus* spores.

REFERENCES

Akers, M.J., Boand, A.V., and Binkley, D.A. (1984). Preformulation method for parenteral preservative efficacy evaluation. *J. Pharm. Sci.,* 73, 903–905.

British Pharmacopoeia. (1993). Efficacy of antimicrobial preservation. *British Pharmacopoeia*, Vol. II. Her Majesty's Stationery Office, London, pp. A191–A192.

CTFA. (1973). Preservation Subcommittee of the CTFA Microbiological Committee. Evaluation of methods for determining preservative efficacy. *CTFA J.,* 5, 1–4.

CTFA. (2001). Preservation Subcommittee of the CTFA Microbiological Committee. Determination of preservative adequacy in cosmetic formulations. *CTFA Microbiological Guidelines*. CTFA, Washington, D.C.

European Pharmacopoeia. (2005). Efficacy of antimicrobial preservation. *European Pharmacopoeia*. 5th ed.. Council of Europe, Strasbourg, France, pp. 447–449.

Geis, P.A. and Hennessy, R.T. (1997). Preservative development. In *Cosmetic Microbiology: A Practical Handbook*. Brannan, D.K., Ed. CRC Press, Boca Raton, FL, pp. 143–161.

Leak, R.E., Morris, C., and Leech, R. (1996). Challenge tests and their predictive ability. In *Microbial Quality Assurance in Cosmetics, Toiletries and Pharmaceuticals*. Baird, R.M. and Bloomfield, S.F., Eds. Ellis Horwood, Chichester, U.K., pp. 199–216.

Levy, E. (1987). Insights into microbial adaptation to cosmetic and pharmaceutical products. *Cosmet. Toil.,* 102(12), 69–74.

Li, R.C., Nix, D.E., and Schentag, J.J. (1994). Pharmacodynamic modelling of bacterial kinetics: β-lactam antibiotics against *Escherichia coli. J. Pharm. Sci.,* 83, 970–975.

Orth, D.S. (1979). Linear regression method for rapid determination of cosmetic preservative efficacy. *J. Soc. Cosmet. Chem.,* 30, 321–332.

Orth, D.S. (1984). Evaluation of preservatives in cosmetic products. In *Cosmetic and Drug Preservation. Principles and Practice*. Kabara, J.J., Ed. Marcel Dekker, New York, pp. 403–421.

Orth, D.S. (1991). Standardizing preservative efficacy test data. *Cosmet. Toil.,* 106(3), 45–48, 51.

Orth, D.S. (1993). *Handbook of Cosmetic Microbiology*. Marcel Dekker, New York.

Orth, D.S. (1997). Inactivation of preservatives in surfactants. In *Surfactants in Cosmetics*. 2nd ed. Rieger, M.M. and Rhein, L.D., Eds. Marcel Dekker, New York, pp. 583–603.

Orth, D.S. (1999). *An Introduction to Cosmetic Microbiology*. IFSCC Monograph no. 5. Micelle Press, Weymouth, Dorset.

Orth, D.S. and Brueggan, L.R. (1982). Preservative efficacy testing of cosmetic products. Rechallenge testing and reliability of the linear regression method. *Cosmet. Toil.,* 97(5), 1–65.

Orth, D.S. and Eck, K.S. (2001). Preservative efficacy testing of aqueous cosmetics and drugs without counting colonies. *Cosmet. Toil.,* 116(4), 41–44, 46, 48, 50.

Orth, D.S. and Enigl, D.C. (1993). Preservative efficacy testing by a rapid screening method for estimation of D-values. *J. Soc. Cosmet. Chem.,* 44, 329–336.

Orth, D.S. and Milstein, S.R. (1989). Rational development of preservative systems for cosmetic products. *Cosmet. Toil.,* 104(10), 91, 92, 94–100, 102, 103.

Orth, D.S., Lutes, C.M., Milstein, S.R., and Allinger, J.J. (1987). Determination of shampoo preservative stability and apparent activation energies by the linear regression method of preservative efficacy testing. *J. Soc. Cosmet. Chem.,* 38, 307–319.

Orth, D.S., Barlow, R.F., and Gregory, L.A. (1992). The required *D*-value: evaluating product preservation in relation to packaging and consumer use/abuse. *Cosmet. Toil.,* 107(12), 39–43.

Orth, D.S., Dumatol, C., and Zia, S. (1996). House organisms: dealing with the "bug in the plant." *Cosmet. Toil.,* 111(6), 59–66, 68–70.

Orth, D.S., Delgadillo, K.S., and Dumatol, C. (1998). Maximum allowable *D*-values for Gram-negative bacteria: determining killing rates required in aqueous cosmetics. *Cosmet. Toil.,* 113(8), 53–56, 58, 59.

United States Pharmacopeia. (2004). *United States Pharmacopeia 28.* U.S. Pharmacopeial Convention, Rockville, MD, pp. 2149–2150.

19 Preservatives, Antimicrobial Agents, and Antioxidants: Registration and Regulatory Affairs*

R. Keith Greenwood and Robin J. Harman

CONTENTS

---

* Regulations develop on a regular basis. The reader is advised to consult the most recent official documents for the most up-to-date position.

19.1 INTRODUCTION

This chapter provides an understanding of the legal framework governing the use of chemicals with preservative, antimicrobial, or antioxidant properties in medicinal products. Although, strictly speaking, the antimicrobial chemical represents the only "true" antimicrobial preservative agent, previous chapters (Chapters 13 through 15 and Chapter 17) have shown the integrated nature of preservation. Thus, product-stabilizing preservatives and antioxidants have been included in this regulatory review because these will undoubtedly contribute, even if only indirectly, to the security of the product (see Chapter 13). Even though the preservatives, antimicrobial agents, and antioxidants are not the most important ingredients in a pharmaceutical formulation (the active ingredient that provides the therapeutic effect has this role), the regulatory authorities that approve the sale and supply of a pharmaceutical formulation place great importance on having as complete a knowledge as possible of the function and properties of the preserving agent(s).

The pharmaceutical industry is probably the most highly regulated of all industries, especially in the requirement for legislative approval obtained prior to a company being able to market its products. All ingredients of a medicinal product are scrutinized during the regulatory process. The important role of a preservative, antimicrobial, or antioxidant in a pharmaceutical formulation requires submission of comprehensive data on the ingredient, its effectiveness, and its influence, if any, upon the product and container. This is in addition to extensive data on the active ingredient(s), excipients, and the container as a whole.

Owing to its membership in the European Community, the U.K. legal framework for registration of medicinal products has been based upon European Community

(EC) Directives and legislation. In 2001, the European Parliament initiated a process consolidating all major procedural directives and their amendments, since 1965, for human and veterinary medicines and on December 18, 2001, two new Community Codes came into force. These were Directive 2001/83/EC Community Code relating to medicinal products for human use and Directive 2001/82/EC Community Code relating to veterinary medicinal products (Anonymous 2001a, 2001b). All earlier directives were revoked but the exercise did not change the legislation. Consequently, much of the legislative detail given in this chapter also applies to the sale and supply of medicinal products in all Member States (MSs) of the EC. More recent advances in harmonization of legislative requirements extend many of those requirements to all major markets in the world.

Any medicinal product, for which a therapeutic, preventative, or diagnostic effect is claimed when administered to humans or animals, has to undergo detailed scientific scrutiny in each country where that product is to be marketed. Data are provided by the company wishing to market the product, in a marketing authorization application (MAA). This is assessed by regulatory authorities to ensure that the medicinal product conforms to the necessary criteria of "quality, safety, and efficacy," as laid down in the EC in various forms of EC legislation (Sections 19.3 and 19.4). This assessment must be completed satisfactorily before permission is granted by the country's regulatory authority for the sale or supply of the medicinal product in its territory.

Preservatives, antimicrobials, or antioxidants are found in many different products, from multidose eye drops and oral liquid preparations to topical creams. The objective of their presence is to minimize the degradation with time, which may occur in a pharmaceutical formulation. Therefore, they are of vital importance in maintaining the quality, safety, and efficacy of a medicinal product throughout the period of its shelf life by:

- Ensuring stability of ingredients (quality)
- Preventing degeneration of any ingredient into potentially harmful by-products (safety)
- Ensuring that the formulated product remains therapeutically effective (efficacy)

19.2 DEFINING REGULATORY AFFAIRS AND REGISTRATION

Regulatory affairs is the gathering and application of all kinds of intelligence that may be used to ensure that new or established medicinal products are researched, developed, manufactured, and supplied in accordance with current (and possibly future) legislation in any country where they may be marketed (Greenwood 1985). The process by which a company collects, collates, records, and presents its data to the authorities and obtains a valid marketing authorization (MA) for a medicinal product is termed *registration*.

The regulatory affairs and registration function is normally handled by a regulatory professional, a scientist, and often a pharmacist, with postgraduate training in the topic. Their work encompasses the briefing of fellow scientists, doctors, and marketers on the regulations and the collecting, collating, and recording of data on all aspects of the medicinal product, including:

- Details on the creation, research, development, manufacture, assembly, storage and distribution, and the quality assurance systems applied to both the raw materials and the finished medicinal product
- Preclinical pharmacological, toxicological, pharmacodynamic, and phar-macokinetic assessments of the active ingredient(s) and where relevant, other components of the formulation (excipients)
- Clinical pharmacological assessment and clinical research to demonstrate the safety and efficacy of the product, and that it has the therapeutic effects for which claims are being made by the marketing company
- Postmarketing surveillance (pharmacovigilance), in which the use of the product is monitored for adverse effects in the general population after the regulatory authority has issued a marketing authorization for the product

An efficient regulatory operation is vital for all companies marketing medicinal products. Even when a product is on the market, it must continue to meet changing regulatory requirements throughout its lifetime. As with all specialties, a separate vocabulary has evolved. A brief glossary of some of the terms used in regulatory affairs and registration is given in Table 19.1.

19.3 REGULATORY PROCESS IN THE EUROPEAN UNION: AN OVERVIEW

19.3.1 THE ROLE OF REGULATORY AUTHORITIES

National regulatory authorities advise their respective licensing authorities in each country, usually government ministers, who are responsible for granting a MA. The European Agency for the Evaluation of Medicinal Products (EMEA) similarly advises the European Commission, which is the licensing authority for the European Union. The functions of regulatory authorities vary between countries, but tasks that are common to most are listed in Table 19.2.

Pharmaceutical companies can choose varying routes by which to obtain approval to market a medicinal product but, in essence, the assessment of applications by each route is essentially similar. These different routes are described in Section 19.4.

19.3.2 PURPOSES OF REGULATORY PROCEDURES

Procedures for the authorization and supervision of medicinal products in the EC are described in the Community Codes, which came into force on December 18, 2001 (Anonymous 2001a, 2001b). The codification exercise consolidated but did not change earlier legislation.

TABLE 19.1
Glossary of Terms Used in Regulatory Affairs and Registration

Active Substance
The ingredient in a medicinal product with therapeutic, diagnostic, or prophylactic activity.
Synonyms: drug substance; active ingredient; new chemical entity (NCE), superseded by new active substance (NAS).

Appeals Procedure
Provision for a marketing authorization application (MAA) to be reconsidered by the regulatory authority when notice is given that the MAA is to be refused. An appeal may also be allowed if a marketing authorization (MA) is revoked or is subject to compulsory variation.

Assessment Report
A summary of the assessment of the application data prepared by the regulatory authority. Can be for internal authority purposes only, or may be released, on request, to the applicant or to other regulatory authorities.
Synonyms: summary basis for approval; evaluation report.

Dosage Form
Formulation of active ingredient(s) for administration to a patient in a specified quantity or strength. Examples include tablets, capsules, injection solutions, syrups, ointments, and suppositories.
Synonyms: pharmaceutical form; finished product.

Generic Product
A pharmaceutical product sold under a nonproprietary name of the active substance, often by reference to a pharmacopoeial monograph.
Synonyms: generic drug; standard product; nonproprietary product.

Good Laboratory Practice (GLP)
Code for the conduct of nonclinical investigations and experiments in the laboratory, during the research and development stages of a new product.

Good Manufacturing Practice (GMP)
A code of quality assurance procedures that must be followed throughout the manufacture of a pharmaceutical product to ensure the quality of the finished product.

Guidelines
Recommendations, which are not legally binding, issued by a regulatory authority or other organization. A guideline can relate to the MAA format and content, to the assessment procedure, or to the scope and manner of conducting studies.
Synonym: notes for guidance.

Hearing
An opportunity for company representatives or their regulatory advocates to appear in person before a regulatory authority, often related to an appeals procedure. Carried out to clarify specific points of issue raised during the assessment of the MAA.

Manufacturer's Authorization
A permit issued by government allowing a company to manufacture pharmaceutical products. It is normal for the premises to be inspected by the regulatory authorities before the issue of the authorization and to be checked by inspectors periodically to ensure compliance with the code of GMP, which has been adopted throughout the EC.
Synonym: manufacturer's license.

Marketing Authorization (MA)
Authorization from the licensing authority (usually the government or its agency) for a medicinal product to be made available under stated conditions, for sale or supply to the marketplace. This is usually in the form of a letter or certificate, which defines the terms of the authorization, normally by reference to the application data and subsequent amendments.
Synonyms: product license, registration.

(continued)

TABLE 19.1 (continued)
Glossary of Terms Used in Regulatory Affairs and Registration

Marketing Authorization Application (MAA)	The data provided to the regulatory authority in support of a request for a marketing authorization. Synonyms: submission, registration dossier, registration file.
Pharmaceutical or Medicinal Product	Any substance or combination of substances presented for treating or preventing disease in human beings or animals. Any substance or combination of substances that may be administered to human beings or animals with a view to making a medical diagnosis or to restoring, correcting, or modifying physiological functions in human beings or animals (Anonymous 2001a, 2001b). Note: For the purpose of product registration, the term normally has a legal definition to specify the categories of product, which fall within the scope of the registration procedure. Definitions will vary between regulatory systems. Synonyms: medicinal product, drug, pharmaceutical, medicine.
Prescription Products	Pharmaceutical products that may only be sold or supplied in accordance with an order (prescription) from a qualified practitioner. Synonyms: ethicals, proprietary products.
Product Evaluation	The process of determining whether or not a MA should be granted, in accordance with an application. Synonyms: product assessment, drug evaluation or assessment.
Product Registration	The procedure by which the regulatory authority processes and documents information on a product and authorizes its marketing. Synonym: drug registration.
Product Regulation	The system for the control of the sale and supply of medicinal products, including manufacture, distribution, marketing, labeling, promotion, and supply for clinical investigation. Control is normally exercised through the issue of authorizations to market, manufacture, and so on, and enforcement of the conditions of those authorizations is through inspection, sampling, testing, legal sanctions, and so on. Synonyms: drug control, drug regulation.
Proprietary Product	A medicinal product sold or supplied under a special name (a brand or trade name) rather than the generic name of the ingredient alone. Synonyms: pharmaceutical speciality, branded product.
Publicly Promoted Medicines	Products that are advertised directly to the general public for the purpose of self-medication. These are, by definition, nonprescription, and may be sold through pharmacies or other retail outlets. Synonym: OTC (over-the-counter) products.
Regulatory Authority	Government department or agency responsible for the control of the sale and supply of medicinal products through a system of product registration or authorization. Synonyms: licensing authority, registration authority, drug regulatory agency (DRA)
Variation	An alteration made to the terms of a MA as a result of changes made to the product. This may either be at the request of the holder of the MA or at the instigation of the authorities on the basis of new data. Synonyms: modification, amendment, change.

TABLE 19.2

Functions of a Regulatory Authority for Human and Veterinary Medicinal Products

Safeguarding the public health by ensuring that all marketed medicines are of an appropriate standard of quality, safety, and efficacy through:

Licensing of Medicinal Products
Assessing applications for marketing authorizations for medicinal products
Dealing with parallel imports of medicinal products
Approving and monitoring clinical trials

Carrying Out Postmarketing Activities
Pharmacovigilance
    Variations to marketing authorizations
    Renewals of marketing authorizations
Changes of legal classification of medicinal products
Regulation of product information
    Regulation of advertising and promotion of medicinal products

Inspection and Enforcement
Ensuring quality assurance in manufacture of medicinal products
Licensing manufacturers and wholesalers
Monitoring good manufacturing practice and good distribution practice
Monitoring good laboratory practice
Monitoring good clinical practice
Monitoring the quality of marketed medicinal products
Monitoring the quality of foreign manufacturing operations exporting to the country
Investigating unlawful activities in medicinal products

The dual objectives of current procedures are:

1. To ensure that effective therapeutic products from the pharmaceutical industry gain rapid access to the citizens of the EC in all its MSs. This is achieved either by obtaining a single "centralized" community-wide MA based on an assessment by the EMEA or by a process of "mutual recognition," whereby second and further MSs accept the assessment carried out by the first MS.
2. To rationalize the system of assessment, by keeping resources to a minimum and avoiding costly and time-consuming duplication.

No criteria other than those of quality, safety, and efficacy are permitted to prevent the granting of a MA (for instance, the price of products cannot be considered).

The most common system of obtaining a MA is based on data assessment conducted by national assessment bodies whose responsibilities were increased in 1995 with the introduction of the new mutual recognition procedure (MRP).

The EMEA, located in London, acts as the focus for scientific evaluation of medicinal products submitted through the centralized procedure (CP), which is used to gain a MA throughout the European Union based on one application. The scientific assessment of a MAA submitted through the centralized procedure is carried out by a scientific committee, the Committee for Human Medicinal Products (CHMP) or the Committee for Veterinary Medicinal Products (CVMP). The EMEA only becomes involved in the MRP if mutual recognition cannot be agreed between the MSs. In this event, Article 11 or Article 12 procedures (of Council Regulation C2309/93) are triggered, in which the CHMP or the CVMP acts as arbitrator.

Unlike some national regulatory authorities, the EMEA has no authority to issue MAs. For products approved through the centralized procedure, the European Commission is the licensing authority. For products approved through mutual recognition, each national regulatory authority acts as the licensing authority in its territory, hence the pivotal role of national bodies.

The full details of the procedures for registration of human and veterinary medicinal products are given in a series of publications from the European Commission entitled *The Rules Governing Medicinal Products in the European Union* (Anonymous 1998). Full, updated details of these publications can be obtained from EC national regulatory authorities or on the Enterprise Directorate's Pharmacos Web site.

19.3.3 TYPES OF EC LEGISLATIVE INSTRUMENTS

The EC legislation affecting the pharmaceutical industry comes in various forms. A clear understanding of their differences is essential.

Regulations are passed by the European Commission or the European Council and, upon ratification, are immediately effective in all MSs. A binding legal effect in each MS is attained without individual national legislation having to be promulgated. If a conflict exists between pre-existing national legislation and the new regulation, the regulation has legislative priority. Examples of regulations affecting the pharmaceutical industry are:

- Council Regulation 2309/93, July 22, 1993 (European Community 1993) — Outlined procedures for the authorization and supervision of medicinal products for human and veterinary use, and established the EMEA.
- Council Regulation 297/95, February 10, 1995 (European Community 1995a) — Outlined fees payable to the EMEA.
- Commission Regulation 540/95, March 10, 1995 (European Community 1995b) — Outlined the arrangements for the reporting of suspected, unexpected adverse reactions which are not serious, whether arising in the community or a third country, to medicinal products for human or veterinary use authorized in accordance with the provisions of Council Regulation 2309/93.

Directives are legislative instruments whose objectives are binding upon MSs but, to become operational, have to be implemented into each MS's national legislation within a defined time. However, the method by which the MS implements the directive is for the MS to decide. EC directives have no legal foundation in individual MSs, but the European Commission, the European Court of Justice, and other

European bodies ensure that the directive has been implemented in each MS, and that the "spirit" of the directive has also been transferred to national legislation. Sanctions may be imposed upon MSs that do not do so within the required time.

Directives are by far the most common instrument used within pharmaceutical legislation. Apart from Community Codes 2001/83/EC (The Human Use Directive) and 2001/82/EC (The Veterinary Use Directive) (Anonymous 2001a, 2001b) the most important are:

Council Directive 91/356/EEC—Good Manufacturing Practices (GMP) Code, June 13, 1991—Outlined the principles and guidelines of good manufacturing practice for medicinal products for human use (Anonymous 1991)

Council Directive 1999/11/EC—Harmonized the laws, regulations, and administrative provisions relating to the application of the principles of good laboratory practice (GLP) and the verification of their applications for the tests on chemical substances (Anonymous 1999).

Council Directive 2001/20/EC, April 4, 2001—Discussed the approximation of the laws, regulations, and administrative provisions of the member states relating to the implementation of good clinical practice (GCP) in the conduct of clinical trials on medicinal products for human use (Anonymous 2001c)

Guidelines (also referred to as *Notes for Guidance*) do not have any legal authority, but their use by companies is critical in satisfying the requirements of the regulatory authorities. Guidelines provide an important means of helping ensure that all regulatory authorities interpret scientific and technical issues from the volumes of *The Rules Governing Medicinal Products in the European Union* (Anonymous 1998) in a similar way. (Note that published volumes are now updated on the Pharmacos Web site.) Knowing that guidelines are being applied in a particular way by all regulatory authorities assists companies in compiling the MAAs in a way that should satisfy all quality, safety, and efficacy issues.

19.4 PROCEDURES FOR REGISTRATION OF MEDICINAL PRODUCTS IN THE EC

A company's regulatory affairs department must ensure that development scientists are fully aware of requirements at the time of providing data in support of a MAA. Reference to all original and amended legislative documents is essential prior to ensuring compliance with their requirements.

19.4.1 THE CENTRALIZED PROCEDURE

The attraction for the pharmaceutical industry of the centralized procedure is that only a single MA is required for a medicinal product to be approved throughout all MSs of the European Community.

In the centralized procedure, the MAA is submitted to the EMEA. Once the application has been validated (i.e., checked that it contains all the required information) and the registration fee has been paid by the applicant, it is passed to the CHMP (for human medicines) or the CVMP (for veterinary medicines). The committee has 210 days (or 150 days under special circumstances) from the date on which the application was submitted to assess it and reach an *Opinion*. The assessment is coordinated by a rapporteur appointed by the C(H/V)MP from among its members. The applicant company can suggest who it would like to be rapporteur, and most applicants list three or four possibilities. By doing so, one of its choices is more likely to be selected than if a single choice is suggested. It is the responsibility of the rapporteur to assemble a team of assessors who are most suited to the type of product undergoing assessment. Usually, for logistical reasons, the assessors come predominantly from the rapporteur's own regulatory authority.

At the end of the period of assessment of the MAA, three outcomes are possible:

1. If the MAA does not satisfy the requirements of Council Regulation 2309/93, a *negative Opinion* will be given. In such an event, the applicant can lodge an appeal to the C(H/V)MP within 60 days. In turn, the C(H/V)MP has 60 days to assess the appeal, usually having been given additional data by the applicant company in support of its application.
2. A *conditional Opinion* may be given. This may require amendments to the Summary of Product Characteristics (the definitive statement of the product's uses) before a positive Opinion will be granted. The applicant has two weeks after a conditional Opinion to appeal against the changes demanded.
3. A *positive Opinion* may be given, in which case the C(H/V)MP has approved the MAA.

Once a positive Opinion is reached, the EMEA has 30 days within which to send its Opinion to the European Commission, the licensing authority. Within a further 30 days, the European Commission must prepare a draft Decision on the Opinion and send it to all EC MSs and the applicant. The final Decision, granting the marketing authorization, is taken by the Standing Committee on Medicinal Products for Human Use when it has received the draft Decision. The approval is then published in the *Official Journal of the European Communities*. There is no defined time limit for this to be completed.

19.4.2 THE MUTUAL RECOGNITION PROCEDURE

The principle of mutual recognition is well founded in European Community law; it is not restricted to the pharmaceutical sector. The principle is based upon the recognition by one MS of the legislative and other administrative decisions taken in another MS. A second MS can refuse to apply the principle, which triggers an "arbitration" process. Resolution of arbitration will ultimately lead to a decision taken by the European Community and with which all MSs have to comply.

Following Directive 2001/83/EC (Anonymous 2001b), a company can seek approval of a MAA within one MS. The MS has 210 days in which to assess the application. If the assessment is positive, it issues a marketing authorization. The

MS in which the company has obtained its primary authorization is referred to as the "reference member state" (RMS). The company has complete freedom of choice as to which MS it approaches to act as RMS. A regulatory authority with which a pharmaceutical company has forged a positive working relationship over several years is an obvious choice as a proponent. Once approved in one MS, the company can submit the same MAA in another MS and ask for mutual recognition, within 90 days of submission, of the first MS's approval.

A second MS has to agree to the scientific assessment carried out by the RMS covering each of the following: indications, contraindications, dosage and method of administration, and shelf life and storage requirements. Should the second MS dispute any of these issues, it has to raise objections in writing with the first MS within the first 60 days of the 90-day period. The remaining 30 days are used for dialogue to lead to the (hopefully) successful resolution of the second MS's concerns. If the issues cannot be satisfactorily resolved, a complicated, and potentially long drawn-out process of arbitration will ensue. Companies try to avoid reaching this stage whenever possible.

Currently, applications for all medicinal products can proceed via this MRP *except* for those products required to be subject to the centralized procedure (Regulation EC 2309/93; European Community 1993) and those products that the company has opted to authorize through the centralized route.

19.4.3 THE ICH PROCESS

Above and beyond the national (and for the EC, supranational) processes for registration of medicinal products lies the ICH process. The acronym stands for International Harmonisation of the Technical Requirements for the Registration of Human Medicinal Products, later shortened to International Conference on Harmonisation. The ICH process has been one of the major successes in global registration. Although the process itself does not lead to issuing of a MA, it has led to many guidelines being standardized throughout the world, reducing the number of tests that have to be carried out by companies in supplying data in different regions.

The ICH process comprises six primary partners: the regulatory authorities and trade associations of each of the European Communities, the United States, and Japan. The first major ICH conference was held in 1991; "consolidation" meetings have been held subsequently.

The immense importance of the ICH process for both the pharmaceutical industry and the regulatory authorities has been the willingness of both sides to attempt to minimize the differences in regulatory requirements between their regions. Significant progress has been made, especially in the "quality" aspects of the MAA in the form of the common technical document. The involvement of the regions' respective pharmacopoeial authorities in the discussions since 1993 has given further impetus to the harmonization of quality standards (Table 19.3, see part II), and particularly those relating to preservation.

Due to the ICH process, differences between the requirements of the authorities in the European Community, United States, and Japan have closed considerably but not comprehensively. It must not be assumed that data in compliance with the EC

TABLE 19.3
EC *Notice to Applicants;* Typical Table of Contents of a Marketing Authorization Application

Foreword
Introduction

Part I	Summary of the Dossier
Part I A	Administrative data
Part I B1	Summary of product characteristics (SPC)
Part I B2	Proposal for packaging, labeling, and package insert
Part I B3	SPCs already approved in the member states
Part I C	Expert reports:
	On chemical, pharmaceutical, and biological documentation
	On toxicopharmacological (preclinical) documentation
	On clinical documentation
Part II	Chemical, Pharmaceutical, and Biological Documentation for Chemical Active Substances
Part II A	Composition
Part II B	Method of preparation
Part II C	Control of starting materials
Part II D	Control tests on intermediate products
Part II E	Control tests on the finished product
Part II F	Stability
Part II G	Bioavailibility/bioequivalence
Part II H	Data related to the environmental risk assessment for products containing genetically modified organisms (GMOs)
Part II Q	Other information

There are variants of Part II for radiopharmaceutical products, biological medicinal products, and vegetable medicinal products.

Part III	Toxicopharmacological Documentation
Part III A	Toxicity
Part III B	Reproductive function
Part III C	Embryofetal and perinatal toxicity
Part III D	Mutagenic potential
Part III E	Carcinogenic potential
Part III F	Pharmacodynamics
Part III G	Pharmacokinetics
Part III H	Local tolerance
Part III Q	Other information
Part III R	Environmental risk assessment
Part IV	Clinical Documentation
Part IVA	Clinical pharmacology
Part IVB	Clinical experience
Part IV Q	Other information
Annex	Table of contents for remainder of the dossier

requirements will automatically satisfy the other authorities. Many differences in interpretation are still in evidence. Consequently, consultation with regulatory professionals having specific and up-to-date knowledge in each of these major markets is an essential prerequisite for the design of protocols for studies submitted to these authorities. This is particularly true at this time as both scientific and legal requirements in all markets are undergoing rapid change.

19.4.4 The Format of an EC Application for a Marketing Authorization

All EC MAAs must be submitted to the regulatory authority in a defined format, ideally incorporating the form of the ICH common technical document (ICH CTD). Immediately upon receipt, a check is made to ensure that all the required documents are present, and that all are in the required style and format. If they are not, the MAA is immediately rejected and returned to the applicant company. Much goodwill and benefit are gained in providing a well-organized, appropriately indexed, and complete dossier at this stage.

The Rules Governing Medicinal Products in the European Union (Anonymous 1998) give full instructions on the format of the MAA. The overall general format is shown in Table 19.3, but a more detailed composition incorporating the ICH CTD is issued as Volume 2B of the Notice to Applicants on the Pharmacos Web site to which reference should be made for the latest requirements.

19.5 PRESERVATIVES, ANTIMICROBIALS, AND ANTIOXIDANTS IN FORMULATED PRODUCTS: LEGISLATIVE CONSIDERATIONS

During the course of a number of years, guidelines have been drawn up by the EC to deal with the problem of pharmaceutical products subject to potential microbial contamination. Most guidelines have been based upon the principle of preservative efficacy testing.

The main sources of standards for preservatives and their efficacy have been national pharmacopoeias. They too, however, like other aspects of pharmaceutical regulatory issues, have become increasingly harmonized between different regions (e.g., the European Community, the United States, and Japan), especially because of their participation in the quality aspects of ICH (see Section 19.4.3).

Whether the use of a preservative, antimicrobial, or antioxidant is necessary will depend upon the product's intended clinical use, the intended clinical properties of the pharmaceutical product, and the design and composition of the container. By the very nature of their ingredients, many pharmaceutical formulations are ideal environments for microbial growth (see Chapter 2). The pharmaceutical development scientist must show the regulatory authority that every effort has been made to prevent such contamination.

Chemical and microbial preservation must often be employed for formulated products that are to be used on more than one occasion once opened. Degradation by oxygen is a particular problem for many active ingredients, particularly when light and moisture are also present. The addition to the formulation of an ingredient that is oxidized more easily than the active ingredient is one approach in such cases.

Ascorbic acid is an example of a frequently used, relatively inert, antioxidant but whatever the choice, the regulatory authority will wish to see adequate evidence that the formulator has proved that the system works.

Microbial contamination is often more problematic than oxidative degradation. However, not all formulations require the addition of an antimicrobial agent; some products contain ingredients that already fulfill that function. A product with a high proportion of glycerol (or a related substance) is one example of a potentially self-preserving formulation (see Chapter 13). Bacterial contamination is also less likely in products containing a high proportion of sucrose (e.g., 70% or more) due to the high osmotic pressure (low water activity) of the solution (Morris 1994, see also Chapter 15).

Nevertheless, the regulatory authorities assume problems will occur in the development phase of a new pharmaceutical formulation; therefore, they should be anticipated rather than assumed to be unlikely. Examples of microbial contamination problems that have occurred in pharmaceuticals, and in particular in topical pharmaceuticals, include steroid creams contaminated with *Pseudomonas aeruginosa; P. aeruginosa* and coliform organisms found in a mouthwash; and microbial contamination of ophthalmic ointments and eye drops (Matthews 1993).

Absence of microbial contamination cannot be guaranteed after the patient has started to use the medicinal product. However, by the application of pharmaceutical development skills, possibly employing the use of preservatives, the formulator can demonstrate that the product created is appropriate for the purpose and conditions of use intended. It must be shown that any accidental contamination of the product during normal use will be minimized and does not present a significant danger to the user. The regulatory authorities will want to see an intrinsically robust product; any form of label disclaimer designed to overcome inadequate preservation or formulation would be quite unacceptable to the authorities.

A preservative must also remain effective throughout any changes to the product that are required before administration. Oral antibiotic solutions are inherently unstable, as a result of which many have to be reconstituted upon dispensing immediately prior to handing to the patient. The solution is then administered over a 5- or 7-day period. Any antioxidant or preservative added to the dry preparation must be shown to remain effective and nonharmful for the full period of treatment. Another example of a preparation similarly reconstituted is powder for injection to which sterile water is added. The effectiveness of the preservative system in such instances must also be demonstrated when the reconstituted injections are added to parenteral delivery systems.

More sophisticated products that must equally demonstrate that they can withstand microbial challenge include the more patient-friendly pen-injection devices. Not only must the active ingredient (e.g., insulin) be protected and remain stable throughout its shelf life, the injecting device must be shown to be able to prevent deliberate or accidental contamination of the contents (see Chapter 16).

19.5.1 Tests Used to Determine the Effectiveness of a Preservative System

Antimicrobial preservative efficacy (APE) testing systems are standardized tests (see Chapter 17) whose results can be incorporated into regulatory submissions. Such tests are described in current monographs in official publications such as the *British Pharmacopoeia* (BP) (2005), the *European Pharmacopoeia* (EP) (2004), and the *United States Pharmacopeia 28* (USP) (2005). If a formulator chooses to use a test system other than that specified in one of the official publications, full justification for this must be given in the regulatory submission. This should be discussed beforehand with a regulatory professional.

The basis of each of the APE tests is to challenge the pharmaceutical formulation, ideally in its final container, with a prescribed quantity of inoculum of microorganisms. The inoculated preparation is incubated at a defined temperature for a specified period, with samples being taken at preset intervals. The number of surviving organisms in the extracted samples is then determined, and these values compared against the requirements laid down in the pharmacopoeias.

The pharmacopoeias give guidance on the ways in which APE tests should be carried out, with reference to test organisms to be used, the maintenance of cultures, inoculum preparation, test procedures, and the interpretation of results (see Chapter 17). Any formulator planning to market a product in the United States as well as Europe must be aware of the differences between the tests for preservative efficacy in the two regions. These differences in applicability of the APE tests in the EP and the USP are of significance for the regulatory professional when justifying results presented in submissions and can sometimes cause problems in gaining approval.

When the data have been gathered, it is not sufficient to state in the regulatory submission that the APE tests have been carried out and satisfactory results obtained. The complete details of the test procedures must be described, making clear what variables were applied to the testing and sampling process. The results should also be supported by data from stability studies used to justify the shelf life of the product. APE data on two freshly made batches of the product should be obtained and submitted, together with test results from an older batch of product. Ideally, the older batch material should be derived from a production-scale sample, not a pre-scale-up process. In discussing the data submitted, its relevance to the actual use of the product by the patient should be considered, plus any contamination problems that might have occurred in smaller-scale use of the product (e.g., during clinical trials).

Advances in harmonization of the APE test have occurred, following a comparison of the tests required by the EP, the USP, and the *Japanese Pharmacopoeia* (JP) (Anonymous 1996, Matthews 1996). The EP test was described as applying primarily to the development product phase as opposed to a routine postmanufacture test. Agreement on some variation in the level of preservative activity was possible in Europe. In the United States, however, the test criteria had to be met for all products at all times—it was a mandatory, definitive requirement. The JP test was intended to be a guidance test rather than an absolute test. More positively, various criteria for the APE test were agreed to be the same in all three pharmacopoeias. The criteria include incubation times and temperatures, number of samples, and validation procedures (see Chapter 17).

The greatest obstacle to harmonization, however, remains the interpretation of the data from the APE tests. It is worthwhile reminding the reader that liaison with a regulatory professional helps to keep laboratory work in line with the latest regulatory thinking and minimizes the risk of rejection of data at the time of registration.

19.5.2 EC GUIDELINE ON INCLUSION OF ANTIOXIDANTS AND ANTIMICROBIAL PRESERVATIVES IN MEDICINAL PRODUCTS

The EC note for guidance entitled "Inclusion of Antioxidants and Antimicrobial Preservatives in Medicinal Products" (CPMP/CVMP/QWP/115/95) (Anonymous 1997) became effective in January 1998. The guideline stated that the MAA should explain why each antioxidant and antimicrobial preservative had been used and provide proof of efficacy of each. The finished product method of control should be described; the preservative should be listed in the finished product labeling, and its safety profile also described. Rather than naming specific antimicrobial or preservative compounds on the labeling, the agent(s) should be referred to by their pharmacopoeial monograph name, or in the absence of one, the Chemical Abstract Service (CAS) registry numbers should be declared. In addition to the chemicals' name, its intended uses must be stated.

The guideline goes on to explain the use of preservatives in development pharmaceutics and reminds the development scientist that the chemicals are in fact excipients and should therefore also conform to *The Rules Governing Medicinal Products in the European Union,* volume 3 (Anonymous 1998). Stability issues and correct labeling must also be confirmed for the use of the antioxidents and/or antimicrobial preservative.

The spectrum of antimicrobial preservatives traditionally recognized by regulatory authorities was summarized in the first edition of this book (Greenwood 1990) and the most commonly used agents are reviewed in Chapter 14 of this edition. Occasionally the choice of preservative system will not be previously known to the authorities. The more novel the concept, the more extensive the laboratory work will have to be in support of its registration. At the extreme, the preservative may be treated as a new chemical entity. In such cases, an authority's demands may be as stringent as those for registration of a completely new "active" drug substance. Tests for a new preservative will have to be provided in the same detail and possibly at a similar cost to that for a new "active" pharmaceutical. Until the authorities involved are convinced that all tests for safety, efficacy and quality, both for the new raw material per se and in the formulated product, are satisfactory, the product will not be permitted a license for sale or supply to the public. Under these circumstances a critical path program for the whole development project will probably be worthwhile.

19.5.3 TYPICAL PROBLEM AREAS FOR REGISTRATION

It is not practical to consider all potential regulatory problems that might arise when attempting to register a product containing either a well-known or a new preservative system. Table 19.4 lists some of the pharmaceutical parameters likely to be relevant to preservation and to require investigation or scientific comment for inclusion in the registration application. These are considered below and some in further detail in Chapter 14 and Chapter 15. For a complete list, it would be appropriate to consult a regulatory professional.

TABLE 19.4
Typical Pharmaceutical Parameters Relevant to Preservation That Require Attention for the Registration Application

Type of product
Toxicity
Type of container
Solubility
Stability
Volatility
Spectrum of activity
pH
Color
Odor
Taste
Preservative ability and efficacy
Interaction with ingredients
Scale-up and production
Raw materials
Analytical methods
Results of tests and interpretation
Validation of methods

19.5.3.1 Type of Product

The type of product indicates the kind of preservative system to be chosen. The authorities need evidence that the choice of preservative is appropriate for the purposes for which the product will be used. This is usually taken to mean that the degree of preservative activity should be related to the intended use and frequency of use of the product.

19.5.3.2 Toxic Potential of Preservatives

In formulating a medicine, preservatives themselves may be a source of toxicity for the patient. Adverse drug reactions (ADRs) are constantly under scrutiny by regulatory authorities. It is reported in the United Kingdom that preservatives attract attention as a cause of ADRs and the authorities welcome the increasing trend toward the development of unit-dose, preservative-free packaging for suitable products (Adams 1987, see also Chapter 16).

19.5.3.3 Type of Container

Even if the container is ideal for use with the product it may not be compatible with the preservative system to be used (see Chapter 15). The MAA will need to address any details of incompatibility with the containers in which the product is stored. If the containers used in stability tests differ from those proposed for the marketed product, the significance of the differences should be discussed. Regulatory authorities will consider the following parameters and how they might influence stability of the product:

- Fabric of the container—for example, the type of glass, plastic, composition of strip packaging
- Nature of closures—for example, details of liners, caps, and crimps

Large containers intended for bulk dispensing may require more stringent tests to allow for the effects upon the contents and the preservative system of continual opening and closing and, over a period of time, the increasing ullage and ingression of possible contaminants.

19.5.3.4 Solubility

In order to be effective, preservatives must usually be dissolved in the product, and be present at appropriate concentrations in the various phases of the formulation. Reassurance is needed that preferential solubility of the preservative in one or other of the phases does not detract from its efficacy throughout all of the product (see Chapters 14 and 15).

If a preservative is dissolved in a small portion of the aqueous phase (perhaps with the aid of heat) before it is added to the bulk, then it is necessary to demonstrate that the preservative does not precipitate from solution when added to the bulk.

19.5.3.5 Stability

A full stability program for the finished product is required with a minimum of three batches studied over a 3-year period under elevated, reduced, temperate, and humidity controlled climatic conditions as laid down in the ICH guidelines. The stability program is primarily aimed at the finished product packaged in the way proposed for marketing. During development and stability testing, it will be necessary to demonstrate, using suitable antimicrobial challenge tests, the efficacy of the preservative system used. Products containing a preservative system should also be assayed for levels of preservative during the storage tests. An assurance should be provided (where necessary) that ongoing data will be generated to show that the product, at the end of the shelf life, is satisfactorily preserved (Anonymous 2001a, 2001b).

19.5.3.6 Volatility

For preservative action to be sustained, it is essential that the preservative remains available at the correct strength within the product throughout the shelf life. Volatility might be relevant to its action but it can also mean that the preservative is lost from the product through evaporation. If this occurs at ambient temperature, it is likely to be even more pronounced at elevated temperatures. Consequently, regulatory authorities will wish to see evidence that the formulator is aware of the problem and has taken steps to overcome the loss of ingredients particularly under hot climatic conditions. Most regulatory bodies will require evidence showing that the ingredients are not escaping from a poorly sealed container. Loss of a volatile preservative degradation product, a particular problem when using plastic containers, can shift the equilibrium to further degradation of the preservative, as happens with thiomersal, for example (Hepburn 1987). Under user conditions, once the pack is opened, volatility is likely to result in a loss of preservative action, making user-challenge tests an item for close scrutiny.

19.5.3.7 Spectrum of Activity

A variety of pathogenic and nonpathogenic bacteria, molds, and yeasts may contaminate and spoil a product. The capability of the chosen preservative system to meet all eventualities throughout the shelf life is a very important regulatory consideration. Practical laboratory studies demonstrating that the system works when challenged by a variety of organisms must be provided in the MAA (Hepburn 1987), particularly if the system is new. If the product is liable to attack by a specific organism peculiar to the conditions in a particular market, then this aspect is likely to be scrutinized by the authority in that area. They would expect to see laboratory work demonstrating effective resistance to such contamination included in the application.

19.5.3.8 pH

The chosen preservative should be appropriate for the pH of the product; if not, a reasoned argument should be given for its selection. For most ionizable preservatives, the antimicrobial activity and the degree of dissociation are closely linked and, therefore, both are likely to be influenced by pH. If antimicrobial activity operates best at a specific pH, departure from that optimum will produce a less effective system. This clearly must be addressed during laboratory work on the proposed formulation. The formulator must show the regulator that the chosen formulation will not impair the preservative system either by being unfavorable to the stability of the preservative(s) or by causing changes within the basic formulation over time.

19.5.3.9 Color

Most preservative systems create no color problems for the formulator. Naturally, the appearance of color during the lifetime of the product would call for close scrutiny as it might indicate a loss of preservative efficacy. The authorities would require to be convinced that there was no problem. Any new preservative system that sought to provide a change in color, for example, when preservative levels became ineffective would require substantial experimental data demonstrating its reliability.

19.5.3.10 Odor

As long as it is acceptable to patients, any preservative system that imparts an odor to a formulation is not likely to be a cause for rejection of a MAA. However, if the odor is distinctive, problems for the blinding of clinical trials may arise if the placebos and competitor products are noticeably different. In addition, an offensive odor could lead to questions by regulatory authorities about potential patient non-compliance.

19.5.3.11   Taste

As most people have sensitive palates, taste will be an important consideration for the formulator of an oral product. Preservatives, although present in small quantities, often impart an aftertaste that some patients find unacceptable. From a regulatory viewpoint, taste is unlikely to be a contentious issue unless, like odor (see above), it invalidates test results (e.g., in a double-blind clinical study or by causing patient dosage noncompliance). The formulator that can provide matching placebos or, with more difficulty, matching competitor product for the clinical study will probably have no problems.

19.5.3.12   Preservative Ability and Efficacy

In designing a preservative efficacy test, consideration should be given to the proposed use of the product (Hepburn 1987). Factors such as reconstitution, dilution before use, duration of treatment, and frequency of opening the pack in use may be relevant. Bulk dispensing packs may require a more stringent test. Both antibacterial and antifungal efficacy should be demonstrated and the tests should include suitable positive and negative controls. The EP (2004) provides a full description of the current procedure to be used in the test for the efficacy of antimicrobial preservatives (see Chapter 17). However, reliance cannot be placed upon any pharmacopoeial method as being definitive in terms of safety and efficacy testing of a particular product (Matthews 1987). For the purposes of registration, one must always consider whether or not a test reflects the state of the art in this regard. Registration authorities have a duty when considering safety to apply the latest standards. This makes reliance upon pharmacopoeial and other published texts always a matter for circumspection by the formulator.

19.5.3.13   Interactions with Ingredients

Martindale (1999) offers ample evidence that preservatives interact with other chemicals, and this is covered in detail in Chapter 15. The formulator must demonstrate to the regulatory authorities that ingredients in the formulation are compatible with both the pack and with the chosen preservative system. Literature references will give some guidance, but physical tests and ultimately chemical and biological tests must be conducted. The results must be interpreted in terms of compatibility as well as confirmation that the formulation is satisfactory.

19.5.3.14   Scale-Up and Production

The difference between laboratory preparation and full-scale manufacture can be marked. For example, in a liquid preparation containing hydrocortisone or aluminium hydroxide in suspension, a homogeneous mixture can be maintained in a beaker prior to dispensing into bottles by using a simple paddle stirrer. In a 1000-L tank piped to the bottling machine, keeping the bulk homogeneous is not so easy. Settling in pipes and corners can become a problem. Continuous circulation between the bottling machine and the bulk may overcome the problem, but the introduction of flexible plastic tubing to achieve this may result in the preservative system being extracted by adsorption. Soaking the tubing in preservatives prior to use may slow or suitably change the rate of uptake. It may also result in preservative leaching from the tube into the product. The authorities will look for evidence that all the variables have been checked and considered.

In processes that use heat, the influence of temperature on stability is important. As batch size increases, the time for which preservatives are subjected to elevated temperature increases (unless forced cooling is used) and this can lead to greater degradation.

19.5.3.15   Raw Materials

Preservatives, like any ingredient in a product, must be subject to the full rigor of a raw material specification and classification procedure. Where conventional supplies of European Pharmacopoeia materials are used, it is likely there will be a full and complete specification for the preservative. However, if less conventional sources are used, for whatever reason, then a full characterization of the material will be needed. If a proprietary system is employed, its manufacturer must be prepared to release a full characterization and specification for the raw material that will satisfy the authorities that the material will be consistent and remain within the stated limits on all occasions (Anonymous 1998). Typically, the specification might include:

- Characteristics—Odor, appearance, color
- Identity tests—Melting point, boiling point, spectrophotometric and chromatographic characterization against a pure reference sample
- Physicochemical tests—Solubility in a variety of solvents
- Purity tests—Thin-layer chromatography, high-pressure liquid chromatography, ultraviolet spectroscopy, loss on drying, incineration, heavy metals
- Assay—Against a pure reference sample

Regulatory authorities will seek assurances that the material is not intrinsically toxic at the levels intended for use.

19.5.3.16   Analytical Methods

All experimental studies must take into account the latest thinking. If the published method, in the light of scientific advance, has been found wanting, then a more reliable method will have to be developed.

Details are required of the analytical methods used to monitor stability during the studies. Where these methods are the same as those described under the finished product specification, cross-reference can be made. Where other methods are used, they must be described in full. It is important that analytical assay methods used in the stability trials should be stability indicating, that is, sufficiently specific and sensitive to detect deterioration. Reporting results obtained using nonspecific methods without supporting data on levels of degradation products will not be sufficient. Similarly, if the method is not sufficiently precise, the early results of the long-term stability study will be of limited reliability as an indication of long-term stability (Anonymous 1998).

19.5.3.17 Results of Tests and Interpretation

Results—Details of the actual results obtained when the samples were tested should be given in a tabulated form. In some instances, a graphical presentation may also be helpful.

Initial assay results should be expressed in the same way as on the product label. Assay results for subsequent checkpoints should be given in the same way (Anonymous 1998).

Discussion of results—Reference should be made to any special precautions, including storage conditions and user information required for the product. Comment should be made on any assay or other test results that are near to, or outside, the check assay limits (Anonymous 1998). From a regulatory viewpoint, assessment of the preservative system is of critical importance. It should demonstrate the correct level of preservatives over the shelf life of the product, as well as effectiveness of preservatives using preservative efficacy tests.

19.5.3.18 Validation of the Methods Used

Data should be presented to show precision, accuracy, and sensitivity of the methods used.

The analytical methods and assay procedures selected for routine control of the formulation should be discussed. This should include evidence to show the validity of the methods employed, for example, standard error of assay methods.

Clear copies of spectra, gas–liquid chromatography traces or other recordings should be provided where these are used for assay or identification purposes. The emphasis here should be on demonstrating that the proposed specifications and methods are adequate to ensure batch-to-batch uniformity of the product (Anonymous 1998).

19.6 CONCLUSIONS

Three cardinal points are critical to a successful registration and must be remembered at all times.

First, one's paperwork is the shop front of one's science. No matter how good the scientific work done, if it is not collected, collated, and presented both clearly and concisely it will jeopardize the chance of gaining registration for the product. Poorly presented or incomplete data simply bring a lack of confidence to bear upon all work, inviting rejection of the whole registration application.

Second, regulations are the key to the marketplace. Any failure to understand the rules of the game simply invites disqualification regardless of the quality of the science.

Third, at the time of the application, one should ensure that both the regulatory and scientific aspects of the submission coincide and are comprehensive. Bearing in mind that both regulations and science are continuously changing, and not always in unison, early monitoring of both is a vital part of the success of any project.

The importance of obtaining adequate and scientifically based data on preservatives, antimicrobial agents, and antioxidants in pharmaceutical formulations cannot and should not be underestimated. Recognition of their importance is given in clear guidelines that have been promulgated in the EC; these are ignored at great risk. Close adherence to these guidelines should be considered as important as the attention paid to the detailed analysis and investigation of the active ingredients in a formulation. Indeed, failure to take seriously the presentation of data requirements for preservatives can be just as disastrous for the approval of the MAA as a failure to obtain data for any other ingredient in the formulation.

Moreover, the increasingly international nature of the pharmaceutical market and the benefits of reaching as many markets as possible as quickly as possible mean that all regulatory authority requirements should be treated with equal importance. Liaison with a regulatory professional throughout development work is crucial if that work is to be used at any time for a submission to the authorities. The cost of having to repeat work that is not regulatory compliant far outweighs the cost of obtaining an up-to-date regulatory opinion prior to conducting a study.

REFERENCES

Adams, P.N. (1987). Clinical issues in the regulation of eye care products: a view from Market Towers. *BIRA J.*, 6, 10–11.

Anonymous. (1991). Directive 91/356/EC of the European Parliament: Laying down the principles and guidelines of good manufacturing practice for medicinal products for human use. *Official Journal of the European Communities*, L193 (July 17), 30.

Anonymous. (1996). Inclusion of antioxidant and antimicrobial preservatives in medicinal products. *The Regulatory Affairs Journal*, 7, 1033–1036.

Anonymous. (1997). Notes for guidance on the inclusion of antioxidants and antimicrobials in medicinal products. *MCA Euro Direct Publication Number QWP/115/95*. July 8, 1997.

Anonymous. (1998). The Rules Governing Medicinal Products in the European Union. Vols. 1–3C. The Stationery Office, London. As updated on http://pharmacos.eudra.org/.

Anonymous. (1999). Directive 1999/11/EC of the European Parliament. *Official Journal of the European Communities* L77 (March 23), 8.

Anonymous. (2001a). Directive 2001/82/EC of the European Parliament and of the Council of 6 November 2001 on the Community code relating to veterinary products. *Official Journal of the European Communities*, L311 (November 28), 1.

Anonymous. (2001b). Directive 2001/83/EC of the European Parliament and of the Council of 6 November 2001 on the Community code relating to medicinal products for human use. *Official Journal of the European Communities*, L311 (November 28), 67.

Anonymous. (2001c). Directive 2001/20/EC of the European Parliament and of the Council of 4 April 2001 on the approximation of the laws, regulations, and administrative provisions of the Member States relating to the implementation of good clinical practice in the conduct of clinical trials on medicinal products for human use. *Official Journal of the European Communities*. L121 (May 1), 34.

British Pharmacopoeia. (2005). *British Pharmacopoeia.* The Stationery Office, London.

European Community. (1993). Council Regulation 2309/93 of July 22, 1993.

European Community. (1995a). Council Regulation 297/95 of February 10, 1995.

European Community. (1995b). Commission Regulation 540/95 of March 10, 1995.

European Pharmacopoeia. (2004). *European Pharmacopoeia.* 5th ed. and Suppl. Council of Europe, Strasbourg.

Greenwood, R.K. (1985). Chairman's report. *BIRA J.,* 4, 1–5.

Greenwood, R.K. (1990). Preservatives: registration and regulatory affairs. In *Guide to Microbiological Control in Pharmaceuticals.* Denyer, S.P. and Baird, R.M., Eds. Ellis Horwood, Chichester, U.K., pp. 313–340.

Hepburn, D. (1987). Achieving the complete Part II. *BIRA J.,* 6, 12–15.

Martindale: The Extra Pharmacopoeia. (1999). 32nd ed. Pharmaceutical Press, London.

Matthews, B.R. (1987). Regulation of eye care products: a view from Market Towers, pharmaceutical issues. *BIRA J.,* 6, 8–9.

Matthews, B.R. (1993). *The Regulatory Affairs Journal,* 4, 455–461.

Matthews, B.R. (1996). *The Regulatory Affairs Journal,* 7, 455–463.

Morris, J.M. (1994). Pharmaceutical development. In *International Pharmaceutical Product Registration: Aspects of Quality, Safety and Efficiency.* Matthews, B.R and Cartwright, A.C., Eds. Ellis Horwood, Chichester, U.K.

Pharmacos Web site. http://pharmacos.eudra.org/ and http://dg3.eudra.org/.

United States Pharmacopeia. (2005). *United States Pharmacopeia 29th Revision* and *National Formulary 23rd Revision.* U.S. Pharmacopeial Convention, Rockville, MD.

20 Microbial Standards for Pharmaceuticals

*Marie L. Rabouhans**

CONTENTS

20.1 INTRODUCTION

Microbial quality assurance is an essential feature of pharmaceutical product development, informing decisions on formulation, manufacture, container, storage, and shelf life. Microbial contamination, whether derived from source materials or arising during manufacture, storage, or use, may result in product spoilage or in patient infection. Spoilage may cause a reduction in potency or a loss of patient acceptability and compliance (Denyer 1988). Regulatory authorities expect microbiological aspects to be fully covered in the data submitted by manufacturers applying for a marketing authorization (product license). In the European Union, the regulatory requirements for medicinal products for both human and veterinary use are laid down in directives and are amplified by notes for guidance. A collection of these documents is published at intervals by the European Commission as volumes in the series *The Rules Governing Medicinal Products in the European Union* (e.g., Anonymous 1998a). New European pharmaceutical legislation was published in 2004 (Anonymous 2004a). Current documentation can be found on the Web site of the European Medicines Agency (http://www.emea.eu.int).

In addition to assessing the microbiological data provided by the manufacturer, the authorities inspect manufacturing facilities for compliance with good manufacturing practice and may take samples for analysis. These regulatory activities are

*When this chapter was written, it reflected the pharmacopoeial requirements of that time. However, for the most recent guidance, the reader is advised to check pharmacopoeial and regulatory Web sites for current information.

underpinned by the published standards of the pharmacopoeias. Pharmacopoeial texts provide specifications comprising mandatory requirements together with, in some cases, nonmandatory recommendations or guidance.

The pharmacopoeial standards applicable in the United Kingdom are those of the British and European Pharmacopoeias. As a monograph of the European Pharmacopoeia takes precedence over any former corresponding national monographs, all the monographs and test methods of the European Pharmacopoeia are incorporated into the British Pharmacopoeia to provide a comprehensive collection of U.K. requirements (Introduction and supplementary chapter IV B, BP 2005).

20.1.1 INTERNATIONAL HARMONIZATION

In recent years, the regulatory authorities and the major research-based companies within the pharmaceutical industry in Europe have been working with their counterparts in the United States and Japan under the auspices of the International Conference on Harmonisation (ICH) to harmonize regulatory requirements and expectations in the three regions. Once agreement is reached on a particular topic, regional guidance documents are replaced by a harmonized guideline. The ICH process relates primarily to market authorization for products containing new active substances and is thus prospective in its application.

In parallel with the ICH regulatory activity, the three regional pharmacopoeias [the European Pharmacopoeia (EP), the Japanese Pharmacopoeia (JP), and the United States Pharmacopeia (USP)] are also engaged in a process of international harmonization. Under the guidance of the Pharmacopoeial Discussion Group (PDG), the three pharmacopoeias are working on a voluntary, consensual basis to harmonize certain key pharmacopoeial requirements. In carrying out this work, the pharmacopoeias recognize that harmonization exists in terms of degree rather than as an absolute, all-or-nothing phenomenon. The goal is to harmonize as far as possible but the process may not always achieve full agreement between the pharmacopoeias.

One constraint on harmonization of pharmacopoeias stems from their wide sphere of application; the shelf life compliance standards of pharmacopoeias apply to all relevant products, both established and new.

A general text on international harmonization is included as section 5.8 in the fifth edition (2004) of the EP. The method texts underpinning the microbiological standards of the pharmacopoeias are among the key texts to which priority for harmonization has been assigned.

20.2 MICROBIAL STANDARDS

20.2.1 STERILE PHARMACEUTICALS

Certain categories of pharmaceutical products are always required to be sterile; these include parenteral preparations, ophthalmic preparations, and preparations for the irrigation of body cavities. The relevant general monographs of the European and British Pharmacopoeias (BP) define these preparations as sterile and require them to comply with the test for sterility. Other categories of products may be required to be sterile, depending on the circumstances in which they are intended to be used. For example, topical semisolid preparations such as creams, gels, and ointments are required to be sterile when intended for use on large open wounds or severely injured skin. Peritoneal dialysis solutions and hemofiltration solutions are always required to be sterile, whereas concentrated solutions for hemodialysis may be required to be sterile in certain circumstances but are otherwise required to have as low a degree of microbial contamination as possible (EP 2004, BP 2005).

As far as bulk raw materials are concerned, EP/BP requirements for sterility are usually restricted to active substances that are intended for use in the manufacture of nonterminally sterilized parenteral preparations. Compliance with the test for sterility is specifically included in monographs for active substances that represent a risk with respect to microbial contamination (for example, materials of biological origin and antibiotics produced by fermentation). A more general statement covering the circumstances in which both active ingredients and excipients are required to be sterile is included in the EP general monograph for Substances for Pharmaceutical Use (monograph number 2034) (EP 2004, BP 2005).

A general text on methods of preparation of sterile products (5.1.1) together with associated texts on biological indicators (5.1.2) and application of the F_0 concept to steam sterilization (5.1.5) are included in the fifth edition (2004) of the EP and are reproduced in Appendix XVIII of the BP (2005). Informational chapter 1211 in USP 28 (2005) provides similar coverage. These general pharmacopoeial texts are intended to be used in conjunction with the relevant good manufacturing practice (GMP) guidance (Anonymous 1998a, MCA 2002). In the EP, emphasis is firmly placed on the use of terminal sterilization, wherever possible. It should be noted that the EP definition of terminal sterilization does not extend to filtration, unlike that in the USP. The established overkill reference cycle of 121°C for 15 min for steam sterilization of aqueous preparations provides a wide safety margin for the bioburdens typically encountered for pharmaceutical products. Other cycles may be chosen provided that a sterility assurance level of 10^{-6} or better is achieved (see Chapters 10 and 12).

Compliance with the test for sterility is mandatory for all products that are required or that are claimed to be sterile. Performance of the test is not, however, necessarily a prerequisite for a manufacturer in assessing compliance with the pharmacopoeia before release of a product. The purpose of this and other pharmacopoeial tests is to provide an independent control analyst with the means of verifying that a particular article meets the requirements of the European Pharmacopoeia. As is made clear in the general notices of the EP (2004), a manufacturer is neither obliged to carry out such tests nor precluded from using modifications of, or alternatives to, the stated method provided it is satisfied that, if tested by the official method, the material in question would comply. The relevance of this notice to sterility testing is highlighted in the EP (2004) text on methods of preparation of sterile products (5.1.1). This text indicates that, when a fully validated terminal sterilization method by steam, dry heat, or ionizing radiation is used, parametric release, that is the release of a batch of sterilized items based on process data rather than on the basis of submitting a sample to sterility testing, may be carried out, subject to the approval of the competent authority. For a more detailed discussion on parametric release, the reader is referred to Chapters 11 and 12 and Bill (2000) in the companion *Handbook*.

In general, pharmacopoeial tests for sterility are based on the absence of growth following culture in media that have been shown to support the growth of a range of microorganisms (typically fluid thioglycollate and soya bean casein digest broths, intended primarily for the detection of anaerobic bacteria and aerobic bacteria or fungi, respectively). The media are incubated at 30 to 35°C for bacteria and at 20 to 25° for fungi for a defined time. Tests using membrane filtration and direct inoculation are described; both the EP (2004) and the USP (2005) require the former technique to be used whenever the nature of the product permits. For further details, the reader is referred to Millar (2000) and Bill (2000) in the companion *Handbook*.

Proposals for the international harmonization of the test for sterility were published in 1994 and 1995 and good agreement on many aspects of test methodology and interpretation was reached in 1996 (Anonymous 1994, 1995, 1996). Placing greater emphasis on the user validating the test as performed was expected to allow a degree of flexibility on aspects such as choice of organisms and size of inoculum. A move away from retesting would be reflected in a shift in emphasis from an assumption of test invalidity to one of product failure in interpreting the test. This, in turn, would require tight control of the test environment. When revised tests were prepared for publication in the 1998 Supplement to the EP 1997 and in the eighth Supplement (1998) to USP 23, it was suggested that the difference between the two texts would be minimal and that no methodological differences of note would exist (Dabbah and Knapp 1997). However, differences of some significance did remain. For example, although the EP adopted a 14-day incubation period for both membrane filtration and direct inoculation, this was qualified in the USP by allowing a 7-day incubation for products terminally sterilized by moist heat when tested by membrane filtration. Revision of the EP text continued in response to comments received following publication of the 1998 text and an interim revision was published in the 2000 EP Supplement. Interpharmacopoeial discussion continued, meanwhile, with a view to refining the texts to improve harmonization and clarity. A consensus text was "signed off" by the three pharmacopoeias in 2002 and was published in the fifth edition of the EP (2004). Although residual differences remain between the pharmacopoeial tests, it is considered possible to carry out one test that will meet the requirements of the three pharmacopoeias (Anonymous 2003).

20.2.2 NONSTERILE PHARMACEUTICALS

The situation with respect to standards for the microbiological quality of nonsterile pharmaceutical products and their constituents is more complex than that of sterile products. Setting appropriate specifications requires consideration of a number of factors. The nature and origin of the material and the method of production will influence the potential for microbial contamination, both in terms of the type and level of contaminants. Materials of plant, animal, or soil origin and aqueous formulations have been identified as having the highest potential for contamination (Baird 1985). The pathogenicity of the microorganisms and the intended use of the material or preparation must also be taken into account. Certain microorganisms can be considered "always objectionable" but others are "usually objectionable" (Bruch 1972, Spooner 1985; see also Chapter 13). Standards may be set for a specific organism because it falls into one of these categories or because it serves as an indicator organism, either for a particular type of contamination or for a breakdown in GMP.

Control of microbial contamination has been a feature of pharmacopoeias for many years but is an area where policy with respect to application and status is still evolving and is currently under active debate. Mandatory requirements are now included in specific EP monographs for certain materials of natural origin such as acacia, kaolin, and pancreatin powder (Table 20.1). These requirements may comprise both absence of certain specific organisms such as *Escherichia coli* in 1 g and *Salmonella* in 10 g and limits for total viable counts (TVCs). Before 1995, reference to microbiological testing was included in discretionary statements appended as footnotes to the relevant EP monographs in recognition of the different approaches pursued within Europe. In the United Kingdom, for example, TVCs were regarded as useful for in-house process control but, for regulatory purposes, reliance had been placed on testing for the absence of pathogens. In contrast, the Scandinavian approach had been to use TVC as the main control and to require absence of specific pathogens only when the TVC exceeded 10^2 colony-forming units (CFU) per g or mL. The inclusion of mandatory requirements followed a policy decision taken by the European Pharmacopoeia Commission in 1994 [Introduction, BP (1993) and Addendum (1995)]. In the interests of European harmonization, it was agreed that the way in which microbial contamination was controlled would no longer be left to the discretion of national authorities.

The mandatory requirements of the EP are supported by method texts for specified microorganisms (section 2.6.13, EP 2004; Appendix XVI B1, BP 2005) and for TVC (section 2.6.12, EP 2004; Appendix XVI B2, BP 2005). These method texts are also used to support the relevant nonmandatory recommendations made within the guidelines on the microbiological quality of pharmaceutical preparations (section 5.1.14, EP 2004; Appendix XVI D, BP 2005). Since the revision included in BP 2001, the text for TVC testing distinguishes between the two different uses of these quantitative tests. The text indicates that the tests are designed primarily to determine whether or not a substance that is the subject of a monograph in the pharmacopoeia complies with the mandatory requirements therein. An important feature is the inclusion of a sampling plan based on taking five separate samples together with appropriate interpretation (a three-class system). This scheme is given as an example of a sampling plan applicable when homogeneity with respect to the distribution of microorganisms may be a problem. Additional guidance is provided in BP (Supplementary chapter I M, BP 2005), and the reader is also referred to the companion *Handbook* (Baird 2000). Control laboratories are advised to consider adoption of the three-class system; this would provide a more reliable basis for binding decisions on bulk materials that can be assumed to be heterogeneous with respect to potential microbial contamination.

As noted above, the EP now provides guidance with respect to the microbiological quality of pharmaceutical preparations according to their designated category. Reference is made to this guidance within the "production" sections of relevant general monographs, for example, that for "Liquids for Oral Use." Further explanatory guidance concerning the application of the advisory limits for nonsterile

TABLE 20.1
Microbial Limits of the EP (2004)

Material	Total Viable Aerobic Count CFU per g	Fungi CFU per g	Absence of Organisms in Specified Quantity of Product		Other Organisms
			E. coli	Salmonella	
Acacia	10^4		1 g		
Agar	10^3		1 g	10 g	
Alginic acid	10^2		1 g	10 g	
Aluminium hydroxide	10^3		1 g[a]		Enterobacteria and certain other Gram-negative bacteria 1 g
Bentonite	10^3				
Calcium gluconate	10^3				
Calcium stearate	10^3		1 g[a]		
Cellulose microcrystalline[a]	10^3	10^2	1 g	10 g	Pseudomonas aeruginosa 1 g, Staphylococcus aureus 1 g
powdered	10^3	10^2	1 g	10 g	P. aeruginosa 1 g, S. aureus 1 g
Charcoal activated	10^3				
Dextran 1/40/60/70 for injection	10^2		1 g		
Ferrous gluconate	10^3				
Galactose	10^2				
Gelatin	10^3		1 g	10 g	
Guar	10^4		1 g	10 g	
Guar galactomannan	10^4		1 g	10 g	
Kaolin	10^3				
Lactose	10^2		1 g		

(continued)

TABLE 20.1 (continued)
Microbial Limits of the EP (2004)

Material	Total Viable Aerobic Count CFU per g	Fungi CFU per g	Absence of Organisms in Specified Quantity of Product		Other Organisms
			E. coli	Salmonella	
Lactulose	10^2		1 g		
Maize starch	10^3 bacteria	10^2	1 g		
Pancreatic extract	10^4		1 g	10 g	
Pepsin powder	10^4		1 g	10 g	
Potato starch	10^3 bacteria	10^2	1 g		
Rice starch	10^3 bacteria	10^2	1 g		
Tapioca starch			1 g		
Sodium alginate	10^3		1 g	10 g	
Sodium starch glycollate type A and type B			1 g	10 g	
Starch pregelatinized	10^3 bacteria	10^2	1 g		
Talc					
if intended for topical administration	10^2 aerobic bacteria and fungi				
if intended for oral administration	10^3 bacteria	10^2			
Tragacanth	10^4		1 g	10 g	
Trypsin	10^4		1 g	10 g	
Wheat starch	10^3 bacteria	10^2	1 g		
Xanthan gum	10^3	10^2	1 g		

a Internationally harmonized monograph.

products (categories 2 to 4) is given in BP (supplementary chapter I M, BP 2005). The availability of authoritative guidelines is expected to provide a useful yardstick for official control laboratories when conducting surveys of products on the market (Alexander et al. 1997). The advisory limits may also be used by manufacturers for process validation or as part of a general monitoring program. Their inclusion in the pharmacopoeia is not, however, intended to imply the need for routine end-product testing.

The need to rationalize microbial standards within the EP is recognized as is the desirability of harmonizing standards between the pharmacopoeias. Issues that need to be considered include the reason why some monographs for materials of natural origin (see Table 20.1) have microbial requirements and others, for example, those for herbal materials do not. Differences in TVC limits (for example, 10^4 CFU per g for acacia, 10^3 for agar, and 10^2 for alginic acid) also need to be examined and compared with those in, for example, the USP NF (no TVC requirement for acacia and agar; alginic acid set at 200). As a first step, proposals were made for harmonization of the methodology (Anonymous 1998b) and suggestions made concerning the application of water activity measurement (Friedel and Cundell 1998) and the use of a decision tree (Anonymous 1998c). Current practice in the testing of raw materials by some European manufacturers has been surveyed (Anonymous 2003). With respect to the methodology, discussion is proceeding within the PDG process (Anonymous 2003) and had reached harmonization stage (official enquiry) by October 2004 (Anonymous 2004b).

20.2.3 WATER

Water is a major pharmaceutical excipient (see Chapter 3) and two grades have been defined for many years within the EP, water for injections (monograph number 0169) and purified water (monograph number 0008). Both these monographs have been revised in recent years (EP 2004, BP 2005). The key difference between the two monographs is in the method of manufacture and the control of microbial contamination. Water for injections may be produced by distillation only, whereas the method of production of purified water is unrestricted. The restriction on the method of production of water for injections EP contrasts with the open definition in the corresponding USP monograph. More recently, a third EP water monograph, highly purified water (monograph number 1927), has been published, covering a grade for which the primary method of production is reverse osmosis (EP 2004). A European Community note for guidance concerning the use of the different grades of water was adopted in 2002 (Anonymous 2002). For a more detailed discussion of methods of water production, the reader is referred to Chapter 3.

Once it is distributed in containers and sterilized, water for injections complies with the test for sterility and with a limit of 0.25 IU.mL^{-1} for bacterial endotoxins. A recommended action limit for the bulk water before sterilization is given in the "production" section of the revised monograph. A TVC limit of 10 microorganisms per 100 ml (determined using membrane filtration, a low nutrient agar medium, and a test volume of at least 200 ml) is given as appropriate under normal conditions. The monograph states that for aseptic processing, stricter alert limits may need to be applied. For purified water, a TVC of $100\,CFU.mL^{-1}$ (determined using the same method and medium as for water for injections) is given both as a recommended action limit under production of the bulk water. In addition, a compliance limit of $100\ CFU.mL^{-1}$ is given for the water distributed in containers. This aspect of the monograph should, perhaps, be reexamined and the need for a mandatory shelf life limit reconsidered. As it stands, for a manufacturer to be confident of consistently meeting a $100\ CFU.mL^{-1}$ shelf life limit for packaged water, he would need to adopt a tighter action limit because proliferation of microorganisms such as pseudomonads is possible. A limit of $0.25\ IU.mL^{-1}$ for bacterial endotoxins applies to purified water intended for use in the manufacture of dialysis solutions.

20.3 CONCLUDING REMARKS

The fundamental principles of microbiological quality assurance are well established and are reflected in the published standards of the pharmacopoeias. The importance of these standards is shown by the high priority that has been given to the microbiological method texts within the international harmonization program. Harmonization of the methods and their application began in the late 1990s and is currently still in progress. In order to keep abreast of developments, it is recommended not only to refer to the current edition of the pharmacopoeias, but also to consult the relevant publications, such as *Pharmeuropa* and the pharmacopoeial and regulatory Web sites.

REFERENCES

Alexander, R.M., Wilson, D.A., and Davidson, A.G. (1997). Medicines Control Agency investigation of the microbial quality of herbal products. *Pharm. J.*, 259, 259–261.

Anonymous. (1994). Harmonisation of sterility tests. *Pharmeuropa*, 6(1), (March).

Anonymous. (1995). Harmonisation of sterility tests. *Pharmeuropa*, 7(2), (June), 256–264.

Anonymous. (1996). International harmonisation of sterility tests. *Pharmeuropa*, special issue (October).

Anonymous. (1998a). *The Rules Governing Medicinal Products in the European Union*. Eudralex Office for Official Publications of the European Communities, Luxembourg.

Anonymous. (1998b). Microbiological examination of non-sterile products. *Pharmeuropa*, 10(3), (September), 418–447.

Anonymous. (1998c). Decision-tree for determining what objectionable micro-organism test(s) is appropriate. *Proceedings of the USP Open Conference on Microbiology for the 21st Century*. U.S. Pharmacopeial Convention, Rockville, MD.

Anonymous. (1999). General monographs. Active substances and excipients. *Pharmeuropa*, 11(3), (September), 532–536.

Anonymous. (2002). *Note for Guidance on Quality of Water for Pharmaceutical Use*. European Agency for the Evaluation of Medicinal Products, London.

Anonymous. (2003). *Proceedings of the International Symposium on Microbiological Control Methods in the European Pharmacopoeia: Present and Future*. Copenhagen, May 2003, Council of Europe, Strasbourg, France.

Anonymous. (2004a). *Official Journal of the European Union*, 47, L136.

Anonymous. (2004b). State of work of international harmonization. *Pharmeuropa*, 16(4), (October), 315–316.

Baird, R.M. (1985). Microbial contamination of pharmaceutical products made in a hospital pharmacy: a nine year survey *Pharm. J.*, 231, 54–55.

Baird, R.M. (2000). Sampling: principles and practice. In *Handbook of Microbiological Quality Control: Pharmaceuticals and Medical Devices*. Baird, R.M., Hodges, N.A., and Denyer, S.P., Eds. Taylor & Francis, London, pp. 38–53.

Bill, A. (2000). Microbiology laboratory methods in support of the sterility assurance system. In *Handbook of Microbiological Quality Control: Pharmaceuticals and Medical Devices*. Baird, R.M., Hodges, N.A.. and Denyer, S.P., Eds. Taylor & Francis, London, pp. 120–143.

British Pharmacopoeia. (1993). *British Pharmacopoeia*. Her Majesty's Stationery Office, London.

British Pharmacopoeia. (1995). *Addendum* (to *British Pharmacopoeia* 1993). Her Majesty's Stationery Office, London.

British Pharmacopoeia. (2001). *British Pharmacopoeia*. The Stationery Office, London.

British Pharmacopoeia. (2005). *British Pharmacopoeia*. The Stationery Office, London.

Bruch, C.W. (1972). Objectionable micro-organisms in non-sterile drugs and cosmetics. *Drug. Cosm. Ind.*, 10, 51–54, 150–156.

Dabbah, R. and Knapp, J.E. (1997). Harmonization of microbiological methods—a status report. *Pharmacopeial Forum*, 23(6), 5334–5340.

Denyer, S.P. (1988). Clinical consequences of microbial action on medicines. In *Biodeterioration 7*. Houghton, D.R., Smith, R.N., and Eggins, H.O.W., Eds. Elsevier Applied Science, London, pp. 146–151.

European Pharmacopoeia. (1998). *Supplement 1998* (to *European Pharmacopoeia*, 3rd ed.). Council of Europe, Strasbourg, France.

European Pharmacopoeia. (2000). *Supplement 2000* (to *European Pharmacopoeia*, 3rd ed.). Council of Europe, Strasbourg, France.

European Pharmacopoeia. (2004). *European Pharmacopoeia*. 5th ed. Council of Europe, Strasbourg, France.

Friedel, R.R. and Cundell, A.M. (1998). The application of water activity measurement to the microbiological attributes testing of nonsterile over-the-counter drug products. *Pharmacopeial Forum*, 24(2), 6087–6090.

MCA (2002). *MCA Rules and Guidance for Pharmaceutical Manufacturers and Distributors*. The Stationery Office, London.

Millar, R. (2000). Enumeration of micro-organisms. In *Handbook of Microbiological Quality Control: Pharmaceuticals and Medical Devices*. Baird, R.M., Hodges, N.A., and Denyer, S.P., Eds. Taylor & Francis, London, pp. 54–68.

Spooner, D.F. (1985). Microbiological criteria for non-sterile pharmaceuticals. *Manufact. Chem.*, 4, 41–45.

United States Pharmacopeia. (1998). *United States Pharmacopeia 23*. 8th Suppl. U.S. Pharmacopeial Convention, Rockville, MD.

United States Pharmacopeia. (2005). *United States Pharmacopeia 28—National Formulary 23*. U.S. Pharmacopeial Convention, Rockville, MD.

21 Risk Management and Microbiological Auditing

Robert A. Pietrowski

CONTENTS

21.1 INTRODUCTION

No activity or pursuit in life is totally risk free. Simple everyday activities, such as crossing the road or playing a sport, involve an element of risk. Furthermore, risk cannot be eliminated from life; it is an inevitable part of living.

In pharmaceutical manufacture, as in life, risk-free activities do not exist. However, medicines offered for sale must be fit for their intended use and, as far as is possible, should not put patients at risk due to inadequate safety, quality, or efficacy. The achievement of this ideal requires strong emphasis on risk management, incorporating, where possible and practicable, risk elimination.

21.2 RISK MANAGEMENT

Embodied within risk management is the acknowledgment of the concept of acceptable risk. This is based upon the realization that risk cannot be completely removed from pharmaceutical manufacture and thus the extent to which risk can be minimized must be influenced and guided primarily by patient safety but also by manufacturing cost.

Patient safety would appear, superficially at least, to be an absolute requirement that cannot under any circumstances be compromised. Unfortunately, however, things are rarely so simple. For example, in the case of an aseptically prepared injectable product, the overwhelmingly important safety criterion is sterility; however, this is the one attribute that cannot be ensured or confirmed, because confirmation requires performing the test for sterility on every container, leaving nothing to supply to the patient!

Special precautions, therefore, must be taken during the manufacture of such products to provide a high degree of confidence that microbiological contamination has been excluded from every container. This requires heavy emphasis on technology as well as the adoption and operation of an effective quality management system, but this is not limitless. At a certain point, the law of diminishing returns dictates that additional costs do not contribute significantly to patient safety. Beyond this point, there is a danger of simply pricing the medicine out of the market place and this in itself becomes a significant patient safety issue.

Thus, risk management should constitute a critical element of any pharmaceutical manufacturer's quality assurance system as it forms the basis for the very system itself. Essential to effective risk management are the following activities: identification of areas of risk or vulnerability, assessment of risk, implementation of actions to reduce or eliminate risk, adoption of in-process controls to confirm risk control, and audit of the system for ongoing effectiveness.

All these steps are crucial to the management of risk, but this chapter will concentrate on just two aspects—risk assessment and audit.

21.3 APPROACHES TO RISK ASSESSMENT

The assessment and quantification of risk are key parts of any strategy of risk management.

Quantification of risk may not extend to a numerical assignment that can be scientifically justified, but it is important to be able to rank different vulnerabilities in terms of their significance for patient safety in an objective manner. In doing this, better use of limited resources can be made to address the greatest areas of risk first.

Within the pharmaceutical industry, there are two fundamental approaches to risk assessment, namely intuitive or structured. Intuitive methods of risk assessment rely upon an individual or a group identifying and quantifying risk purely on the basis of intuition, with no formal structure or agreed criteria to guide the process. Such an approach is fraught with problems and disadvantages because not only is the process dependent upon the experience of the participants, but also there is no mechanism to ensure that each potential risk is assessed in the same manner. Furthermore, on the basis of objective criteria, the process is not amenable to justification.

For these and other reasons, the pharmaceutical industry almost universally adopts structured approaches to risk assessment, as exemplified by failure mode and effect analysis (FMEA) and hazard analysis of critical control points (HACCP).

21.3.1 FAILURE MODE AND EFFECT ANALYSIS

FMEA is a technique that has been used successfully for many years to assess risk associated with activities and operations. Although it has its roots in engineering, FMEA has been successfully used for risk assessment in numerous industries, including the pharmaceutical industry. FMEA involves analyzing each step in a process or activity by the following criteria:

- The severity of the consequence of failure (e.g., for the patient) (S)
- The probability of occurrence of the failure (O)
- The probability of detection of the failure, should it occur (D)

Each of these criteria can then be assigned a score, commonly from 1 to 10. Thus, if the consequences of failure are considered to be very severe, a high number will be assigned; if the probability of occurrence of failure is also considered to be high, this too will be assigned a high number. If, however, the probability of detection of the failure is considered to be high, a low number will be assigned to this criterion.

From these, scores can be calculated leading to an overall risk probability number (RPN) for the process step, where

$$RPN = S \cdot O \cdot D \tag{21.1}$$

Thus, for the step described above where S is considered high, O is considered high, and D is considered high, the calculated risk probability number may, for example, be

$$RPN = 8 \times 8 \times 2 = 128$$

Similarly, for a step where the severity of failure is considered high, the probability of occurrence is, however, considered to be low and the probability of detection is also considered to be low, the risk probability number may be

$$RPN = 9 \times 3 \times 8 = 216$$

In this way, the risk associated with each step in a process, or with any given activity, can be objectively assessed and quantified. This approach permits the ranking of steps in terms of risk associated with them and allows the user to assign priorities for action in order to eliminate or minimize the risk by, for example, modification of the process or targeted in-process monitoring.

21.3.2 HAZARD ANALYSIS AND CONTROL OF CRITICAL POINTS

HACCP is a structured approach to risk assessment and risk management that was developed from the principles of FMEA. HACCP has its origins in the food industry, where it was developed as a system of food safety assurance. The system was formulated in the 1960s by the Pillsbury food company in the United States, the United States Army Laboratories at Natrick, and the National Aeronautics and Space Administration (NASA) in their joint development of foods for the American space program. In this project, food production processes had to be designed to ensure the elimination of pathogens and toxins from the foods being produced. End-product testing was not considered to be a reliable measure of microbiological quality and so other methods to ensure the safety of the food had to be developed. Thus the concept of HACCP was born.

Since Pillsbury first presented its experiences with HACCP in 1971, the food industry has widely adopted the approach and developed it further. Thus the U.S. Food and Drug Administration (1973) incorporated HACCP into its low acid canned foods regulations. Both the World Health Organization (1996) and the International Commission on Microbiological Specifications for Foods (1988) have advocated the use of HACCP. HACCP is also recommended to help demonstrate "due diligence" under the U.K. Food Safety Act of 1990. Within Europe, systems based upon the principles of HACCP have been incorporated into the 1993 EC directive on the hygiene of foodstuffs.

21.3.2.1 Applicability to the Pharmaceutical Industry

Like FMEA, HACCP seeks to ensure quality not only by reducing emphasis on end-product testing (testing for failure), but also by increasing emphasis on process design and control (preventing failure). However, unlike FMEA, HACCP is an approach that has been designed primarily to identify and eliminate microbiological risks. As such, the technique is of immediate and direct value to the pharmaceutical industry, where microbiological contamination is a major concern.

In recent years, the pharmaceutical industry has become aware of the many advantages that HACCP offers in the systematic approach to microbiological quality assurance. In particular, its application to the control of aseptic manufacture has been beneficial in providing a logical framework for the many layers of control and monitoring necessary in the quality assurance of such products (Ljungqvist and Reinmüller 1995, Jahnke 1997). In this regard, it is interesting to note that during the development of the international standard on aseptic processing (International

Organization for Standardization ISO TC 209), HACCP was included as a means of risk assessment, only to be excluded from the final document in favor of a more general statement that risk assessment should be undertaken. Most notably, the U.S. Food and Drug Administration, which has long adopted HACCP for food applications, is considering widening the approach to medical devices and possibly also to pharmaceuticals as a means of "streamlining inspections."

21.4 HACCP TO ENHANCE MICROBIOLOGICAL CONTROL OF MEDICINES

The application of HACCP involves, in its simplest form, a series of linked steps. These include definition of the product and process, identification of potential hazards and potential measures for their control, determination of critical control points, establishment of critical limits for each critical control point, establishment of a monitoring system for each critical control point, implementation of corrective action plans to reestablish control when necessary, and establishment of verification procedures to demonstrate compliance.

21.4.1 DEFINITION OF THE PRODUCT AND PROCESS

As far as the product is concerned, the following considerations may have an important impact upon risk: raw materials and components, product composition, dosage form, the pack, storage conditions, shelf life, and patient target group (e.g., age, health status; see Chapter 2).

In the case of the process, the following should be considered: all process steps, process equipment, major utilities, manufacturing environment, holding times and conditions, cleaning and environmental hygiene, product segregation, and routes of potential contamination or cross-contamination (see Chapters 3 through 6).

21.4.2 IDENTIFICATION OF POTENTIAL HAZARDS AND POTENTIAL CONTROL MEASURES

Potential hazards can arise from acknowledged sources of contamination, including raw materials, equipment, personnel, and the environment. Less obviously, certain combinations of product-processing conditions, such as, pH, a_w, and temperature, may also represent a potential hazard. Furthermore, poor management systems may indirectly be identified as creating hazards; this may arise, for example, through inadequate production planning or from frequent product changeovers.

Each potential hazard should be analyzed using the following typical criteria: likelihood of occurrence, severity of hazard, magnitude of hazard (e.g., batch size affected), age and vulnerability of those exposed to the hazard (and hence the likely impact), the survival or multiplication potential of contaminating microorganisms, the potential for production of and persistence of microbial toxins or other potentially detrimental materials, the existence of conditions conducive to microbial contamination or toxin production.

Finally, potential control measures, if they exist, should be reviewed for their usefulness in eliminating, preventing, or reducing such identified potential hazards. At the same time, both the product and process should be examined to ascertain whether they may be redesigned to eliminate, prevent, or reduce risk.

21.4.3 DETERMINATION OF CRITICAL CONTROL POINTS

Figure 21.1 represents a decision tree for the identification of critical control points (CCP) for processes.

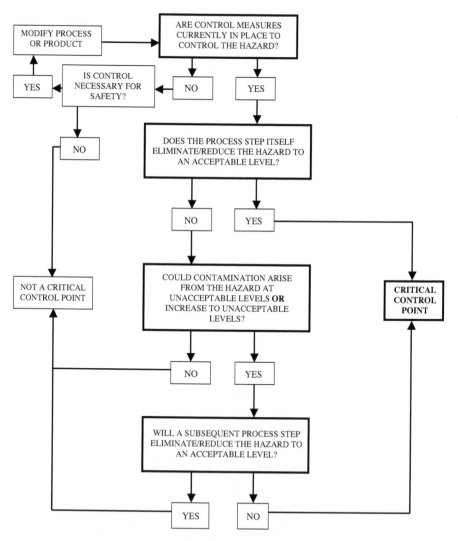

FIGURE 21.1 Decision tree for the identification of critical control points. (Adapted from Leaper, S., Ed. (1997). *HACCP: A Practical Guide.* Technical Manual no. 38. Campden and Chorleywood Food Research Association, Chipping Campden, U.K. With permission.)

21.4.4 Establishment of CCP Limits

For each CCP identified, a critical limit should be established. This should be discriminatory in that it should distinguish between what is acceptable and what is unacceptable. It may therefore represent an accept or reject limit or an action limit; where the action may be reprocessing of the product, additional testing, or action to prevent a recurrence or re-establishment of process control.

21.4.5 Establishment of a CCP Monitoring System

The establishment of a truly effective monitoring scheme for each CCP is an essential part of risk management by HACCP. The monitoring scheme must therefore be able to detect a loss of control and provide timely information that permits appropriate corrective action to be taken, preferably before product rejection becomes the only option. In this regard, the design of monitoring schemes and their operation become critical to the total risk management program.

In many cases, monitoring will involve sampling and analysis, but this is by no means the only approach open to the manufacturer. Collection of physical data (e.g., temperature, pressure, mixing speed) and comparison to pre-agreed upper and lower set points, either continuously or periodically, represent an effective monitoring method for many parameters.

Whichever approach to monitoring is adopted, its effectiveness will be influenced not only by the sample size but also by the monitoring frequency or number of datum points. The sample size should be based upon the accuracy of the monitoring method and limit to be met. For example, if it is necessary to show that a water system consistently complies with a bioburden limit of <10 CFU per 100 mL, then the volume of water to be sampled and tested will be greater than if the bioburden limit is <100 $CFU.mL^{-1}$. The monitoring frequency will be influenced by the ranking of the CCP, the probability of occurrence of the potential hazard, and the confidence in the monitoring method used. Thus the frequency of microbiological monitoring and the number of samples taken for a raw material of natural origin to be used in the manufacture of an aseptically prepared injectable will be greater than for an inorganic salt to be used in the manufacture of a terminally sterilized small-volume parenteral product.

21.4.6 Establishment of a Corrective Action Plan

If monitoring data indicate a loss of control, it is imperative that appropriate action is taken to regain control. Wherever possible, this action should be defined in writing at the time that the monitoring system is approved. It should include the following: what action is to be taken and when; who is to act; and how the effectiveness of the action will be verified.

21.4.7 Establishment of Verification Procedures to Demonstrate Compliance

The effectiveness of the control measures described above should be challenged and verified. This can be achieved in several ways. Using trend analysis of data over extended time periods, overall control and the effectiveness of corrective actions performed to regain control will be demonstrated. Data trended and reviewed need not necessarily be specific to an individual CCP, but may include data from other CCPs and the final product to demonstrate the contribution of a specific control strategy to overall product quality assurance. Reviewing deviations will demonstrate consistency of process control and also the effectiveness of corrective actions employed. Through audit of the integrated systems, they will remain well-targeted, relevant, effective, and complied with at all levels.

If such reviews show that the integrated system is not exerting satisfactory control over CCPs and the product, then the system must be modified to improve control. In this regard, it is important to ensure that any risk management system is linked closely to the company's change control system so that, if necessary, the risk management system is adapted to take account of changes to such things as raw materials and components, processing equipment, procedures, environments, methods of sampling and analysis, and regulatory limits.

21.5 USING RISK ANALYSIS TO ENHANCE THE EFFECTIVENESS OF MICROBIOLOGICAL AUDITING

Before discussing how risk analysis can enhance the value of microbiological auditing, it is important to define the purpose of auditing. The EC Directive 91/356/EEC (1991) on good manufacturing practice (GMP) states that "self-inspections shall be conducted in order to monitor the implementation and compliance with Good Manufacturing Practice principles and to propose necessary corrective measures." Thus, audits are seen by regulators as a critical component of quality assurance and GMP. In essence, audits are the principal means by which the effectiveness of the quality management system is assessed and improved.

The audit should therefore challenge aspects of the quality management system to ensure that it is comprehensive, understood, relevant to the activities undertaken, effective in exerting control, and its objectives are met. Should problems be identified by the audit process, the audit must make provision for the execution of timely and effective corrective action.

From the foregoing, it should be clear that the audit is a key element of any risk management system. The major role of any audit is to identify areas of vulnerability. These may arise through noncompliance with regulatory requirements, with license submissions, or with the written quality management system as well as from inadequate product quality assurance. In the latter case, this may stem from a failure to identify areas of risk, from inadequate monitoring and control procedures, from a failure to take effective corrective actions to regain control, or perhaps from inadequate staff training or awareness.

In assessing vulnerability, structured techniques such as FMEA and, in particular, HACCP are invaluable tools for the auditor. The principles of risk assessment can be used for all phases of the audit—planning the audit, conducting the audit, communicating concerns, and in making recommendations for corrective actions.

For a more detailed discussion of hazard analysis and auditing in practice, the reader is referred to the companion *Handbook* (Lush 2000).

21.5.1 PLANNING THE AUDIT

As part of the essential preparation for any audit, the auditor must review in detail the activity to be audited. This will ensure not only adequate understanding of the activity but also identify areas of vulnerability (CCPs) within the activity. In turn, this will allow the effectiveness of the auditee's control measures to be challenged and assessed. This process provides the structure for the audit and determines how much time and emphasis should be allocated to different aspects of the activity.

The identification of areas of vulnerability, and thus the focus of the audit process, will be greatly enhanced by the application of a formal risk assessment as this will allow objective identification of CCPs in the activity and will permit the auditor to explain why these points are considered to be of critical importance. As always, time invested in proper planning pays dividends.

21.5.2 CONDUCTING THE AUDIT

During the audit itself, the auditor should challenge each CCP identified during the planning phase and attempt to answer the following questions:

- Does the auditee recognize this as an area of risk (and hence a CCP)?
- Has the auditee attempted to "design out" the risk at this stage of the activity?
- Have appropriate limits been set for this CCP?
- Does the auditee monitor the activity at this point, and if so is the monitoring strategy sufficient in terms of monitoring frequency, number of samples, sample size, means of analysis, and in the communication of results?
- Is the system capable of identifying loss of control or movement toward loss of control?
- Is there a clear, effective corrective plan in place to regain control?
- Is there a system in place to demonstrate and confirm the adequacy of all these control measures through trend analysis of data, follow-up on corrective actions, change control, and periodic review or audit?

The following examples are intended to illustrate specific areas of risk in different situations and how the audit questions should be framed to reflect such perceived risks.

21.5.2.1 Tablet Manufacture

It is tempting to assume that because tablet manufacture is essentially a dry process there is little or no attendant microbiological hazard. In many respects this is true, but there are aspects of the process where potential for microbiological contamination of the final product exists. These include the supply of contaminated raw materials, wet granulation and the preparation and storage of aqueous solutions, or equipment washing, if accompanied by inadequate drying. These therefore represent critical control points that should be assessed during the audit.

By way of illustration, and with regard to raw materials, questions to be asked will include:

- Has the manufacturer acknowledged and identified those raw materials considered to be a potential contamination risk?
- Are they obtained from suppliers with a history of satisfactory supply?
- Have appropriate bioburden limits been set for raw materials?
- For those "at risk" raw materials are the number of samples, sampling sites, sampling frequency, and tests performed sufficient to ensure adequate control?
- Is trend analysis undertaken on cumulative data to demonstrate on-going control?
- In the event of loss of control are appropriate corrective action strategies in place and are they used?
- Is the effectiveness of the overall control strategy reviewed periodically?
- Is there a satisfactory change control procedure in place for raw materials, suppliers, and storage and manufacturing processes to ensure that changes are adequately identified, evaluated, and appropriate actions taken when necessary?

21.5.2.2   Small Volume Parenteral Manufacture by Aseptic Processing

The manufacture of sterile products by an aseptic process perhaps represents the most hazardous pharmaceutical manufacturing activity, requiring rigorous control at numerous points in the process if acceptable product quality is to be ensured. The application of HACCP to such a manufacturing process has been the subject of an excellent review (Jahnke 1997) and, therefore, will not be covered in detail here. Rather, just one aspect of the process will be discussed, namely the exclusion of microbiological contamination at the point of fill.

Here the key control measure is the provision of a localized grade A environment (see Chapter 6) that provides air free of microorganisms at the point of fill. The auditor should therefore check that:

- The control measure, a grade A environment, exists
- The unidirectional air flow environment is subject to appropriate qualification and ongoing maintenance, calibration, and performance monitoring
- Procedures are in place to minimize operator intrusion into this critical zone
- Cleaning and disinfection procedures exist for the critical zone
- The exclusion of contamination is monitored by appropriate microbiological methods
- Appropriate alert and action limits for microbiological contamination are in place

- The microbiological monitoring program is sufficient to detect contamination in terms of sampling locations, sampling methods, number and frequency of sampling, and procedures for incubation of media and enumeration of colonies
- Results of monitoring are compared with approved alert and action limits and appropriate action taken in the event of loss of control
- Results are subject to trend analysis, which may indicate a move toward loss of control
- Records of all activities are maintained and subject to periodic review
- Changes to procedures, processes, equipment, components, and people are identified and evaluated to determine whether modifications to the control strategy are required

21.5.2.3 The Microbiological Quality Control Laboratory

Microbiological monitoring activities of the laboratory rely heavily on the supply of culture media of the required quality, that is, both sterile and nutritious. The preparation of media from a dry powdered medium is a critical control point in the laboratory. Thus, any audit of the laboratory should focus on this activity and involve the following lines of questions:

- Is there a clear purchasing policy for dry powdered medium?
- Are storage conditions for dry powdered medium defined and maintained?
- Is growth medium prepared according to defined procedures?
- What controls, if any, are exerted on the water used for preparation?
- Is the effectiveness of the sterilization process ensured?
- Are procedures for aseptic dispensing of sterilized medium in place?
- Is the prepared, sterilized medium assessed not only for sterility but also for retention of growth promotion properties?
- Are the sampling and testing procedures used to ensure sterility and growth promotion adequate?
- Are both limits and corrective action procedures in place?
- Are results subject to trend analysis?
- Are all results documented and traceable?
- Are appropriate change control procedures in place?

For a more detailed discussion on auditing the microbiology laboratory, the reader is referred to the companion *Handbook* (Lush 2000).

As can be seen from the preceding examples, effective planning and the use of risk analysis can bring structure to the audit process and improve the overall efficiency of the audit. It must be borne in mind, however, that during any audit, however well planned, certain practices or activities may require an "on-the-spot" assessment of risk. In many cases, the auditor may not have sufficient knowledge of the activity or sufficient time to carry out a formal risk assessment. Under these circumstances, the auditor should ask what are the likely implications, severity and probability of failure, as well as the probability of detection. By doing this, the auditor can assess the extent of risk and hence the necessity for control measures. It is worth stressing at this point that most problems identified during microbiological audits stem not from a failure to identify, monitor, and control critical control points, but from a failure to act upon trended data indicating loss of control or the inadequacy of corrective actions to eliminate the problem. In this regard, there is a strong tendency to treat symptoms of problems rather than the root cause of the problem itself. Such failures may manifest themselves in many ways, for example, as intermittent high bacterial counts or endotoxin results in pharmaceutical water systems, sporadic high microbiological counts on environmental monitoring plates, or variable bioburden results on raw materials and in-process solutions.

21.5.3   COMMUNICATING CONCERNS

It is not enough simply to identify problems during an audit. The auditee must understand and share the auditor's concerns; otherwise, the auditee may not be sufficiently motivated to assign the necessary urgency and diligence to rectify the problem. Failure to communicate the reasons for concerns is perhaps the single most common cause of inadequate follow-up to audits. The structured approach to identification of areas of vulnerability afforded by FMEA and HACCP and the objective criteria by which the effectiveness of control measures may be judged provide the auditor with an excellent means of discussing concerns and can enable the auditor and auditee to find a common basis for understanding and hopefully agreement.

21.5.4   RECOMMENDATIONS FOR CORRECTIVE ACTIONS

Once there is a clear understanding of the vulnerability and its scale, the task of making recommendations for corrective action becomes much simpler and more objective. For example, the vulnerability may arise from issues such as failure to: identify a CCP; set appropriate control limits; perform appropriate monitoring; identify loss of control or a move toward loss of control; take effective corrective actions that prevent recurrence; modify practices in the event of change.

If the structured approach to risk management afforded by FMEA or HACCP is used during audit, the precise nature of the vulnerability can be identified, thus providing the auditor with an objective basis for recommending corrective actions. Wherever appropriate, the auditor should suggest ways in which the effectiveness of the corrective actions can be confirmed. For some unaccountable reason, some auditors see their role solely in terms of identification of problems and not in their rectification. If the auditor can offer recommendations for action, it is his or her duty to do so.

21.6   OVERVIEW

There is a tendency to believe that pharmaceutical auditing is all about products and processes, facts and figures, and documents and records. Indeed, all that has gone before in this chapter would seem to support that contention. It should not be forgotten, however, that an audit is made up of a series of human interactions between the auditor and auditee. The success of any audit thus depends in no small part on the ability of the auditor to build a relationship with the auditee, based upon mutual trust and respect. If the auditee considers the auditor to be arrogant, self-opinionated, and unsympathetic, then the auditor's findings are less likely to be accepted or advice put into practice. There is a valuable lesson here to be learned by any auditor.

It is the role of the auditor to leave any department, company, or activity having improved awareness of regulations and standards, current expectations, and of strengths and weaknesses. Furthermore the organization should be more motivated to build on its strengths and correct its weaknesses.

If the auditor does not achieve this, then he or she has failed.

REFERENCES

EC Directive. (1991). Council Directive 91/356/EEC 1. Laying down the principles and guidelines of good manufacturing practice for medicinal products for human use. *Official Journal of the European Communities,* L193, 30–33.

EC Directive. (1993). Council Directive 93/43/EEC. Hygiene of foodstuffs. *Official Journal of the European Communities*, L175, 1–11.

Food and Drug Administration. (1973). Thermally processed low-acid foods packed in hermetically sealed containers GMP. *Federal Register,* 38, 2398–2410.

International Commission on Microbiological Specifications for Foods. (1988). *Microorganisms in Foods 4—Application of the Hazard Analysis Critical Control Point System to Ensure Microbiological Safety and Quality.* Blackwell Scientific Publications, Oxford.

International Organization for Standardization. (1999). *ISO/TC 209—Cleanrooms and Associated Controlled Environments.* Institute of Environmental Sciences and Technology, Rolling Meadows, IL.

Jahnke, M. (1997). Use of the HACCP concept for the risk analysis of pharmaceutical manufacturing processes. *Eur. J. Parent. Sci.,* 2, 113–117.

Leaper, S., Ed. (1997). *HACCP: a Practical Guide.* Technical Manual no. 38. Campden and Chorleywood Food Research Association, Chipping Campden, U.K.

Ljungqvist, B. and Reinmüller, B. (1995). Hazard analyses of airborne contamination in cleanrooms: application of a method for limitation of risks. *PDA J. Pharm. Sci. Technol.,* 49, 239–243.

Lush, M. (2000). Microbiological hazard analysis and audit: the practice. In *Handbook of Microbiological Quality Control: Pharmaceuticals and Medical Devices.* Baird, R.M., Hodges, N.A., and Denyer, S.P., Eds. Taylor & Francis, London, pp. 221–238.

U.K. Food Safety Act. (1990). Her Majesty's Stationery Office, London.

World Health Organization. (1996). *Training Aspects of the Hazard Analysis Critical Control Point System (HACCP).* WHO/FNU/FOS/96.3. World Health Organization, Geneva.

Index

A

Acacia, 33, 354, 442
Acanthamoeba, 42, 350
Acceptable quality level (AQL), 158
Acetylsalicylic acid, 31
Acholeplasma laidlawii, 97
Achromobacter, 25, 27
Acinetobacter anitratus, 37
Acinetobacter lwoffi, 31
Acinetobacter spp., 25, 27, 36, 56, 57
Activated carbon, 56, 442
Activation energy for inactivation, 209–210, 216, 217
Adaptation index (AI), 313
Adsorption, preservatives and, 352–353, 357
Aerobacter, 40
Aerobic microorganisms, 8–9
Aerosol challenge testing, 377
Aerosol medications, 39, 368
Aflatoxins, 20, 32
Agar, 30, 33, 442
Air ballasting, 284
Airborne particulates, 73
  cell culture contamination, 94
  contaminant deposition estimation, 146–147
  defining cleanroom air quality, 78, 123–124
  dispersion of, 143–153, *See also* Air supply
    and ventilation system
    general mathematical principles, 143–145
    limitation of risks, 151–153
    still air and turbulent mixing, 145–148
    unidirectional air flow, 148–149
    wakes and vortices, 149–150
  limitation of risks, 151–153
  mathematical estimation, 78
  microbiological monitoring guidelines, 147–148
  microorganism size, 147
  monitoring manufacturing areas, 94
  sampling methods, 176–177
Airlock systems, 133
Air pressure differences, 82, 133, 145
Air reservoir tanks, 64
Air sampling, 176–177
Air–steam mixture cycles, 284
Air supply and ventilation system, 65, 74–82, 133

airborne contaminant dispersion, *See* Airborne
    particulates
  biotechnological production facilities, 94
  filtration, 74–82
    diffusive, 75
    electrostatic, 75
    HEPA, 65, 70, 74, 75–76, 145
    inertial impaction, 75
    prefilters, 75–76
  good manufacturing practice, 131
  pressurization, 77
  side walls, 150
  standards and guidelines, 70–71, 81
  turbulent flow, 78
  unidirectional flow, 70, 79, 145, 148–149
  validation and monitoring, 82
  vertical and horizontal flows, 81
Alcaligenes, 25, 56, 57
Alcaligenes faecalis, 39
Alcohols, 14
  edetic acid interactions, 334
  microbial growth and, 35
  neutralization in culture media, 161
  preservation spectrum, 332
  preservative applications (table), 325
  solubility and partitioning effects, 330
  temperature effects on preservative activity, 329
Alginic acid, 442
Alkaloids, 31
Alkylating gaseous sterilization agents, 220–223
Alkyltrimethylammonium bromide, 408
Aluminum hydroxide, 442
Aluminum tubing, 136
Amines, 15
Amino acid metabolism, 15, 30
α-Amylase, 30
Anaerobic microorganisms, 8
Animal droppings, 60
Antacid suspensions, 353
Antibiotics
  antineoplastic drugs, 361
  culture media, 95, 102, 103, 105, 159–160
  inactivating, 177, 389
  intravenous administration, 350
  microbial contamination, 157
  microbial degradation of, 31

C